Conservation Physiology

Conservation Physiology

https://academic.oup.com/conphys
@conphysjournal

Conservation Physiology is an online only, fully open access journal published on behalf of the Society for Experimental Biology. It publishes research on all taxa (microbes, plants, and animals) focused on understanding and predicting how organisms, populations, ecosystems and natural resources respond to environmental change and stressors. Physiology is considered in the broadest possible terms to include functional and mechanistic responses at all scales. *Conservation Physiology* also welcomes research towards developing and refining strategies to rebuild populations, restore ecosystems, inform conservation policy, and manage living resources.

Conservation Physiology

Applications for Wildlife Conservation and Management

EDITED BY

Christine L. Madliger
Department of Biology, Carleton University, Canada

Craig E. Franklin
School of Biological Sciences, University of Queensland, Australia

Oliver P. Love
Department of Integrative Biology, University of Windsor, Canada

Steven J. Cooke
Department of Biology, Carleton University, Canada

OXFORD
UNIVERSITY PRESS

OXFORD
UNIVERSITY PRESS

Great Clarendon Street, Oxford, OX2 6DP,
United Kingdom

Oxford University Press is a department of the University of Oxford.
It furthers the University's objective of excellence in research, scholarship,
and education by publishing worldwide. Oxford is a registered trade mark of
Oxford University Press in the UK and in certain other countries

© Oxford University Press 2021

The moral rights of the authors have been asserted

First Edition published in 2021
Impression: 1

Published in the United States of America by Oxford University Press
198 Madison Avenue, New York, NY 10016, United States of America

British Library Cataloguing in Publication Data
Data available

Library of Congress Control Number: 2020948787

ISBN 978–0–19–884361–0 (hbk.)
ISBN 978–0–19–884362–7 (pbk.)

DOI: 10.1093/oso/9780198843610.001.0001

Printed and bound by
CPI Group (UK) Ltd, Croydon, CR0 4YY

Contents

**Chapter 17   Using applied physiology to better manage and conserve the
white rhinoceros (*Ceratotherium simum*)   287**

Anna J. Haw, Andrea Fuller, and Leith C.R. Meyer

Part VI Increasing the Reach of Conservation Physiology   301

**Chapter 18   Communication in conservation physiology: linking diverse
stakeholders, promoting public engagement, and encouraging
application   303**

Taryn D. Laubenstein and Jodie L. Rummer

**Chapter 19   Optimism and opportunities for conservation physiology in the
Anthropocene: a synthesis and conclusions   319**

Steven J. Cooke, Christine L. Madliger, Jordanna N. Bergman, Vivian M. Nguyen,
Sean J. Landsman, Oliver P. Love, Jodie L. Rummer, and Craig E. Franklin

Preface

When Rachel Carson raised the alarm about the effects of dichlorodiphenyltrichloroethane (DDT) on raptor populations in her 1962 book *Silent Spring*, she was unlikely aware of her significant pioneering role in establishing the nascent discipline of conservation physiology. Her work and that of others in the scientific community went beyond simply documenting the declining raptor populations and reproductive failures (i.e. describing patterns) to identifying the mechanistic basis for the declines. Efforts to establish cause–effect relationships between DDT and eggshell quality served as the basis for evidence-based regulatory change and eventually the implementation of an environmental policy (the 1998 Aarhus Protocol on Persistent Organic Pollutants). This serves as a classic example of the power of using physiological tools, knowledge, and concepts to understand and solve complex conservation problems and effect change.

The term 'conservation physiology' was formally coined in 2006 by Martin Wikelski (then at Princeton University) and Steven Cooke (Carleton University), and immediately gained traction within the scientific community as a novel discipline. Scientists began to self-identify as 'conservation physiologists' along with a growing number of opportunities to come together at conferences and workshops to advance the field. Fast forward to 2013, the Society for Experimental Biology and Oxford University Press collaborated to launch the journal *Conservation Physiology* with leadership from Tony Farrell, Craig Franklin, and Steven Cooke. The journal quickly became a trusted outlet for those working on using physiological techniques to tackle conservation-related questions, and further catalysed the community and codified the discipline.

Our reference book represents a logical next step in the evolution and growth of conservation physiology as an integrative, proactive, and evidence-based discipline. There are of course many ways to organize a book, but in this case the structure was immediately apparent to us. We recruited scientists working on diverse topics, using various tools, and spanning different taxa and ecosystems, who had achieved success in generating knowledge relevant to disentangling underlying mechanisms and helping to solve conservation problems. The resultant chapters represent a rich series of case studies with emergent and interconnected themes and lessons that are summarized at the end of the book in a concluding chapter.

It is our hope that readers will find that these case studies are illuminating and provide tangible examples of both successes and failures when working in conservation physiology. Our goal was to create a text that will help drive the further evolution of this discipline, and by extension, its ability to contribute to resolving conservation problems worldwide.

We are grateful to the authors of the various chapters for sharing their candid experiences. We also wish to acknowledge and celebrate the broader community of practice working on defining and operationalizing conservation physiology.

**Christine L. Madliger, Craig E. Franklin,
Oliver P. Love, and Steven J. Cooke**

List of contributors

Alejandro Fernández Ajó Northern Arizona University, USA; Instituto de Conservación de Ballenas, Argentina

Cedric Alaux INRAE, France

Lesley A. Alton Monash University, Australia

Jordanna N. Bergman Carleton University, Canada

Nolan N. Bett University of British Columbia, Canada

Ian A. Bouyoucos James Cook University, Australia; PSL Research University, France

Charles A. Brown Louisiana State University, USA

Jean-Luc Brunet INRAE, France

C. Loren Buck Northern Arizona University, USA

Elizabeth A. Burgess New England Aquarium, USA

Nicholas J. Burnett BC Hydro, Canada

Renae Charalambous Western Sydney University, Australia

Christian O. Chukwuka University of Otago, New Zealand

Steven J. Cooke Carleton University, Canada

Rebecca L. Cramp The University of Queensland, Australia

Alison Cree University of Otago, New Zealand

Glenn T. Crossin Dalhousie University, Canada

Kara Dodge New England Aquarium, USA

Benjamin Dubansky University of North Texas, USA

Yvonne A. Dzal University of Winnipeg, Canada

Erika J. Eliason University of California Santa Barbara, USA

Anthony P. Farrell University of British Columbia, Canada

Craig E. Franklin The University of Queensland, Australia

Andrea Fuller University of the Witwatersrand, South Africa

Fernando Galvez Louisiana State University, USA

Kelly M. Hare University of Waikato, New Zealand

Anna J. Haw University of Veterinary Medicine Vienna, Austria

Brian Helmuth Northeastern University, USA

Mickael Henry INRAE, France

Chelsea D. Hess Louisiana State University, USA

Scott G. Hinch University of British Columbia, Canada

Kathleen E. Hunt George Mason University, USA

Nigel E. Hussey University of Windsor, Canada

Charles Innis New England Aquarium, USA

Kenneth M. Jeffries University of Manitoba, Canada

Sean J. Landsman Carleton University, Canada

Taryn D. Laubenstein James Cook University, Australia

Oliver P. Love University of Windsor, Canada

Carley Lowe Northern Arizona University, USA

Daniel J. Madigan Harvard University, USA

Christine L. Madliger Carleton University, Canada

Eduardo G. Martins University of Northern British Columbia, Canada

Leith C.R. Meyer University of Pretoria, South Africa

Kristina M. Miller Fisheries and Oceans Canada, Canada

Christopher Myrick Colorado State University, USA

Edward J. Narayan The University of Queensland, Australia

Nicola J. Nelson Victoria University of Wellington, New Zealand

Vivian M. Nguyen Carleton University, Canada

Michel E.B. Ohmer University of Pittsburgh, USA

David A. Patterson Fisheries and Oceans Canada, Simon Fraser University, Canada

Graham D. Raby Trent University, Canada

Essie M. Rodgers University of Canterbury, New Zealand

Jodie L. Rummer James Cook University, Australia

James Sakker Department of Primary Industries Fisheries, Australia

Christina A.D. Semeniuk University of Windsor, Canada

Oliver N. Shipley Stony Brook University, USA

Amy K. Teffer University of British Columbia, Canada

Marisa L. Trego University of California Davis, USA

Jo Virens University of Otago, New Zealand

Andrew Whitehead University of California Davis, USA

Tony D. Williams Simon Fraser University, Canada

Craig K.R. Willis University of Winnipeg, Canada

Nathan Young University of Ottawa, Canada

CHAPTER 1

The history, goals, and application of conservation physiology

Christine L. Madliger, Oliver P. Love, Steven J. Cooke, and Craig E. Franklin

1.1 The history of combining animal physiology and conservation science

Although conservation physiology is often cited as one of the newest branches of conservation science (for an overview of conservation science writ large, see Soulé 1985), its foundation is formed by nearly 200 years of comparative animal physiology research. As that primarily laboratory-based discipline expanded into natural settings, the field of ecological (or environmental) physiology began to take shape. By the mid-1900s, scientists were characterizing how physiological adaptations allow organisms to prosper in extreme environments like deserts, the depths of the oceans, high altitudes, and the poles (Feder et al. 1997). Ecological physiology became increasingly interdisciplinary, drawing on molecular biology, evolutionary and life history theory, behavioural ecology, and natural history to characterize physiological diversity and adaptations across all environmental types and scales (Willmer et al. 2009). Given the capacity of this knowledge base, it is unsurprising that, as environmental movements gained traction in the 1960s, some scientists turned to physiology to discern the underlying mechanistic basis of widespread conservation issues. One prominent example is that of dichlorodiphenyltrichloroethane (DDT) causing reproductive failure in avian species, particularly large raptors such as bald eagles (*Haliaeetus leucocephalus*) and peregrine falcons (*Falco peregrinus*). It was the discovery of altered eggshell deposition dynamics—an interplay of reproductive

physiology and toxicology (see Bitman et al. 1969; Jefferies 1969)—that provided some of the key evidence leading to pesticide bans in the United States, forming one of the earliest success stories in the field of conservation physiology.

Despite this success, it would be decades before researchers began to formally frame the process of integrating physiological tools into conservation science as a discipline in its own right. Much of the first published literature discussing the potential applications of physiology in conservation focused on endocrinology, in particular reproductive and stress physiology. For example, building off knowledge gained in captive breeding scenarios monitoring reproductive hormone levels, Berger et al. (1999) discussed the application of 'conservation endocrinology' for wild populations. By monitoring faecal progesterone levels, the authors determined that a low frequency of juvenile moose (*Alces alces*) in the Greater Yellowstone Ecosystem of the United States was not the result of increased predation by recolonizing wolves (*Canis lupus*) or grizzly bears (*Ursus arctus*), but instead was due to low pregnancy rates. It was therefore clear that physiology could help identify the underlying cause of population instability in a wild setting. In relatively close succession, Millspaugh and Washburn (2004), noting the increasing use of faecal glucocorticoid levels as indicators of 'stress' in wild animals, outlined numerous considerations for sample collection, processing, and interpretation that were specific to using this tool in conservation biology research. This work stressed just

Christine L. Madliger, Oliver P. Love, Steven J. Cooke, and Craig E. Franklin, *The history, goals, and application of conservation physiology*
In: *Conservation Physiology: Applications for Wildlife Conservation and Management*. Edited by: Christine L. Madliger, Craig E. Franklin, Oliver P. Love, and Steven J. Cooke, Oxford University Press (2021). © Oxford University Press. DOI: 10.1093/oso/9780198843610.003.0001

how important it is to have detailed knowledge of the role of any aspect of physiology in maintaining homeostasis (i.e. the value of validating tools for specific species and contexts) to employ a physiological metric as a conservation biomarker.

Conservation physiology became much more solidified in the mid-2000s as researchers came together to define its purpose and scope. In 2004, a symposium titled 'Ecophysiology and Conservation: The Contribution of Endocrinology and Immunology' at the Society for Integrative and Comparative Biology's annual meeting specifically showcased physiological research with conservation implications (Stevenson et al. 2005). Many of the associated papers from this symposium covered endocrine disruption in aquatic environments (Stevenson et al. 2005), reflecting the history of using physiological techniques to understand the effects of environmental toxins and pollutants. Soon after, Wikelski and Cooke (2006, p. 38) formally defined and described in detail the scope and goals of conservation physiology as an emerging, cogent discipline. Their definition stated that conservation physiology is, 'the study of physiological responses of organisms to human alteration of the environment that might cause or contribute to population declines'. Importantly, they stressed that one of the most valuable characteristics of conservation physiology is that it reaches beyond a description of patterns to provide information on the mechanism(s) underpinning a conservation issue.

The next major leap in the field came in 2013 with the launch of the dedicated journal *Conservation Physiology* by Oxford. For scientists sometimes facing difficulty fitting their research into the scope of existing journals, this became an outlet where studies could be published without having to tailor to either a solely physiological or conservation science-focused venue. The inaugural paper of the journal refined the definition of conservation physiology to be, 'an integrative scientific discipline applying physiological concepts, tools, and knowledge to characterizing biological diversity and its ecological implications; understanding and predicting how organisms, populations, and ecosystems respond to environmental change and stressors; and solving conservation problems across the broad range of taxa (i.e. including microbes, plants, and animals)' (Cooke et al. 2013, p. 2). This is an inherently broad definition that considers the diversity of physiological traits available, which span immunology/epidemiology, endocrinology, bioenergetics, cardiorespiratory physiology, physiological genomics, neuro- and sensory physiology, and toxicology (Cooke et al. 2013). The authors envisioned the discipline contributing to diverse conservation goals, including identifying strategies to rebuild populations, ecosystem restoration, conservation policy development, and

Table 1.1 The scope of conservation physiology. Adapted from Cooke et al. (2013).

Monitoring and identifying threats	Predicting change	Integrating with diverse disciplines	Achieving conservation success
Providing a mechanistic understanding of the effect of environmental change on organisms	Developing mechanistic models for species distributions	Integration of physiology with conservation behaviour, conservation medicine, conservation toxicology, conservation genetics, and other relevant sub-disciplines	Exploiting knowledge of organismal physiology to control invasive species and restore threatened habitats and populations
Understanding the influences of anthropogenic disturbance and variation in habitat quality on organism condition, health, and survival	Developing mechanistic relationships between population declines and physiological processes	Understanding the relevance of ecology and evolution of physiological diversity to conservation	Understanding the optimal environmental conditions for ex situ preservation of endangered species (e.g. captive breeding)
Understanding the physiological mechanisms involved in changes in community, ecosystem, and landscape structure, as well as individual species, in response to environmental change		Understanding the relevance of acclimatization and adaptation of physiological processes to environmental variation to management and conservation	Integrating physiological knowledge into ecosystem management and development of tools to solve complex conservation problems
Evaluating stress responsiveness and environmental tolerances relative to environmental change (including climate change and ocean acidification)	Developing predictive models in conservation practices that include physiological parameters	Applications of contemporary genomic and post-genomic technologies to conservation physiology	Evaluating and improving the success of various management and conservation interventions
Applying physiological biomarkers as part of long-term environmental monitoring programmes			Understanding the policy implications of conservation physiology research

Figure 1.1 Conservation physiology successes cover a diversity of taxa, ecosystems, landscape scales, and physiological systems. (A) Birds of prey, such as osprey, have rebounded following regulations on DDT. (B) Plague is being combated in the endangered black-footed ferret via a targeted vaccination programme. (C) Caribou and wolf populations are being effectively managed via physiological monitoring of scat. (D) Nutrition programmes support successful breeding in the critically endangered kakapo. (E) Ecotourism feeding practices are regulated for stingrays in the Cayman Islands. (F) Sensory physiology has informed shoreline lighting regulations for nesting sea turtles. (G) Recovery chambers decrease the stress associated with bycatch in salmonids. (H) Physiological monitoring is improving translocation success in white rhino. Photo credits: (A) Randy Holland; (B) USGS NWHC; (C) Wayne Sawchuk, Samuel Wasser; (D) Kakapo Recovery; (E) Christina Semeniuk; (F) Sea Turtle Conservancy; (G) Cory Suski, Jude Isabella; and (H) Andrea Fuller. Reproduced with modification from Madliger et al. (2016).

natural resource management (Cooke et al. 2013; Table 1.1).

Over the past 15 years, conservation physiology has accrued measurable successes (Madliger et al. 2016, Cooke et al. 2020; Figure 1.1). It has not done so in isolation, but is often used in conjunction with tools and techniques from conservation behaviour, genetics, and/or social sciences. In this way, the discipline is becoming increasingly integrated into the broader conservation toolbox, enabling initiatives that support conservation actions for imperilled species (reactive approaches) and those that are part of sustainable management (proactive approaches) (Cooke et al. 2020). In the rest of this chapter, we briefly outline some of the benefits and applications conservation physiology has to offer, provide an overview of the current toolbox being used by researchers and practitioners, and discuss the layout and goals of the text.

1.2 What can conservation physiology offer?

1.2.1 An increasingly expanding and validated toolbox

The toolbox currently available to conservation physiologists is diverse and ever-expanding (Table 1.2). Endocrine tools have been the most heavily investigated in the context of conservation applications (Lennox et al. 2014; Madliger et al. 2018), with stress hormones (glucocorticoids such as cortisol and corticosterone) drawing the greatest attention. However, there has been a disconnect between the propensity with which stress hormones are measured for conservation purposes (e.g. inferring stress or disturbance in wild populations) and the number of resulting success stories (i.e. examples where measurement of the physiological trait resulted in a change in conservation policy, practice, or behaviour) (Madliger et al. 2016). As a result, we urge readers to consider conservation physiology in the most diverse manner possible, taking note that there are dozens of physiological traits in the toolbox (Madliger et al. 2018). To illustrate this diversity, Table 1.2 provides examples of many of the available tools partitioned by physiological sub-discipline, and the subsequent chapters in the text illustrate how many of these tools have been

put into practice across a variety of animal taxa. Indeed, many metrics have the potential to fulfil a role in generating robust knowledge that can be integrated as decision-support tools for conservation science.

The potential for application of a vast array of traits becomes more realistic as researchers continue to validate tools across species and contexts and work to find techniques that are non-invasive or as minimally invasive as possible. For example, although the 'traditional' sample medium for a variety of reproductive, stress, and energetic hormones has been blood, there are now techniques to assess many hormone levels in saliva, urine, faeces, fur, feathers, claws, scales, shed skin, eggs, baleen, whale blow, ear wax, and water. Some of these sample types can be acquired without any handling or disturbance of an individual animal. For example, researchers obtained snags of hair from grizzly bears (*Ursus arctos*) to assess responses in reproductive and stress physiology to changing nutritional quality (Bryan et al. 2013). Faeces can similarly be collected entirely non-invasively from many organisms in the field. Hunninck et al. (2020) analysed glucocorticoid levels in faecal samples from wild ranging impala (*Aepyceros melampus*) paired with an indicator of forage quality (normalized difference vegetation index—NDVI) and determined that climate-induced changes in vegetation represent a larger disturbance than human-related land-use changes. Excitingly, some sample media also give us retrospective glimpses into physiological functioning long before the sample is collected; for example, whale baleen provides a cross-section of reproductive and stress hormone level changes across the decades in which it was grown (Hunt et al. 2014; Hunt et al., Chapter 12, this volume). This ability to reconstruct a species past history can even reveal where an animal has lived and its broad-scale movement patterns. Madigan et al. (Chapter 5, this volume) highlight the value of harnessing stable isotope analyses of various tissues (e.g. muscle, liver, erythrocytes, and plasma) that have different turnover rates and the geographical variation of isotopic signatures (i.e. isoscapes) to reveal migration patterns and habitat use of overexploited Pacific bluefin tuna (*Thunnus orientalis*).

Table 1.2 Physiological sub-disciplines, ways each can contribute to conservation science, and examples of common tools/techniques. Adapted from Cooke et al. (2013) and Madliger et al. (2018).

Physiological sub-discipline	Examples of contributions to conservation physiology	Examples of tools
Bioenergetics, metabolic, and nutritional physiology	Assessing whole-organism response to environmental change; improving captive breeding and rehabilitation through adequate nutrition; monitoring conservation management scenarios; identifying mechanisms behind population decline	Body condition indices Daily energy expenditure Lipid and fatty acid concentrations Metabolic rate Plasma glucose Plasma lactate Stable isotopes Ucrit (prolonged swimming speed) Vitellogenin
Cardiorespiratory physiology	Predicting and monitoring responses to environmental change; predicting invasive species spread; predicting species distributions under climate change scenarios	Aerobic scope Haematocrit Haemoglobin concentration Heart rate Respiratory rate EPOC (excess post-exercise oxygen consumption)
Immunology/epidemiology	Predicting spread of diseases; design of control and vaccination programmes; determining sub-lethal consequences of environmental change	Disease state (e.g. serum total protein) Humoral and cell-mediate immune response Cytokines
Neurophysiology/sensory biology	Determining guidelines/optimal designs to reduce human–wildlife conflicts; understanding mechanisms behind behavioural responses to environmental change	Neural activity Electrical excitability Pheromones Sensory sensitivity/tolerance
Reproductive physiology	Identifying mechanisms behind population declines; improving captive breeding success; monitoring success of reintroduction programmes	Developmental rate Fecundity Reproductive hormone levels (e.g. oestrogen, testosterone, progesterone) Sperm motility
Stress physiology	Predicting and monitoring responses to environmental change; monitoring success of restoration programmes; identifying best practices for translocation; improving welfare in captive scenarios	Electrolyte balance Glucocorticoids (e.g. cortisol, corticosterone) Oxidative status/stress pH Telomere shortening
Thermal physiology	Determining organismal capacity to respond to climate change (e.g. thermal plasticity, range expansion); determining thermal dependence of performance and tolerances across various environmental conditions; predicting potential limitations of invasive species spread	Thermal tolerance (e.g. CTMax, CTMin) Thermal performance curves Q_{10} Enzyme kinetics
Toxicology	Determining sources of population declines; delineating regulatory guidelines for chemicals/pollutants; designing remediation protocols	Pollutant/chemical contaminant concentration in various tissues Trace element/metal concentrations

There is also a great deal of capacity to assess stress, health, and physiological functioning at a variety of biological scales using the conservation physiology toolbox (Table 1.2), including 'gene expression (e.g. physiological genomics), gene products (e.g. physiological proteomics), cells (e.g. sperm physiology), individual tissues (e.g. muscle oxygenation), organs (e.g. heart rate) and the whole-organism (e.g. daily energy expenditure)' (Madliger et al. 2018, p. 5). As a result of the different scales that various physiological traits capture, their suitability for conservation questions and planning will also vary. For example, repeated individual measures of reproductive hormones can be useful for designing and improving captive breeding programmes focused around encouraging copulation, assessing reproductive potential, or improving artificial insemination practices in captivity (e.g. Swanson 2003; Dehnhard et al. 2008). Indeed, acquiring fundamental knowledge on reproductive physiology through non-invasive endocrine monitoring in killer whales (*Orcinus orca*) was necessary to develop the artificial insemination technology that resulted in the first successful conceptions (i.e. resulting in live offspring) in any cetacean species (Robeck et al. 2004). For highly vulnerable and endangered groups, like sea turtles, physiological biomarkers can also be used to understand physiological dysfunction and direct veterinary treatment options that aid in the rehabilitation of individuals (Innis and Dodge, Chapter 14, this volume; Narayan and Charalambous, Chapter 11, this volume). When conservation actions require capture and translocation of individual animals, the quantification of physiological biomarkers can ensure the health and safety of animals, such as the monitoring of arterial blood gases in white rhinos (*Ceratotherium simum*) that have been immobilized and anaesthetized (Haw et al., Chapter 17, this volume).

Other traits can be harnessed to understand conservation-relevant consequences within populations, such as how differences in immune function or energetics influence disease susceptibility and potential or realized population decline (e.g. Rohr et al. 2013; McCoy et al. 2017). Development and use of sensitive biomarkers and physiological approaches in the fight against emerging novel diseases is strongly advocated by Dzal and Willis (Chapter 9, this volume) and Ohmer et al. (Chapter 10, this volume). Determining the susceptibility of bats and amphibians to cutaneous fungal pathogens has been dependent on understanding the underlying physiology of the disease state, with applications for conservation strategy design to limit and treat disease spread within and across populations. Furthermore, other traits can be used for broader, macrophysiological applications (Chown and Gaston 2008). Assessments of meta-

bolic scope have the power to determine environmental tolerance and therefore predict population ranges under climate change and/or the spread of invasive species across large spatial scales (e.g. Deutsch et al. 2015; Marras et al. 2015; Winwood-Smith et al. 2015).

In some conservation-relevant systems, researchers and practitioners can combine multiple physiological traits to solve problems across biological and spatial scales. In assessing the migration failure of Pacific salmon (*Oncorhynchus* genus), Cooke et al. (Chapter 3, this volume) highlight the value of the measurement of a suite of biomarkers that span levels of biological organization, from genomic to organismal, that are derived from both field- and laboratory-based studies. In this case, the measurement of multiple biomarkers provided a comprehensive understanding of physiological function and organismal performance, which in turn assists in better evaluating and predicting migration success or failure in salmon. The following sections and the chapters contained in this text will further illustrate many ways that physiology can provide both snapshots and longer-term assessments of animal stress, function, and health across biological, spatial, and temporal scales.

1.2.2 Sensitive biomarkers of organismal condition and health

Traditionally, physiological biomarkers have often been used as indicators of health, diagnosing the presence of a pathogen, and assessing severity of a disease state or pathological condition. Beyond this, physiological biomarkers are frequently used to determine the general welfare and condition of animals, with blood diagnostic tests quantifying immune function, endocrine function, oxygen transport capacity, nutritional/metabolic state, and electrolyte balance being common. As discussed above, conservation physiologists have developed an innovative and growing array of biomarkers to assess the health of organisms, identify threats, and ultimately aid in conservation efforts (Madliger et al. 2018; Cooke et al. 2020; Table 1.2). For biomarkers to be useful in evaluating the condition of animals, assessing and predicting how species perform in response to changing environmental conditions, or providing an indication of the health of an ecosystem, they must be sensitive to environmental changes of interest (Cooke et al. 2013, 2017a, 2020; Madliger et al. 2017). There are countless publications that investigate the sensitivity of various metrics of physiology to environmental conditions or the internal state of animals (e.g. reproductive status, development, age). Just in the most recent issue of the journal *Conservation Physiology* (Volume 8, 2020), we can find articles that link ocean acidification and warming to changes in the transcriptome of an Antarctic pteropod (*Limacina helicina antarctica*) (Johnson and Hofmann 2020); indicate zinc concentrations in walrus (*Odobenus rosmarus divergens*) teeth can reflect the onset of female reproductive maturity (Clark et al. 2020); show oxygen-carrying capacity is compromised under exposure to nitrate and low pH in perch (*Leiopotherapon unicolor*) (Gomez Isaza et al. 2020); and describe faecal cortisol and oestradiol concentrations varying with human disturbance in Asian elephants (*Elaphas maximus*) (Tang et al. 2020).

In every chapter of this book, sensitive biomarkers that assess physiological function, animal condition, and performance are described with concrete conservation applications. Specifically, biomarkers can help discriminate acute from chronic stress and identify specific stressors, with some of the more novel and ingenious techniques being able to assess physiological and health status of an organism in the absence of any physical sample taken from an animal. For example, Hunt et al. (Chapter 12, this volume) describe how droplets in the respiratory vapour of whales ('blow') can be collected by using a small aerial drone at the moment a whale exhales. The blow samples have been found to contain all major steroid and thyroid hormones, making them useful for health and reproductive monitoring across seasons and years.

Use of OMICS technologies (e.g. transcriptomics, gene expression) and generation of genomic/transcriptomic biomarkers can provide detailed mechanistic information that can cover an array of biological/physiological systems, including metabolic, digestive, and immune (Ge et al. 2013; McMahon et al. 2014; Bahamonde et al. 2016). In particular, information gained from genome-wide gene expression profiling (that generates large datasets) has been invaluable in diagnosing the health impacts of toxicants and chemical pollution (Trego et al., Chapter 7, this volume). These OMICS biomarkers can often be highly sensitive, detecting perturbations not easily revealed by more traditional biomarkers like stress hormones.

Sensitive physiological biomarkers can also provide red flags that show condition and reproductive health are changing in response to environmental variability. Crossin and Williams (Chapter 2, this volume) describe how yolk precursors, the primary sources of protein and lipid in egg yolk (e.g. vitellogenin), in the plasma of breeding birds can provide an assessment of the reproductive status of free-living birds. This approach is highly valuable in long-lived Arctic-breeding seabirds that would normally require large longitudinal datasets to ascertain a decline, resulting in delays that could make conservation approaches aimed at reversing declines much less effective. Further, because vitellogenin predictably and sensitively changes in relation to the environmental conditions that drive breeding decisions, a small blood sample can be taken prior to the commencement of egg laying and therefore avoid disturbing birds at times when they would abandon their nests. Alaux et al. (Chapter 4, this volume) further note that vitellogenin levels found in bees can provide an indication of tolerance to oxidative stress, an evaluation of immune defence, and ultimately could be used to provide an assessment of the effects of habitat enhancement and protection on bee health. These examples from birds and bees illustrate how a ubiquitous protein can provide useful conservation insights across vastly different taxonomic groups.

Organismal locomotor and performance traits (e.g. fish swimming speed, Wilson et al. 2001; Kern et al. 2018; frog jumping ability, Hudson & Franklin, 2002) are often not viewed as biomarkers, yet fundamentally they can provide an important measure of condition and health, but also locomotor capability. For example, Cramp et al. (Chapter 6, this volume) detail how measures of swimming performance can be effective in predicting the ability of fish to successfully transverse man-made structures, like culverts and fish ladders. Organismal performance traits can also elucidate impacts of environmental change. For example, determining the relationship between body temperature and performance in ectotherms can provide valuable insights into the impact of climate warming and the sensitivity of organisms to temperature increases, whether they are intertidal organisms (Helmuth, Chapter 13, this volume) or reptiles (Cree et al., Chapter 16, this volume).

1.2.3 Identification of underlying mechanisms of decline

Monitoring the abundance and vital rates of animal populations is fundamental to management and conservation (Krebs 1989). However, such information tends to only reveal that a problem exists (e.g. a population decline) rather than identifying the drivers or mechanistic basis for a decline. Associations (e.g. coincident with a decline in population was an increase in a given stressor) can be identified; yet, it is not possible to determine with any certainty if an association or correlation is spurious (Mayr 1961). In contrast, cause-and-effect relationships bring a level of certainty (Mayr 1961) that allows managers to identify optimal strategies and know that their efforts are focused on identifying threats that have a direct negative impact on organisms (Carey 2005; Cooke and O'Connor 2010; Seebacher and Franklin 2012). In this way, managers avoid mitigation strategies being put in place at later-than-optimal time points or wasting precious resources on efforts that will not result in any benefit to wildlife (Sutherland and Wordley 2017).

Conservation physiology is particularly well suited to understanding mechanisms and pathways of effect because, as also outlined above and throughout the subsequent chapters of this text, it sensitively links an organism's internal state with its external environment, providing an objective measure of how animals respond to or cope with changes (Wikelski and Cooke 2006; Tracy et al. 2006; Cooke and O'Connor 2010; Cooke et al. 2013). In particular, physiological studies focused on individual organisms allow scientists to use experimental approaches or comparisons across disturbed and undisturbed sites that can lead to the identification of cause-and-effect relationships (Cooke et al. 2017a) that can then be scaled up to the level of the population and even ecosystem (see Section 1.2.4). Such cause-and-effect relationships are so compelling

that they are used as the gold standard in courts of law (Cooke and O'Connor 2010), and physiological measurements have been used to allow decision-makers to better target their conservation strategies. For example, monitoring multiple physiological traits (faecal reproductive, thyroid, and adrenal hormones) in Puget Sound killer whales allowed researchers to tease apart the impacts of boat traffic and nutritional stress, leading to the identification of protecting the whales' salmon prey as a more effective conservation strategy compared with limiting vessel disturbance (Ayres et al. 2012). Similarly, by combining faecal corticosterone and thyroid hormone monitoring in woodland caribou (*Rangifer tarandus caribou*), researchers have been able to determine how wolf predation and extraction of petroleum products from the Canadian oil sands development differentially impact a population, which has led to the deemphasizing of wolf removal programmes and greater effort in preservation of habitat that provides lichen as a food source (Wasser et al. 2011; Joly et al. 2015).

Being able to ascertain the mechanisms underlying conservation challenges can also provide compelling quantifiable measures for decision-makers, providing the evidence necessary to move forward with conservation action. For example, Semeniuk (Chapter 8, this volume) promotes the use of biomarkers of animal health in the management of wildlife provisioning tourism. Together with other physiological and biochemical indicators, non-esterified fatty acid profiles were used as a biomarker for assessing diet composition, lipid requirements, and nutritional status of tourist-fed stingrays to determine that tourism was imposing an ecological trap on the animals. This information provided concrete evidence to the Caymanian government that ecotourism guidelines should be updated. Beyond policy-makers, physiological mechanisms can also be effective tools in communicating conservation messages to stakeholders and the general public to draw support for conservation action (Bouyoucos and Rummer, Chapter 11, this volume; Laubenstein and Rummer, Chapter 18, this volume).

Finally, monitoring physiology to determine underlying mechanisms of decline can provide greater resolution on the complexity of some conservation challenges, which can in turn also benefit the development of evidence-informed mitigation strategies. Studying elk (*Cervus elaphus*) in the Greater Yellowstone Ecosystem, Creel et al. (2007) found that wolf predation pressure leads to greater vigilance, decreased foraging, and altered habitat selection and diet. These behavioural responses are associated with decreased levels of progesterone in adult females that is linked to lower calving rates and declining population size. By further measuring glucocorticoid (faecal cortisol) levels, Creel et al. (2009) showed that the negative influence of wolves on calf recruitment is not due to chronic stress, but is instead much more likely to be the result of the increased predation risk resulting in changes in foraging and nutrition. Declines in body mass and fat resources then result in a lowered ability of females to maintain pregnancies (Creel et al. 2009). From a management perspective, this type of information creates a clearer picture of the direct versus indirect effects of wolf predation, and how this can be intertwined with habitat change and availability.

1.2.4 Proactive and predictive capacity

Harnessing the predictive capacity to forecast the performance and fitness responses of organisms in complex systems is a primary motivation behind incorporating any mechanism (i.e. gene transcription, behaviour, physiology) into conservation studies (Madliger et al. 2015, 2018). Quantifying variation in physiological traits can allow us to link larger-scale abiotic processes to the organismal responses that influence fitness, population demography, and even ecosystem functioning (Bergman et al. 2019; Ames et al. 2020). With knowledge of the structure and strength of these linkages, we can then use predictive modelling techniques (McClane et al. 2011) to more effectively forecast expected outcomes for individuals and populations under expected future environmental scenarios (Semeniuk et al. 2012a; Cooke et al. 2013; Madliger et al. 2015). As outlined below, this type of scaled, predictive approach uses previous findings to cement mechanistic linkages to build strong predictive modelling capacity, and as a collective result has the power to better inform conservation decisions across a diversity of systems.

Since physiology links the organism to its environment, we can use a multitude of traits to quantify how both small-scale (e.g. variation in habitat/resource quality) and large-scale (e.g. variation in weather, climate) environmental variation influences organismal functioning (Madliger et al. 2015). Examples of emerging success stories range from the finer-scale examination of how human-induced increases in turbidity from sediment dredging in the Great Barrier Reef impact respiratory physiology and gill microbiome responses in larval reef fish (Hess et al. 2015, 2017; Illing and Rummer 2017), to using energetic physiology and stable isotopes in wide-ranging oceanic top predators such as seabirds, tuna, and sharks to monitor how global change is impacting resource acquisition and key biological processes (Hennin et al. 2016; Ferguson et al. 2017; Pethybridge et al. 2018; Descamps et al. 2019; Lorrain et al. 2020; Madigan et al., Chapter 5, this volume). Bringing this field together is the unified framework of 'macro-physiology'—the investigation of variation in physiological traits over large geographic and temporal scales and the ecological implications of this variation (Chown and Gaston 2016)—which is enabling researchers to predict and test a diversity of environment–physiology relationships (Lennox et al. 2018).

The next step is to quantify how environmentally related changes in centrally regulated physiological traits affect performance and fitness (Madliger and Love 2016b). Variation in breeding phenology/investment/success and survival are all being predicted by traits as diverse as energetic and glucocorticoid physiology (Madliger and Love 2014, 2015; Hennin et al. 2016, 2018, 2019; Sorenson et al. 2017; Minke-Martin et al. 2017; Crossin and Williams, Chapter 2, this volume), oxidative stress (Guindre-Parker et al. 2013; Costantini and Dell'Omo 2015), and cardiorespiratory physiology (Brownscombe et al. 2017). For example, Crossin and Williams (Chapter 2, this volume) outline how yolk precursors, the primary sources of protein and lipid in developing follicles and egg yolk, can predict breeding propensity and reproductive success in at-risk bird species. Although these types of linkages are often some of the most difficult to establish regardless of the system (Madliger et al. 2016), they are imperative for being able to scale these relationships up to quantify impacts on populations and ecosystems (Cooke and O'Connor 2010; Bergman et al. 2019; Ames et al. 2020).

Despite the expected management benefits of scaling individual variation in physiological traits up to predict population-level demographic responses to environmental variation or disturbance (Cooke and O'Connor, 2010; Madliger et al. 2018; Bergman et al. 2019), accomplishing this has been challenging in the wild (Bergman et al. 2019). However, recent successful case studies include using energetic traits to predict abrupt depopulation of managed honey bee colonies in Europe (Dainat et al. 2012; Alaux et al. 2017; López-Uribe et al. 2020; Alaux et al., Chapter 4, this volume); using resource-induced changes in glucocorticoids to predict population success in seabirds (Kitaysky et al. 2007); using environmentally induced changes in telomere length to predict negative changes in relative abundance and risk of extinction in lizard (*Zootoca vivipara*) populations across Europe (Dupoué et al. 2017); and using glucocorticoid and androgen hormones to link weather, habitat stressors, and altered social structure to predict declines in fecundity and population size in at-risk zebra (*Equus zebra zebra*) populations in South Africa (Lea et al. 2018). In other systems, a growing body of research is beginning to form a clearer picture of the links between environmental stressors, physiology, and fitness. For example, Dzal and Willis (Chapter 9, this volume) outline how a skin infection of hibernating bats (white-nose syndrome) leads to the disruption of physiological homeostasis and mortality. Research is also showing that the same degree of environmental change can result in different, individually flexible physiological decisions and therefore different fitness outcomes (Love et al. 2014; Madliger and Love 2016a), making our ability to predict larger-scale effects at the population level more complex. Ultimately, scaling processes up to determine ultimate impacts on ecosystem functioning is difficult, but progress is already being made (Schimel et al. 2007; Jungblut et al. 2017) and advances will increase with our ability to model this complexity effectively.

Indeed, the inclusion of physiological traits within the field of predictive modelling—the ability to model future outcomes of individuals, populations, and species based on variation in underlying phenotypic responses to environmental change—under future environmental scenarios is already a reality (e.g. Pirotta et al. 2018). Models spanning multiple physiological systems are being used to predict the tolerance and spread of invasive species (Kolbe et al. 2010; Seebacher and Franklin 2011; Higgins and Richardson 2014; Marras et al. 2015; Winwood-Smith et al. 2015), determine broad-scale host–parasite interactions (Rohr et al. 2013), predict the spread of novel diseases and disease dynamics (Legagneux et al. 2014; Ceccato et al. 2016; Becker et al. 2018), temporally and spatially define source vs. sink populations (Whitlock et al. 2015), examine how industrial activity will impact at-risk species (Muhly et al. 2011; Semeniuk et al. 2012a, b, 2014), and assess how the impacts of tourism and human decision-making will influence population viability (Semeniuk et al. 2010; Semeniuk, Chapter 8, this volume). Recently, studies using physiological traits across a diversity of taxa are helping to predict species responses to the effects of climate change (e.g. Farrell et al. 2008; Kearney and Porter 2009; Wilczek et al. 2010; Dey et al. 2017, 2018). The strengthening of this predictive capacity will enable practitioners to quantify how further changes to habitat, ecosystem, and climatic functioning will impact species continuance over vast spatiotemporal scales (Madliger et al. 2017).

1.3 Layout of the book: what to expect

We view this book as an opportunity to convey the current status of the field of conservation physiology, while also providing examples of research and practice that will be relevant into the future. We have structured the book as a series of 'case studies'—overviews of bodies of research that illustrate the variety of taxa, tools, and conservation issues that conservation physiologists are addressing. The case study chapters vary in their scope. Some cover topics that are relatively new to the discipline's research space or just gaining traction, such as the landscape ecology and physiology of bees (Alaux et al., Chapter 4, this volume) or the use of physiology in social-ecological models to mitigate human wildlife conflict (Semeniuk, Chapter 8, this volume). As a result, some chapters will detail a single case study, with developed sections on how to build the theory and implement the tools more extensively moving forwards. Other topics have been addressed using a conservation physiology approach more widely, and such chapters incorporate multiple case studies. For example, Crossin and Williams (Chapter 2, this volume) highlight work in four different avian systems to illustrate how physiological measurements can provide information on reproductive status that is otherwise nearly impossible to obtain. Cramp et al. (Chapter 6, this volume) similarly cover four case studies where physiology has contributed to understanding and rectifying barriers to fish passage in freshwater ecosystems. We believe this type of chapter structure is an ideal illustration of how the field of conservation physiology continues to build from a strong foundation of fundamental and applied work.

Through the 16 case study chapters, the text will cover a diversity of taxa (reptiles, amphibians, birds, fish, mammals, insects, crustaceans), ecosystems (terrestrial, freshwater, marine), and conservation questions (monitoring environmental stress, predicting the impact of climate change, understanding disease dynamics, improving captive breeding, reducing human–wildlife conflict). The tools and techniques included are also highly varied, spanning stress, energetic, immune, nutritional, cardiorespiratory, and reproductive physiology. They range from metrics that are considered relatively 'simple' (e.g. body condition assessed through photography) to those requiring more complex experimental set-ups (e.g. respirometry). They also vary in the degree to which an animal must be handled for collection, with some (like faecal samples for hormone analysis) being quite noninvasive, and others involving capture and handling (e.g. swabs to assess skin infection). We view the great variety of tools available in the conservation physiology toolbox as a benefit, as researchers and practitioners can choose the techniques that best suit their questions and constraints. By acting as practical roadmaps across a diversity of subdisciplines, we hope the case studies can serve to increase the accessibility of this discipline to new researchers, illustrate the far-reaching nature of the field, and allow readers to gain an appreciation of the purpose, value, and status of the field of conservation physiology. We would also like to draw readers' attention to the take-home messages that are included at the start of each chapter. These are designed to provide a one- or two-sentence overview of what the chapter will cover, but also to illustrate the broader goals that can be accomplished with a conservation physiology approach.

We acknowledge that the book is focused only on animals, as we felt that this arm of conservation physiology is the most established. Indeed, only 3 per cent of the papers published in the journal *Conservation Physiology* in the 5 years after its launch (2013–2018) focused on plants (Madliger et al. 2018). However, we recognize the value and success of such research (Madliger et al. 2016) and hope to see its continued growth.

Following the case studies in the main section of the book, we have included two concluding chapters. In the first, Laubenstein and Rummer (Chapter 18, this volume) cover some of the important goals of conservation physiology that extend beyond acquiring data and publishing results. Here, readers will find information on establishing relationships with stakeholders, carrying research through to application, and sharing their findings with the broader public. Finally, our synthesis and conclusion chapter (Cooke et al., Chapter 19, this volume) outlines 12 themes that came to light through the series of case studies, the challenges and gaps the discipline currently faces, and a final message of optimism for future growth and cohesion with the ultimate aim of ensuring the conservation of earth's remarkable biodiversity.

References

Alaux, C., Allier, F., Decourtye, A. et al., 2017. A 'Landscape physiology' approach for assessing bee health highlights the benefits of floral landscape enrichment and semi-natural habitats. *Scientific Reports*, 7(1), 1–10.

Ames, E.M., Gade, M.R., Nieman, C.L. et al., 2020. Striving for population-level conservation: integrating physiology across the biological hierarchy. *Conservation Physiology*, 8(1), coaa019.

Ayres, K.L., Booth, R.K., Hempelmann, J.A. et al., 2012. Distinguishing the impacts of inadequate prey and vessel traffic on an endangered killer whale (*Orcinus orca*) population. *PLoS ONE*, 7(6), e36942.

Bahamonde, P.A., Feswick, A., Isaacs, M.A. et al., 2016. Defining the role of omics in assessing ecosystem health:

perspectives from the Canadian environmental monitoring program. *Environmental Toxicology and Chemistry*, 35(1), 20–35.

Becker, D.J., Hall, R.J., Forbes, K.M. et al., 2018. Anthropogenic resource subsidies and host–parasite dynamics in wildlife. *Philosophical Transactions of the Royal Society of London*, 373(1745), 20170086. 10.1098/rstb.2017.0086.

Berger, J., Testa, J.W., Roffe, T., and Monfort, S.L., 1999. Conservation endocrinology: a noninvasive tool to understand relationships between carnivore colonization and ecological carrying capacity. *Conservation Biology*, 13(5), 980–9.

Bergman, J.N., Bennett, J.R., Binley, A.D. et al., 2019. Scaling from individual physiological measures to population-level demographic change: case studies and future directions for conservation management. *Biological Conservation*, 238, 108242.

Bitman, J., Cecil, H.C., Harris, S.J., and Fries, G.F., 1969. DDT induces a decrease in eggshell calcium. *Nature*, 224(5214), 44–6.

Brownscombe, J.W., Cooke, S.J., Algera, D.A. et al., 2017. Ecology of exercise in wild fish: integrating concepts of individual physiological capacity, behavior, and fitness through diverse case studies. *Integrative and Comparative Biology*, 57, 281–92.

Bryan, H.M., Darimont, C.T., Paquet, P.C. et al., 2013. Stress and reproductive hormones in grizzly bears reflect nutritional benefits and social consequences of a salmon foraging niche. *PLoS ONE*, 8(11), e80537.

Carey, C., 2005. How physiological methods and concepts can be useful in conservation biology. *Integrative and Comparative Biology*, 45(1), 4–11.

Ceccato, E., Cramp, R.L., Seebacher, F., and Franklin, C.E., 2016. Early exposure to ultraviolet-B radiation decreases immune function later in life. *Conservation Physiology*, 4, cow037.

Chown, S.L. and Gaston, K.J., 2008. Macrophysiology for a changing world. *Proceedings of the Royal Society B: Biological Sciences*, 275(1642), 1469–78.

Chown, S.L. and Gaston, K.J., 2016. Macrophysiology—progress and prospects. *Functional Ecology*, 30, 330–44.

Clark, C.T., Horstmann, L., and Misarti, N., 2020. Zinc concentrations in teeth of female walruses reflect the onset of reproductive maturity. *Conservation Physiology*, 8(1), coaa029.

Cooke, S.J. and O'Connor, C.M., 2010. Making conservation physiology relevant to policy makers and conservation practitioners. *Conservation Letters*, 3, 159–66.

Cooke, S.J., Birnie-Gauvin, K., Lennox, R.J. et al, 2017a. How experimental biology and ecology can support evidence-based decision-making in conservation: avoiding pitfalls and enabling application. *Conservation Physiology*, 5(1), cox043.

Cooke, S.J., Hultine, K.R., Rummer, J.L., and Franklin, C.E., 2017b. Reflections and progress in conservation physiology. *Conservation Physiology*, 5(1), cow071.

Cooke, S.J., Madliger, C.L., Cramp, R.L. et al., 2020. Reframing conservation physiology to become more inclusive, integrative, relevant and forward-looking: reflections and a horizon scan. *Conservation Physiology*, 8(1), coaa016.

Cooke, S.J., Sack, L., Franklin, C.E. et al., 2013. What is conservation physiology? Perspectives on an increasingly integrated and essential science. *Conservation Physiology*, 1, cot001.

Costantini, D. and Dell'Omo. G., 2015. Oxidative stress predicts long-term resight probability and reproductive success in Scopoli's shearwater (*Calonectris diomedea*). *Conservation Physiology*, 3(1), cov024.

Creel, S., Christianson, D., Liley, S., and Winnie, J.A., 2007. Predation risk affects reproductive physiology and demography of elk. *Science*, 315(5814), 960.

Creel, S., Winnie, J.A., and Christianson, D., 2009. Glucocorticoid stress hormones and the effect of predation risk on elk reproduction. *Proceedings of the National Academy of Sciences*, 106(30), 12388–93.

Dainat, B., Evans, J.D., Chen, Y.P. et al., 2012. Predictive markers of honey bee colony collapse. *PLoS ONE*, 7, 10.1371.

Dehnhard, M., Naidenko, S., Frank, A. et al., 2008. Non-invasive monitoring of hormones: a tool to improve reproduction in captive breeding of the Eurasian lynx. *Reproduction in Domestic Animals*, 43, 74–82.

Descamps, S., Ramírez, F., Benjaminsen, S. et al., 2019. Diverging phenological responses of Arctic seabirds to an earlier spring. *Global Change Biology*, 25, 4081–91.

Deutsch, C., Ferrel, A., Seibel, B. et al., 2015. Climate change tightens a metabolic constraint on marine habitats. *Science*, 348(6239), 1132–5.

Dey, C.J., Richardson, E., McGeachy, D. et al., 2017. Increasing nest predation will be insufficient to maintain polar bear body condition in the face of sea-ice loss. *Global Change Biology*, 23, 1821–31.

Dey, C.J., Semeniuk, C.A., Iverson, S.A. et al., 2018. Forecasting the outcome of multiple effects of climate change on northern common eiders. *Biological Conservation*, 220, 94–103.

Dupoué, A., Rutschmann, A., Le Galliard, J.F. et al., 2017. Shorter telomeres precede population extinction in wild lizards. *Scientific Reports*, 7, 16976.

Farrell, A.P., Hinch, S.G., Cooke, S.J. et al., 2008. Pacific salmon in hot water: applying aerobic scope models and biotelemetry to predict the success of spawning migrations. *Physiological and Biochemical Zoology*, 81(6), 697–708.

Feder, M.E., Bennett, A.F., Burggren, W.W., and Huey, R.B., 1987. *New Directions in Ecological Physiology*. Cambridge University Press, Cambridge.

Ferguson, S.H., Young, B.G., Yurkowski, D.J. et al., 2017. Demographic, ecological, and physiological responses of ringed seals to an abrupt decline in sea ice availability. *PeerJ*, 5, e2957.

Ge, Y., Wang, D.Z., Chiu, J.F. et al, 2013. Environmental OMICS: current status and future directions. *Journal of Integrated OMICS*, 3(2), 75–87.

Gomez Isaza, D.F., Cramp, R.L., and Franklin, C.E., 2020. Simultaneous exposure to nitrate and low pH reduces the blood oxygen-carrying capacity and functional performance of a freshwater fish. *Conservation Physiology*, 8(1), coz092.

Guindre-Parker, S., Baldo, S., Gilchrist, H.G. et al., 2013. The oxidative costs of territory quality and offspring provisioning. *Journal of Evolutionary Biology*, 26, 2558–65.

Hennin, H.L., Dey, C., Bety, J. et al., 2018. Higher rates of pre-breeding condition gain positively impacts clutch size: a mechanistic test of the condition-dependent individual optimization model. *Functional Ecology*, 32, 2019–28.

Hennin, H.L., Legagneux, P., Bêty, J. et al., 2016. Energetic physiology mediates individual optimization of breeding phenology in a migratory Arctic seabird. *American Naturalist*, 188, 434–45.

Hennin, H.L., Legagneux, P., Gilchrist, H.G. et al., 2019. Plasma mammalian leptin analogue predicts reproductive phenology, but not reproductive output in a capital-income breeding seaduck. *Ecology and Evolution*, 9, 1512–22.

Hess, S., Prescott, L.J., Hoey, A.S. et al., 2017. Species-specific impacts of suspended sediments on gill structure and function in coral reef fishes. *Proceedings of the Royal Society of London*, 284, 10.1098/rspb.2017.1279

Hess, S., Wenger, A.S., Ainsworth, T.D., and Rummer, J.L., 2015. Exposure of clownfish larvae to suspended sediment levels found on the Great Barrier Reef: impacts on gill structure and microbiome. *Scientific Reports*, 5, 10561.

Higgins S.I. and Richardson, D.M., 2014. Invasive plants have broader physiological niches. *Proceedings of the National Academy of Sciences of the USA*, 111, 10610–10614.

Hudson, N.J. and Franklin, C.E., 2002. Effect of aestivation on muscle characteristics and locomotor performance in the green-striped burrowing frog, *Cyclorana albogutta*. *Journal of Comparative Physiology B—Biochemical Systemic and Environmental Physiology*, 172, 177–82.

Hunninck, L., May, R., Jackson, C.R. et al., 2020. Consequences of climate-induced vegetation changes exceed those of human disturbance for wild impala in the Serengeti ecosystem. *Conservation Physiology*, 8(1), coz117.

Hunt, K.E., Stimmelmayr, R., George, C. et al., 2014. Baleen hormones: a novel tool for retrospective assessment of stress and reproduction in bowhead whales (*Balaena mysticetus*). *Conservation Physiology*, 2, cou030.

Illing, B. and Rummer, J.L., 2017. Physiology can contribute to better understanding, management, and conservation of coral reef fishes. *Conservation Physiology*, 5, cox005.

Jefferies, D.J., 1969. Induction of apparent hyperthyroidism in birds fed DDT. *Nature*, 222(5193), 578–9.

Johnson, K.M. and Hofmann, G.E., 2020. Combined stress of ocean acidification and warming influence survival and drives differential gene expression patterns in the Antarctic pteropod, *Limacina helicina antarctica*. *Conservation Physiology*, 8(1), coaa013.

Joly, K., Wasser, S.K., and Booth, R., 2015. Non-invasive assessment of the interrelationships of diet, pregnancy rate, group composition, and physiological and nutritional stress of barren-ground caribou in late winter. *PLoS ONE*, 10(6), e0127586.

Jungblut, S., Boos, K., McCarthy, M. et al., 2017. Respiration physiology and ecosystem impact of European and Asian shore crabs in a temperate European habitat. The Crustacean Society Mid-Year Meeting, Barcelona, Spain, 19–22 June 2017.

Kearney M. and Porter, W., 2009. Mechanistic niche modelling: combining physiological and spatial data to predict species' ranges. *Ecology Letters*, 12, 334–50.

Kern P., Cramp R.L., Gordos M.A. et al., 2018. Measuring Ucrit and endurance: equipment choice influences estimates of fish swimming performance. *Journal of Fish Biology*, 92(1), 237–47.

Kitaysky, A., Piatt, J., and Wingfield, J., 2007. Stress hormones link food availability and population processes in seabirds. *Marine Ecology Progress Series*, 352, 245–58.

Kolbe, J.J., Kearney, M., and Shine, R., 2010. Modeling the consequences of thermal trait variation for the cane toad invasion of Australia. *Ecological Applications*, 20, 2273–85.

Krebs, C.J., 1989. *Ecological Methodology*. Harper & Row, New York.

Lea, J.M.D., Walker, S.L., Kerley, G.I.H. et al., 2018. Noninvasive physiological markers demonstrate link between habitat quality, adult sex ratio and poor population growth rate in a vulnerable species, the Cape mountain zebra. *Functional Ecology*, 32, 300–12.

Legagneux, P., Berzins, L.L., Forbes, M. et al., 2014. No selection on immunological markers in response to a highly virulent pathogen in an Arctic breeding bird. *Evolutionary Applications*, 7(7), 765–73. 10.1111/eva.12180.

Lennox, R. and Cooke, S.J., 2014. State of the interface between conservation and physiology: a bibliometric analysis. *Conservation Physiology*, 2, cou003.

Lennox, R.J., Suski, C.D., and Cooke, S.J., 2018. A macrophysiology approach to watershed science and management. *Science of the Total Environment*, *626*, 434–40.

López-Uribe, M.M., Ricigliano, V.A., and Simone-Finstrom, M., 2020. Defining pollinator health: a holistic approach based on ecological, genetic, and physiological factors. *Annual Review of Animal Biosciences*, *8*, 269–94.

Lorrain, A., Pethybridge, H., Cassar, N. et al., 2020. Trends in tuna carbon isotopes suggest global changes in pelagic phytoplankton communities. *Global Change Biology*, *26*, 458–70.

Love, O.P., Madliger, C.L., Bourgeon, S. et al., 2014. Evidence for baseline glucocorticoids as mediators of reproductive investment in a wild bird. *General and Comparative Endocrinology*, *199*, 65–9.

Madliger, C.L., Cooke, S.J., Crespi, E.J. et al., 2016. Success stories and emerging themes in conservation physiology. *Conservation Physiology*, *4*, cov57.

Madliger, C.L. Franklin, C.E., Hultine, K.R. et al., 2017. Conservation physiology and the quest for a 'good' Anthropocene. *Conservation Physiology*, *15*, cox003.

Madliger, C.L. and Love, O.P., 2014. The need for a predictive, context-dependent approach to the application of stress hormones in conservation. *Conservation Biology*, *28*, 283–7.

Madliger, C.L. and Love, O.P., 2015. The power of physiology in changing landscapes: considerations for the continued integration of conservation and physiology. *Integrative and Comparative Biology*, *55*, 545–53.

Madliger, C.L. and Love, O.P., 2016a. Employing individual measures of baseline glucocorticoids as population-level conservation biomarkers: considering within-individual variation in a breeding passerine. *Conservation Physiology*, *4*, cow048.

Madliger, C.L. and Love, O.P., 2016b. Conservation implications of a lack of relationship between baseline glucocorticoids and fitness in a wild passerine. *Ecological Applications*, *26*, 2732–45.

Madliger, C.L., Love, O.P., Hultine, K., and Cooke, S.J., 2018. The conservation physiology toolbox: status and opportunities. *Conservation Physiology*, *6*, coy029.

Madliger, C.L., Semeniuk, C.A.D., Harris, C.M. and Love, O.P., 2015. Assessing baseline stress physiology as an integrator of environmental quality in a wild avian population: implications for use as a conservation biomarker. *Biological Conservation*, *192*, 409–17.

Marras, S., Cucco, A., Antognarelli, F. et al., 2015. Predicting future thermal habitat suitability of competing native and invasive fish species: from metabolic scope to oceanographic modelling. *Conservation Physiology*, *3*(1), cou059.

Mayr, E., 1961. Cause and effect in biology. *Science*, *134*(3489), 1501–6.

McClane, A.J., Semeniuk, C.A.D., and Marceau, D., 2011. The role of agent-based models in wildlife ecology and management: the importance of accommodating individual habitat-selection behaviours and spatially explicit movement for conservation planning. *Ecological Modelling*, *222*, 1544–56.

McCoy, C.M., Lind, C.M., and Farrell, T.M., 2017. Environmental and physiological correlates of the severity of clinical signs of snake fungal disease in a population of pigmy rattlesnakes, *Sistrurus miliarius*. *Conservation Physiology*, *5*(1), cow077.

McMahon, B.J., Teeling, E.C., and Höglund, J., 2014. How and why should we implement genomics into conservation? *Evolutionary Applications*, *7*(9), 999–1007.

Millspaugh, J.J. and Washburn, B.E., 2004. Use of fecal glucocorticoid metabolite measures in conservation biology research: considerations for application and interpretation. *General and Comparative Endocrinology*, *138*, 189–199.

Minke-Martin, V., Hinch, S.G., Braun, D.C. et al., 2017. Physiological condition and migratory experience affect fitness-related outcomes in adult female sockeye salmon. *Ecology of Freshwater Fish*, *27*, 296–309.

Muhly, T.B., Semeniuk, C.A.D., Massolo, A. et al., 2011. Human activity helps prey win the predator-prey space race. *PLoS ONE*, *6*, e17050.

Pethybridge, H., Choy, C.A., Logan, J.M. et al., 2018. A global meta-analysis of marine predator nitrogen stable isotopes: relationships between trophic structure and environmental conditions. *Global Ecology and Biogeography*, *27*, 1043–55.

Pirotta, E., Mangel, M., Costa, D.P. et al., 2018. A dynamic state model of migratory behavior and physiology to assess the consequences of environmental variation and anthropogenic disturbance on marine vertebrates. *American Naturalist*, *191*, E40–56.

Robeck, T.R., Steinman, K.J., Gearhart, S. et al., 2004. Reproductive physiology and development of artificial insemination technology in killer whales (*Orcinus orca*). *Biology of Reproduction*, *71*(2), 650–60.

Rohr, J.R., Raffel, T.R., Blaustein, A.R. et al., 2013. Using physiology to understand climate-driven changes in disease and their implications for conservation. *Conservation Physiology*, *1*(1), cot022.

Schimel, J., Balser, T.C., and Wallenstein, M., 2007. Microbial stress-response physiology and its implications for ecosystem function. *Ecology*, *88*, 1386–94.

Seebacher, F. and Franklin, C.E., 2011. Physiology of invasion: cane toads are constrained by thermal effects on physiological mechanisms that support locomotor performance. *Journal of Experimental Biology*, *214*(9), 1437–44.

Seebacher, F. and Franklin, C.E., 2012. Determining environmental causes of biological effects: the need for a mechanistic physiological dimension in conservation biology. *Philosophical Transactions of the Royal Society B: Biological Sciences*, 367(1596), 1607–14.

Semeniuk, C.A.D., Haider, W., Cooper, A., and Rothley, K.D., 2010. A linked model of animal ecology and human behaviour for the management of wildlife tourism. *Ecological Modelling*, 221, 2699–713.

Semeniuk, C.A.D., Musiani, M., Birkigt, D.A. et al., 2014. Identifying non-independent anthropogenic risks using a behavioral individual-based model. *Ecological Complexity*, 17, 67–78.

Semeniuk, C.A.D., Musiani, M. Hebblewhite, M. et al., 2012a. Evaluating risk effects of industrial features on woodland caribou habitat selection in west central Alberta using agent-based modelling. *Procedia Environmental Sciences*, 13, 698–714.

Semeniuk, C.A.D., Musiani, M., Hebblewhite, M. et al., 2012b. Incorporating behavioural-ecological strategies in pattern-oriented modelling of caribou habitat use in a highly industrialized landscape. *Ecological Modelling*, 243, 18–32.

Sorenson, G.H., Dey, C., Madliger, C.L., and Love, O.P., 2017. Effectiveness of baseline corticosterone as a monitoring tool for fitness: a meta-analysis in seabirds. *Oecologia*, 183, 353–65.

Soulé, M.E., 1985. What is conservation biology? *BioScience*, 35(11), 727–34.

Sutherland, W.J. and Wordley, C.F., 2017. Evidence complacency hampers conservation. *Nature Ecology & Evolution*, 1(9), 1215–16.

Stevenson, R.D., Tuberty, S.R., DeFur, P.L. and Wingfield, J.C., 2005. Ecophysiology and conservation: the contribution of endocrinology and immunology—introduction to the symposium. *Integrative and Comparative Biology*, 45(1), 1–3.

Swanson, W.F., 2003. Research in nondomestic species: experiences in reproductive physiology research for conservation of endangered felids. *ILAR Journal*, 44(4), 307–16.

Tang, R., Li, W., Zhu, D. et al., 2020. Raging elephants: effects of human disturbance on physiological stress and reproductive potential in wild Asian elephants. *Conservation Physiology*, 8(1), coz106.

Tracy, C.R., Nussear, K.E., Esque, T.C. et al., 2006. The importance of physiological ecology in conservation biology. *Integrative and Comparative Biology*, 46(6), 1191–205.

Wasser, S.K., Keim, J.L., Taper, M.L., and Lele, S.R., 2011. The influences of wolf predation, habitat loss, and human activity on caribou and moose in the Alberta oil sands. *Frontiers in Ecology and the Environment*, 9(10), 546–51.

Whitlock, R.E., Hazen, E.L., Walli, A. et al., 2015. Direct quantification of energy intake in an apex marine predator suggests physiology is a key driver of migrations. *Science Advances*, 1, e1400270.

Wikelski, M. and Cooke, S.J., 2006. Conservation physiology. *Trends in Ecology & Evolution*, 21(1), 38–46.

Wilczek, A.M., Burghardt, L.T., Cobb, A.R. et al., 2010. Genetic and physiological bases for phenological responses to current and predicted climates. *Philosophical Transactions of the Royal Society of London Biological Sciences*, 365, 3129–47.

Willmer, P., Stone, G., and Johnston, I., 2009. *Environmental Physiology of Animals*, second edition. Blackwell Science Ltd, Oxford.

Wilson, R.S., Franklin, C.E., Davison, W., and Kraft, P., 2001. Stenotherms at sub-zero temperatures: thermal dependence of swimming performance in Antarctic fish. *Journal of Comparative Physiology B—Biochemical Systemic and Environmental Physiology*, 171, 263–9.

Winwood-Smith, H.S., Alton, L.A., Franklin, C.E., and White, C.R., 2015. Does greater thermal plasticity facilitate range expansion of an invasive terrestrial anuran into higher latitudes? *Conservation Physiology*, 3(1), cov010.

PART I

Monitoring and Managing Wild Populations

CHAPTER 2

Using physiology to infer the reproductive status and breeding performance of cryptic or at-risk bird species

Glenn T. Crossin and Tony D. Williams

⊃ **Take-home message**

The measurement of yolk precursors in the plasma of breeding birds can provide a useful means for assessing the reproductive status of threatened and endangered birds, which can provide those tasked with the conservation and management of rare and cryptic bird species important information about their phenology and population dynamics.

2.1 Introduction

Spatial and temporal variation in breeding parameters (e.g. date of breeding initiation or 'lay date' in birds, number of offspring, etc.) are key determinants of population ecology (Williams 2012). For species considered at-risk (special concern, threatened, endangered), declines in breeding productivity could foretell population declines, and so documenting patterns of reproductive investment is critical for determining conservation status (Newton 1979; Newton 1998; Krebs 1985). Historically, the estimation of breeding parameters for most bird populations has relied on conventional methods where researchers locate nests and count the numbers of eggs or chicks produced (Williams 2012). From such data, population size and breeding productivity can be estimated directly

(Perrins and Moss 1975; Dunn et al. 2011). However, for cryptic, rare, and at-risk populations these simple methods can be problematic if such species are difficult to locate and study, or are sensitive to disturbance. For example, marbled murrelets (*Brachyramphus marmoratus*), a threatened seabird, typically nest in the canopies of large trees within old-growth forests of the Pacific Northwest, often in isolated, remote locations. Similarly, harlequin ducks (*Histrionicus histrionicus*) are difficult to study during the breeding season as they nest in the riparian zones of remote boreal, sub-Arctic streams. Even when nests of rare or cryptic species can be located, there are often concerns about negative effects of nest disturbance and handling on breeding outcomes (Moran-Lopez et al. 2006; Ellenberg et al. 2013; Jorgensen et al. 2016).

Glenn T. Crossin and Tony D. Williams, *Using physiology to infer the reproductive status and breeding performance of cryptic or at-risk bird species*
In: *Conservation Physiology: Applications for Wildlife Conservation and Management.* Edited by: Christine L. Madliger, Craig E. Franklin, Oliver P. Love, and Steven J. Cooke, Oxford University Press (2021). © Oxford University Press. DOI: 10.1093/oso/9780198843610.003.0002

Therefore, for cryptic and at-risk species, accurately determining even very basic breeding parameters can be very difficult. Population estimates might then have to rely on less invasive methods like point-counts, or on capture of breeding individuals away from their nests. However, birds opportunistically captured away from nests will most often be of unknown breeding status, and so lethal examination of reproductive tract development, or behavioural observation for traits associated with breeding activity, has been used in the past to estimate a bird's reproductive status (Ankney 1977; Vézina and Williams 2003). These approaches are often both undesirable (e.g. lethal collection) and very time-intensive (e.g. behavioural observations). In such cases physiological analysis of blood samples can be a powerful means for determining individual sex and breeding status, as well as providing estimates of breeding phenology and effective population size. Hormonal and physiological changes associated with the avian reproductive cycle have been well characterized in a wide range of birds (Wingfield and Farner 1978; Dawson 1983; Mays et al. 1991; Cockrem and Seddon 1994; Christians and Williams 1999). Here we will focus on analysis of yolk precursors in female birds, which we argue is an especially useful, relatively non-invasive, and simple technique for identifying the stage-specific breeding profiles differentiating the reproductive state of individuals in a mixed population (Vanderkist et al. 2000).

2.2 What are yolk precursors and why are they useful?

When we speak of yolk precursors in birds and other oviparous vertebrates, we refer specifically to vitellogenin (VTG) and yolk-targeted very-low-density lipoprotein (VLDLy), which are the primary sources of protein and lipid in developing follicles and egg yolk, ultimately providing all of the nutritional and energetic needs of a developing embryo (Deeley et al. 1975; Wallace 1985; Walzem et al. 1999; Williams 2012). In egg-producing females, the synthesis of these compounds occurs in the liver, in

response to rising plasma levels of oestrogen (e.g. 17ß-oestradiol), which are in turn a response to seasonally increasing levels of gonadotropins (luteinizing hormone [LH], and follicle stimulating hormone [FSH]). VLDLy can be distinguished from generic VLDL, which is synthesized constitutively in non-breeding females (and males) and which functions to meet their metabolic needs (Walzem 1996). VLDLy has a smaller particle size compared with generic VLDL, a different apolipoprotein composition, and is resistant to breakdown by lipase enzymes (Williams 2012). Synthesis of VLDLy, at the onset of egg formation, thus marks a significant shift in the lipid metabolism of a female; once lipids are modified and targeted for deposition into yolky follicles, VLDLy cannot be drawn upon as an energy source by the female—it is committed to the egg follicle (Williams 2012). Measurement of generic VLDL itself, outside the period of rapid follicle growth, has been related to energetic management (e.g. Hennin et al. 2015; Hennin et al. 2016) and can be used to infer fattening rates (e.g. Williams et al. 2007; Evans Ogden et al. 2013). In egg-producing females, VLDLy and VTG are secreted from the liver into circulation and taken up by ovarian follicles via receptor-mediated endocytosis to form yolk prior to ovulation (Williams 2012). Normally, VTG is not present in males or non-egg-producing females so it is undetectable in both males and females outside of the breeding season. However, during the breeding season, both VTG and VLDLy are useful biomarkers of egg production in female birds, with VTG generally regarded as the more robust plasma indicator of follicle development and vitellogenesis, although VLDL has the advantage of requiring smaller plasma volumes for analysis (10 μl vs. 75 μl; Challenger et al. 2001).

Measurement of yolk precursors has proven to be a valuable technique for assessing reproduction in cryptic and at-risk species because elevated plasma levels of VLDL and VTG are very tightly coupled to egg production, that is, they only occur in fecund females (Figure 2.1). Early studies made use of simple, colourimetric assays developed by poultry researchers (Mitchell and Carlisle 1991) for

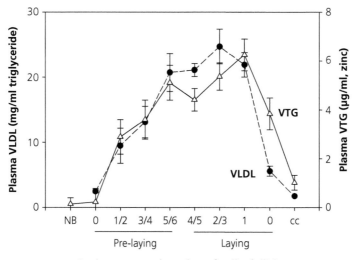

Figure 2.1 Relationship between plasma levels of the two yolk precursors, very-low-density lipoprotein (VLDL) and vitellogenin (VTG), laying stage, and the number of yolky follicles present in the ovary in female European starlings (*Sturnus vulgaris*). NB = non-breeding birds (sampled 3 weeks before first egg), CC = clutch completion. Reproduced from Challenger et al. (2001) with the permission of the University of Chicago Press.

assessing reproduction in egg-laying vertebrates, which require only small volumes of plasma. Specifically, these assays involved the measurement of plasma zinc, which provides a robust index of VTG, and of plasma triglyceride, which provides a measure of VLDL (either generic VLDL in non-breeders or VLDLy in females undergoing egg formation). Importantly, critical validation of this technique for use in wild or other non-domesticated bird species relied on laboratory studies of 'model' species (e.g. zebra finches, *Taeniopygia guttata*; Salvante and Williams 2002) or tractable study species in the wild (nest-box breeding, invasive European starlings, *Sturnus vulgaris*; Challenger et al. 2001). It is also important to note that none of the early studies that were critical to the development and application of yolk precursor analysis as a conservation tool were actually initiated with an applied, conservation-related question in mind, nor were they funded by conservation dollars: the studies of Challenger et al. (2001), Salvante and Williams (2002), and Salvante et al. (2007) represented basic research aimed at elucidating physiological mechanisms underpinning egg production in free-living female birds.

In this chapter, we will highlight several case studies in which measurement of yolk precursors provided a means for assessing the breeding status of threatened and endangered birds. We also highlight studies wherein yolk precursors are measured in free-living birds that are not yet threatened, but which are of special concern due to dramatic changes in environmental conditions, especially in the Arctic and Antarctic. Finally, we present examples where variation in yolk precursor levels has been used in toxicological studies to trace the maternal transfer of organic contaminants into eggs.

2.3 Case study 1: marbled murrelets

Marbled murrelets (*Brachyramphus marmoratus*) are marine seabirds, but they are atypical members of the alcid or auk family in that, although they forage in marine habitats, they nest almost exclusively in coastal, coniferous forests along the west coast of North America up to 30–50 km inland (Nelson 1997). Marbled murrelets require mossy nesting 'platforms' located on large tree limbs, and these occur mainly in old-growth coniferous trees that are >30 m tall and at least 140–200 years old (Burger 2002). Consequently, the marbled murrelet was identified early on as an indicator species of coastal old-growth forests in the Pacific Northwest (along with the spotted owl, *Strix occidentalis*; Abate 1992; COSEWIC 2012). Loss of old-growth forests was estimated at over 20 per cent between 1970 and 2000, which represents a direct loss of essential breeding habitat for the species, and marbled murrelets face additional threats from proposed shipping routes, increased habitat fragmentation, fisheries bycatch, and changing at-sea conditions (COSEWIC 2012). This led to marbled murrelets being International Union for Conservation of Nature (IUCN) Red Listed as 'Endangered', Federally listed under the Endangered Species Act as a threatened species in Washington, Oregon, and California, and (in 1990) designated as threatened in Canada (listed in Schedule 1 of the Species at Risk Act [SARA]).

Ralph et al. (1995) highlighted the fact that a lack of basic information on population and breeding biology was hampering the development of management and conservation protocols for marbled murrelets. Consequently, obtaining data on breeding biology and population demography of marbled murrelets became a major conservation priority in the 1990s, and one that continues to this day (e.g. in British Columbia there are still no surveys that allow a good assessment of population trends over the past 30 years, COSEWIC 2012). Due to its secretive behaviour and the comparative inaccessibility of its nesting areas, marbled murrelets were called the 'enigma of the Pacific' (Guiguet 1956). Indeed, the first active nests of this species were not found until 1974 in the USA (Binford et al. 1975) and 1993 in Canada (Nelson 1997). The largest sample of nests (*n* = 156) was obtained in British Columbia using radio-telemetry and ground searching between 1998 and 2002 (Zharikov et al. 2006) and even here basic metrics such as breeding success had to be inferred indirectly, via analysis of radio-telemetry patterns of nest attendance rather than visual surveys due to concerns about disturbance.

Consequently, it proved easier to capture and study marbled murrelets off the nest rather than via traditional methods of direct observation of nesting birds, and two technical developments were critical

here. Kaiser et al. (1995) developed a floating mist-net system that caught marbled murrelets as they flew low over the water through a narrow coastal channel between marine foraging areas and inland nesting sites at Theodosia Inlet, British Columbia; in the first year alone this method caught >220 birds. Large numbers of murrelets could therefore be caught off-nest. However, all birds were of unknown breeding status, and because this species is sexually monomorphic, not even the sex of birds was known. Molecular sexing (Griffiths et al. 1996) resolved the first problem, allowing males and females to be identified in the capture sample. Interestingly, this also highlighted a strong male bias (1.8:1) in birds mist-netted during the daytime, whereas murrelets captured using an alternative night-lighting technique (Whitworth et al. 1997) were not male-biased (Vanderkist et al. 1999), indicating sex differences in diurnal activity patterns at the capture location.

The next problem was how reproductive state could be determined for females caught off-nest. Application of the simple physiological assay for measurement of plasma VTG and VLDL (Mitchell and Carlisle 1991) allowed females to be classified as either 'egg-producing' or 'non-egg-producing'. Vanderkist et al. (2000) validated measurement of plasma levels of these yolk precursors as indices of egg production for the characterization of fecund females. Data for a species where breeding chronology could be directly assessed (Cassin's auklet, *Ptychoramphus aleuticus*) confirmed the validity of this approach: plasma VTG levels were highest during the defined egg-laying period, and the ighest proportion of females were defined as egg-producing in this period. Analysis of samples for marbled murrelets caught off-nest clearly identified a putative egg-laying phase (mid-May to early July), with a single, protracted breeding season and no evidence of a bimodal distribution, which would be associated with multiple broods (i.e. second clutches). Furthermore, analysis of body mass showed that egg-producing females, with elevated plasma VTG, were on average 40 g heavier than other females, equivalent to the mass of the single egg (36–41 g; Vanderkist et al. 2000).

Subsequently, Lougheed et al. (2002) and McFarlane Tranquilla et al. (2003a) extended the use of this yolk precursor technique in marbled murrelets. Lougheed et al. (2002) compared several methods for studying annual variation in chronology and synchrony of breeding in marbled murrelets over 3 years (1996–1998) at Desolation Sound, British Columbia. They confirmed that physiological analysis of the yolk precursor VTG could produce a complete distribution of breeding events if sampling was initiated early and conducted throughout laying. McFarlane Tranquilla et al. (2003a) confirmed that breeding in marbled murrelets is highly asynchronous but similar in both years (Figure 2.2).

They also showed that the predicted timing of chick fledging based on plasma VTG analyses was within 1 day of the first sightings of fledglings at sea, confirming that the yolk precursor technique provides accurate information on breeding chronology. At this study site, the percentage of captured females defined as 'egg producers' was relatively high and similar in 2 years: 54 per cent in 1999 and 56 per cent in 2000. McFarlane Tranquilla et al. (2003a) included a detailed discussion of how to determine a 'threshold' value for classifying egg-roducing from non-egg-producing birds. Together, data from these two studies confirmed the hypothesis that marbled murrelets are more asynchronous

in breeding than other alcids, perhaps because of their solitary nesting habits (Hamer and Nelson 1995), or as a strategy to reduce the risk of nest predation. Identification of a more prolonged, asynchronous nesting season was important in terms of planning when forestry or logging operations could proceed without (directly) impacting nesting success.

McFarlane Tranquilla et al. (2003b) combined at-sea capture of marbled murrelets with radio-tracking and yolk precursor analysis. They found that 92 per cent of the birds that were paired at capture were male–female pairs, and that paired females were more likely (73 per cent) to be producing eggs than were single females (8 per cent). These data confirmed the assumption that the majority of marbled murrelets caught or observed together as pairs early in the breeding season are mated pairs; critical information given that at-sea surveying is a commonly used technique to census marbled murrelets to estimate local productivity, population sizes, and population trends.

Peery et al. (2004a, b) subsequently applied the yolk precursor technique to marbled murrelets in California under a very different conservation scenario. They combined demographic, behavioural, and physiological data and used a 'multiple com-

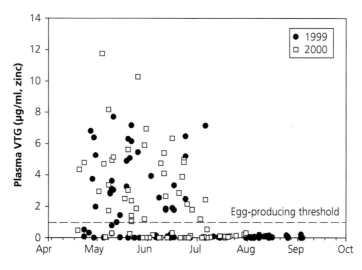

Figure 2.2 Plasma vitellogenin (VTG) levels in female marbled murrelets (*Brachyramphus marmoratus*) caught off-nest in relation to date during two breeding seasons (1999 and 2000) at Desolation Sound, British Columbia. Values above the dashed line (0.96 µg/ml, zinc) indicate egg-producing birds. Redrawn from McFarlane Tranquilla et al. (2003a) using original data provided by Dr Laura McFarlane Tranquilla. Reproduced with the permission of Oxford University Press.

peting hypothesis' approach, to determine the importance of three limiting factors—food availability, nest site availability, and nest predation—in explaining low productivity and population decline of this species in California. Although the proportion of potential breeders, with elevated plasma VTG, was similar among captured females between California and British Columbia, the percentage of known breeders (31 per cent) was half that in British Columbia, and nest success (16 per cent) was almost three times lower than in British Columbia (Bradley et al. 2004). This suggested that low reproductive performance of marbled murrelets in central California was not related to nest site availability. Rather, low food availability and/or high nest predation likely caused low reproductive success below that required to sustain the population (Peery et al. 2004a). Finally, Janssen et al. (2009) used plasma VTG levels to identify egg-producing females and combined this with diet data inferred using stable carbon (δ13C) and nitrogen (δ15N) analysis to test the hypothesis that the quality of the pre-breeding trophic feeding level positively influences breeding success in this species. There were significant differences in δ15N (reflecting trophic level of the diet) in the pre-breeding diet among egg-producing and non-egg-producing female murrelets in 2 years, but in opposite directions, with no differences in 2 other years (Figure 2.3). This

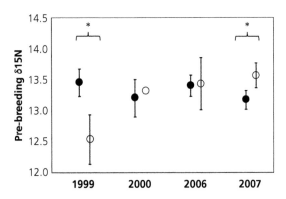

Figure 2.3 Levels of nitrogen stable isotopes (δ15N) in the red blood cells of egg-producing (solid circles) and non-egg-producing female (open circles) marbled murrelets (*Brachyramphus marmoratus*) during the pre-breeding period at Desolation Sound, British Columbia. Asterisks indicate statistically significant differences within a year (α =0.05). Error bars are ± SE. Reproduced from Janssen et al. (2009) with the permission of the Pacific Seabird Group.

work contributed to our understanding of factors underpinning long-term declines in murrelet populations.

As this case study has shown, yolk precursor analysis has shown great utility for population monitoring, and determining the breeding phenology of a cryptic, endangered seabird species. In the subsequent case studies, we build on these findings to showcase yolk precursor analysis as a means for assessing the status of bird species breeding in the rapidly changing boreal, Arctic, and Antarctic environments.

2.4 Case study 2: Arctic-nesting waterfowl

During the past 20 years, there has been increasing concern about the population status of Arctic-nesting waterfowl in North America, especially in light of the rapid, broad-scale changes occurring in the Arctic environment, including warming temperatures, earlier springs, reduced sea-ice extent, etc. (Canadian Wildlife Service Waterfowl Committee 2013). Waterfowl tend to be fairly inconspicuous in the Arctic, breeding in generally low densities and at remote locales across a wide geographic area, making it difficult to collect data for some species on basic aspects of their ecology and population dynamics. Here, we focus on three waterfowl species (scaup, harlequin duck, and common eider, each of which is of conservation concern), where yolk precursor analysis has proven valuable in improving our understanding of reproductive and population biology of these otherwise hard to study species.

2.4.1 Scaup

Greater scaup (*Aythya marila*) is one of the few diving duck species with a circumpolar distribution (Kessel et al. 2002). Breeding widely throughout the tundra of North America, Europe, and Asia during the boreal summer, usually adjacent to large lakes, scaup then migrate to the Pacific and Atlantic coasts in winter, where hundreds to thousands aggregate. Since the 1980s there has been concern about the decline of North American scaup populations (e.g. greater scaup), especially in the boreal areas of northern Canada and eastern Alaska (Austin et al. 2000; Afton and Anderson 2001). To inform the conservation status of greater scaup, a species that has particularly high energetic and nutritional egg production costs relative to other waterfowl (Flint and Grand 1999; Alisauskas and Ankney 1994), Gorman et al. (2007, 2008, 2009) examined ovarian follicle dynamics and yolk precursor production. First, by sampling females on the breeding grounds, they examined follicle dynamics among individuals and developed predictive models describing the temporal pattern of follicle growth. This allowed them to precisely define the period of rapid follicular growth (RFG), during which yolk precursors are synthesized and deposited into developing follicles, relative to individual nest initiation dates and clutch sizes. These data provided important baselines that could then be used to non-lethally predict the reproductive status of female scaup of unknown breeding status captured away from nests via yolk precursor analysis.

In highly seasonal environments like the Arctic, it is generally assumed that the birds arriving at breeding areas with the highest energetic reserves will have the best chances of breeding successfully and thus maximizing fitness (Drent and Daan 1980; Stearns 1992; Rowe et al. 1994; McNamara and Houston 1996). Generally, the earliest arriving birds are in good body condition, and early arrival and laying increases the survival probability of offspring (Drent and Daan 1980; Rowe et al. 1994; Lepage et al. 2000; Bêty et al. 2003). Gorman et al. (2008) used proximate analysis to glean insights to the energetic condition of breeding female scaup, in relation to egg production. Interestingly, endogenous lipid, protein, and mineral reserves among individuals did not show a temporal decline throughout the period of egg production and were constant irrespective of the date of onset of rapid follicle growth. This contrasts with the condition dependence of egg production and breeding success observed in most other waterfowl species, which rely to some extent on endogenous, somatically stored energy (e.g. Warren et al. 2013). Gorman et al. (2009) identified a condition threshold for the initiation of RFG, as non-breeding females had lower lipid, protein, and mineral levels relative to breeders. The high investment of scaup towards

egg production relative to other waterfowl species likely relates to the presumed 'income' strategy used by the species, via the acquisition of local area resources prior to breeding. However, in the Arctic where phenologies and environmental conditions are rapidly changing, this income strategy could lead to an increased incidence of breeding deferral in years when food availability is low.

Having established an understanding of follicle growth dynamics and the energetic requirements for breeding in female scaup, Gorman et al. (2008) then sought to relate these traits to the production of the yolk precursors VTG and VLDL at discrete stages of the annual cycle. By comparing circulating levels of both precursors to the follicle sizes within the ovary (via dissection and preservation of ovaries), changes in VTG and VLDL were, as predicted, found to be tightly coupled to follicle development. Furthermore, by sampling birds during the non-breeding season in winter and comparing with the non-developed ovary, both yolk precursors could differentiate 'non-developed' or 'non-breeding' females from active breeders, with the former having virtually no VTG or VLDL (or VLDLy) in circulation (Figure 2.4).

From a conservation perspective, these empirical relationships between yolk precursors and follicle dynamics provide a non-lethal, physiological means for determining the reproductive state of female waterfowl caught off-nest. Additionally, yolk precursor analysis can be used to discern the breeding success or failure of individuals. For example, those with low or basal levels of VTG or VLDL might be found in deferring or non-breeding individuals, or could be active breeders who failed to reproduce successfully, e.g. their eggs failed to hatch (*sensu* Peery et al. 2004a; Bond et al. 2008).

2.4.2 Harlequin ducks

Harlequin ducks (*Histrionicus histrionicus*) are small, diving ducks that breed in remote, high latitude areas, along cold, fast-moving streams in north-western and north-eastern North America, Greenland, Iceland, and eastern Russia. After breeding, they migrate to nearby areas along the Atlantic and Pacific coasts, characterized by

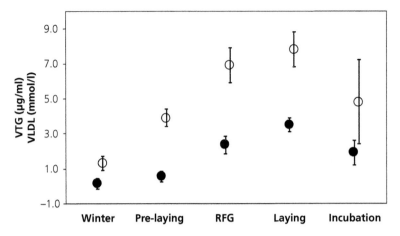

Figure 2.4 Mean concentration of vitellogenin (VTG, solid circles) and total very-low-density lipoprotein (VLDL, open circles) measured in female greater scaup (*Aythya marila*) at discrete stages of the annual cycle. RFG = rapid follicle growth stage. Values are means ± SE. Reproduced from Gorman et al. (2008) with the permission of Oxford University Press.

pounding surf and productive intertidal zones. Harlequin ducks are another highly cryptic waterfowl species of conservation concern, due in part to low and variable levels of breeding productivity (Smith et al. 2001). Like other Arctic-nesting waterfowl, harlequin ducks are thought to exhibit a high degree of non-breeding propensity (i.e. birds that choose not to breed in a given year) (Bond et al. 2008), which has implications for population-level productivity irrespective of climate change scenarios. In North America, harlequin ducks form two geographically separated groupings or populations. The eastern population consists of only a few thousand individuals (COSEWIC 2012), which are found in small breeding colonies in remote headland areas in Labrador, Quebec, and Greenland. Although the conservation status of the eastern population is listed federally in Canada as 'special concern', provincial listings include 'endangered' and 'vulnerable' (COSEWIC 2012). The western population in Canada and the United States is not listed, and is much larger and more stable, numbering upwards of 400 000 individuals (Canadian Wildlife Service Waterfowl Committee 2013), and can be found in coastal areas around urban environments. It is thought however that their breeding productivity is too low to compensate for observed mortality rates, and so there has been research directed at examining population demographics (Smith et al. 2001). Bond et al. (2008) used VTG and

VLDL to estimate the proportion of sexually mature females initiating egg production (i.e. breeding propensity). By capturing females in breeding areas, collecting blood samples, and attaching radio-transmitters to individual females to monitor nesting activity, they found that the proportion of breeding females was 92 per cent. When examining VTG and VLDL concentrations in females telemetrically determined as breeders and non-breeders, distinct yolk precursor clusters were observed, which could be used in future studies as a means for identifying birds of unknown breeding status (Figure 2.5).

Collectively, these results dispelled long-standing ideas that the species had low breeding propensity, and instead suggest that declining population trends might be related to influences on other phases of their annual cycle.

2.4.3 Common eiders

Common eiders (*Somateria mollissima*) are the largest duck species in the northern hemisphere, and perhaps the best studied of all Arctic waterfowl. They form very large breeding assemblages in the tens of thousands, which are important to Inuit communities for egg harvest and the collection of eider down (Henri et al. 2010). They are also extensively harvested by subsistence hunters throughout their wintering areas in Greenland and Newfoundland (Canadian Wildlife Service

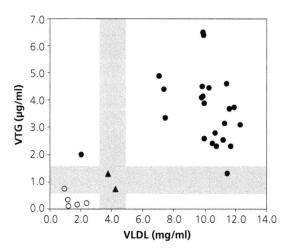

Figure 2.5 The relationship between circulating vitellogenin (VTG) and very-low-density lipoprotein (VLDL) can be used to infer the breeding status of female harlequin ducks (*Histrionicus histrionicus*). Dark circles represent VTG and VLDL levels in known egg-producing females, open circles are non-egg-producing females, while dark triangles are birds whose breeding status was unknown. Grey areas are zones of uncertainty determined by lower and upper cut-off values for egg-producing and non-egg-producing individuals. Data are from years 2003 and 2004. Reproduced from Bond et al. (2008) with the permission of John Wiley and Sons.

Waterfowl Committee 2013). Currently listed as near-threatened (BirdLife International 2018), the colony-wide breeding failures of common eiders in some colonies of the Canadian Arctic in recent years, due to indirect effect of climate change and reduced sea-ice coverage via increased polar bear predation, mark a unique, Arctic-specific conservation issue. Recently, yolk precursor measurements have been applied in a study of common eiders at a large breeding colony in Nunavut, Canada. Hennin et al. (2015) examined the partitioning of energy in pre-breeding eiders, which are a mixed-strategy breeder (i.e. a combination of income and capital breeding). A large body of work from this colony has detailed the importance of pre-breeding condition for the optimization of lay date and increased fitness probability (Descamps et al. 2011). Characterizing pre-breeding condition is also the only means for obtaining physiological insights to breeding eiders, as females are prone to abandon reproduction when handled on the nest. More broadly though, understanding pre-breeding energetic management in highly seasonal environments is important for discerning the importance of local area resources in a changing climate, and how local resources influence reproductive success. By measuring VTG and VLDL, in addition to corticosterone (a hormone that at baseline, non-stress-induced levels mediates daily and annual energetic management [Romero 2002; Landys et al. 2006] and reproductive investment [Love et al. 2014]), it was shown that baseline corticosterone increased significantly from the time eiders arrived at the colony breeding grounds to the initiation of RFG (Hennin et al. 2015). Baseline corticosterone also correlated positively with body mass, which suggests a role in fattening rates in advance of egg production (Hennin et al. 2015).

Corroborating these findings are recent studies that used experimental corticosterone increases in a captive eiders, which identified a potential causal effect on fat storage (Hennin et al. 2016). VLDL also increased throughout the pre-breeding period, reaching maximal values at the initiation of RFG. Unexpectedly, VTG levels were unprecedentedly high in arriving females, which should theoretically be low prior to RFG. Although it remains to be examined more carefully, this was interpreted as a pre-emptive strategy for having VTG in storage, ready for deposition into developing follicles at arrival, assuming that females meet the energetic threshold for breeding. These results suggest that in the seasonal Arctic environment, where the window of opportunity for breeding is narrow, a store of circulating VTG during the pre-laying fattening period optimizes the rate of egg production once a female has met the energetic threshold for laying, thereby ensuring an early lay date. Finally, Hennin et al. (2016) show that those individuals with higher VLDL at arrival lay earlier and produce larger clutches (via its direct effect on lay date; Hennin et al. 2018).

2.5 Case study 3: threatened penguins and albatrosses

According to the IUCN Red List criteria, pelagic seabirds are among the most threatened of all groups of birds (Croxall et al. 2012). The albatrosses (*Diomedeidae*) and the penguins (*Spheniscidae*) are particularly at risk, with 15 of 22 albatross species listed as vulnerable (i.e. IUCN classification of high risk of endangerment in the wild) or threatened (IUCN classification of endangered or critically endangered) (68 per cent; Phillips et al. 2016), as well as 10 of 18 penguin species (56 per cent; IUCN Red List online). Both families are characterized by a long lifespan, high adult survival rate, delayed sexual maturity, and low fecundity. The albatrosses all lay a single-egg clutch, while all penguins lay a two-egg clutch, except for those in the genus *Aptenodytes* (king and emperor penguins), which lay single-egg clutches. Given the slow life histories of both albatrosses and penguins, population trajectories tend to be particularly sensitive to changes in adult mortality (Croxall and Rothery 1991; Trathan et al. 2014; Phillips et al. 2016). This is especially a concern in the Antarctic and sub-Antarctic, where climate change is having profound effects on sea-ice coverage and recruitment of prey species important to many Antarctic predators (Flores et al. 2012).

Crossin et al. (2012a) examined the variation in pre-breeding physiology of black-browed albatrosses (*Thallasarche melanophrys*) at a colony in South Georgia, where a changing Antarctic environment could affect prey recruitment and result in carryover effects via variation in foraging success during the winter non-breeding period (O'Connor et al. 2014). Variation in foraging success could broaden the variability of physiological markers indicative of laying probability, via effects on circulating levels of yolk precursors. By sampling blood plasma in migratory female albatrosses upon their return to the colony, prior to laying, and then monitoring breeding activity throughout the season, Crossin et al. (2012a) showed that the decision to reproduce was made prior to their arrival. Breeding females arrived with high condition-related markers (e.g. body mass, and indicators of aerobic condition—haematocrit and haemoglobin

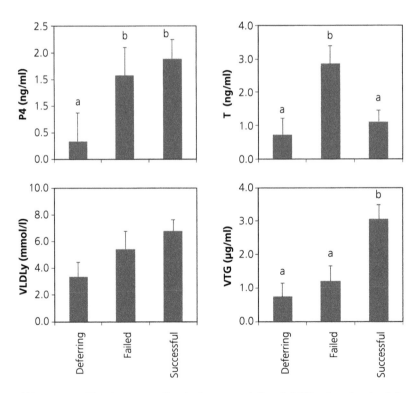

Figure 2.6 Sex steroid (progesterone P4, and testosterone T) and yolk precursor vitellogenin (VTG) and very-low-density lipoprotein (VLDLy) profiles of female black-browed albatrosses (*Thalassarche melanophris*) upon their arrival at a breeding colony at Bird Island, South Georgia. Birds are grouped according to breeding outcome, and bars represent least square means ± SEM. Differing letters indicate statistically significant contrasts (α = 0.05). Reprinted from *General & Comparative Endocrinology*, 176, Crossin et al., Migratory carryover effects and endocrinological correlates of reproductive decisions and reproductive success in female albatrosses, pp.151–157, Copyright (2012) with permission from Elsevier.

concentrations), high sex steroid levels (progester-one, testosterone), and high VTG and VLDL con-centrations. Non-breeding or 'deferring' females have lower levels of all of these markers (Crossin et al. 2012a; Figure 2.6).

Regarding reproductive success (whether a lay-ing female successfully fledged a chick), body con-dition markers showed little association (Crossin et al. 2012a). However, the steroidogenic processes underlying follicle development, specifically sex steroid production and VTG production, differed significantly between successful and failed breeders (Crossin et al. 2012a). Breeding failure was charac-terized by high testosterone and low VTG, whereas conversely, success was characterized by low tes-tosterone and high VTG (Figure 2.6). These results suggested that, independent of body condition, breeding success was mediated by factors affecting the enzymatic processes governing the conversion of testosterone to 17ß-oestradiol, and the subse-quent production of VTG. Failed females did not appear to produce VTG at sufficient levels to gener-ate egg yolks that would sustain embryonic devel-opment (Crossin et al. 2012a).

Crossin et al. (2010, 2012b) also applied yolk pre-cursor measurement to investigate the unusual pat-tern of egg production in macaroni penguins (*Eudyptes chrysolophus*) and rockhopper penguins (*E. chrysocome*), which involves an extreme degree of egg-size dimorphism unobserved in any other bird species. This egg-size dimorphism is a unique feature of their fixed two-egg clutch size, and is characterized by a first laid A-egg, which rarely ever hatches, and which is among the genus, 18–57 per cent smaller than the second laid B-egg (Williams 1995; Crossin et al. 2013). Such extreme dimorphism is thought to be related to a carryover effect in which VTG production is constrained when the physiological processes regulating egg production occur simultaneously with migratory activity (Crossin and Williams 2016). By sampling newly arrived females after long-distance migra-tion, it was shown that variation in VTG levels were lower, and egg-size dimorphism greater, when RFG was initiated in females before they arrived at the colony, during homeward migration. In contrast, females arriving at the colony and spending nearly all of the RFG period on land had higher VTG levels

and lower levels of egg-size dimorphism (Crossin et al. 2010).

2.6 Case study 4: toxicology studies—maternal transfer of contaminants to developing eggs

Prior to the formal recognition of 'conservation physiology' as a scientific discipline (*sensu* Wikelski and Cooke 2006), toxicology studies provided the most obvious, and important, application of physi-ology to management of species impacted by human activities (such as effects of pesticides), and yolk precursor analysis has been a useful tool in eco-toxicological studies. Jimenez et al. (2007) developed diagnostic tools for assessing the impact of endocrine-disrupting agents, such as organo-chlorine compounds, in peregrine falcons (*Falco per-egrinus*) collected in the wild. In males exposed to such agents, they detected increased VTG and zona radiata proteins (proteins found in the membranes of egg follicles and important for successful fertil-ization), neither of which are normally expressed in males. This suggests that when male falcons are exposed to organochlorine compounds in the wild, they experience disruptive pro-oestrogenic effects. VTG has also been used to study the maternal trans-fer of organohalogenated contaminants to eggs in rockhopper penguins (Dehnhard et al. 2017). Because of the highly lipophilic nature of organic contaminants, such compounds can be actively transported from the maternal circulation to devel-oping egg follicles via a 'piggy-back' attraction to similarly lipophilic VTG and subsequent receptor-mediated deposition during RFG. In rockhopper penguins, there was little evidence that elevated VTG played a significant role in maternal transfer of contaminants to eggs, and individual variation in female VTG concentrations did not significantly correlate with any measured organohalogenated compounds in either female plasma or eggs. Eng et al. (2013) experimentally dosed captive zebra finches (*Taeniopygia guttata*) with a brominated flame retardant (e.g. BDE-99) and measured mater-nal VLDL concentrations during RFG and at clutch completion, as well as contaminant levels in the eggs. In contrast to Dehnhard et al. (2017), individual variation in circulating maternal VLDL was significantly related to BDE-99 transfer to egg yolks when BDE-99 was at low background levels, but unexpectedly, the relationship was not detected when contaminants were elevated in the dosed birds. Collectively, these studies suggest that relationships between maternal lipoprotein status, reproduction and contaminant levels in eggs can be complex and highly variable, something that needs to be considered when using eggs as bio-indicators of environmental variation in contaminant levels (for a review of maternal influences on egg quality, see Williams and Groothuis 2015).

2.7 Conclusions and future directions

We have highlighted a series of case studies to demonstrate how the indirect, physiological assessment of reproductive state via the measurement of yolk precursors has provided valuable information on the breeding biology of cryptic and at-risk birds, which would have been difficult to obtain in any other way (e.g. proportion of fecund females, breeding phenology, single vs. multiple-clutch breeding pattern). Yolk precursor analysis has now been applied more broadly in a range of avian studies, for example to study (1) local adaptation of timing of breeding in blue tits (*Cyanistes caeruleus*; Caro et al. 2009), and (2) analysis of breeding condition and phenology in gyrfalcons (*Falco rusticolus*; Lamarre and Franke 2017). This technique should be broadly applicable to any oviparous vertebrate population where essential information on breeding biology cannot be obtained by more traditional methods, especially for threatened or endangered species as the technique is non-lethal and minimally invasive, requiring only a small blood sample.

The coupling of physiological sampling methods with other techniques like electronic tracking has allowed researchers to make unprecedented insights to the ecology of species and the mechanisms underlying fitness-related processes (Crossin et al. 2014). Studies like McFarlane Tranquilla et al. (2003b) and Bond et al. (2008), described herein, used radio-telemetry and yolk precursor analysis to identify breeding activity and status in otherwise unknown-status individuals. Janssen et al. (2009) also used a multi-disciplinary approach to inform the breeding and trophic ecology of breeding birds by measuring VTG levels and linking these to variation in δ13C and δ15N levels. Many other techniques, such as remote sensing, electronic tracking, genomics and transcriptomics, etc., can be linked to physiological techniques. These will ultimately benefit the study of avian reproduction and conservation and management efforts by resolving the mechanisms underlying individual variation in fitness or proxies thereof. As this chapter has shown, the measurement of yolk precursors can provide a useful means for assessing the breeding status of threatened and endangered birds, which can provide important information about phenology and population dynamics to those tasked with the management of rare and cryptic bird species.

Acknowledgements

Work reported in this chapter was supported by multiple NSERC Discovery Grants (T.D.W., G.T.C.), and long-term Environment Canada (now Environment Climate Change Canada) funding to T.D.W. through the Centre for Wildlife Ecology at Simon Fraser University. Thanks to the many students who have contributed to these studies over the years. We thank Dr Laura McFarlane Tranquilla for providing data for Figure 2.2.

References

Abate, T., 1992. Which bird is the better indicator species for old-growth forest? *Bioscience*, *42*, 8–9.

Afton, A.D. and Anderson, M.G., 2001. Declining scaup populations: a retrospective analysis of long-term population and harvest survey data. *Journal of Wildlife Management*, *65*, 781–96.

Alisauskas, R.T. and Ankney, C.D., 1994. Nutrition of breeding female ruddy ducks: the role of nutrient reserves. *Condor*, *96*, 878–97.

Ankney, C.D., 1977. Feeding and digestive organ size in breeding lesser snow geese. *Auk*, *94*, 275–82.

Austin, J.E., Afton, A.D., Anderson, M.G., and Clark, R.G., 2000. Declining scaup populations: issues, hypotheses, and research needs. *Wildlife Society Bulletin*, *28*, 254–63.

Bêty, J., Gauthier, G., and Giroux, J.-F., 2003. Body condition, migration, and timing of reproduction in snow geese: a test of the condition-dependent model of optimal clutch size. *American Naturalist*, *162*, 110–21.

Binford, L.C., Elliott, B.G., and Singer, S.W., 1975. Discovery of a nest and the downy young of the marbled murrelet. *Wilson Bulletin*, 87, 303–19.

BirdLife International, 2018. *Somateria mollissima*. The IUCN Red List of Threatened Species 2018: e. T22680405A132525971.   https://dx.doi.org/10.2305/IUCN.UK.2018-2.RLTS.T22680405A132525971.en

Bond, J.C., Esler, D., and Williams, T.D., 2008. Breeding propensity of female harlequin ducks. *Journal of Wildlife Management*, 72, 1388–93.

Bradley, R.W., Cooke, F., Lougheed, L.W., and Boyd, W.S., 2004. Inferring breeding success through radiotelemetry in the marbled murrelets. *Journal of Wildlife Management*, 68, 318–31.

Burger, A.E., 2002. Conservation assessment of marbled murrelets in British Columbia: review of the biology, populations, habitat associations, and conservation. Part A of Marbled Murrelet Conservation Assessment. Tech. Rep. Ser. 387. Canadian Wildlife Service, Delta, BC.

Canadian Wildlife Service Waterfowl Committee, 2013. Population status of migratory game birds in Canada: November 2013. CWS Migratory Birds Regulatory Report Number 40.

Caro, S.P., Charmantier, A., Lambrechts, M.M. et al., 2009. Local adaptation of timing of reproduction: females are in the driver's seat. *Functional Ecology*, 23, 172–9.

Challenger, W.O., Williams, T.D., Christians, J.K., and Vezina, F., 2001. Follicular development and plasma yolk precursor dynamics through the laying cycle in the European starling (*Sturnus vulgaris*). *Physiological and Biochemical Zoology*, 74, 356–65.

Christians, J.K. and Williams, T.D., 1999. Effects of exogenous 17β-estradiol on the reproductive physiology and reproductive performance of European starlings (*Sturnus vulgaris*). *Journal of Experimental Biology*, 202, 2679–85.

Cockrem, J.F. and Seddon, P.J., 1994. Annual cycle of sex steroids in the yellow-eyed penguin (*Megadyptes antipodes*) on South Island, New Zealand. *General and Comparative Endocrinology*, 94, 113–21.

COSEWIC, 2012. COSEWIC assessment and status report on the marbled murrelet *Brachyramphus marmoratus* in Canada. Committee on the Status of Endangered Wildlife in Canada, Ottawa, ON.

Crossin, G.T., Cooke, S.J., Goldbogen, J.A., and Phillips, R.A., 2014. Tracking fitness in marine vertebrates: current knowledge and opportunities for future research. *Marine Ecology Progress Series*, 496, 1–17.

Crossin, G.T., Phillips R.A., Trathan P.N. et al., 2012a. Migratory carryover effects and endocrinological correlates of reproductive decisions and reproductive success in female albatrosses. *General & Comparative Endocrinology*, 176, 151–7.

Crossin, G.T., Poisbleau, M., Demongin, L. et al., 2012b. Migratory constraints on yolk precursor production limit egg androgen deposition and underlies a brood reduction strategy in rockhopper penguins. *Biology Letters*, 8, 1055–8.

Crossin, G.T., Trathan, P.N., and Crawford, R.J.M., 2013. The macaroni penguin (*Eudyptes chrysolophus*) and the royal penguin (*E. schlegeli*). In P. Garcia-Borboroglu and P.D. Boersma, eds. *Penguins: Natural History and Conservation*, pp. 185–208. University of Washington Press, Seattle, WA.

Crossin, G.T., Trathan, P.N., Phillips, R.A. et al., 2010. A carry-over effect of migration underlies individual variation in reproductive readiness and extreme egg size dimorphism in macaroni penguins (*Eudyptes chrysolophus*). *American Naturalist*, 176, 357–66.

Crossin, G.T. and Williams, T.D., 2016. Migratory lifehistories explain the extreme egg size dimorphism of Eudyptes penguins. *Proceedings of the Royal Society of London B*, 283, 1413–18.

Croxall, J.P., Butchart, S.H.M., Lascelles, B., and Stattersfield, A., 2012. Seabird conservation status, threats and priority actions: a global assessment. *Bird Conservation International*, 22, 1–34.

Croxall, J.P. and Rothery, P., 1991. Population regulation of seabirds: implications of their demography for conservation. In C.M. Perrins, J.D. Lebreton, and G.J.M. Hirons, eds. *Bird Population Studies: Relevance to Conservation and Management*, pp. 272–96. Oxford University Press, New York.

Dawson, A., 1983. Plasma gonadal steroid levels in wild starlings (*Sturnus vulgaris*) during the annual cycle and in relation to the stages of breeding. *General and Comparative Endocrinology*, 49, 286–94.

Deeley, R.G., Mullinix, K.P., Weterkam, W. et al., 1975. Vitellogenin synthesis in the liver: vitellogenin is the precursor of the egg yolk phosphoproteins. *Journal of Biological Chemistry*, 250, 9060–6.

Dehnhard, N., Jaspers, V.L.B., Demongin, L. et al., 2017. Organohalogenated contaminants in plasma and eggs of rockhopper penguins: does vitellogenin affect maternal transfer? *Environmental Pollution*, 226, 277–87.

Descamps, S., Bêty, J., Love, O.P., and Gilchrist, H.G., 2011. Individual optimization of reproduction in a long-lived migratory bird: a test of the condition-dependent model of laying date and clutch size. *Functional Ecology*, 25, 671–81.

Drent, R.H. and Daan, S., 1980. The prudent parent: energetic adjustments in avian breeding. *Ardea*, 68, 225–52.

Dunn, P.O., Winkler, D.W., Whittingham, L.A. et al., 2011. A test of the mismatch hypothesis: how is timing of reproduction related to food abundance in an aerial insectivore? *Ecology*, 92, 450–61.

Ellenberg, U., Mattern, T., and Seddon, P.J., 2013. Heart rate responses provide an objective evaluation of human disturbance stimuli in breeding birds. *Conservation Physiology, 1*, doi: 10.1093/conphys/cot013

Eng, M.L., Elliott, J.E., Letcher, R.J., and Williams, T.D., 2013. Individual variation in body burdens, lipid status, and reproductive investment is related to maternal transfer of a brominated diphenyl ether (BDE-99) to eggs in the zebra finch. *Environmental Toxicology and Chemistry, 32*, 345–52.

Evans Ogden, L., Martin, K., and Williams, T.D., 2013. Elevational differences in estimated fattening rates suggest high elevation sites are high quality habitats for fall migrants. *Auk, 130*, 98–106.

Flint, P.L. and Grand, J.B., 1999. Patterns of variation in size and composition of greater scaup eggs: are they related? *Wilson Bulletin, 111*, 465–71.

Flores, H., Atkinson, A., Kawaguchi, S. et al., 2012. Impact of climate change on Antarctic krill. *Marine Ecology Progress Series, 458*, 1–19.

Gorman, K.B., Esler, D., Flint, P.L., and Williams, T.D., 2008. Nutrient reserve dynamics during egg production by female greater scaup (*Aythya marila*): relationships with timing of reproduction. *Auk, 125*, 384–94.

Gorman, K.B., Esler, D., Walzem, R.L., and Williams, T.D., 2009. Plasma yolk precursor dynamics during egg production by female greater scaup (*Aythya marila*): characterization and indices of reproductive state. *Physiological and Biochemical Zoology, 82*, 372–81.

Gorman, K.B., Flint, P.L., Esler, D., and Williams, T.D., 2007. Ovarian follicle dynamics of female greater scaup during egg production. *Journal of Field Ornithology, 78*, 64–73.

Griffiths, R., Daan, S., and Dijkstra, D.C., 1996. Sex identification in birds using two CHD genes. *Proceedings of the Royal Society of London B, 263*, 1251–6.

Guiguet, C.J., 1956. Enigma of the Pacific. *Audubon, 58*, 164–7.

Hamer, T.E. and Nelson, S.K., 1995. Characteristics of marbled murrelet nest trees and nesting stands. In C.J. Ralph, G.L. Hunt, Jr, M.G. Raphael, and J.F. Piatt, eds. *Ecology and Conservation of the Marbled Murrelet*, pp. 69–82. USDA Forest Service General Technical Report PSW-GTR-152, Albany, CA.

Hennin, H.L., Bêty, J., Legagneux, P. et al., 2016. Energetic physiology mediates individual optimization of breeding phenology in a migratory Arctic seabird. *American Naturalist, 188*, 434–45.

Hennin, H.L., Dey, C., Bêty, J. et al., 2018. Higher rates of pre-breeding condition gain positively impacts clutch size: a mechanistic test of the condition-dependent individual optimization model. *Functional Ecology, 32*, 2019–28.

Hennin, H.L., Legagneux, P., Bêty, J. et al., 2015. Pre-breeding energetic management in a mixed-strategy breeder. *Oecologia, 177*, 235–43.

Henri, D., Gilchrist, H.G., and Peacock, E., 2010. Understanding and managing wildlife in Hudson Bay under a changing climate: some recent contributions from Inuit and Cree ecological knowledge. In S.H. Ferguson, L.L. Loseto, and M.L. Mallory, eds. *A Little Less Arctic*, pp. 267–89. Springer, Dordrecht.

Janssen, M.H., Arcese, P., Kyser, T.K. et al., 2009. Pre-breeding diet quality and timing of breeding in a threatened seabird, the marbled murrelet. *Marine Ornithology, 37*, 33–40.

Jimenez, B., Mori, G., Concejero, M.A. et al., 2007. Vitellogenin and zona radiata proteins as biomarkers of endocrine disruption in peregrine falcon (*Falco peregrinus*). *Chemosphere, 67*, S375–8.

Jorgensen, J. Dinan, G.L.R., and Brown, M.B., 2016. Flight initiation distances of nesting piping plovers (*Charadrius melodus*) in response to human disturbance. *Avian Conservation and Ecology, 11.* http://dx.doi.org/10.5751/ACE-00826-110105

Kaiser, G.W., Derocher, A.E., Crawford, S. et al., 1995. A capture technique for marbled murrelets in coastal inlets. *Journal of Field Ornithology, 66*, 321–33.

Kessel, B., Rocque, D.A., and Barclay, J.S., 2002. Greater scaup (*Aythya marila*), version 2.0. In A.F. Poole and F.B. Gill, eds. *The Birds of North America*. Cornell Lab of Ornithology, Ithaca, NY.

Krebs, C.J., 1985. *Ecology: The Experimental Analysis of Distribution and Abundance*, thirrd edition. Harper & Row, New York.

Lamarre, V. and Franke, A., 2017. Body condition and reproductive phenology. In D.L. Anderson, C.J.W. McClure, and A. Franke, eds. *Applied Raptor Ecology: Essentials from Gyrfalcon Research*, pp. 234–64. The Peregrine Fund, Boisie, ID.

Landys, M.M., Ramenofsky, M., and Wingfield, J.C., 2006. Actions of glucocorticoids at a seasonal baseline as compared to stress-related levels in the regulation of periodic life processes. *General and Comparative Endocrinology, 148*, 132–49.

Lepage, D., Gauthier, G., and Menu, S., 2000. Reproductive consequences of egg-laying decisions in snow geese. *Journal of Animal Ecology, 69*, 414–27.

Lougheed, C., Vanderkist, B.A., Lougheed, L.W., and Cooke, F., 2002. Techniques for investigating breeding chronology in marbled murrelets, Desolation Sound, British Columbia. *Condor, 104*, 319–30.

Love, O.P., Bourgeon, S., Madliger, C.L. et al., 2014. Evidence for baseline glucocorticoids as mediators of reproductive investment in a wild bird. *General and Comparative Endocrinology, 199*, 65–9.

Mays, N.A., Vleck, C.M., and Dawson, J., 1991. Plasma luteinizing hormone, steroid hormones, behavioral role, and nest stage in cooperatively breeding Harris' hawks (*Parabuteo unicinctus*). *Auk, 108*, 619–37.

McFarlane Tranquilla, L., Williams, T., and Cooke, F., 2003a. Using vitellogenin to identify interannual variation in breeding chronology of marbled murrelets. *Auk, 120*, 512–21.

McFarlane Tranquilla, L.A., Yen, P.P-W., Bradley, R.W. et al., 2003b. Do two murrelets make a pair? Breeding status and behavior of marbled murrelet pairs captured at sea. *Wilson Bulletin, 115*, 374–81.

McNamara, J.M. and Houston, A.I., 1996. State-dependent life-histories. *Nature, 380*, 215–21.

Mitchell, M.A. and Carlisle A.J., 1991. Plasma zinc as an index of vitellogenin production and the reproductive status in the domestic fowl. *Comparative Biochemistry and Physiology, 100A*, 719–24.

Moran-Lopez, R., Guzman, J.M.S., Borrego, E.C., and Sanchez, A.V., 2006. Nest-site selection of endangered cinereous vulture (*Aegypius monachus*) populations affected by anthropogenic disturbance: present and future conservation implications. *Animal Conservation, 9*, 29–37.

Nelson, S.K., 1997. Marbled murrelet (*Brachyramphus marmoratus*), version 2.0. In A.F. Poole and F.B. Gill, eds. *The Birds of North America*. Cornell Lab of Ornithology, Ithaca, NY.

Newton, I., 1979. *Population Ecology of Raptors*. Buteo Books, Vermillion, SD.

Newton, I., 1998. *Population Limitation in Birds*. Academic Press, San Diego, CA.

O'Connor, C.M., Norris, N.R., Crossin, G.T., and Cooke, S.J., 2014. Biological carryover effects: linking common concepts and mechanisms in ecology and evolution. *Ecosphere, 5*, 1–11.

Peery, M.Z., Beissinger, S.R., Newman, S.H. et al., 2004a. Applying the declining population paradigm: diagnosing causes of low reproductive success in marbled murrelets. *Conservation Biology, 18*, 1088–98.

Peery, Z., Beissinger, S.R., Newman, S.H. et al., 2004b. Individual and temporal variation in inland flight behavior of marbled murrelets: implications for population monitoring. *Condor, 106*, 344–53.

Perrins, C.M. and Moss, D., 1975. Reproductive rates in the great tit. *Ibis, 44*, 695–706.

Phillips, R.A., Gales, R., Baker, G.B. et al., 2016. The conservation status and priorities for albatrosses and large petrels. *Biological Conservation, 201*, 169–83.

Ralph, C.J., Hunt, G.L.J., Raphael, M.G., and Piatt, J.F., 1995. Ecology and conservation of the marbled murrelet in North America: an overview. In C.J. Ralph, L.H. George Jr, M.G. Raphael, and J.F. Piatt, eds. *Ecology and Conservation of the Marbled Murrelet, General Technical Report PSW-152*, pp. 3–22. US Department of Agriculture, Albany, CA.

Romero, L.M., 2002. Seasonal changes in plasma glucocorticoid concentrations in free-living vertebrates. *General and Comparative Endocrinology, 128*, 1–24.

Rowe, L., Ludwig, D., and Schluter, D., 1994. Time, condition, and the seasonal decline of avian clutch size. *American Naturalist, 143*, 698–722.

Salvante, K.G., Lin, G., Walzem, E.R.L., and Williams, T.D., 2007. Characterization of very-low density lipoprotein particle diameter dynamics in relation to egg production in a passerine bird. *Journal of Experimental Biology, 210*, 1064–74.

Salvante, K.G. and Williams, T.D., 2002. Vitellogenin dynamics during egg-laying: daily variation, repeatability and relationship with egg size. *Journal of Avian Biology, 33*, 391–8.

Smith, C.M., Goudie, R.I., and Cooke, F., 2001. Winter age ratios and the assessment of recruitment of harlequin ducks. *Waterbirds, 24*, 39–44.

Stearns, S.C., 1992. *The Evolution of Life Histories*. Oxford University Press, Oxford.

Trathan, P.N., García-Borboroglu, P., Boersma, P.D. et al., 2014. In a changing climate, pollution, habitat loss and fishing remain the top threats to the world's penguins. *Conservation Biology, 29*, 31–41.

Vanderkist, B.A., Williams, T.D., Bertram, D.F. et al., 2000. Indirect, physiological assessment of reproductive state and breeding chronology in free-living birds: an example in the marbled murrelet (*Brachyramphus marmoratus*). *Functional Ecology, 14*, 758–65.

Vanderkist, B.A., Xue, X-H., Griffiths, R.A. et al., 1999. Evidence of male-bias in capture samples of marbled murrelets (*Brachyramphus marmoratus*) from genetic studies in British Columbia. *Condor, 101*, 398–402.

Vézina, F. and Williams, T.D., 2003. Plasticity in body composition in breeding birds: what drives the metabolic costs of egg production? *Physiological and Biochemical Zoology, 76*, 713–30.

Wallace, R.A., 1985. Vitellogenin and oocyte growth in non-mammalian vertebrates. In L.W. Browder, ed. *Developmental Biology*, pp. 127–77. Plenum, New York.

Walzem, R.L., 1996. Lipoproteins and the laying hen: form follows function. *Poultry and Avian Biology Review, 7*, 31–64.

Walzem, R.L., Hansen, R.J., Williams, D.L., and Hamilton, R.L., 1999. Estrogen induction of VLDLy assembly in egg-laying hens. *Journal of Nutrition, 129*, 467S–72.

Warren, J.M., Cutting, K.A., and Koons, D.N., 2013. Body condition dynamics and the cost-of-delay hypothesis in a temperate-breeding duck. *Journal of Avian Biology, 44*, 575–82.

Whitworth, D.L., Takekawa, J.Y., Carter, H., and McIver, W.R., 1997. A night-lighting technique for at-sea capture of Xantus' murrelets. *Colonial Waterbirds, 20,* 525–31.

Wikelski, M. and Cooke, S.J., 2006. Conservation physiology. *Trends in Ecology and Evolution, 21,* 38–46.

Williams, T.D., 1995. *The Penguins*. Oxford University Press, Oxford.

Williams, T.D., 2012. *Physiological Adaptations for Breeding in Birds*. Princeton University Press, Princeton, NJ.

Williams, T.D. and Groothuis, T.G.G., 2015. Egg quality, embryonic development and post-hatching phenotype: an integrated perspective. In D.C. Deeming and S.J. Reynolds, eds. *Nests, Eggs and Incubation: New Ideas about Avian rReproduction*, pp. 114–26. Oxford University Press, Oxford.

Williams, T.D., Warnock, N., Takekawa, J., and Bishop, M.A., 2007. Flyway scale variation in plasma triglyceride levels as an index of refueling rate in spring migrating western sandpipers. *Auk, 124,* 886–97.

Wingfield, J.C. and Farner, D.S., 1978. The endocrinology of a natural breeding population of the white-crowned sparrow (*Zonotrichia leucophrys pugetensis*). *Physiological Zoology, 51,* 188–205.

Zharikov, Y., Lank, D.B., Huettmann, F. et al., 2006. Habitat selection and breeding success in a forest-nesting Alcid, the marbled murrelet, in two landscapes with different degrees of forest fragmentation. *Landscape Ecology, 21,* 107–20.

CHAPTER 3

On conducting management-relevant mechanistic science for upriver migrating adult Pacific salmon

Steven J. Cooke, Graham D. Raby, Nolan N. Bett, Amy K. Teffer,
Nicholas J. Burnett, Kenneth M. Jeffries, Erika J. Eliason,
Eduardo G. Martins, Kristina M. Miller, David A. Patterson,
Vivian M. Nguyen, Nathan Young, Anthony P. Farrell,
and Scott G. Hinch

> ➲ **Take-home message**
>
> Diverse physiological tools and approaches have revealed the mechanisms that underpin migration failure of Pacific salmon in the Fraser River of British Columbia, Canada, providing fisheries managers with predictive tools and management options for balancing fisheries opportunities with conservation of salmon populations.

3.1 Defining the conservation challenge—Pacific salmon in the Anthropocene

By any and all measures, Pacific salmon (*Oncorhynchus* genus) are among the most iconic of all migratory species. These anadromous species engage in impressive migrations that are cyclical and predictable in time and space (Groot and Margolis 1991). As adults, Pacific salmon have to migrate from open-ocean feeding grounds to spawning grounds. Except for steelhead (*Oncorhynchus mykiss*), Pacific salmon are exclusively semelparous whereby failure to reproduce

means zero lifetime fitness. Migration for salmon is not easy—along the way they experience dynamic and challenging environmental conditions (i.e. water velocity, turbulence, and temperature, as well as pathogens and predators). Pacific salmon are therefore especially vulnerable to environmental changes that make a difficult migration even more challenging (Crozier et al. 2008). Moreover, these predictable migrations make Pacific salmon particularly vulnerable to capture fisheries (commercial, recreational, and Indigenous) that use a diversity of gear types (ranging from hook-and-line to nets). While most fisheries are conducted to harvest salmon, a component of the catch is released for conservation, either because of fishery regulations or the conservation ethic of fishers in the context of catch-and-release angling. Some individual salmon also encounter and escape from fishing gear (Patterson et al. 2017a, b) by engaging in high-intensity exercise, perhaps with injury, which introduces additional challenges for fish that need to reach spawning grounds to reproduce. Even catch-and-release fishing involves intense exercise, handling, and perhaps exposure to air.

Collectively, the natural and anthropogenic stressors encountered by Pacific salmon represent dramatic and important selective forces such that in some years and for some populations only a small proportion of individuals survive to reach spawning grounds (Cooke et al. 2004). Indeed, many Pacific salmon populations are in decline as a result of multiple stressors that are experienced by fish at various times during their life history (Lichatowich et al. 1999; Portley et al. 2014; also see International Union for Conservation of Nature [IUCN] Red List for individual stocks that are listed). As such, there is a dire need for science-based tools that consider the mechanistic basis behind mortality or fitness impairments to identify opportunities for mitigating threats and informing management options.

Historically research on Pacific salmon focused on characterizing trends and patterns in population size (e.g. Shea and Mangel 2001) and vital rates (e.g. Bradford 1995)—*how many* fish are there—*how many* fish were harvested—*how* many fish died—*when* did they die—and *where* did they die? While these are essential aspects of stock assessment, much less effort was devoted to understanding *why* fish died. This is not surprising given the stock assessment and research tools available, which tended to involve use of external markers at best and typically considered the fate of individuals. Today we have modern electronic tagging (see Figure 3.1) and genomics tools combined with laboratory cardio-respiratory assessments and other multi-disciplinary approaches to understand how salmon interact with each other, their environment, and humans. Our team has used a diversity of approaches, tools, and endpoints that span the lab and field to understand how Pacific salmon

complete (or attempt to complete) their complicated life history. Indeed, among all animal taxa, some of the most impressive and clear stories (see Madliger et al. 2016) related to conservation physiology exist among Pacific salmon (e.g., Cooke et al. 2012). Therefore, we are much closer to understanding when, where, and why a Pacific salmon dies during its river migration.

Here we use a case study approach to explore how our collective work on Pacific salmon migration biology has advanced our understanding of Pacific salmon biology and in doing so generated policy- and management-relevant science. Our approach involves providing an overview of how eight specific scientific impacts of environmental/anthropogenic stress have been informed by multiple measures of physiology to generate new knowledge that has been used by practitioners and policy makers for conservation gains. Our story is unique in that it is long term and we have previously synthesized some of our findings (e.g. Cooke et al. 2012; Hinch et al. 2012; Patterson et al. 2016). Rather than focusing solely on the biological science, we have included both social scientists and practitioners on our team. We consider how the diverse contemporary tools available to the practising conservation physiologist can be applied more broadly to other taxa. The tools that underpin the examples presented here both serve to advance the field and constrain what is possible. This juxtaposition represents the frontier of animal biology and is where major investments are needed.

3.2 Elevated temperatures can kill salmon

3.2.1 The science

For over half a century, researchers have explored the role of water temperature on the biology of adult Pacific salmon (e.g. seminal work by Rolly Brett). Early work focused on swimming performance and respiratory aspects of energetics (Brett 1971, 1973). Over the decades that work has expanded to include extensive lab and field studies that range from understanding the effects of water temperature (especially warming temperatures) on

(a)

(b)

Figure 3.1 Research focused on wild Pacific salmon engaged in upstream migration inherently requires consideration of behaviour and physiology while combining research from the lab and the field. A core component of such work is biotelemetry (1A), which for upstream migrating Pacific salmon often means the gastric implantation of a telemetry transmitter to enable the tracking of behaviour and fate. Photo credit: S. Cooke. Another key tool is non-lethal biopsy (1B) of various tissues (e.g. gill, muscle, blood), which can be done on fish that are tagged with telemetry transmitters. Photo credit: S. Cooke.

gene expression (Jeffries et al. 2012a, b, 2014a, b; Akbarzadeh et al. 2018), field energetics (Rand et al. 2006), and fate (i.e. whether it survives to spawn or dies prematurely; Farrell et al. 2008; Martins et al. 2012a, b).

One of the core messages with that collective body of work is that elevated water temperatures can kill salmon. Furthermore, that acute (*sensu* Hinch et al. 2012) warm water can lead to exceptionally high levels of salmon mortality when they are faced with challenges of upstream migration even at temperatures well below the upper lethal limits for the species (Servizi and Jensen 1977). Yet, the mechanisms by which hot water kills salmon are diverse. For example, detailed swimming per-

formance studies have suggested that aerobic scope (see Figure 3.2) collapses at high temperatures (which are stock specific; Farrell et al. 2008; Eliason et al. 2011), presumably underpinned by constraints on cardiac function (Eliason et al. 2013a, b). Indeed, maximum heartrate in non-salmonid fish is typically reached below the upper acute lethal limits (Farrell 2016). These laboratory studies (e.g. Eliason et al. 2011, 2013a, b) align well with field-observed patterns of mortality using telemetry across a range of upriver migration temperatures (e.g. Farrell et al. 2008).

Another mechanism associated with mortality can be additional energy depletion at elevated water temperatures. In warm years, it is possible for fish, which have ceased feeding before entering the river, to run out of energy due to enhanced costs of swimming, brought about by an increase in standard metabolic rate amplifying the overall cost of swimming, rather than an increase in net cost of swimming (Brett 1995). Consequently, a migration delay in warmer water as fish hold waiting for preferable migration conditions (Cooke et al. 2004) will burn up valuable energy stores at a faster rate without the fish moving upstream. Hinch et al. (2005) proposed that sockeye salmon need an energy density of at least 4 MJ/kg when they reach the spawning area to successfully spawn. With rising water temperatures in the Fraser basin, it is anticipated that fish will deplete energy resources more quickly and thus we anticipate seeing more instances of condition-dependent pre-spawn mortality (Martins et al. 2012a).

What is particularly valuable about this overall body of work is that it spans lab, field, and modelling realms, and combines observational (including comparative physiology and genomics) and experimental approaches. In particular, the experimental approaches have helped to establish causation. For example, Crossin et al. (2008) captured upriver migrating sockeye, held them in the lab at two temperatures (an optimal temperature and a super-optimal temperature) for several weeks; then released them back into the wild to complete their migration. Such studies linked laboratory findings to field-relevant patterns of mortality and definitively show that hot water kills upriver migrating salmon.

(a)

(b)

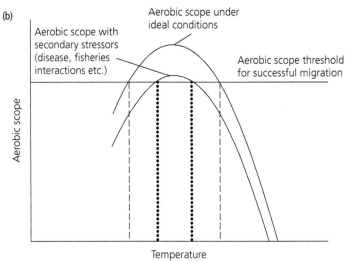

Figure 3.2 To simulate strenuous swimming conditions experienced during migration, laboratory swim flume studies (2A) can be used to assess swimming performance and cardio-respiratory physiology. Photo credit: S. Cooke. Aerobic scope curves (2B) under ideal conditions and under the influence of secondary stressors, as a function of temperature. For a given salmon population, there is a minimum aerobic scope threshold for successful migration to reach the spawning ground. This threshold will vary yearly depending on environmental conditions (e.g. may increase or decrease due to varying river flow, etc.). The optimal range of temperatures is restricted when fish are physiologically compromised due to secondary stressors (dotted lines) compared with under ideal conditions (dashed lines).

3.2.2 The conservation implications

In the Fraser River, peak summer water temperatures have been steadily rising by an average ~0.5°C over the past half of each century (Morrison et al. 2002). Because Pacific salmon are locally adapted, even slight increases in water temperature can push fish from their optimal temperatures to those where aerobic scope is constrained, which could alter migration success (Hague et al. 2011; Farrell 2016), not to mention alterations in pathogen exposure and pathogenicity (Miller et al. 2014). Because of the manifold effects of water temperature on fish along with these specific examples for upriver migrating

Pacific salmon, managers have accepted that hot water kills salmon (although admittedly it took some time; Cooke et al. 2012). This knowledge has translated to changes in the management adjustment model—a model that is used by fisheries managers to determine the levels of catch that can be permitted in the face of predicted en route loss associated with adverse environmental conditions to achieve the desired level of spawning escapements (i.e. the number of fish that reach spawning grounds; Macdonald et al. 2010; Patterson et al. 2016).

3.3 Intraspecific variation is the norm

3.3.1 The science

Pacific salmon, and especially sockeye salmon, return faithfully to their natal areas to reproduce, resulting in hundreds of reproductively isolated populations across a diverse range of environments (Groot and Margolis 1991). Morphological, physiological, behavioural, and life history traits vary extensively among populations and much of this diversity is a result of local adaptation (Taylor 1991; Crossin et al. 2004; Fraser et al. 2011; Eliason et al. 2013a, b). For example, considerable research has focused on the adult upriver spawning migration, where populations can encounter highly variable migratory conditions depending on when and where they spawn (e.g. varying in temperature, flow, distance travelled, elevation). Fraser River sockeye salmon populations with more difficult migrations (e.g. longer distances, greater elevation gains) start their migration with more somatic energy, have a more streamlined body shape, fewer eggs, larger hearts with more compact myocardium and its associated coronary circulation, and greater cardiac sarco(endo)plasmic reticulum Ca^{2+}-ATPase activity (which means Ca^{2+} can cycle faster to increase heartrate). They also have higher aerobic capacity, greater cardiac performance, and enhanced swimming performance compared with populations with shorter, easier migrations (Lee et al. 2003a, b; Crossin et al. 2004; Eliason et al. 2011; Eliason et al. 2013; Anttila et al. 2019). Taken together, variation in how challenging the upriver migration is has

selected for specific morphological and physiological traits that enable an energetically conservative yet high-performance athletic phenotype.

Thermal biology also varies across populations. The optimal range of temperatures for aerobic scope (thermal dependence curve) generally corresponds to the typical range of temperatures encountered for adult Fraser River sockeye salmon (Lee et al. 2003a, b; Farrell et al. 2008; Eliason et al. 2011). Similarly, a warm-temperature migrating population of chum salmon (*Oncorhynchus keta*) in Japan had a thermal dependence curve for aerobic scope shifted 3°C higher compared with a cool-temperature migrating population (Abe et al., 2019). In contrast, the thermal performance (i.e. swimming) of two coastal populations of adult autumn-migrating Coho salmon (*Oncorhynchus kisutch*) displayed divergent patterns, matching historical conditions in one case (Lee et al. 2003a, b) and extending above historical temperatures in another case (Raby et al. 2016). Thermal tolerance also varies across populations at other life stages. Egg thermal tolerance differed substantially for nine Fraser River sockeye salmon populations reared in a common garden experiment at three temperatures (Whitney et al. 2013). The emergent juveniles showed population-specific differences in their upper acute lethal limits, but this difference may have been a result of thermal plasticity or the juveniles emerging at different body size (caused presumably by differences in egg size among populations) because these population differences disappeared when test were performed at a common body mass of ~1 g (Chen et al. 2013). Importantly, these studies reveal that thermal tolerance is a flexible trait and is constrained at different life stages.

Finally, over the past decade or so, our team has identified an alarming trend where adult female salmon have elevated mortality compared with males, especially when exposed to secondary stressors (e.g. Roscoe et al. 2011; Jeffries et al. 2012a, b; Martins et al. 2012a, b). This trend is consistent across both field studies (e.g. tagging studies and passage studies through dammed watersheds; Roscoe et al. 2011; Martins et al. 2012a, b) and lab studies (e.g. experimental holding studies while exposing fish to various temperature regimes; Jeffries et al. 2012a, b). However, the physiological mechanisms driving these sex-specific differences in mortality are currently unknown.

3.3.2 The conservation implications

Managers and stakeholders cannot assume Pacific salmon species, populations, or sexes will respond similarly to environmental stressors. Pacific salmon display huge variability in traits, largely as a result of local adaptation (Taylor 1991; Fraser et al. 2011). This trait variability makes fish differentially susceptible to natural and anthropogenic perturbations (Martins et al. 2011, 2012a, b; Eliason et al. 2011; Donaldson et al. 2014). Greater population diversity has been shown to buffer against the impact of climate variability and change (Anderson et al. 2015), so preserving genetic diversity and a broad range of thermal tolerances is critical to support a robust metapopulation. Given that female salmon govern the fecundity of a reproducing population, a differential elevation in adult female mortality is a serious concern. Despite recent work in this area, the underlying mechanisms of this phenomenon still need identifying so effective mitigation can occur.

3.4 Stressors rarely act alone

3.4.1 The science

Bringing wild large-bodied fish into a laboratory to conduct controlled studies is challenging. Yet, this approach already has extensively characterized the responses of Pacific salmon to relevant environmental stressors (e.g. Wagner et al. 2005; Crossin et al. 2008; Clark et al. 2011; Eliason et al. 2011; Jeffries et al. 2012a). Like most lab studies they typically study one specific stressor at a time, whereas the reality in nature is that fish face multiple, simultaneous environmental stressors (Figure 3.3). Therefore, interactions among stressors can make interpretations of performance (e.g. migration success, fitness) difficult (Johnson et al. 2012). For example, an adult salmon moving from the saltwater to freshwater in summer face dramatic and simultaneous changes in their water temperature and salinity. In addition, fish become exposed

to freshwater-specific pathogens (e.g. Tierney and Farrell 2004; Wagner et al. 2005; Bradford et al. 2010). Furthermore, pre-existing pathogen infections can potentially influence the response to traumatic handling stress related to fisheries (Teffer et al. 2018). Therefore, the cumulative impact on wild fish populations of pathogens and abiotic stressors is likely pervasive but poorly understood.

One possible way to tease apart the types of stressors that a salmon is responding to is to develop molecular biomarkers (i.e. gene expression) of a response to a certain type of stressor (Akbarzadeh et al. 2018; Houde et al. 2018a, b; Miller et al. 2017), including an immune response to a range of pathogens (Miller et al. 2014; Jeffries et al. 2014b). This approach is challenged, however, by the fact that individual genes often respond to multiple stressors (e.g. certain heat shock proteins) as part of a generalized stress response (Feder and Hofmann 1999; Kultz 2005). As a result, directly linking a shift in expression of a single biomarker to a specific stress response and/or to a performance- or fitness-level consequence may not always be possible. A new approach to stressor prediction developed by our team relies on the resolution of panels of biomarkers that, when co-expressed, are indicative of individual stressors. For example, temperature holding studies that utilized transcriptome-wide screening approaches across multiple salmon

Figure 3.3 Visualization of the multiple stressors that often act together to influence the fate of migratory Pacific salmon. Photo credit: S. Cooke.

species (Jeffries et al. 2012b, 2014a; Tomalty et al. 2015) were mined for the development of biomarkers for thermal stress response in Pacific salmon (Akbarzadeh et al. 2018). The activity and specificity of this, and other panels of biomarkers activated in gill tissue, were then validated in a multi-stressor challenge study (Houde et al. 2018a), which not only resolved the most robust panels of biomarkers capable of identifying individual stressors, but also revealed biomarkers within those panels that were predictive of ensuing mortality. Simultaneous assessments of multiple biomarker panels are enabled through use of high-throughput microfluidics quantitative (q)PCR techniques. Host biomarker panels can be combined with assays for a range of pathogens (e.g. Miller et al. 2014; Jeffries et al. 2014b) to characterize the interplay between stress and infectious disease development. Overall, we anticipate that this approach will bridge the gap in our understanding of cumulative stressor/disease impacts from the laboratory to the field.

3.4.2 The conservation implications

Establishing approaches to tease apart responses to multiple stressors will aid in our ability to prioritize efforts to minimize the effects of certain stressors on Pacific salmon. Some stressors may be beyond our control (i.e. natural pathogen communities; but for an example of experimental surface flushing flows where extra water is released to reduce *Ceratonova shasta* concentrations in the river and address disease concerns, see Voss et al., 2018), however regulating activities that can exacerbate the effects of other stressors (i.e. fisheries encounters, industrialization) is one potential strategy. Additionally, the lower reaches and estuaries of major salmon-producing rivers (e.g. Fraser River, Columbia River, Sacramento River) often have an accumulation of municipal, industrial, and agricultural contaminants that can interact with other abiotic stressors affecting fish in a system. Contaminants in aquatic systems generally occur in mixtures and the direct and indirect effects on anadromous fishes are often complex (Laetz et al. 2008; Tierney et al. 2008). Understanding how adaptive responses to natural stressors and homing during migrations are impacted

by chemical contaminants will be a challenge for salmon researchers in the future (Ross et al. 2013).

3.5 A range of indices associated with compromised health can predict migration fate

3.5.1 The science

Stress and health biomarkers can be linked to migration failure of Pacific salmon at both the outbound juvenile (Olivier 2002; Tucker et al. 2016) and the inbound adult life stages (Young et al. 2006; Miller et al. 2011Cooke et al. 2012; Hinch et al. 2012). These indices can be measured at molecular, cellular, suborganismal, and whole-animal levels. Evaluations that comprise responses at multiple levels (e.g. genomic, metabolic, behavioural, disease-associated) provide a comprehensive assessment of the mechanisms of migration failure. Our understanding of how fish health impacts migration capacity has benefited from the use of physiological indices, such as hormone and metabolite levels in blood (Barton 2002; Cook et al. 2014), and molecular biomarkers described in previous sections (e.g. thermal and osmotic stress), as well as those relating to host immunity (Miller et al. 2017) and pathogen presence (Miller et al. 2014, 2017). Genomic tools for assessing wild salmon health are quickly becoming an invaluable method for detecting and describing infectious agents in wild salmon populations (Miller et al. 2014; Bass et al. 2017; Figure 3.4) because of the advent of high-throughput methods. Molecular approaches have also been used to evaluate pathogen impacts on health, longevity, and migration success of wild salmon, often within the context of environmental or fisheries stressors (Jeffries et al. 2014a, b; Teffer et al. 2017, 2018; Bass et al. 2019).

Infectious disease processes are increasingly being investigated as mechanisms of migration failure and mortality of wild salmon at both juvenile (Jeffries et al. 2014a, b; Miller et al. 2014) and adult life stages (Miller et al. 2011; Teffer et al. 2018; Bass et al. 2019). For juvenile salmon, indices that identify abnormal physiological processes involved in smoltification (i.e. osmoregulatory and immunological shifts from fresh- to saltwater) or starvation in the marine environment can be used to predict migration failure as fish fail to meet the energetic demands of seaward migration (Maule et al. 1987; Groot and Margolis 1991; Houde et al. 2019). Physiological or disease-associated impairment during this critical life stage may also enhance predation risk, which is a major factor influencing juvenile survival and population productivity (Miller et al. 2014; Furey 2016; Tucker et al. 2016). During the adult spawning migration, tremendous physiological and metabolic changes comprise osmoregulatory, reproductive, and immunological components (Shrimpton et al. 2005; Miller et al. 2009; Dolan et al. 2016). Deviations from these trajectories may not only impact migratory fate, but also reproductive potential and survival (Cooperman et al. 2010; Evans et al. 2011; Jeffries et al. 2011; Miller et al. 2011; Teffer et al. 2017). Molecular and metabolic indices that identify physiological impairment due to injury, osmotic imbalance, or aerobic collapse

are predictive of migration failure and often found in association with migratory stressors like high temperature or fishery interactions (Eliason et al. 2013a, b; Donaldson et al. 2014; Raby et al., 2014, 2015a, b; Bass et al. 2018). Like all organisms, Pacific salmon respond to their environment and exhibit an array of indices that can predict migration fate.

3.5.2 The conservation implications

As we learn more about genomic and physiological responses of wild salmon to environmental stressors and become better equipped to measure molecular biomarkers in large numbers of fish in the wild, we can combine this information with environmental parameters to inform management decisions that regulate human activities (e.g. land use, fisheries). If wild migrating fish are found to be exhibiting a particular response (e.g. thermal or osmotic stress), management actions can be taken towards short-term (e.g. reduce additional stressors like fisheries) and long-term solutions (e.g. improve habitat quality, river temperature regulation at dams). Great strides have been made in comprehending the molecular signatures of thermal stress (Akbarzadeh et al. 2018), osmotic impairment (Houde et al. 2018b), and viral disease development (Miller et al. 2017), which can be used to derive physiological and disease-associated biomarkers for reduced survival and reproductive success of wild salmon. Incorporating molecular biomarker data and other indices of unhealthy fish from in-season sampling into population dynamics models may improve the precision of productivity estimates under different ecological conditions.

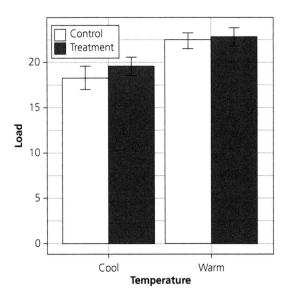

Figure 3.4 Loads of *Parvicapsula minibicornis* in the gills of adult Coho salmon from the Chilliwack River, British Columbia, Canada. Samples were taken after fish were held for 1 week at cool (10°C) or warm (15°C) temperature and a subset of fish from each temperature group received a simulated bycatch release treatment (gill net entanglement and air exposure). Loads were derived using qPCR and then subtracting the quantification cycle from 40 (maximum). Data from Teffer et al. (2019).

3.6 Simple reflex indicators can be used to refine fishing practices

3.6.1 The science

RAMP (Reflex Action Mortality Predictors), first pioneered in the laboratory (Davis 2005, 2007; Davis and Ottmar 2006), is a simple technique for assessing the vitality of a captured fish (Davis 2010). A RAMP assessment typically incorporates five or more reflexes consistently exhibited when a fish is vigorous and healthy. These reflexes, however, progressively disappear in fish that are exhausted, in poor condition, or moribund (Davis 2010; Raby et al. 2012). RAMP assessments are rapid, taking just 10–20 s per fish; the RAMP score is the proportion of reflexes that were absent (impaired; Davis 2007) and so higher RAMP scores indicate low vitality and a higher probability of delayed mortality. Although aspects reflexes are behavioural, the mechanisms underpinning reflexes are entirely physiological. The original promise of RAMP for field application was an objective, inexpensive, and rapid predictor of mortality.

We adapted RAMP for Pacific salmon assessment in the wild (Raby et al. 2012). Our motivation for doing so was to evaluate whether RAMP could be applied across Pacific salmon fisheries, many of which release non-target salmon species, to rapidly generate estimates of post-release mortality in fish discards (bycatch released overboard; Davis 2010). Five reflexes were scored, such as whether the fish responded to having its caudal fin held ('tail grab') and whether it righted itself when turned upside-down in the water ('orientation'). Given the ease of performing RAMP assessments, they were used in many studies examining post-release mortality in simulated adult salmon fisheries (Donaldson et al. 2012; Raby et al. 2012, 2013, 2014, 2015a, b; Robinson et al. 2013; Gale et al. 2014; Nguyen et al. 2014; Cook et al. 2018, 2019). Collectively, our work showed that RAMP responds well to variation in the severity of stressors like handling and air exposure that cause exhaustion (e.g. Raby et al. 2013; Nguyen et al. 2014) and therefore is a useful indicator of adult salmon vitality (Figure 3.5). RAMP is a powerful assessment tool because it conceivably integrates individual fish stress, independent of a prior condition or a stress imposed during the actual fishery encounter (e.g. its position within a crowded net; Raby et al. 2014, or its unique responsiveness to stress; Cook et al. 2014).

In some situations, RAMP predicted a fish's likelihood of surviving after release (Raby et al. 2014), but not all situations had clear relationships between stressor severity and rates of mortality (Donaldson et al. 2012; Raby et al. 2013; Robinson et al. 2013),

especially ones where injury was a major contributor to survival probability (Donaldson et al. 2012; Nguyen et al. 2014). Therefore, in fisheries where visible injuries are common, injury metrics should be integrated with RAMP into vitality assessments (Raby et al. 2015a; Uhlmann et al. 2016; Meeremans et al. 2017; Cook et al. 2018, 2019). Importantly, in addition to repeatedly showing that RAMP is sensitive to differences in fishing techniques, RAMP is effective at reflecting underlying physiological disturbances based on established, but much harder to determine, indices of exhaustion based on blood samples (e.g. lactate build-up, disturbance to ion concentrations; Raby et al. 2013, 2015a; Cook et al. 2019, and see McArley and Herbert 2014).

3.6.2 The conservation implications

Some of our recommendations for fish handling have passed on to the fishers by fisheries managers (Raby et al. 2014) with the intention of RAMP becoming an integral part of a 'rapid assessment' of the survival probability whenever a salmon was released from a specific fishery (Patterson et al. 2017a, b). In British Columbia alone, there are over 100 such fisheries with different combinations of fishing gear, species, location, and environmental conditions. Some existing fisheries observer programmes are already using vitality assessments on discarded bycatch (e.g. Benoît et al. 2012) and were incorporated (more crudely as a ranking scale) in earlier Pacific salmon bycatch research (e.g. Farrell et al. 2001a, b). Clearly, using RAMP moves these types of vitality evaluations to a new and more objective level and provides the research community with a cheap and simple biomarker of complex underlying processes that are often difficult to measure. Although more RAMP work has been done on Pacific salmon than any other fish species, the technique is being employed across a diverse range of freshwater and marine fish and even other taxa (e.g. turtles).

3.7 Facilitated recovery of exhausted salmon is context-dependent

3.7.1 The science

When Pacific salmon encounter various fishing gears (rod and reel, seine net, gill net) a component

(a)

(b)

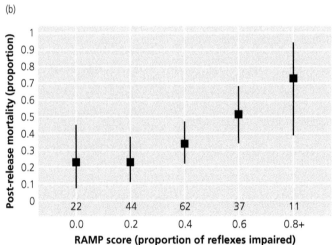

Figure 3.5 Reflex Action Mortality Predictors (RAMP) are obtained from fish to assess vitality. One of the more important and responsive indicators is the righting response (5A) where fish are turned upside-down to determine if they can right themselves within 3 s. Photo credit: V. Nguyen. (5B) Post-release mortality increases with increasing levels of reflex impairment (RAMP score) for adult Coho salmon released from a beach seine fishery in the Fraser River. Sample sizes for each level of RAMP score are given along the bottom of the figure. Symbols represent observed level of mortality (failure of telemetry tagged fish to reach terminal radio telemetry receivers near spawning areas), while error bars indicate 95 per cent confidence intervals calculated using the Clopper–Pearson Exact Method. The RAMP assessment upon which was the score (x-axis) was based is described in detail by Raby et al. (2012). Figure created using data for Raby et al. (2014), which are publicly archived alongside the paper.

of the catch is often released to comply with regulations or because of voluntary conservation ethic. The condition of live fish following capture varies greatly from extremely vigorous to moribund. Because fishing and associated handling (including air exposure) induce physiological disturbances (e.g. depletion of tissue energy stores, elevation of metabolic rate, accumulation of metabolic by-products), fish that are to be released may be exhausted. This state of exhaustion can lead directly to mortality or do so indirectly whereby exhausted fish are susceptible to post-release predation, displacement (e.g. being washed downstream), or simply holding until suitably recovered to continue the migration. We have conducted experiments in the lab and field using tools such as blood and muscle physiology, RAMP (as described above), biotelemetry, and even swimming studies to determine if it is possible to facilitate recovery of exhausted fish.

Early work in the marine environment revealed that it was possible to take adult Coho salmon that were fully exhausted (i.e. unable to maintain equilibrium) after capture by gill net or trolling; by holding them in a 'fish-sized' compartment with a jet that forced water towards the mouth of the fish, most individuals would be vigorous within 15 min after recovering their physiological status (as assessed by blood and muscle biochemistry; Farrell et al. 2001a, b). Even fish classed as 'moribund' were revived and, after an overnight recovery in a net pen, performed a swim test just as well as fish classed as 'vigorous' at the time of capture. Indeed, the short-term mortality rate on incidentally caught Coho salmon could be reduced to as little as 6 per cent with the use of the recovery box, modified gear, short net soak times, and careful handling of fish on removal from the gill net, compared with the more typical Coho mortality rate of 35–70 per cent associated with gillnetting (Buchanan et al. 2002).

Because of the apparent success of this method of facilitated recovery in the marine migration phase (assessed using blood and muscle biochemistry; Farrell et al. 2001a, b), we have also tested the value of using a recovery box during the upriver migration phase, while recognizing the impracticality of using such a box designed for a large vessel with shore-based fishers or small vessels. The recovery of migratory adult fish was certainly facilitated with the recovery box, but it did not translate to higher levels of post-release survival assessed with biotelemetry, likely because dermal injuries enable opportunistic pathogens to take hold in freshwater more so than seawater (Nguyen et al. 2013). Further tests with portable and inexpensive fish holding bags that were more suited to river fishing came to a similar conclusion (Donaldson et al. 2013; Raby et al. 2014); short-term holding suppressed the cortisol stress response, but unless the salmon was completely exhausted, there was little benefit in terms of post-release survival (Donaldson et al. 2013). Lastly, efforts to facilitate recovery by hand

(just by holding fish into the flow) also seemed to generate minimal benefits except for the most exhausted individuals (Robinson et al. 2013, 2015).

3.7.2 The conservation implications

We concluded that it is best to immediately release a salmon if it is not completely exhausted, able to maintain equilibrium, and not severely injured; facilitated recovery benefited salmon in the very worst condition, that is, unable to maintain equilibrium. Of course, RAMP is a type of vitality assessment that can distinguish whether a salmon is fully exhausted. This simple tool can be used by fishers to triage captured fish that must be released to comply with harvest regulations.

3.8 The smell has to be the right one to get home

3.8.1 The science

Pacific salmon rely on their highly developed sense of smell (olfaction) to locate their natal streams during the spawning migration. Early research found that salmon imprint on the odour of their natal water as juveniles and are progressively guided by those imprinted cues as adults get closer to the natal stream using their sensory physiology apparatus (Hasler and Scholz 1983). Yet, water resource developments such as dams can alter the odour cues if they involve altering the relative contribution of different water sources to flow. Imprinting occurs during the juvenile parr–smolt transformation stage (Dittman and Quinn 1996), although recent studies indicate it may even occur prior to this life stage (Bett et al. 2016; Havey et al. 2017). Regardless of the timing, successful migration is nonetheless reliant on detection of imprinted cues, which can be difficult for returning salmon to detect if the natural movement or composition of water in their natal system is disrupted. For example, recent work in a regulated watershed in British Columbia, Canada found that an altered water flow pattern and composition associated with a power-generating station caused confusion and a migration delay in adult sockeye salmon (Middleton et al. 2018). In this system, natal water is partially diverted from the migratory route for power generation, and the remaining natal water is diluted with make-up water from another watershed. In these systems we have used biotelemetry and controlled behavioural choice experiments to show that normal hydro-operational procedures thath would reduce the amount of dilution appeared to enhance migration speed and migration success through the area (Bett et al. 2018; Drenner et al. 2018).

Olfactory toxicity, or a reduction in olfactory sensitivity following exposure to pollutants, is another concern for salmon homing. Electrophysiological and behavioural studies indicate that chemicals such as pesticides or heavy metals impair a salmonid's ability to detect and respond to olfactory cues (Tierney et al. 2008; Sovová et al. 2014). Urban areas release these and other pollutants into rivers, potentially affecting adult salmon navigation to natal spawning areas. In addition, because the olfactory abilities of juvenile Pacific salmon are adversely affected by the projected increases in freshwater CO_2 (Ou et al. 2015), a similarly negative effect may occur in returning adults. Whether or not pollutants or elevated CO_2 levels encountered along the migratory route also affect pheromone detection is unknown. Pheromones have long been considered to play a role in salmon homing and are now believed to be particularly important directional cues when salmon stray to seek alternate, non-natal spawning habitat (Bett and Hinch 2015).

3.8.2 The conservation implications

Navigation problems could be compounded by issues related to dam passage, making careful management of regulated systems a significant concern for salmon populations. Hydroelectric development that best emulates the natural flow patterns along migration routes will limit the negative consequences to salmon navigation. Increased urbanization and rising CO_2 levels could also negatively affect navigation, although restrictions on industrial effluents, agricultural runoff, or other sources of pollution could mitigate these effects. More research that integrates field observations (via telemetry) with behavioural and physiological

experiments will be needed to determine the extent to which these factors can affect olfactory ability, and as a consequence navigation and migration success.

3.9 Not too much, not too little

3.9.1 The science

Burst swimming (i.e. when fish swim at near maximal speeds using white muscle fibres that are fuelled by anaerobic metabolism) is used by fish to negotiate and pass through areas of high water flow. Swimming anaerobically cannot be sustained indefinitely and requires oxygen uptake elevated for a substantial period post-exercise (known as excess post-exercise oxygen consumption, or EPOC; Lee et al. 2003a, b). If a fish has fatigued, the time to recovery from EPOC is up to 4 h (Brett 1973; Milligan 1996). But if a fish is exhausted and no longer able to maintain a righting reflex, recovery is much longer (12–15 h to recovery from EPOC; Zhang et al. 2018) to re-establish metabolic homeostasis. Thus, anaerobic swimming, if excessive, has the potential to influence subsequent behaviour, survival, and spawning success of salmon.

Salmon are often exposed to high water flows in dam tailraces (i.e. area downstream of dams). Yet, the potential for carryover effects (i.e. when an individual's previous history and experience explain their current performance; O'Connor et al. 2014) associated with dam passage is generally underappreciated and overlooked by researchers and natural resource managers because few studies have made a direct link between an individual's experience during dam passage and their ability to fulfil other life history requirements. Recent research on sockeye salmon passage at Seton Dam in British Columbia, Canada found reduced survival to spawning grounds following dam passage (Roscoe et al. 2011). Burnett et al. (2014) aimed to understand the mechanism(s) underlying poor post-passage survival, with a focus on characterizing high-flow-induced burst swimming. Using acoustic accelerometer transmitters to estimate swimming speed and oxygen consumption, Burnett et al. (2014) identified a trade-off where sockeye salmon required burst swimming to successfully pass through fishway attraction flows, but this behaviour had significant consequences for survival to spawning grounds. Individuals that spent more time swimming anaerobically were less likely to reach spawning grounds, perhaps because they became fatigued or, worse still, exhausted, which resulted in a greater and more prolonged EPOC (Lee et al. 2003a, b; Zhang et al. 2018). With an understanding of the physiological mechanism of reduced survival, Burnett et al. (2017) conducted a large-scale management experiment of fishway attraction flows to maximize dam passage and survival to spawning grounds. Fish tagged with acoustic accelerometer transmitters and Passive Integrated Transponder (PIT) tags were released and tracked to spawning grounds under two flow conditions: (1) a baseline condition that has been used by managers since the 1950s; and (2) an alternative condition that attempted to reduce the exposure of fish to high flows during their approach to the fishway entrance. Fish exposed to alternative flow conditions required significantly more time to pass Seton Dam but showed improved survival to spawning grounds—a finding that scaled up from individual tagged fish to the population level. Ultimately, the benefits of reducing high flows to improve survival to spawning grounds outweighed the costs associated with increased passage time.

3.9.2 The conservation implications

Taken together, this research has pointed to a less impactful, cost-effective flow management strategy that maximizes the survival of sockeye salmon to spawning grounds without impacting power production and downstream flow requirements. Fishway attraction flows at Seton Dam are operated in a manner consistent with facilities worldwide, where attraction flows are released directly adjacent to the fishway entrance. Dams are ideal model systems for testing management options in the field (Memmott et al. 2010) through large-scale experiments (Walters and Holling 1990), highlighting that there are opportunities elsewhere to understand and minimize the effects of dams and dam operations on aquatic animals. Research of this nature highlights the value of integrating applied ecological research and adaptive management to provide natural resource managers and conservation practitioners with the information required to make responsible environmental management decisions.

3.10 Reflections on physiology in practice

The above examples of applied conservation physiology research have improved our scientific understanding of how environmental stressors mechanistically impact salmonid performance and fitness. As a result, the incorporation of physiology has helped to reduce uncertainty in the scientific advice presented to managers on a broad range of topics related to Pacific salmon migration. However, researchers simply presenting a better mechanistic understanding of salmon physiology in relation to individual fish survival does not automatically translate into management actions directed primarily at the population level. Working with managers for over 20 years has afforded our group with the opportunity to reflect on both reasons for success and failure in management effectively using science information in their decision-making processes (Hinch et al. 2012; Patterson et al. 2016). The main reasons for failing to adopt scientific advice include scientific uncertainty, science management integration, and institutional resistance of management agencies.

The lack of adoption of science information by management can result from uncertainty in scientific results, especially results that either appear contradictory or challenge current beliefs. For example, the lack of universal benefit to facilitated recovery has likely delayed clear management responses. Results that challenge the current management paradigm and/or have a high cost to implement make the adoption of the advice more difficult and increase the scrutiny of the uncertainty in the results. One of the key successes of the research at Seton Dam was that the changes in flow regime were cost-neutral. In contrast, the initial attempts to reduce salmon harvest levels based on forecasts of salmon exposure to prolonged high water temperatures were met with scepticism and resistance, even though there was good physiological research on temperature impacts and robust correlational models between high temperature, freshwater residence, and in-river mortality (Macdonald et al. 2010). One of the main advantages of using physiologically based mechanisms to predict survival is the ability to generate consistent results over time (Horodysky et al. 2015), unlike many other studies that simply link environmental covariates to fish survival, which break down over time (Myers 1998). More recently, management has accepted the results from the environment-based correlation models, in large part due to the model not breaking down since it was first proposed 20 years ago! In the interim, we continued to add physiological research support of a problem we believed in. Challenges still remain in communicating uncertainty, especially for the challenging multiple stressor work, but these can be overcome with better integration with management.

Science working in isolation of management, as well as being detached from the groups that are potentially affected by the results, can create impediments to use of science information. For example, our work to research and understand sex-specific differences in adult survival was done in isolation from management. This isolation was primarily because initially we did not expect sex-specific differences, but began to notice the difference when we tried to match the numbers of males and females in our experimental designs. Neither did management expect the differences that emerged. As such, there is no current mechanism within the management structure to use the unexpected information; spawning objectives would have to be changed to explicitly consider sex-specific targets. Similarly, failure to integrate/communicate at the beginning of the fisheries interaction work did lead to suspicions and active attempts to discredit the work before results had been generated by some user groups that could be impacted by changes to post-release mortality rates. We have now learnt to share study designs and preliminary results with managers and interested parties during our regular workshops to build awareness, answer questions, communicate uncertainty, and build trust in the work. For example, the innovative work on biomarkers is being shared with relevant branches within Fisheries and Oceans Canada (DFO) to maximize the potential utility of these powerful tools.

Institutional resistance within management organizations is often overlooked when examining the failure to adopt science advice. Resistance can result from the relatively low importance of science in the organization's existing decision-making framework and the lack of organizational support (expertise, workload, and funding) to integrate science. Science, of course, is only one of many factors considered by decision makers, and so science inevitably competes with social factors that also influence the decision process (Rice et al. 2011). Therefore, the best hope to affect positive change is to increase the scientific expertise within the organization. We have been particularly fortunate to work with fisheries management staff who have science backgrounds. Indeed, many of our former graduate students now work for the regulatory agencies. However, while many managers may have the necessary science literacy, a common refrain is that they may not have time to properly evaluate the research due to workload issues; as such the information becomes tabled.

Many funding opportunities in Canada aimed at improving management decisions through better science actually restrict any money going back to the regulator (i.e. government agency). This then restricts participation of both the managers that are responsible for understanding and processing the science information being generated by the funds, and government scientists who can help with interpretation and implementation of the work. Where possible we have tried to include funding for management to deal with extra work associated with working with scientists; this includes funds to help organize and/or reanalyse their own data and participate in workshops. In addition, we have social scientists within our research groups to help understand the successes and failures of science from the perspective of management and different groups affected by the work.

3.11 Insights from social science

In parallel to the physiological work, and in order to better understand the potential applications of this research, we have also collaborated with social scientists to develop an interdisciplinary research programme focusing on the socio-ecological dimension of fisheries governance in the Fraser River. The social-ecological approach sees environmental conditions (biological, physical, and ecosystem) and human actions (perceptions, decisions, and relationships) as intertwined, with each factor deeply affecting the other (Ostrom 2009). Human beings impact the natural environment via their individual and collective actions, and the environment affects humans by providing natural resources and imposing material constraints, as well as by providing cultural symbols and shaping lived experiences (Berkes et al. 2003). Therefore, our social science research programme included in-depth interviews with the fishers (Indigenous and non-Indigenous), the leaders of stakeholder and rights-holder groups, the participants in fisheries co-management, the scientists (inside and outside of government), the regulators, and the policy makers (see Nguyen et al. 2016; Young et al., 2016a, b; 2018).

Findings from the social science research reinforced the often-heard argument that salmon have deep economic and cultural significance in British Columbia (Scarce 2000). The cultural importance of salmon to the province's Indigenous (First Nations) people cannot be overstated: salmon movements and physiology play a key role in the cosmology of many groups and communities in both coastal and inland regions (Harris 2001). Salmon also have deep cultural significance to non-Indigenous stakeholders, and to fisheries managers in government, who are personally committed to the animals and their proper management. The economic and cultural prominence of salmon can, however, complicate the integration of new science—including physiological science—into fisheries governance. The social science research found that stakeholders and government employees alike are reluctant to take risks in salmon management, and modifying the evidence base for decision making is perceived as risky (Young et al. 2013). The high profile of salmon issues in the province means that all participants in fisheries governance feel under the microscope, and believe that a broad consensus on the validity of new knowledge is required prior to adjusting policy. Participants from all backgrounds reported that the complexity of the issue made it difficult to evaluate the veracity of new knowledge

and evidence (Young et al. 2016a, b, 2018). Government employees stressed the importance of having new knowledge vetted by the DFO, in the form of a Canadian Science Advice Secretariat report, before being comfortable using that knowledge. Stakeholders, on the other hand, expressed scepticism about the reductionist or decontextualized knowledge that can result from such reviews. Instead, they often stressed the importance of researchers' personal knowledge and understanding of the Fraser River as key markers of valid knowledge, because researchers who understand the social-ecology of human–salmon interactions are more likely to make responsible recommendations and less likely to be careless in their thinking. This gulf in preferences about the form and format of new knowledge and evidence is a hidden but significant barrier to the uptake of new science into practice.

3.12 Conclusions and future directions

Starting in the mid-1990s, a team of academic and government scientists engaged in mission-oriented research focused on understanding the factors that influence migration success in Pacific salmon. Along the way we incorporated diverse expertise, disciplines, and techniques that told us about animal–environment and animal–human interactions. Extensive interaction and collaboration with resource managers through a co-production model ensured that the research was relevant to knowledge users. Nonetheless, challenges remain with respect to incorporation of findings into management. Through collaboration with social scientists and ongoing reflection, we identified barriers to knowledge mobilization and worked actively to overcome them. What is clear is that there was no magic bullet—rather all of the advances in management that occurred as a result of this collective body of research were because of application of multiple tools and approaches (spanning the lab and field and including modelling, observation, and experimentation) in a co-production framework. We adeptly and vigorously incorporated emerging tools and technologies (e.g. genomics, telemetry) into our research programme and more innovations are certainly on the horizon. Yet, it is not a single tool or study that matters—it is about how the individual tools combine and how the individual studies compound to generate nuanced, comprehensive, and evidence-based knowledge. The single most important advice that we have for others engaged in conservation physiology is to engage in frequent and meaningful collaboration with knowledge users—that is the formula for success in conservation physiology. Specific to Pacific salmon, most of the research on migration biology thus far has focused on sockeye salmon, so there is need for similar research on other species to understand the extent to which the observations presented here represent general phenomena. In addition, there is need for mechanistic research on migration that extends across all life stages—not just the adult migration phase. That type of work is ongoing but it is too early to attempt synthesis.

References

Abe, T.K., Kitagawa T., Makiguchi, Y., and Sato, K., 2019. Chum salmon migrating upriver adjust to environmental temperatures through metabolic compensation. *Journal of Experimental Biology*, 222(3), jeb186189.

Akbarzadeh, A., Günther, O.P., Houde, A.L. et al., 2018. Developing specific molecular biomarkers for thermal stress in salmonids. *BMC Genomics*, 19, 749.

Anderson, S.C., Moore, J.W., McClure, M.M. et al., 2015. Portfolio conservation of metapopulations under climate change. *Ecological Applications*, 25, 559–72.

Anttila, K., Farrell, A.P., Patterson, D.A. et al., 2019. Cardiac SERCA activity in sockeye salmon populations: an adaptive response to migration conditions. *Canadian Journal of Fisheries and Aquatic Sciences*, 76, 1–5.

Barton, B.A., 2002. Stress in fishes: a diversity of responses with particular reference to changes in circulating corticosteroids. *Integrative and Comparative Biology*, 42, 517–25.

Bass, A., Hinch, S.G., Casselman, M.T. et al., 2018. Visible gill net injuries predict migration and spawning failure in adult sockeye salmon. *Transactions of the American Fisheries Society*, 147, 1085–99.

Bass, A.L., Hinch, S.G., Teffer, A.K. et al., 2017. A survey of microparasites present in adult migrating Chinook salmon (*Oncorhynchus tshawytscha*) in south-western British Columbia determined by high-throughput quantitative polymerase chain reaction. *Journal of Fish Diseases*, 40, 453–77.

Bass, A.L., Hinch, S.G., Teffer, A.K. et al., 2019. Fisheries capture and infectious agents were associated with travel rate and survival of Chinook salmon during spawning migration through a natal river. *Fisheries Research*, 209, 156–66.

Benoît, H.P., Hurlbut, T., Chassé, J., and Jonsen, I.D., 2012. Estimating fishery-scale rates of discard mortality using conditional reasoning. *Fisheries Research*, 125–126, 318–30.

Berkes, F., Colding, J., and Folke, C., 2003. *Navigating Social-Ecological Systems: Building Resilience for Complexity and Change*. Cambridge University Press, New York.

Bett, N.N. and Hinch, S.G., 2015. Attraction of migrating adult sockeye salmon to conspecifics in the absence of natal chemical cues. *Behavioral Ecology*, 26, 1180–7.

Bett, N.N., Hinch, S.G., and Casselman, M.T., 2018. Effects of natal water dilution in a regulated river on the migration of Pacific salmon. *River Research and Applications*, 34, 1151–7.

Bett, N.N., Hinch, S.G., Dittman, A.H., and Yun, S.S., 2016. Evidence of olfactory imprinting at an early life stage in Pacific salmon (*Oncorhynchus gorbuscha*). *Scientific Reports*, 6, 36393.

Bradford, M.J., 1995. Comparative review of Pacific salmon survival rates. *Canadian Journal of Fisheries and Aquatic Sciences*, 52(6), 1327–38.

Bradford, M.J., Lovy, J., Patterson, D.A. et al., 2010. *Parvicapsula minibicornis* infections in gill and kidney and the premature mortality of adult sockeye salmon (*Oncorhynchus nerka*) from Cultus Lake, British Columbia. *Canadian Journal of Fisheries and Aquatic Sciences*, 67(4), 673–83.

Brett, J.R., 1971. Energetic responses of salmon to temperature. A study of some thermal relations in the physiology and freshwater ecology of sockeye salmon (*Oncorhynchus nerka*). *American Zoologist*, 11, 99–113.

Brett, J.R., 1973. Energy expenditure of sockeye salmon, *Oncorhynchus nerka*, during sustained performance. *Journal of the Fisheries Board of Canada*, 30, 1799–809.

Brett, J.R., 1995. Energetics. In C. Groot, L. Margolis, and W.C. Clarke, eds. *Physiological Ecology of Pacific Salmon*, pp. 3–68. University of British Columbia Press, Vancouver, BC.

Buchanan, S., Farrell, A.P., Fraser, J. et al., 2002. Reducing gill-net mortality of incidentally caught Coho salmon. *North American Journal of Fisheries Management*, 22(4), 1270–5.

Burnett, N.J., Hinch, S.G., Bett, N.N. et al., 2017. Reducing carryover effects on the migration and spawning success of sockeye salmon through a management experiment of dam flows. *River Research and Applications*, 33, 3–15.

Burnett, N.J., Hinch, S.G., Braun, D.C. et al., 2014. Burst swimming in areas of high flow: delayed consequences of anaerobiosis in wild adult sockeye salmon. *Physiological and Biochemical Zoology*, 87, 587–98.

Chen, Z., Anttila, K., Wu, J. et al., 2013. Optimum and maximum temperatures of sockeye salmon (*Oncorhynchus nerka*) populations hatched at different temperatures. *Canadian Journal of Zoology*, 91, 265–74.

Clark, T.D., Jeffries, K.M., Hinch, S.G., and Farrell, A.P., 2011. Exceptional aerobic scope and cardiovascular performance of pink salmon (*Oncorhynchus gorbuscha*) may underlie resilience in a warming climate. *Journal of Experimental Biology*, 214(18), 3074–81.

Cook, K. V., Crossin, G.T., Patterson, D.A. et al., 2014. The stress response predicts migration failure but not migration rate in a semelparous fish. *General and Comparative Endocrinology*, 202, 44–9.

Cook, K.V., Hinch, S.G., Drenner, S.M., et al., 2018. Population-specific mortality in Coho salmon (*Oncorhynchus kisutch*) released from a purse seine fishery. *ICES Journal of Marine Science*, 75, 309–18.

Cook, K.V., Hinch, S.G., Drenner, S.M. et al., 2019. Dermal injuries caused by purse seine capture result in lasting physiological disturbances in Coho salmon. *Comparative Biochemistry and Physiology -Part A Molecular and Integrative Physiology*, 227, 75–83.

Cooke, S.J., Hinch, S.G., Donaldson, M.R. et al., 2012. Conservation physiology in practice: how physiological knowledge has improved our ability to sustainably manage Pacific salmon during up-river migration. *Philosophical Transactions of the Royal Society Biological Science*, 367, 1757–69.

Cooke, S.J., Hinch, S.G., Farrell, A.P. et al., 2004. Early-migration and abnormal mortality of late-run sockeye salmon in the Fraser River, British Columbia. *Fisheries*, 29(2), 22–33.

Cooperman, M.S., Hinch, S.G., Crossin, G.T. et al., 2010. Effects of experimental manipulations of salinity and maturation status on the physiological condition and mortality of homing adult sockeye salmon held in a laboratory. *Physiological and Biochemical Zoology*, 83, 459–72.

Crossin, G.T., Hinch, S.G., Cooke, S.J. et al., 2008. Exposure to high temperature influences the behaviour, physiology, and survival of sockeye salmon during spawning migration. *Canadian Journal of Zoology*, 86(2), 127–40.

Crossin, G. T., Hinch, S.G., Farrell, A.P. et al., 2004. Energetics and morphology of sockeye salmon: effects of upriver migratory distance and elevation. *Journal of Fish Biology*, 65(3), 788–810.

Crozier, L.G., Hendry, A.P., Lawson, P.W. et al., 2008. Potential responses to climate change in organisms with complex life histories: evolution and plasticity in Pacific salmon. *Evolutionary Applications*, 1(2), 252–70.

Davis, M.W., 2005. Behaviour impairment in captured and released sablefish: ecological consequences and possible substitute measures for delayed discard mortality. *Journal of Fish Biology*, 66, 254–65.

Davis, M.W., 2007. Simulated fishing experiments for predicting delayed mortality rates using reflex impairment in restrained fish. *ICES Journal of Marine Science*, 64, 1535–42.

Davis, M.W., 2010. Fish stress and mortality can be predicted using reflex impairment. *Fish and Fisheries*, 11, 1–11.

Davis, M.W. and Ottmar, M.L., 2006. Wounding and reflex impairment may be predictors for mortality in discarded or escaped fish. *Fisheries Research*, 82, 1–6.

Dittman, A. and Quinn, T., 1996. Homing in Pacific salmon: mechanisms and ecological basis. *Journal of Experimental Biology*, 199, 83–91.

Dolan, B.P., Fisher, K.M., Colvin, M.E. et al., 2016. Innate and adaptive immune responses in migrating spring-run adult Chinook salmon, *Oncorhynchus tshawytscha*. *Fish Shellfish Immunology*, 48, 136–44.

Donaldson, M.R., Hinch, S.G., Jeffries, K.M. et al., 2014. Species- and sex-specific responses and recovery of wild, mature Pacific salmon to an exhaustive exercise and air exposure stressor. *Comparative Biochemistry and Physiology— Part A Molecular and Integrative Physiology*, 173, 7–16.

Donaldson, M.R., Hinch, S.G., Raby, G.D. et al., 2012. Population-specific consequences of fisheries-related stressors on adult sockeye salmon. *Physiological and Biochemical Zoology*, 85, 729–39.

Donaldson, M.R., Raby, G.D., Nguyen, V.N. et al., 2013. Evaluation of a simple technique for recovering fish from capture stress: integrating physiology, biotelemetry, and social science to solve a conservation problem. *Canadian Journal of Fisheries and Aquatic Sciences*, 70(1), 90–100.

Drenner, S.M., Harrower, W.L., Casselman, M.T. et al., 2018. Whole-river manipulation of olfactory cues affects upstream migration of sockeye salmon. *Fisheries Management and Ecology*, 25, 488–500.

Eliason, E.J., Clark, T.D., Hague, M.J. et al., 2011. Differences in thermal tolerance among sockeye salmon populations. *Science*, 332(6025), 109–12.

Eliason, E.J., Clark, T.D., Hinch, S.G., and Farrell, A. P., 2013a. Cardiorespiratory collapse at high temperature in swimming adult sockeye salmon. *Conservation Physiology*, 1(1), cot008. doi:10.1093/conphys/cot008

Eliason, E.J., Wilson, S.M., Farrell, A.P. et al., 2013b. Low cardiac and aerobic scope in a coastal population of sockeye salmon *Oncorhynchus nerka* with a short upriver migration. *Journal of Fish Biology*, 82(6), 2104–12.

Evans, T.G., Hammill, E., Kaukinen, K.H. et al., 2011. Transcriptomics of environmental acclimatization and survival in wild adult Pacific sockeye salmon (*Oncorhynchus nerka*) during spawning migration. *Molecular Ecology*, 20, 4472–89.

Farrell, A.P., 2016. Pragmatic perspective on aerobic scope: peaking, plummeting, pejus and apportioning. *Journal of Fish Biology*, 88(1), 322–43.

Farrell, A.P., Gallaugher, P.E., Fraser, J. et al., 2001b. Successful recovery of the physiological status of Coho salmon on board a commercial gillnet vessel by means of a newly designed revival box. *Canadian Journal of Fisheries and Aquatic Sciences*, 58(10), 1932–46.

Farrell, A.P., Gallaugher, P.E., and Routledge, R., 2001a. Rapid recovery of exhausted adult Coho salmon after commercial capture by troll fishing. *Canadian Journal of Fisheries and Aquatic Sciences*, 58(12), 2319–24.

Farrell, A.P., Hinch, S.G., Cooke, S.J. et al., 2008. Pacific salmon in hot water: applying aerobic scope models and biotelemetry to predict the success of spawning migrations. *Physiological and Biochemical Zoology*, 81, 697–708.

Feder, M.E. and Hofmann, G.E., 1999. Heat-shock proteins, molecular chaperones, and the stress response: evolutionary and ecological physiology. *Annual Review of Physiology*, 61(1), 243–82.

Fraser, D., Weir, L., Bernatchez, L. et al., 2011.Extent and scale of local adaptation in salmonid fishes: review and meta-analysis. *Heredity*, 106, 404–20.

Furey, N.B., 2016. *Migration Ecology of Juvenile Pacific Salmon Smolts: The Role of Fish Condition and Behaviour across Landscapes*. University of British Columbia, Vancouver, BC.

Gale, M.K., Hinch, S.G., Cooke, S.J. et al., 2014. Observable impairments predict mortality of captured and released sockeye salmon at various temperatures. *Conservation Physiology*, 2, 1–15.

Groot, C. and Margolis, L., 1991. *Pacific Salmon Life Histories*. University of British Columbia Press, Vancouver, BC.

Hague, M.J., Ferrari, M.R., Miller, J.R. et al., 2011). Modelling the future hydroclimatology of the lower Fraser River and its impacts on the spawning migration survival of sockeye salmon. *Global Change Biology*, 17(1), 87–98.

Harris, D.C., 2001. *Fish, Law, and Colonialism: The Legal Capture of Salmon in British Columbia*. University of Toronto Press, Toronto, ON.

Hasler, A.D. and Scholz, A.T., 1983. *Olfactory Imprinting and Homing in Salmon*. Springer-Verlag, Berlin.

Havey, M.A., Dittman, A.H., Quinn, T.P. et al., 2017. Experimental evidence for olfactory imprinting by sockeye salmon at embryonic and smolt stages. *Transactions of the American Fisheries Society*, 146, 74–83.

Hinch, S.G., Cooke, S.J., Farrell, A.P. et al., 2012. Dead fish swimming: a review of research on the early migration and high premature mortality in adult Fraser River sockeye salmon *Oncorhynchus nerka*. *Journal of Fish Biology*, 81, 576–99.

Hinch, S.G., Cooke, S.J., Healey, M.C., and Farrell A.P., 2005. Behavioural physiology of fish migrations: salmon as a model approach. In K.A. Solomon, R.W. Wilson, and S. Balshine, eds. *Fish Physiology Series, vol. 24,*

Behaviour and Physiology of Fish, pp. 239–95. Academic Press, New York.

Horodysky, A.Z., Cooke, S.J., and Brill, R.W., 2015. Physiology in the service of fisheries science: why thinking mechanistically matter. *Reviews in Fish Biology and Fisheries*, 25, 425–47.

Houde, A.L.S., Akbarzadeh, A., Günther, O.P. et al., 2018a. Salmonid gene expression biomarkers indicative of physiological responses to changes in salinity, temperature, but not dissolved oxygen. *Journal of Experimental Biology*, 222, jeb198036. doi:10.1242/jeb.198036

Houde, A.L.S., Gunther, O.P., Strohm, J. et al., 2018b. Discovery and validation of candidate smoltification gene expression biomarkers across multiple species and ecotypes of Pacific salmonids. *bioRxiv*, 474692.

Houde, A.L.S., Schulze, A.D., Kaukinen, K.H. et al., 2019. Transcriptional shifts during juvenile Coho salmon (*Oncorhynchus kisutch*) life stage changes in freshwater and early marine environments. *Comparative Biochemistry and Physiology—Part D Genomics and Proteomics*, 29, 32–42.

Jeffries, K.M., Hinch, S.G., Donaldson, M.R. et al., 2011. Temporal changes in blood variables during final maturation and senescence in male sockeye salmon *Oncorhynchus nerka*: reduced osmoregulatory ability can predict mortality. *Journal of Fish Biology*, 79, 449–65.

Jeffries, K.M., Hinch, S.G., Gale, M.K. et al., 2014b. Immune response genes and pathogen presence predict migration survival in wild salmon smolts. *Molecular Ecology*, 23(23), 5803–15.

Jeffries, K.M., Hinch, S.G., Martins, E.G. et al., 2012a. Sex and proximity to reproductive maturity influence the survival, final maturation, and blood physiology of Pacific salmon when exposed to high temperature during a simulated migration. *Physiological and Biochemical Zoology*, 85(1), 62–73.

Jeffries, K.M., Hinch, S.G., Sierocinski, T. et al., 2012b. Consequences of high temperatures and premature mortality on the transcriptome and blood physiology of wild adult sockeye salmon (*Oncorhynchus nerka*). *Ecology and Evolution*, 2(7), 1747–64.

Jeffries, K.M., Hinch, S.G., Sierocinski, T. et al., 2014a. Transcriptomic responses to high water temperature in two species of Pacific salmon. *Evolutionary Applications*, 7(2), 286–300.

Johnson, J.E., Patterson, D.A., Martins, E.G. et al., 2012. Quantitative methods for analysing cumulative effects on fish migration success: a review. *Journal of Fish Biology*, 81(2), 600–31.

Kültz, D., 2005. Molecular and evolutionary basis of the cellular stress response. *Annual Review of Physiology*, 67, 225–57.

Laetz, C.A., Baldwin, D.H., Collier, T.K. et al., 2008. The synergistic toxicity of pesticide mixtures: implications for risk assessment and the conservation of endangered Pacific salmon. *Environmental Health Perspectives*, 117(3), 348–53.

Lee, C.G., Farrell, A.P., Lotto, A. et al., 2003a. The effect of temperature on swimming performance and oxygen consumption in adult sockeye (*Oncorhynchus nerka*) and Coho (*O. kisutch*) salmon stocks. *Journal of Experimental Biology*, 206(18), 3239–51.

Lee, C.G., Farrell, A.P., Lotto, A.G. et al., 2003b. Excess postexercise oxygen consumption in adult sockeye salmon (*Oncorhynchus nerka*) and Coho (*O. kisutch*) salmon following critical speed swimming. *Journal of Experimental Biology*, 206, 3253–60.

Lichatowich, J., Mobrand, L., and Lestelle, L., 1999. Depletion and extinction of Pacific salmon (*Oncorhynchus* spp.): a different perspective. *ICES Journal of Marine Science*, 56(4), 467–72.

Macdonald, J.S., Patterson, D.A., Hague, M.J., and Guthrie, I.C., 2010.Modeling the influence of environmental factors on spawning migration mortality for sockeye salmon fisheries management in the Fraser River, British Columbia. *Transactions of the American Fisheries Society*, 139(3), 768–82.

Madliger, C.L., Cooke, S.J., Crespi, E.J. et al., 2016. Success stories and emerging themes in conservation physiology. *Conservation Physiology*, 4. doi:10.1093/conphys/cov057

Martins, E.G., Hinch, S.G., Cooke, S.J., & Patterson, D.A., 2012a. Climate effects on growth, phenology, and survival of sockeye salmon (*Oncorhynchus nerka*): a synthesis of the current state of knowledge and future research directions. *Reviews in Fish Biology and Fisheries*, 22(4), 887–914.

Martins, E. G., Hinch, S.G., Patterson, D.A. et al., 2011. Effects of river temperature and climate warming on stock-specific survival of adult migrating Fraser River sockeye salmon (*Oncorhynchus nerka*). *Global Change Biology*, 17(1), 99–114.

Martins, E. G., Hinch, S.G., Patterson, D.A. et al., 2012b. High river temperature reduces survival of sockeye salmon (*Oncorhynchus nerka*) approaching spawning grounds and exacerbates female mortality. *Canadian Journal of Fisheries and Aquatic Sciences*, 69(2), 330–42.

Maule, A.G., Schreck, C.B., and Kaattari, S.L., 1987. Changes in the immune system of Coho salmon (*Oncorhynchus kisutch*) during the parr-to-smolt transformation and after implantation of cortisol. *Canadian Journal of Fisheries and Aquatic Science*, 44, 161–6.

McArley, T.J. and Herbert, N.A., 2014. Mortality, physiological stress and reflex impairment in sub-legal *Pagrus auratus* exposed to simulated angling. *Journal of Experimental Marine Biology and Ecology*, 461, 61–72.

Meeremans, P., Yochum, N., Kochzius, M. et al., 2017. Inter-rater reliability of categorical versus continuous scoring of fish vitality: does it affect the utility of the reflex action mortality predictor (RAMP) approach? *PLoS ONE*, 12, 1–22.

Memmott, J., Cadotte, M., Hulme, P.E. et al., 2010. Putting applied ecology into practice. *Journal of Applied Ecology*, 47, 1–4.

Middleton, C.T., Hinch, S.G., Martins, E.G. et al., 2018. Effects of natal water concentration and temperature on the behaviour of up-river migrating sockeye salmon. *Canadian Journal of Fisheries and Aquatic Sciences*, 75, 2375–89.

Miller, K.M., Günther, O.P., Li, S. et al., 2017. Molecular indices of viral disease development in wild migrating salmon. *Conservation Physiology*, 5(1). https://doi.org/10.1093/conphys/cox036

Miller, K.M., Li, S., Kaukinen, K.H. et al., 2011. Genomic signatures predict migration and spawning failure in wild Canadian salmon. *Science*, 331, 214–17.

Miller, K.M., Schulze, A.D., Ginther, N.G. et al. 2009. Salmon spawning migration: metabolic shifts and environmental triggers. *Comparative Biochemistry and Physiology—Part D Genomics and Proteomics*, 4, 75–89.

Miller, K.M., Teffer, A.K., Tucker, S. et al., 2014. Infectious disease, shifting climates, and opportunistic predators: cumulative factors potentially impacting wild salmon declines. *Evolutionary Applications*, 7, 812–55.

Milligan, C.L., 1996. Metabolic recovery from exhaustive exercise in rainbow trout. *Comparative Biochemisty and Physiology Part A Physiology*, 113(1), 51–60.

Morrison, J., Quick, M.C., & Foreman, M.G., 2002. Climate change in the Fraser River watershed: flow and temperature projections. *Journal of Hydrology*, 263(1–4), 230–44.

Myers, R.M., 1998. When do environment-recruitment correlations work? *Reviews in Fish Biology and Fisheries*, 8, 285–305.

Nguyen, V.M., Lynch, A.J., Young, N. et al., 2016. To manage inland fisheries is to manage at the social-ecological watershed scale. *Journal of Environmental Management*, 181, 312–25.

Nguyen, V.M., Martins, E.G., Robichaud, D. et al., 2013. Disentangling the roles of air exposure, gill net injury, and facilitated recovery on the postcapture and release mortality and behavior of adult migratory sockeye salmon (*Oncorhynchus nerka*) in freshwater. *Physiological and Biochemical Zoology*, 87(1), 125–35.

Nguyen, V.M., Martins, E.G., Robichaud, D. et al., 2014. Disentangling the roles of air exposure, gill net injury, and facilitated recovery on the postcapture and release mortality and behavior of adult migratory sockeye salmon (*Oncorhynchus nerka*) in freshwater. *Physiological and Biochemical Zoology*, 87, 125–35.

O'Connor, C.M., Norris, D.R., Crossin, G.T., and Cooke, S.J., 2014. Biological carryover effects: linking common concepts and mechanisms in ecology and evolution. *Ecosphere*, 5, 1–11.

Olivier, G., 2002. Disease interactions between wild and cultured fish—perspectives from the American Northeast (Atlantic Provinces). *Bulletin European Association of Fish Pathologists*, 22, 103–9.

Ostrom, E., 2009. A general framework for analyzing sustainability of social-ecological systems. *Science*, 325(5939), 419–22.

Ou, M., Hamilton, T. J., Eom, J. et al., 2015. Responses of pink salmon to CO_2-induced aquatic acidification. *Nature Climate Change*, 5, 950–5.

Patterson, D.A., Cooke, S.J., Hinch, S.G. et al., 2016. A perspective on physiological studies supporting the provision of scientific advice for the management of Fraser River sockeye salmon (*Oncorhynchus nerka*). *Conservation Physiology*, cow026. doi:10.1093/conphys/cow026

Patterson, D.A., Robinson, K.A., Lennox, R.J., et al., 2017a. Review and evaluation of fishing-related incidental mortality for Pacific salmon. *DFO Canadian Science Advisory* Secretariat Science Advisory Report, Doc. 2017/010. ix + 155pp.

Patterson, D.A., Robinson, K.A., Raby, G.D. et al., 2017b. Guidance to derive and update fishing-related incidental mortality rates for Pacific salmon. DFO Canadian Science Advisory Secretariat Science Advisory Report, Doc. 2017/011. vii + 56pp.

Portley, N., Sousa, P., Lee-Harwood, B. et al., 2014. Global sustainability overview of Pacific salmon fisheries. Sustainable Fisheries Partnership Foundation. 36pp.

Raby, G.D., Casselman, M.T., Cooke, S.J. et al., 2016. Aerobic scope increases throughout an ecologically relevant temperature range in Coho salmon. *The Journal of Experimental Biology*, 219(12), 1922–31.

Raby, G.D., Cooke, S.J., Cook, K.V. et al., 2013. Resilience of pink salmon and chum salmon to simulated fisheries capture stress incurred upon arrival at spawning grounds. *Transactions of the American Fisheries Society*, 142, 524–39.

Raby, G.D., Donaldson, M.R., Hinch, S.G. et al., 2012. Validation of reflex indicators for measuring vitality and predicting the delayed mortality of wild Coho salmon bycatch released from fishing gears. *Journal of Applied Ecology*, 49(1), 90–8.

Raby, G.D., Donaldson, M.R., Hinch, S.G. et al., 2015a. Fishing for effective conservation: Context and biotic variation are keys to understanding the survival of Pacific salmon after catch-and-release. *Integrative and Comparative Biology*, 55, 554–76.

Raby, G.D., Donaldson, M.R., Nguyen, V.M. et al., 2014. Bycatch mortality of endangered Coho salmon: impacts, solutions, and aboriginal perspectives. *Ecological Applications*, 24, 1803–19.

Raby, G.D., Hinch, S.G., Patterson, D.A. et al., 2015b. Mechanisms to explain purse seine bycatch mortality of Coho salmon. *Ecological Applications*, 25(7), 1757–75.

Rand, P.S., Hinch, S.G., Morrison, J. et al., 2006. Effects of river discharge, temperature, and future climates on energetics and mortality of adult migrating Fraser River sockeye salmon. *Transactions of the American Fisheries Society*, 135(3), 655–67.

Robinson, K.A., Hinch, S.G., Gale, M.K. et al., 2013. Effects of post-capture ventilation assistance and elevated water temperature on sockeye salmon in a simulated capture-and-release experiment. *Conservation Physiology*, cot015. doi:10.1093/conphys/cot015

Robinson, K.A., Hinch, S.G., Raby, G.D., 2015. Influence of post-capture ventilation assistance on migration success of adult sockeye salmon following capture and release. *Transactions of the American Fisheries Society*, 144, 693–704.

Roscoe, D.W., Hinch, S.G., Cooke, S.J., and Patterson D.A., 2011. Fishway passage and post-passage mortality of up-river migrating sockeye salmon in the Seton River, British Columbia. *River Research and Applications*, 27(6), 693–705.

Ross, P., Kennedy, C.J., Shelley, L.K. et al., 2013. The trouble with salmon: relating pollutant exposure to toxic effect in species with transformational life histories and lengthy migrations. *Canadian Journal of Fisheries and Aquatic Sciences*, 70(8), 1252–64.

Scarce, R., 2000. *Fishy Business: Salmon, Biology, and the Social Construction of Nature*. Temple University Press, Philadelphia, PA.

Servizi J.A. and Jensen J.O.T., 1977. Resistance of adult sockeye salmon to acute thermal shock. International Pacific Salmon Fisheries Commission, New Westminster, BC. Progress Report 34, pp. 1–11.

Shea, K. and Mangel, M., 2001. Detection of population trends in threatened Coho salmon (*Oncorhynchus kisutch*). *Canadian Journal of Fisheries and Aquatic Sciences*, 58(2), 375–85.

Shrimpton, J.M., Patterson, D.A., Richards, J.G. et al., 2005. Ionoregulatory changes in different populations of maturing sockeye salmon *Oncorhynchus nerka* during ocean and river migration. *Journal of Experimental Biology*, 208, 4069–78.

Sovová, T., Boyle, D., Sloman, K.A. et al., 2014. Impaired behavioural response to alarm substance in rainbow trout exposed to copper nanoparticles. *Aquatic Toxicology*, 152, 195–204.

Taylor, E.B., 1991. A review of local adaptation in Salmonidae, with particular reference to Pacific and Atlantic salmon. *Aquaculture*, 98(1–3), 185–207.

Teffer, A.K., Bass, A.L., Miller, K.M. et al., 2018. Infections, fisheries capture, temperature, and host responses:

multistressor influences on survival and behaviour of adult Chinook salmon. *Canadian Journal of Fisheries and Aquatic Sciences*, 75(11), 2069–83.

Teffer, A.K., Hinch, S.G., Miller, K.M. et al., 2017. Capture severity, infectious disease processes and sex influence post-release mortality of sockeye salmon bycatch. *Conservation Physiology*, 5, cox017.

Teffer, A.K., Hinch, S.G., Miller, K.M., et al., 2019. Cumulative effects of thermal and fisheries stressors reveal sex-specific effects on pathogen development and early mortality of adult Coho salmon (*Oncorhynchus kisutch*). *Physiological and Biochemical Zoology*, 92(5), 505–29.

Tierney, K.B. and Farrell, A.P., 2004. The relationships between fish health, metabolic rate, swimming performance and recovery in return-run sockeye salmon, *Oncorhynchus nerka* (Walbaum). *Journal of Fish Diseases*, 27, 663–71.

Tierney, K.B., Sampson, J.L., Ross, P.S. et al., 2008. Salmon olfaction is impaired by an environmentally realistic pesticide mixture. *Environmental Science & Technology*, 42(13), 4996–5001.

Tomalty, K.M., Meek, M.H., Stephens, M.R., et al., 2015. Transcriptional response to acute thermal exposure in juvenile Chinook salmon determined by RNAseq. *G3: Genes, Genomes, Genetics*, 5(7), 1335–49.

Tucker, S., Hipfner, J.M., Trudel, M., 2016. Size- and condition-dependent predation: a seabird disproportionately targets substandard individual juvenile salmon. *Ecology*, 97, 461–71.

Uhlmann, S., Theunynck, R., Ampe, B. et al., 2016. Injury, reflex impairment, and survival of beam-trawled flatfish. *ICES Journal of Marine Science*, 73, 1244–54.

Voss, A., True, K., and Foott, J., 2018. Myxosporean parasite (*Ceratonova shasta* and *Parvicapsula minibicornis*) prevalence of infection in Klamath River Basin juvenile Chinook salmon, March–August 2018. US Fish and Wildlife Service California—Nevada Fish Health Center, Anderson, CA. Available at: http://www.fws.gov/canvfhc/reports.html (accessed 27 July 2020).

Wagner, G.N., Hinch, S.G., Kuchel, L.J. et al., 2005. Metabolic rates and swimming performance of adult Fraser River sockeye salmon (*Oncorhynchus nerka*) after a controlled infection with *Parvicapsula minibicornis*. *Canadian Journal of Fisheries and Aquatic Sciences*, 62(9), 2124–33.

Walters, C.J. and Holling, C.S., 1990. Large-scale management experiments and learning by doing. *Ecology*, 71, 2060–8.

Whitney, C.K., Hinch, S.G., and Patterson, D.A., 2013. Provenance matters: thermal reaction norms for embryo survival among sockeye salmon *Oncorhynchus nerka* populations. *Journal of Fish Biology*, 82(4), 1159–76.

Young, J.L., Hinch, S.G., Cooke, S.J. et al., 2006. Physiological and energetic correlates of en route mortality for abnormally early migrating adult sockeye salmon (*Oncorhynchus nerka*) in the Thompson River, British Columbia. *Canadian Journal of Fisheries and Aquatic Sciences*, 63, 1067–77.

Young, N., Corriveau, M., Nguyen, V.M. et al., 2016a. How do potential knowledge users evaluate new claims about a contested resource? Problems of power and politics in knowledge exchange and mobilization. *Journal of Environmental Management*, 184(2), 380–8.

Young, N., Corriveau, M., Nguyen et al., 2018. Embracing disruptive new science? Biotelemetry meets co-management in Canada's Fraser River. *Fisheries*, 43(1), 51–60.

Young, N., Gingras, I., Nguyen, V. M. et al., 2013. Mobilizing new science into management practice: the challenge of biotelemetry for fisheries management, a case study of Canada's Fraser River. *Journal of International Wildlife Law and Policy*, 16, 328–48.

Young, N., Nguyen, V.M., Corriveau, M. et al., 2016b. Knowledge users' perspectives and advice on how to improve knowledge exchange and mobilization in the case of a co-managed fishery. *Environmental Science & Policy*, 66, 170–8.

Zhang, Y., Claireaux, G., Takle, H. et al., 2018. Excess post-exercise oxygen consumption in Atlantic salmon (*Salmo salar*): a metabolic recovery in three-phases. *Journal of Fish Biology*, 92, 1385–403.

Integrating physiological and ecological data to increase the effectiveness of bee protection and conservation

Cedric Alaux, Jean-Luc Brunet, and Mickael Henry

> ⮩ **Take-home message**
>
> Linking honeybee physiology to landscape patterns can help identify best-suited environments and effective habitat-restoration schemes for populations. This conservation physiology approach may be transposed to wild bee species in order to monitor the impacts of environmental changes and the effectiveness of local population restoration.

4.1 Introduction

Extensive loss of honeybee colonies (*Apis mellifera*) is a well-established pattern that has emerged over the past few decades around the world, resulting from a range of environmental and biotic constraints in the context of global changes (Potts et al. 2010; Vanbergen et al. 2013). This phenomenon is depicted most spectacularly by the colony collapse disorder syndrome (Oldroyd 2007), which refers to the rapid loss of honeybee workers, leaving behind the queen, stored food, and a few adult and immature honeybees. In addition, there is accumulating evidence that wild bees, which generally have lower resilience capacities, have experienced a severe decline in terms of species diversity (Biesmeijer et al. 2006), abundance (Koh et al. 2016), or distribution (Kerr et al. 2015) in relation to global warming, intensification of agricultural practices, and the

disappearance of natural habitats and floral resources. This, in turn, leads to major concerns about both the sustainability of pollination services for insect-dependent crops, representing around 70 per cent of crops worldwide (Klein et al. 2007), and the maintenance of populations of insect-pollinated wild plants and therefore food webs. Consequently, the conservation of wild bees and the protection of managed honeybees are supported by strong ecosystem-service arguments in addition to intrinsic-value arguments (conserving biodiversity for its own sake).

Thus far, most studies on wild bee declines have focused on species richness and distribution patterns at large (national to continental) scales (e.g. Biesmeijer et al. 2006; Kerr et al. 2015). However, patterns of species range contraction and decreased diversity are a result of extinction processes that were triggered decades or more ago by climate and

Cedric Alaux, Jean-Luc Brunet, and Mickael Henry, *Integrating physiological and ecological data to increase the effectiveness of bee protection and conservation* In: *Conservation Physiology: Applications for Wildlife Conservation and Management.* Edited by: Christine L. Madliger, Craig E. Franklin, Oliver P. Love, and Steven J. Cooke, Oxford University Press (2021). © Oxford University Press. DOI: 10.1093/oso/9780198843610.003.0004 land-use changes. These delayed extinctions are termed extinction debts (Tilman et al. 1994) in that the extinction cost of an environmental disturbance may be paid back only after a long period of time. Although poorly documented, an extinction debt in pollinators is most likely occurring, just as it is in numerous other biological groups (Bommarco et al. 2014). In addition, in response to this wild bee decline and significant honeybee colony losses, there is a need to better understand how new habitat constraints (derived from anthropogenic activity) affect these populations, and as such, to then develop habitats that are as supportive to bees as possible. For that purpose, we collectively require complementary approaches to studies on bee species distribution and abundance. Indeed, mechanisms that allow us to connect the sustainability of bee populations to habitat quality will have great potential for supporting conservation policy.

In this context, conservation biologists urgently need to identify alternative biological metrics suitable as early warning signals of population declines before they become local or regional extinctions, but also as indicators of the potential sustainability of bee populations. Therefore, beyond demographics, those mechanisms should capture the state of key biological functions, such as energetic, reproductive, or immune functions indicative to some extent of the individual health state. Physiology, in particular, offers a promising framework to link organismal responses to alterations in the environment with population-level responses, and to reveal possible mechanisms behind decline processes already operating within species ranges.

When applied on large geographical scales at the community level, conservation physiology is assimilated to 'macrophysiology' (Chown and Gaston 2016), and referred to as 'landscape physiology' when focused on species conservation over narrower spatial scales (Ellis et al. 2012). In response to the growing localized and broad-scale effects of human activities, conservation physiology may therefore be useful both in the context of: (1) protected natural areas to survey pollinator communities facing possible long-term declines (Hallmann et al., 2017), and (2) modern agro-ecosystems to better assess the efficiency of environmentally friendly agricultural practices intended to support pollin-

ator communities (i.e. resource enhancement/restoration) (Kennedy et al. 2013; Scheper et al. 2013; Lichtenberg et al. 2017; Tonietto and Larkin 2018). As an emerging and valuable example of these linkages and outcomes, we report in this chapter a case study outlining how combining information on landscape ecology and honeybee physiology can help to determine the efficiency of habitat enhancement and restoration/protection in an intensive farming system (Alaux et al. 2017). We then suggest transposing this conservation physiological approach to wild bee populations to better assess habitat suitability and effectiveness of current and future conservation strategies.

4.2 Case study: a landscape physiology approach shows that floral landscape enrichment and semi-natural habitats improve honeybee health

In recent decades, anthropogenic land-use changes have largely contributed to the decline in plant abundance and diversity on which bees rely, and are therefore considered as a primary driver of severe honeybee colony losses and the pervasive decline of wild bee populations (Scheper et al. 2014; Goulson et al. 2015; Woodard and Jha 2017). Indeed, floral resource availability has been shown to clearly affect bee nutritional state, survival, and abundance, as well as colony growth of social bees. In response to this resource rarefaction, habitat enhancement in agro-ecosystems (e.g. establishment of grasslands and flower strips) and habitat restoration (management, protection, and reintroduction of plants to historical levels) have become common approaches. In point of fact, two meta-analyses investigating the effects of habitat enhancement and restoration on bees showed overall positive effects on wild bee abundance and richness (Scheper et al. 2013; Tonietto and Larkin 2018). Using these approaches to maintain diverse floral resources is also likely to benefit honeybee colony development and survival (Decourtye et al. 2010). However, before providing recommendations for habitat enhancement and restoration, we need assessment endpoints to evaluate the sustainability of bee populations in these newly designed habitats and to determine not just the attractiveness to bees, but the beneficial effects on targeted bee populations.

Combining physiological and landscape data (Ellis et al. 2012) has the potential to be used for the assessment of the effects of habitat enhancement and restoration/protection on bee health. In particular, in honeybees, assessments made during the pre-wintering period would be essential given that this life-history stage is critical for nutrient storage and, in recent years, they have been experiencing abnormally high winter colony losses.

4.2.1 Honeybee overwintering

In temperate regions, the overwintering state of honeybee colonies is characterized by a cessation of brood rearing (no population renewal), but contrary to most insects, which pass through a diapause stage, honeybees remain slightly active depending on ambient temperature. Below 10°C, bees crowd tightly together in a cluster for efficient thermoregulation (e.g. reduction of heat loss and endothermic heat production), while above 10°C bees disperse throughout the hive and may perform cleansing flights.

As an adaptation for surviving throughout this cold period, long-lived winter bees emerge during the autumn and replace summer bees in the colony. They exhibit a specific winter physiological state, characterized by high levels of nutrient storage in the fat body and an increased tolerance to oxidative stress due to the ubiquitous protein vitellogenin (Amdam et al. 2011; Döke et al. 2015). This protein, produced in the fat body, can constitute up to 40 per cent of the total haemolymph protein fraction (Fluri et al. 1982) and promotes the longevity of bees by acting as an antioxidant (Seehuus et al. 2006) and being involved in immune defence (Salmela et al. 2015). Vitellogenin is present at high levels in both young summer and winter bees, but in winter bees it exhibits a negligible decline over time (Fluri et al. 1982), which likely explains the extreme longevity of winter bees (up to 6 months) as compared with summer bees (4–6 weeks).

The abnormally high winter colony losses recently reported in the United States (around 44 per cent) (Seitz et al. 2015) and in Europe (up to 32 per cent) (Jacques et al. 2017) suggest that preparation for overwintering is becoming especially chal-

lenging for honeybee colonies. Notably, for a successful overwintering, sufficient energetic reserves must be stored within both the colony and individuals during the pre-wintering period (Döke et al. 2015). Nectar, transformed into honey, is used as an energetic resource for colony thermoregulation, while the availability of pollen shapes winter bee physiology since fat body growth and vitellogenin production are both driven by the amount and quality of ingested pollens (Alaux et al. 2010; Di Pasquale et al. 2013; Di Pasquale et al. 2016). The decline of bee pastures due to the current intensification of agriculture might thus contribute to deficient nutrient storage and negatively affect winter preparation.

This case study highlights how the measurement of winter bee physiology can be used to assess whether habitat enhancement via the implementation of melliferous catch crops (i.e. catch crops producing nectar and pollen foraged by honeybees), and increasing amount of semi-natural habitats in the vicinity of the colony, can have beneficial effects on colony survival. Catch crops are quick-growing plants grown between two regular crops in consecutive seasons, with agronomic (nitrogen fixation, reduction of nutrient leaching) and environmental (temporary habitats and resources for invertebrates, birds, or small mammals) benefits. The case study covers the work completed by Alaux et al. (2017), which was made possible thanks to collaborations among researchers, engineers of beekeeping technical institutes, beekeepers, and farmers.

4.2.2 Determining the influence of landscape variables on honeybee physiology and survival

To determine whether floral landscape enrichment and semi-natural habitats provide benefits to honeybee health before overwintering, we exposed different colonies in the autumn to varying amounts of either melliferous catch crops or semi-natural habitats, and measured the gene expression levels of vitellogenin, as well as the fat body mass of bees. We also investigated other metrics, such as the amount and composition of pollen foraged by bees in the studied landscapes, and the infestation levels of the parasitic mite *Varroa destructor*, known to have detrimental effects on colony survival during the winter (Döke et al. 2015). The experimental procedure developed for analysing the influence of landscape variables on winter bee physiology and colony survival is illustrated in Figure 4.1.

The experiment was specifically performed in an intensive farming system in central western France. Landscape enrichment with different amounts of catch crops was performed during the pre-wintering period within a 1.5-km radius area, which corresponds to the honeybee foraging range in autumn (Couvillon et al. 2014). Colonies were installed in the early autumn either inside a catch crop area or 8–10 km farther away with no catch crop areas within the foraging range. The amount of semi-natural habitat (woodlots and hedgerows) was also quantified within the 1.5-km foraging range around the apiaries.

To determine the origin of foraged pollen and ascertain the use of catch crops by honeybees, we sampled pollen from colonies during the flowering of the catch crops and analysed the species composition. Pollen nutritional characteristics were also determined to further investigate the link between

landscape resource and bee health. Finally, at the end of the flowering period, bees were sampled for their winter physiological state (vitellogenin and fat body mass). Colonies were then placed back into their initial apiaries to standardize overwintering conditions and avoid any influence of different micro-climatic conditions. Overwintering was considered unsuccessful for collapsed or weak colonies (i.e. not operational for beekeeping activity at the beginning of spring).

4.2.3 Healthier honeybees around restored floral resources and semi-natural habitats

Model analysis was used to determine which factor contributed most to bee health and their ability to survive the winter. As expected, the overwintering survival of honeybee colonies was negatively influenced by *Varroa* infestation levels recorded in the autumn, and vitellogenin level was linked to fat body mass (Figure 4.2). Landscape quality did not directly influence overwintering survival; however, the

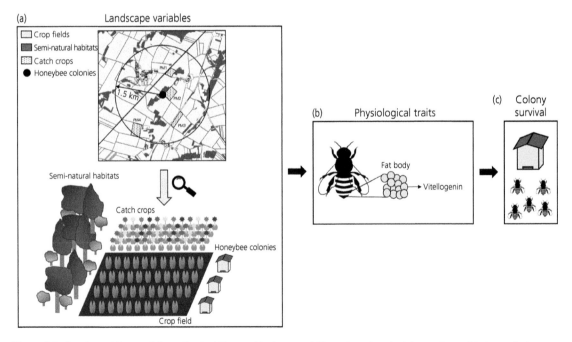

Figure 4.1 Experimental framework for testing the influence of landscape variables on honeybee physiology and overwintering survival. (A) In the autumn, apiaries were either installed near melliferous catch crops or farther away (beyond bee foraging distance). The areas of catch crops and semi-natural habitat (woodlots and hedgerows) were quantified within a foraging range of 1.5 km around each apiary. (B) At the end of the catch crop blooming, bees were sampled from each colony and their winter physiology was analysed (fat body mass and vitellogenin level). (C) Colony overwintering survival was then determined at the end of winter. Figure panels are modified from Alaux et al. (2017).

ecophysiological approach showed that it significantly affected winter bee physiology, with the amount of melliferous catch crops and semi-natural habitats positively influencing honeybee vitality (fat body mass and vitellogenin levels) (Figure 4.2). Furthermore, vitellogenin had a positive influence on overwintering survival, which further confirms the role of vitellogenin as a predictive biomarker for monitoring honeybee population health (Dainat et al. 2012; Smart et al. 2016). For instance, colonies composed of honeybees with low levels of vitellogenin had a winter survival rate of 60 per cent, whereas colonies composed of honeybees with high levels of vitellogenin achieved survival rates of about 90 per cent. Altogether, these results show that landscape quality promoted honeybee health, which in turn improved the survival probability of colonies. Finally, it is important to note that semi-natural habitats exhibited greater effects on both

vitellogenin and fat body than catch crops, indicating a specific effect of landscape features.

The analysis of pollen collected by honeybees revealed that they effectively foraged on catch crops, since half of the pollen diet composition originated from the seeded plant mix. If the presence of catch crops did not modify the daily amount of collected pollen and the overall pollen diet energy and protein contents, it almost doubled pollen diet diversity. This suggests that the designed habitat enhancement improved bee health through higher diet diversity rather than through higher nutritional abundance and quality. However, because diet quality analysis was performed on a limited number of nutritional parameters, there remains the possibility that pollen diets provided by melliferous catch crops were of higher quality in terms of other nutrients (e.g. lipids, amino acids, vitamins).

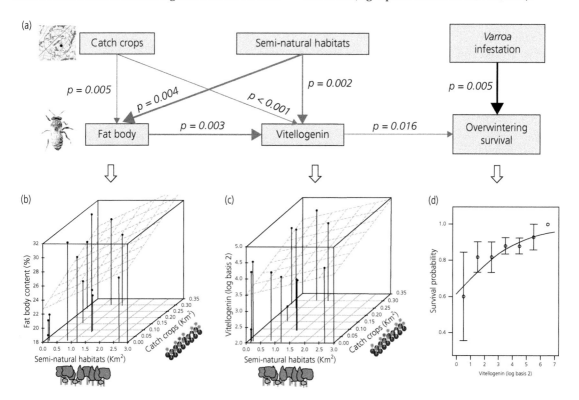

Figure 4.2 Landscape physiology approach showing the influence of landscape variables on honeybee physiology and overwintering survival. (A) Except for *Varroa* infestation levels (black arrow), which negatively influence colony winter survival, all links indicate positive effects (grey arrow). Arrow thickness is proportional to the size of the effect. Significance levels are shown next to each link. The area of catch crop and semi-natural habitats positively influenced the development of fat bodies (B) and vitellogenin level (C). Each dot represents the average value of fat body mass and vitellogenin levels per apiary. The regression planes show the association between landscape variables and bee physiological parameters. Vitellogenin was then found to be positively associated with the probability of colony survival (D). Figure panels are modified from Alaux et al. (2017).

4.2.4 Case study conclusions

Mitigating the negative influence of anthropogenic landscape alteration on bee populations is a priority of bee conservation and protection policy. For that purpose, considerable research efforts have been made to determine the influence of habitat restoration and enhancement schemes on bee abundance and diversity (Kennedy et al. 2013; Scheper et al. 2013; Lichtenberg et al. 2017; Tonietto and Larkin 2018). The highlighted case study moved one step beyond research on species distribution (presence/absence, abundance) to show that landscape enrichment with catch crops, besides being attractive, can promote honeybee health during a critical period of their colony life cycle. Furthermore, the experimental design revealed a greater influence of semi-natural habitats than artificial bee pasture (catch crops) and therefore indicates that the implementation of floral strips should be considered as a complementary management measure of semi-natural habitat preservation or restoration. This case study combining landscape ecology and bee physiology also offers a valuable proof-of-concept for better understanding of the influence of environmental variations on bee health and identifying best-suited habitats. By offering mechanistic insight into population sustainability, this ecophysiological approach will set the stage for more effective conservation of wild bees, which until now has been restricted to the assessment of species distribution.

4.3 Moving beyond honeybee protection: conservation physiology of wild bees

4.3.1 The decline of wild bees and remedial actions

As managed honeybees are extensively monitored through beekeeping surveillance programmes, their colony mortality rates and declining trends are well documented worldwide as well as at smaller regional scales whenever acute mortality events occur (Seitz et al. 2015; Jacques et al. 2017). Unfortunately, such extensive monitoring programmes are historically scarce for wild bees due to the technical difficulty and monetary cost of monitoring them (Lebuhn et al. 2013). Large-scale studies

dedicated to the monitoring of wild bees are usually based on extensive databases of species occurrence records or on meta-analyses of long-term local diversity surveys (Winfree et al. 2009; Kerr et al. 2015). The seminal study by Biesmeijer et al. (2006) comparing bee records pre- and post-1980 reported significant bee species richness decreases in 52 per cent and 67 per cent of studied localities in Britain and the Netherlands, respectively. On the contrary, only 4–10 per cent of localities showed evidence of increases in species richness. Concurrently, the occurrence of bee-pollinated plants has significantly declined in those countries, relative to wind-pollinated plants or predominantly self-pollinating plants. Subsequently, Winfree et al. (2009) reviewed 54 studies on local bee assemblages and established that bee abundance and richness are indeed significantly affected by anthropogenic disturbances, and in particular by habitat loss, with an average decrease effect size estimated at −0.32 and −0.37 units of standard deviation, respectively. The authors however emphasized that those results mainly stand for systems experiencing extreme habitat losses. More recently, Kerr et al. (2015) focused on extensive bumblebee species occurrence records across North America and Western Europe over 110 years and found that their distributional ranges are shrinking as a result of global warming. As heat-sensitive organisms, *Bombus* species have lost on average 300 km on their southern distributional limits, while during the same time period, they failed to expand accordingly into their northern limits.

Mitigating global bee declines thus requires large-scale national policies, such as the European agro-environmental schemes (AES), which provide farmers with financial incentives for adopting environmentally friendly practices to promote diversity in farmlands. Pollinators could benefit from a variety of AES such as sown flower strips in field margins or set-aside areas, extension of pastures and meadows, or organic farming that preserves useful wild-growing species. However, the pollinator-oriented AES are not always effective at mitigating pollinator losses, depending on the focal agricultural system and landscape context (Scheper et al. 2013). Under some circumstances, AES may be more expensive than preserving or restoring natural habitats, and represent less effective strategies for protecting threatened bee species that usually do not persist in agricultural systems (Winfree 2010). In this situation, species-oriented, local restoration programmes may be necessary (Winfree et al. 2009; Winfree 2010), such as floral resource and nesting site restoration, invasive plant removal, or even bee population relocation—though the latter may be controversial (Lozier et al. 2015).

4.3.2 The need for physiological metrics to better support conservation strategies

Understanding and undertaking a comprehensive mitigation on the impact of environmental changes on populations of wild bees continues to be a demanding research challenge due to the large gaps in monitoring capacity and accuracy. As described above, the estimation of insect pollinator population sizes is traditionally reliant on punctual counts of individuals per locality. In this context, the deleterious effects of environmental alteration can only be detected once the population has started to decline. In addition, identifying the drivers that shape pollinator populations requires high-resolution data. Furthermore, the effectiveness of conservation strategies depends on studies of the spatial distribution of species. As such, there is a need to gather complementary information on bee health to assess the potential sustainability of bee populations in habitats designed for their conservation, especially since some practices developed to improve resource access in agrosystems (e.g. wildflower fields) might actually have negative side-effects on bee health, such as increasing the risk of contamination with parasites (Piot et al. 2018) and prolonging the exposure to pesticides (Botias et al. 2015). Finally, similar bee abundance in different habitats does not necessarily mean a similar influence of such habitats on bee life traits, as recently demonstrated by Renauld et al. (2016), who found no effect of landscape quality on the abundance of *Andrena nasonii* but a reduction of individual size with increasing agricultural land uses.

4.3.3 Bee body condition

The first studies of bee decline that moved beyond the recording of spatial distribution used body size

for assessing the response and sensitivity of wild bees to anthropogenic disturbance of habitats. Body size, which is strongly influenced by resource availability and other environmental factors such as temperature, has often been used as a proxy of individual fitness, dispersal ability, and reproductive success in insects (Chown and Gaston 2010). This method appeared useful given that the analysis of museum samples in the north-eastern United States and the Netherlands revealed a negative trend between bee body size and population trend: larger bee species were found to decline at a greater rate than smaller bee species, likely due to their larger pollen requirement (Bartomeus et al. 2013; Scheper et al. 2014). These results therefore highlighted food limitation as a key driver of bee decline. Furthermore, by analysing intraspecific body-size variation of two medium-sized bees in relation to landscape structure, Warzecha et al. (2016) found that body size increased with habitat fragmentation, indicating selection for higher dispersal abilities. However, in a meta-analysis, body size was not consistently linked to species responses and thus appeared inappropriate for predicting bee species response to environmental disturbance (Williams et al. 2010). In addition, fluctuating asymmetry of wings has been used several times in insects as an indicator of stressful conditions, with the underlying reasoning that stressful conditions are liable to disturb the stability of an organism's development, and therefore increase their overall lateral asymmetry at the adult stage. Bumblebee asymmetry, as opposed to overall wing size and shape, did not change upon exposure to stress from natural toxins, parasites, and thermic pressures, nor in inbred individuals (Gerard et al. 2018). These findings suggest that proxies of body condition might provide useful insights into bee sensitivity to environmental changes, but they cannot be considered as an exclusive measure of bee fitness-related traits (Wilder et al. 2016).

4.3.4 Physiological metrics

Assessment of physiological standards in pollinator species might serve as a useful complementary and/or alternative indicator to gauge habitat suitability and effectiveness of conservation strategies, but also in developing and feeding predictive models of population response to global changes. Some studies in honeybees have highlighted the usefulness of physiological biomarkers for assessing environmental quality based on pollution (Badiou-Bénéteau et al. 2013) and resource levels (Alaux et al. 2017; see the case study in Section 4.2). Unfortunately, the study of physiological metrics in the context of wild bee conservation is limited. One study that investigated the nutritional status (lipid content) of several bumblebee species in relation to grassland management practices in agricultural landscapes (Smith et al. 2016) showed that it is a promising surrogate for testing the efficiency of resource enhancement. More recently, in order to predict habitat suitability at the local scale, Tomlinson et al. (2018) projected data of thermal tolerance thresholds and metabolic rates of bees onto high-resolution topoclimatic models. This study highlighted the importance of physiological models to better assess the efficiency of ecological management programmes at local scales, as opposed to species distribution models (e.g. climatic models), which are better adapted to predicting the impact of environmental changes at global, national, and regional scales.

While there is a growing body of literature on the biochemical and molecular characterization of several physiological functions in wild bees, they remain to be integrated into ecological data and tested in a conservation context. Notably it would be interesting to focus on a few essential functions that are critical for the health and the survival of bees and relatively easy to measure. Among the putative functions to target, stored energy reserves would be a good candidate. They are essential to different life-history traits and life cycle stages of insects, including reproduction, flight capacity, and thus dispersal, diapause, and survival. In insects and therefore in bees, the energetic state of individuals is regulated by the fat body through glycogen and triglyceride rates (Arrese and Soulages 2010). These nutrients are mainly acquired through consumption of pollen and nectar, and thus variations in the availability and quality of floral resources are expected to directly affect the bees' energy budget and subsequent health. Finally, the use of physiological metrics as part of bee population management and

conservation could involve the development of species-specific allometric reference standards (combinations of measured physiological metrics to body mass/size). Potential deviations from those standards (e.g. due to poor nutritional environments) would therefore indicate a decreased health status and provide an early indicator of bee sensitivity to environmental quality.

4.4 Conclusions and future directions

Monitoring is essential for the conservation of flora and fauna to better evaluate and counter current biodiversity loss due to anthropogenic perturbation. It can serve as an early warning system for detecting environmental problems, but can also help to determine how well remedial actions are working. In sum, monitoring is the basis for improved decision making. Although the macrophysiological approach for directly assessing population status and trends was conceptualized 10 years ago (Gaston et al. 2009), it has not yet been applied to monitoring bee populations. The use of metrics of an individual's state of health would however represent a complementary and powerful approach to studying species distribution. As we showed in the honeybee case study, it can provide a cause-and-effect relationship between environmental characteristics and bee survival, which could directly contribute to decision making and supporting conservation policy. Indeed, due to the lack of metrics suitable for monitoring the sustainability of bee populations, conservation practitioners and environmental managers often struggle to identify and test the efficiency of remedial actions. Equally important, the identification of physiological metrics suitable as early warning signals could be extremely useful for providing an indication of a stressful environment. This type of assessment would be especially beneficial since placing efforts on preventing or reversing changes before they become catastrophic represents a more proactive approach to conservation, in opposition to the reactive approach consisting of correcting the damages, likely when it is too late.

More studies are needed to investigate the insights provided by physiological conservation in the context of bee decline. There remain some key challenges to be addressed:

1. Dealing with carryover effects. Carryover effects occur when adverse environmental conditions for the studied organism during one season affect an individual's health status or fitness in a subsequent season (Harrison et al. 2011). Ignoring such inter-annual delayed effects can lead to erroneous conclusions about the effectiveness of conservation efforts (O'Connor and Cooke 2015). Carryover effects have been documented in honeybees, with higher winter mortality risks in colonies that have experienced acute nutritional stress due to pollen scarcity during the previous season (Requier et al. 2016). Carryover effects are also indirectly shown in wild bee communities as their abundance is typically influenced by the previous year's nesting conditions, particularly the floral resources available for provisioning nests (Potts et al. 2003; Le Feon et al. 2013). Indeed, in solitary wild bees, females generally emerge in the spring, mate, and then provision their nests with pollen to feed their larvae, which change into adults in the autumn and will emerge the next spring. As a result, the physiological status of adults sampled in one year should mostly be affected by the environmental conditions in the previous year or months (e.g. resources ingested as a larva, climatic conditions during overwintering), and less by the current environmental conditions. Dealing with such carryover effects therefore requires the specificity of each species life cycle to be taken into account.

2. Identifying standard physiological references. As required for any conservation physiology approach, one needs to establish reference standards for a given species. Standard values obtained from reference populations will help determine when adverse conditions trigger a significant deviation from the baseline. In addition, one should identify and validate candidate proxies for physiological metrics that are the least invasive and therefore allow species identification in the field or the lab.

3. Scaling up from populations to communities. Wild bee communities are species rich (up to several hundred species at the landscape to regional scales). Physiological responses of bees to environmental disturbances should therefore

be collated among many species to depict the overall community-level response to environmental stressors. This requires identifying bee functional groups that are more sensitive to the studied stress, as well as standardizing physiological metrics in a consistent manner across species.

4. Assessing the role of physiological metrics as indicators of tipping points. A real challenge for environmental managers and scientists is the ability to anticipate and predict the emergence of tipping points in population dynamics (point of no return). To what extent physiological metrics can be used to predict tipping points in bee populations will need to be investigated. This will notably require the integration of such indicators into surveillance network datasets of bee population size.

Acknowledgements

This work was supported by the French Ministry of Agriculture programme (CASDAR, INTERAPI programme no. 1176) and two grants from INRA SPE department (CLIMBEE and MACROBEE). We are also grateful to Fabrice Allier, Axel Decourtye, Jean-François Odoux, Thierry Tamic, Melanie Chabirand, Estelle Delestra, Florent Decugis, Yves Le Conte, and the volunteer farmers and beekeepers all involved in the described case study.

References

Alaux, C., Allier, F., Decourtye, A. et al., 2017. A 'landscape physiology' approach for assessing bee health highlights the benefits of floral landscape enrichment and semi-natural habitats. *Scientific Reports*, 7, 40568.

Alaux, C., Ducloz, F., Crauser, D., and Le Conte, Y., 2010. Diet effects on honeybee immunocompetence. *Biology Letters*, 6(4), 562–5.

Amdam, G.V., Fennern, E., and Havukainen, H., 2011. Vitellogenin in honey bee behavior and lifespan. In C.G. Galizia, D. Eisenhardt, and M. Giurfa, ed. *Honeybee Neurobiology and Behavior*, pp. 17–29. Springer, Berlin.

Arrese, E.L. and Soulages, J.L., 2010. Insect fat body: energy, metabolism, and regulation. *Annual Review of Entomology*, 55(1), 207–25.

Badiou-Bénéteau, A., Benneveau, A., Géret, F. et al., 2013. Honeybee biomarkers as promising tools to monitor environmental quality. *Environment International, 60,* 31–41.

Bartomeus, I., Ascher, J.S., Gibbs, J. et al., 2013. Historical changes in northeastern US bee pollinators related to shared ecological traits. *Proceedings of the National Academy of Sciences of the United States of America, 110*(12), 4656–60.

Biesmeijer, J.C., Roberts, S.P., Reemer, M. et al., 2006. Parallel declines in pollinators and insect-pollinated plants in Britain and the Netherlands. *Science, 313*(5785), 351–4.

Bommarco, R., Lindborg, R., Marini, L., and Öckinger, E., 2014. Extinction debt for plants and flower-visiting insects in landscapes with contrasting land use history. *Diversity and Distributions, 20*(5), 591–9.

Botias, C., David, A., Horwood, J. et al., 2015. Neonicotinoid residues in wildflowers, a potential route of chronic exposure for bees. *Environmental Science & Technology, 49*(21), 12731–40.

Chown, S.L. and Gaston, K.J., 2010. Body size variation in insects: a macroecological perspective. *Biological Reviews of the Cambridge Philosophical Society, 85*(1), 139–69.

Chown, S.L. and Gaston, K.J., 2016. Macrophysiology— progress and prospects. *Functional Ecology, 30*(3), 330–44.

Couvillon, M.J., Schurch, R., and Ratnieks, F.L.W., 2014. Waggle dance distances as integrative indicators of seasonal foraging challenges. *PLoS ONE, 9*(4), e93495.

Dainat, B., Evans, J.D., Chen, Y.P., Gauthier, L., and Neumann, P., 2012. Predictive markers of honey bee colony collapse. *PLoS ONE, 7*(2), e32151.

Decourtye, A., Mader, E., and Desneux, N., 2010. Landscape enhancement of floral resources for honey bees in agro-ecosystems. *Apidologie, 41,* 264–77.

Di Pasquale, G., Alaux, C., Le Conte, Y. et al., 2016. Variations in the availability of pollen resources affect honey bee health. *PLoS ONE, 11*(9), e0162818.

Di Pasquale, G., Salignon, M., Le Conte, Y. et al., 2013. Influence of pollen nutrition on honey bee health: do pollen quality and diversity matter? *PLoS ONE, 8*(8), e72016.

Döke, M.A., Frazier, M., and Grozinger, C.M., 2015. Overwintering honey bees: biology and management. *Current Opinion in Insect Science, 10,* 185–93.

Ellis, R.D., McWhorter, T.J., and Maron, M., 2012. Integrating landscape ecology and conservation physiology. *Landscape Ecology, 27*(1), 1–12.

Fluri, P., Luscher, M., Wille, H., and Gerig, L., 1982. Changes in the weight of the pharyngeal gland and haemolymph titres of juvenile hormone, protein and vitellogenin in worker honey bees. *Journal of Insect Physiology, 28,* 61–8.

Gaston, K.J., Chown, S.L., Calosi, P. et al., 2009. Macrophysiology: a conceptual reunification. *American Naturalist, 174*(5), 595–612.

Gerard, M., Michez, D., Debat, V. et al., 2018. Stressful conditions reveal decrease in size, modification of shape but relatively stable asymmetry in bumblebee wings. *Scientific Reports, 8*(1), 15169.

Goulson, D., Nicholls, E., Botias, C., and Rotheray, E.L., 2015. Bee declines driven by combined stress from parasites, pesticides, and lack of flowers. *Science, 347*(6229), 1255957.

Hallmann, C.A., Sorg, M., Jongejans, E. et al., 2017. More than 75 per cent decline over 27 years in total flying insect biomass in protected areas. *PLoS ONE, 12*(10), e0185809.

Harrison, X.A., Blount, J.D., Inger, R. et al., 2011. Carryover effects as drivers of fitness differences in animals. *Journal of Animal Ecology, 80*(1), 4–18.

Jacques, A., Laurent, M., Consortium, E. et al., 2017. A pan-European epidemiological study reveals honey bee colony survival depends on beekeeper education and disease control. *PLoS ONE, 12*(3), e0172591.

Kennedy, C.M., Lonsdorf, E., Neel, M.C. et al., 2013. A global quantitative synthesis of local and landscape effects on wild bee pollinators in agroecosystems. *Ecology Letters, 16*(5), 584–99.

Kerr, J.T., Pindar, A., Galpern, P. et al., 2015. Climate change impacts on bumblebees converge across continents. *Science, 349*(6244), 177–80.

Klein, A.M., Vaissiere, B. E., Cane, J.H. et al., 2007. Importance of pollinators in changing landscapes for world crops. *Proceedings of the Royal Society B, 274*(1608), 303–13.

Koh, I., Lonsdorf, E.V., Williams, N.M. et al., 2016. Modeling the status, trends, and impacts of wild bee abundance in the United States. *Proceedings of the National Academy of Sciences of the United States of America, 113*(1), 140–5.

Le Feon, V., Burel, F., Chifflet, R. et al., 2013. Solitary bee abundance and species richness in dynamic agricultural landscapes. *Agriculture Ecosystems & Environment, 166,* 94–101.

Lebuhn, G., Droege, S., Connor, E.F. et al., 2013. Detecting insect pollinator declines on regional and global scales. *Conservation Biology, 27*(1), 113–20.

Lichtenberg, E.M., Kennedy, C.M., Kremen, C. et al., 2017. A global synthesis of the effects of diversified farming systems on arthropod diversity within fields and across agricultural landscapes. *Global Change Biology, 23*(11), 4946–57.

Lozier, J.D., Cameron, S.A., Duennes, M.A. et al., 2015. Relocation risky for bumblebee colonies. *Science 350*(6258), 286–7.

O'Connor, C.M. and Cooke, S.J., 2015. Ecological carryover effects complicate conservation. *Ambio, 44*(6), 582–91.

Oldroyd, B.P., 2007. What's killing American honey bees? *PLoS Biology, 5*(6), 1195–9.

Piot, N., Meeus, I., Kleijn, D. et al., 2018. Establishment of wildflower fields in poor quality landscapes enhances micro-parasite prevalence in wild bumble bees. *Oecologia*, *189*(1), 149–58.

Potts, S.G., Biesmeijer, J. C., Kremen, C. et al., 2010. Global pollinator declines: trends, impacts and drivers. *Trends in Ecology & Evolution*, *25*(6), 345–53.

Potts, S.G., Vulliamy, B., Dafni, A. et al., 2003. Linking bees and flowers: how do floral communities structure pollinator communities? *Ecology*, *84*(10), 2628–42.

Renauld, M., Hutchinson, A., Loeb, G. et al, 2016. Landscape simplification constrains adult size in a native ground-nesting bee. *PLoS ONE*, *11*(3), e0150946.

Requier, F., Odoux, J. F., Henry, M., and Bretagnolle, V., 2016. The carry-over effects of pollen shortage decrease the survival of honeybee colonies in farmlands. *Journal of Applied Ecology*, *54*(4), 1161–70.

Salmela, H., Amdam, G.V., and Freitak, D., 2015. Transfer of immunity from mother to offspring is mediated via egg-yolk protein vitellogenin. *PLoS Pathogens*, *11*(7), e1005015.

Scheper, J., Holzschuh, A., Kuussaari, M. et al., 2013. Environmental factors driving the effectiveness of European agri-environmental measures in mitigating pollinator loss—a meta-analysis. *Ecology Letters*, *16*(7), 912–20.

Scheper, J., Reemer, M., van Kats, R. et al., 2014. Museum specimens reveal loss of pollen host plants as key factor driving wild bee decline in The Netherlands. *Proceedings of the National Academy of Sciences of the United States of America*, *111*(49), 17552–7.

Seehuus, S.C., Norberg, K., Gimsa, U. et al., 2006. Reproductive protein protects functionally sterile honey bee workers from oxidative stress. *Proceedings of the National Academy of Sciences of the United States of America*, *103*(4), 962–7.

Seitz, N., Traynor, K.S., Steinhauer, N. et al., 2015. A national survey of managed honey bee 2014–2015 annual colony losses in the USA. *Journal of Apicultural Research*, *54*(4), 292–304.

Smart, M., Pettis, J., Rice, N. et al., 2016. Linking measures of colony and individual honey bee health to survival among apiaries exposed to varying agricultural land use. *PLoS ONE*, *11*(3), e0152685.

Smith, G.W., Debinski, D.M., Scavo, N.A. et al., 2016. Bee abundance and nutritional status in relation to grassland management practices in an agricultural landscape. *Environmental Entomology*, *45*(2), 338–47.

Tilman, D., May, R.M., Lehman, C.L., and Nowak, M.A., 1994. Habitat destruction and the extinction debt. *Nature*, *371*(6492), 65–6.

Tomlinson, S., Webber, B.L., Bradshaw, S. D. et al., 2018. Incorporating biophysical ecology into high-resolution restoration targets: insect pollinator habitat suitability models. *Restoration Ecology*, *26*(2), 338–47.

Tonietto, R.K. and Larkin, D.J., 2018. Habitat restoration benefits wild bees: a meta-analysis. *Journal of Applied Ecology*, *55*(2), 582–90.

Vanbergen, A.J., Baude, M., Biesmeijer, J.C. et al., 2013. Threats to an ecosystem service: pressures on pollinators. *Frontiers in Ecology and the Environment*, *11*(5), 251–9.

Warzecha, D., Diekötter, T., Wolters, V., and Jauker, F., 2016. Intraspecific body size increases with habitat fragmentation in wild bee pollinators. *Landscape Ecology*, *31*(7), 1449–55.

Wilder, S.M., Raubenheimer, D., and Simpson, S.J., 2016. Moving beyond body condition indices as an estimate of fitness in ecological and evolutionary studies. *Functional Ecology*, *30*(1), 108–15.

Williams, N.M., Crone, E.E., Roulston, T.H. et al., 2010. Ecological and life-history traits predict bee species responses to environmental disturbances. *Biological Conservation*, *143*(10), 2280–91.

Winfree, R., 2010. The conservation and restoration of wild bees. *Annals of the New York Academy of Sciences*, *1195*, 169–97.

Winfree, R., Aguilar, R., Vázquez, D.P. et al., 2009. A meta-analysis of bees' responses to anthropogenic disturbance. *Ecology*, *90*(8), 2068–76.

Woodard, S.H. and Jha, S., 2017. Wild bee nutritional ecology: predicting pollinator population dynamics, movement, and services from floral resources. *Current Opinion in Insect Science*, *21*, 83–90.

CHAPTER 5

Applying isotopic clocks to identify prior migration patterns and critical habitats in mobile marine predators

Daniel J. Madigan, Oliver N. Shipley, and Nigel E. Hussey

⮞ Take-home message

Predator tissue composition can reconstruct animal movement histories on population scales, informing spatial management and predicting distribution shifts with changing climate.

5.1 Migratory marine species: management and conservation challenges

Migration of marine predators challenges effective management and conservation (Lascelles et al. 2014; Runge et al. 2014). Migratory marine taxa that are exploited by fisheries and/or of conservation concern, including teleost, elasmobranch, mammal, turtle, and seabird species, are managed on regional (i.e. state/province), national, and/or international scales, yet their movements do not honour geopolitical boundaries (Harrison et al. 2018; Barkley et al. 2019). Thus, the extent (e.g. trans-oceanic movements) and dynamics (e.g. spatiotemporal aggregations, inter-annual and inter-individual variability) of migrations makes many species differentially susceptible to exploitation based on migration stage (Lascelles et al. 2014; Harrison et al. 2018). Species migration patterns, and variation within those migratory strategies, are also largely unknown for most species throughout ontogeny and on popula-

tion scales, further complicating effective management efforts (Hazen et al. 2012).

General understanding of broad marine predator movements began with simple conventional tag–recapture studies, which provided start- and endpoint location data for linear reconstructions of animal migrations (McFarlane et al. 1990). Marine predator movement data have greatly improved in recent decades, largely owing to technological advances in animal biotelemetry (Hussey et al. 2015), which has also revealed physiological constraints for migration including thermal and energetic requirements (Watanabe et al. 2015, Whitlock et al. 2015). However, electronic tagging technologies broadly share the limitations of high cost, logistics of accessing and placing tags on live animals, limited sample sizes, and the prospective nature of the data: tagged animals provide data on movements after tagging, but no information on prior migrations, often resulting in data gaps for earlier life stages (Hazen et al. 2012). This presents the need for a retrospective tool that can provide

Daniel J. Madigan, Oliver N. Shipley, and Nigel E. Hussey, *Applying isotopic clocks to identify prior migration patterns and critical habitats in mobile marine predators* In: *Conservation Physiology: Applications for Wildlife Conservation and Management.* Edited by: Christine L. Madliger, Craig E. Franklin, Oliver P. Love, and Steven J. Cooke, Oxford University Press (2021). © Oxford University Press. DOI: 10.1093/oso/9780198843610.003.0005 adequately robust movement data for population-wide estimates of species movement patterns.

Chemical tracers in animal tissues provide such a tool. Certain naturally occurring and anthropogenic chemical compounds accumulate in marine animal tissues, either through direct contact with seawater by gills and/or accretionary structures (e.g. otoliths, teeth, baleen), or by ingestion of prey containing local chemical 'signatures', which are incorporated into animal tissues (e.g. muscle, blood, liver, bone). Chemical tracers have historically included trace metals (Sturrock et al. 2012), organic compounds (Dickhut et al. 2009), radioactive isotopes (Madigan et al. 2013), and more recently, stable isotopes (Hobson 1999; Ramos and González-Solís 2012; McMahon et al. 2013b).

Stable isotope analysis (SIA) relies on the premise that isotopes of common elements (e.g. C, N, S, O) occur naturally in variable relative proportions to each other. Early studies recognized that stable isotope ratios of carbon ($\delta^{13}C$) and nitrogen ($\delta^{15}N$) varied with trophic level within ecosystems, allowing for studies of ecosystem structure and nutrient flow (Peterson and Fry 1987). However, ongoing study showed that isotopic signatures also vary widely across oceanographic regions and could be applied to discern migratory patterns (Graham et al. 2010; McMahon et al. 2013b). This premise was often applied to complement other techniques (e.g. biotelemetry) for reconstructing movement patterns of populations, with direct applications to management. One example is the Atlantic bluefin tuna (*Thunnus thynnus*), for which understanding of discrete population (east and west Atlantic) mixing via trans-Atlantic migrations was revolutionized with electronic tag technology (Block et al. 2005). However, the natal origin of tagged fish remained unknown until chemical tracers in otoliths revealed that despite mixing, east and west populations returned to their natal origins to spawn (Rooker et al. 2008). The repeatability of this chemical tracer approach allowed further studies that demonstrated the high inter-annual variability in mixing of these stocks (Rooker et al. 2019). Together, these approaches allowed for improved regional quotas based on stock mixing, which has aided the current positive rebuilding trend of the Atlantic bluefin tuna population (Porch et al. 2019).

A prerequisite to the study above, and any study using an SIA approach to assess movement, is geographical variation of isotopic signatures. In the case above, different oceanographic properties of the Gulf of Mexico and Mediterranean Sea drove differences in seawater $\delta^{13}C$ and $\delta^{18}O$, which was validated using otoliths of larvae from both regions (Rooker et al. 2008). In general, broad oceanographic processes (e.g. nitrogen fixation, nitrate upwelling) and properties (temperature, salinity) of a marine region will dictate the regional isotopic 'baseline': the stable isotope (SI) signatures found at the base of a regional food web, often analysed via particulate organic matter, phytoplankton, or zooplankton (Montoya et al. 2002; Graham et al. 2010; Olson et al. 2010; MacKenzie et al. 2014). In the case of $\delta^{13}C$ and $\delta^{15}N$, two of the most commonly applied isotopes to marine ecology, the sources and assimilation dynamics at the base of the food web dictate regional baseline values. In more oligotrophic regions, uptake of N is dominated by N_2-fixation, resulting in low $\delta^{15}N$ values (due to minimal ^{15}N fractionation) that propagate up regional food webs (Montoya et al. 2002). In contrast, highly productive upwelling regions have higher concentrations of biologically fractionated nitrate (NO_3^-), which can result in baseline differences of >10‰ in some upwelling regions (Montoya et al. 2002; McMahon et al. 2013a). Patterns of $\delta^{13}C$ also correlate with regional productivity, proximity to continental shelves and coastlines, and sea surface temperature (MacKenzie et al. 2011; Magozzi et al. 2017), with $\delta^{13}C$ generally decreasing from coastal to pelagic waters (Carlisle et al. 2012).

Early observations of broad differences between marine ecoregions (e.g. the oligotrophic North Pacific Subtropical Gyre versus the high upwelling California Current Ecosystem; Carlisle et al. 2012) evolved to more spatially explicit 'isoscapes', which provided finer-scale geographic maps of isotopic baselines across oceanographic gradients (Graham et al. 2010; McMahon et al. 2013b; Ohshimo et al. 2019). Such isoscapes provided the baseline mapping necessary for interpretation of predator isotopic signatures, revealed the high correlation of $\delta^{13}C$ and $\delta^{15}N$ values to local oceanography and proximity to land masses, and have been generated using organic particles/phytoplankton (McMahon et al. 2013b; Oczkowski et al. 2016), copepods (Olson et al. 2010), jellyfish (MacKenzie et al. 2014), zooplanktivorous forage fish and squids (Ohshimo et al. 2019), and even higher trophic level species such as small tunas (Graham et al. 2010) and skates (Shipley et al. 2019).

Success of early studies and the growth of available isoscapes drove increased application of isotopic approaches to describe movements of wild marine predators, with standardization of ideal tissue types, tissue treatments, and post-analysis interpretation of SIA data. Accretionary structures, including otoliths, teeth, baleen, and hair were collected for reconstruction of past trophic ecology/ habitat use, and laboratory studies calibrated the extent to which these structures tracked prior diet and/or movement (West et al. 2006; Hobson and Wassenaar 2019; Trueman et al. 2019). In addition, physiologically dynamic tissues (i.e. those that are continually catabolized and anabolized through an organism's life, leading to tissue 'turnover') including muscle, blood, and liver were collected and analysed, as the different turnover rates of these tissues provided complementary temporal windows into past resource use (Fry 2006; Hobson and Wassenaar 2019). Laboratory studies aided in the development of standardized treatments of these tissues to eliminate biases in SI results, including acidification of tissues containing carbonate (teeth, otoliths, elasmobranch muscle) and chemical lipid extraction of lipoidal tissues (e.g. liver, muscle), as carbonate and lipid were demonstrated to bias $\delta^{13}C$ values (Post et al. 2007; Mateo et al. 2008).

Combined field and laboratory research continued to refine the treatment of certain tissues, such as the demonstrable need to extract urea and trimethylamine *n*-oxide from elasmobranch tissues (Carlisle et al. 2016; Li et al. 2016). Finally, algorithms were developed to arithmetically correct for lipid content, eliminating the need for time-consuming and costly lipid extraction approaches in some studies (Sweeting et al. 2006; Post et al. 2007; Logan et al. 2008). With the growth of SIA components to marine predator research came further advancements in the scope and availability of geographical SIA data, including the proposed SIA

data repository IsoBank (Pauli et al. 2017) and global-scale analyses of top predator SI values (Bird et al. 2018; Pethybridge et al. 2018). This set the stage for much more reliable interpretation of field-collected marine predator SI data, and innovation in quantitative approaches to these data, including Bayesian mixing models for diet analysis (Moore and Semmens 2008; Parnell et al. 2010) and isotopic clocks (Phillips and Eldridge 2006; Klaassen et al. 2010) for timing past movements.

This chapter explains how the isotopic clock approach utilizes predator physiology to understand migration and habitat use, illustrates the successful application of this technique to overexploited Pacific bluefin tuna and Pacific salmon species, and promotes increased application of this tool to better understand drivers and dynamics of migration in mobile marine predators. Predator physiology drives many aspects of migratory patterns (Whitlock et al. 2015), and a changing ocean environment will alter physiological restraints, resulting in potential alterations to migration patterns (Lennox et al. 2016). In turn, aspects of predator physiology (here, assimilation of local isotopic signatures into tissues) can be harnessed to better understand predator movements and predict responses to environmental change.

5.2 The isoclock approach

The isotopic clock or 'isoclock' technique extends the retrospective spatiotemporal capacity of SI data interpretation by directly harnessing the physiological dynamics of isotopic shifts in predator tissues. Isoclocks utilize the premise that different metabolically dynamic tissues (muscle, liver, whole blood, red blood cells, blood plasma) change in SI values at predictable rates (isotopic turnover rate) following a diet or migratory shift, and require different timeframes to reach steady-state with the new diet/region (Phillips and Eldridge 2006; Klaassen et al. 2010; Carter et al. 2019). If a metabolically dynamic tissue of a predator is at isotopic steady-state with diet in a certain region, when the animal migrates to a new region that tissue will change according to isotope- and tissue-specific turnover rate until reaching steady-state with the new regional diet. The timing since a movement shift can thus be calculated as a function of the difference between the current SI value and the SI baselines from the prior and current region, if isotopic turnover rate of the analysed tissue is known or can be estimated. This is demonstrated with the simple equation for isoclock measurements (Klaassen et al. 2010):

$$t = \frac{\ln\left(\dfrac{\delta_0 - \delta_f}{\delta_t - \delta_f}\right)}{\lambda}$$

Where δ_0 and δ_f = SI values (measured or estimated) of the study predator at steady-state with the initial (prior) and final (current) region (i.e. reflective of the regional geographic baseline values), λ = turnover rate of the analysed tissue, δ_t = current, measured SI value of an animal's tissue, and t = time in the new region (Figure 5.1). As such, an isoclock estimate of the timing of an animal's past movements requires relatively few measurements and parameters: endmember SI values for given tissue(s) and regions, the current SI value of the focal predator measured at time of sampling, and the isotopic turnover rate of the analysed tissue of the study animal (Figure 5.1). Importantly, if the animal tissue is at isotopic steady-state with the current region ($\delta_t \approx \delta_f$), the isoclock will not result in

reasonable migration timing estimates. However, useful migration timing information can still be derived from tissue(s) SI values that have reached isotopic steady-state with the current region. Such a case indicates the animal has been residential to the region for a period at least as long as the relevant isotope-tissue turnover rate and this can be used to discern recent migrants from relative 'residents' (Madigan et al. 2017a; Tawa et al. 2017).

Predator physiology drives the components of the isoclock. As described above, tissue turnover rates have been shown to follow exponential change after a diet shift (assuming the old and new diet are isotopically distinct), providing the half-life (λ) values that capacitate the temporal element of isoclocks. Controlled laboratory studies have shown that turnover rate is driven by both tissue growth and decomposition processes, as well as diet, animal life stage, and animal ectothermy versus endothermy (see reviews by Thomas & Crowther 2015 and Vander Zanden et al. 2015). Isotopic turnover rate decreases predictably with animal size, allowing for estimates of turnover in large animals that lack lab-derived turnover parameters (Thomas and Crowther 2015 and Vander Zanden et al. 2015). A second, implicit component of isoclocks is the potential need for diet-tissue discrimination

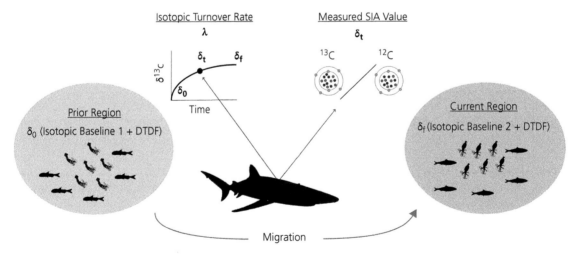

Figure 5.1 Components of an isotopic clock for a migratory marine predator. A predator with tissue isotopic composition reflecting a prior region (δ_0) moves to a new region, in which its tissue isotopic composition changes exponentially, with half-life λ (see 'Isotopic Turnover Rate' plot, upper left) to reflect the new prey baseline (δ_f). Sampling and stable isotope analysis of predator tissue during this transition give a SI value that is intermediate between these two regions (δ_t; see arrows from predator), which with other isotopic parameters (δ_0, δ_f, λ) can be applied to an isotopic clock to back-calculate timing of migration or to discern regional residents from recent migrants.

factors (DTDFs). In brief, metabolic processes mediate the conversion of prey to predator tissue, resulting in higher $\delta^{13}C$ and $\delta^{15}N$ values in predators than in prey, with DTDFs quantifying this difference (Caut et al. 2009). If a prey-based isoscape is used to generate endmember values (δ_0 and δ_f), DTDFs are necessary to generate endmember isoclock values (see Figure 5.1). However, endmember δ_0 and δ_f values can also be approximated using ecologically similar proxy species to the study animal. Both methods have been applied successfully in isoclock studies (see Pacific bluefin tuna and Pacific salmon case studies in Sections 5.3 and 5.4, respectively).

The migration timing estimates provided by isoclocks differentiate the approach from more typical chemical tracer studies that map predator tissue SI values to regional baseline values. The latter approach provides similar information to early conventional tagging studies: endpoint (location of tissue sampling) and migratory origin (discerned from correlating animal and regional isotopic signatures), with no temporal aspect to prior movement patterns. Importantly, the relatively few parameters and measurements needed for isoclock estimates also distinguish this approach from those that reconstruct animal life histories from accretionary structures (e.g. otolith, vertebrae, feather, baleen, eye lens). This use of accretionary tissues requires a transect from the base or core of the structure (representing beginning of life), and sequential sampling along a transect to the outermost part of the structure (representing conditions at each timepoint until death). This is a powerful technique and has been used to reconstruct life-history foraging and movements of teleosts, elasmobranchs, seabirds, and whales (Best and Schell 1996; Baumann et al. 2015; Carlisle et al. 2015; Christiansen et al. 2015; Morra et al. 2018; Quaeck-Davies et al. 2018).

While informative and effective, reconstructing movements from accretionary structures requires the measurement of many SI values from a single individual, as well as subsampling techniques that often require advanced technology (particularly for drilling otoliths and vertebrae), resulting in greater time and resource expense. For example, the cost of analysing a single white shark (*Carcharodon carcharias*) vertebral transect for SIA of $\delta^{13}C$ and $\delta^{15}N$ would be ~US$1000, while the isoclock approach for the same shark, for two tissues, would cost ~US$40. Similar to electronic tagging approaches, the extent of resources exhausted for a single individual can in many cases preclude data collection and analysis on the scale necessary for population-wide estimates of movement or limit the possibility for ongoing studies that capture temporal variability of observed movement trends. The isoclock approach is most valuable when migration timing estimates are needed for large and robust datasets, which may include sampling high numbers of individuals, and capturing movement dynamics across size classes, regions, seasons, and/or years. Below, we present two case studies that have successfully utilized isoclocks to infer migration patterns of highly migratory teleosts, Pacific bluefin tuna (*Thunnus orientalis*) and North Pacific salmon (*Oncorhynchus* spp.). In both cases, isoclocks provided recent migration histories of the study taxa and thus key information for improved contemporary management. These studies provide a blueprint for future use of isoclocks and their efficacy in providing information for management of migratory species.

5.3 Case study 1: Pacific bluefin tuna

The Pacific bluefin tuna was considered the only healthy population of bluefin tunas globally until 2011, though prior evaluations recognized assessments as data-deficient (Collette et al. 2011). Pacific bluefin became a conservation priority overnight in 2012, when population levels were estimated to be at ~4 % of historical, unfished levels (ISC 2012). This pointed to the need for better understanding of movement dynamics for multi-national management and population rebuilding efforts across their exploitation range, including Korea, Japan, China, and Taiwan in the western North Pacific, Mexico and the United States in the eastern North Pacific, and New Zealand in the South Pacific. The species was reviewed for endangered status under the Endangered Species Act in 2017 (NOAA 2016), which ultimately failed.

Pacific bluefin tuna spawn only in the western North Pacific Ocean, in regions around Japan and Taiwan. In their first and second year of life, a previously described 'unknown proportion' of the population migrates to the eastern Pacific Ocean, primarily using California Current waters off California, USA and Baja California, Mexico (Bayliff 1994; Boustany et al. 2010). Early conventional tagging (Bayliff et al. 1991) and more recent electronic tagging studies (Boustany et al. 2010) showed a high degree of residency in the California Current by immature Pacific bluefin tuna, though age composition in the eastern Pacific fluctuated widely. For many years, only age classes 1, 2, and 3 comprised the majority in the eastern Pacific, though there were historical records of much older, larger fish, and the prevalence of age classes 4–8 years has greatly increased in recent years (Madigan et al. 2017a). The population in the eastern Pacific was historically considered a minority (Bayliff 1994), but the paucity of adequate migration data, their overexploited status, and management differences between the East and West Pacific called for a population-level approach to assess the trans-Pacific migratory dynamics of the species.

To apply an isoclock to Pacific bluefin tuna in the East Pacific Ocean, the physiological parameters of isotopic tissue turnover and DTDFs were needed for a relatively large, regionally endothermic tuna. Captive Pacific bluefin tuna in the Tuna Research & Conservation Center (Monterey, CA, USA) allowed the rare opportunity to calculate these parameters from large pelagic fish under controlled conditions. This study of $\delta^{13}C$ and $\delta^{15}N$ turnover rates in captive bluefin found that time for muscle to reach steady-state with diet $\delta^{13}C$ and $\delta^{15}N$ values was approximately 1120 and 720 days, respectively, with respective half-life values (λ) of 255 and 167 days (Madigan et al. 2012b). DTDFs, necessary to calculate isoclock endmember values (δ_0 and δ_j) from West and East Pacific prey $\delta^{13}C$ and $\delta^{15}N$ values, calculated from this study were 1.8 ± 0.3‰ and 1.9 ± 0.4‰, respectively (Madigan et al. 2012b). Consistent isotopic differences were observed between tuna prey in the West and East Pacific (Minami et al. 1995; Mitani et al. 2006; Madigan et al. 2012a), particularly due to large differences in $\delta^{15}N$ between the California Current, an Eastern Boundary Current with high upwelling (high $\delta^{15}N$ values) and the relatively oligotrophic western Pacific (low $\delta^{15}N$ values). These regional differences were used to generate distinct, mean prey SI values

for both regions, based on pelagic forage fish, squid, and crustaceans. By adding the lab-derived bluefin DTDF to these regional prey SI values, the necessary parameters were available for an isoclock application to Pacific bluefin in the East Pacific. Finally, to account for the error associated with multiple isoclock parameters (DTDF and regional prey means used to generate δ_0 and δ_j; reported error around turnover rate λ), an iterative bootstrapping approach was applied, which generated 10^3 estimates of migration timing for each individual, resulting in a distribution of migration timing values for both each individual and the entire sampled population of Pacific bluefin tuna (Madigan et al. 2014; Madigan et al. 2017a).

The first isoclock study of East Pacific bluefin tuna applied SI values of $\delta^{13}C$ and $\delta^{15}N$ in muscle tissue from 130 individuals sampled from 2008–2010 (Madigan et al. 2014). SIA clearly distinguished West Pacific migrants from long-term (>1.5 year) East Pacific residents. These bulk isotope-inferred differences were corroborated by two complementary tracers, Fukushima-derived radiocaesium and amino acid compound-specific isotope analysis (Madigan et al. 2014). Results of this study showed higher residency in the East Pacific with increasing age class, and that most Pacific bluefin tuna migrate to the East Pacific at age 0–1, and a smaller number at age 1–2 (Madigan et al. 2014). The time-of-entry into the East Pacific was also calculated, with an estimated peak centred around May and a second peak in winter, centred around February–March (Figure 5.2). However, that initial study was limited to younger year classes (age 1–3) of Pacific bluefin tuna, which do not begin to spawn in the West Pacific until ages 3.5–4 and may reach sexual maturity several years later.

The subsequent occurrence of larger, older (age class 1–7) Pacific bluefin tuna in the eastern Pacific allowed for a second, more comprehensive isoclock analysis of migration dynamics in Pacific bluefin tuna. SI data were collected from 428 individuals from 2012–2015, representing age classes of 1–6+ years (Madigan et al. 2017a). That study found that all East Pacific bluefin age 3–7 were residents (>1.5 year) to the East Pacific, and that most individuals migrated from the West to East Pacific in their first or second year of life (Figure 5.2). This study also found multiple peaks in arrival time and age-of-entry to the East Pacific. The dual peaks in arrival time, not an expected result based on pre-existing conventional tagging or fisheries catch data, were later corroborated by an extensive electronic tagging study of age-0 Pacific bluefin tuna in the West Pacific (Fujioka et al. 2018). Fujioka et al. retrieved archival electronic tag data from 15 bluefin that had migrated east, providing fine-scale and accurate movement data that corroborated isoclock migration timing estimates, and the dual peaks in arrival, reported by Madigan et al. (2014, 2017a) (Figure 5.2). Finally, a study applying SIA of $\delta^{15}N$ to 155 Pacific bluefin in the Sea of Japan, one of the spawning regions in the West Pacific, found that individuals of age 3–17 in that region were a mix of East Pacific migrants and West Pacific residents (Tawa et al.

2017). By synthesizing SIA studies on both sides of the North Pacific Ocean, and comparing them to trends of historical Pacific bluefin tuna conventional tag–recapture data and regional fisheries catch (Bayliff 1994), it became clear that the proportion of Pacific bluefin tuna that migrate to the East Pacific is not a minority but, at least over certain timeframes, may be the majority of the population (Madigan et al. 2017b).

The results from the studies above have important management/conservation implications and applications to Pacific bluefin tuna. The proportion of the population that migrates to the East Pacific and later returns to the Sea of Japan is demonstrably substantial (Madigan et al. 2017b; Tawa et al. 2017), and stock assessments should integrate fisheries mortality in the East Pacific more fully into

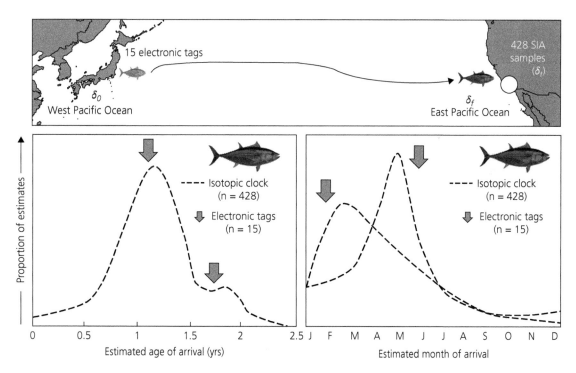

Figure 5.2 Results of isotopic clocks applied to trans-Pacific migration of Pacific bluefin tuna, *Thunnus orientalis*. Pacific bluefin tuna migrate from the West to East Pacific as juveniles (top panel), and stable isotope analysis and isotopic clocks using $\delta^{15}N$ allowed for estimates (dashed lines) of the age (bottom left) and month (bottom right) at which Pacific bluefin migrate to the East Pacific. A subsequent electronic tagging study tracked trans-Pacific migration of 15 individuals, allowing for multi-technique comparison of timing (arrows in bottom panels). Simplified and summarized results reproduced with permission from: (1) John Wiley and Sons, *Ecology*, *95*, Madigan et al., Copyright (2014); (2) Canadian Science Publishing, from *Canadian Journal of Fisheries and Aquatic Sciences*, Madigan et al., *75*, Copyright (2017), permission conveyed through Copyright Clearance Center; (3) Elsevier, *Progress In Oceanography*, *162*, Fujioka et al., Spatial and temporal variability in the trans-Pacific migration of Pacific bluefin tuna (*Thunnus orientalis*) revealed by archival tags, pp. 52–65, Copyright (2018).

assessments, which have traditionally considered the East Pacific population as a negligible proportion. As new potential management and conservation measures are implemented for this overfished species (e.g. Nakatsuka et al. 2017), and climate cycles vary in the North Pacific Ocean (Chavez et al. 2003), migration dynamics will likely change over time (Hazen et al. 2013). Ongoing sampling from captured Pacific bluefin on both sides of the Pacific will facilitate relatively tractable ongoing studies that track temporal change of bluefin movement dynamics in response to implemented management measures and/or climate variability. Finally, these studies demonstrate the utility and feasibility of isoclocks to estimate previously unknown movement dynamics in large sample sets of migratory marine predators that may be of particular commercial, ecological, or conservation concern.

5.4 Case study 2: Pacific salmon

A second study by Moore et al. (2016) demonstrates the applicability of isoclocks not only to ocean basin-wide movements of pelagic predators, but to smaller-scale movements of migratory species that move ontogenetically across freshwater, brackish, and marine habitats. Salmon are anadromous fishes that perform extensive migrations from freshwater habitats where they spawn, to marine habitats where they grow (Groot and Margolis 1991). Estuaries are considered an important habitat for salmon during their migratory cycle and may provide numerous services for different life-history stages (e.g. refuge from predators, feeding opportunities, physiological salinity buffers; Healey 1982; Thorpe 1994; Weitkamp et al. 2014). Understanding the role of estuaries in salmon migration is pertinent considering the ongoing development and reclamation of coastal habitat that may threaten many species and life-history stages (Orth et al. 2006). Management measures for salmon are currently challenged by poor knowledge of estuarine residency and habitat use during early migration phases (Weitkamp et al. 2014). Thus, the degree to which estuaries act as critical stopover habitats versus migration corridors is unknown.

The Skeena River, British Columbia is the second largest salmon-supporting system in Canada, con-

taining five species: Chinook (*Oncorhynchus tshawytscha*), chum (*Oncorhynchus keta*), Coho (*Oncorhynchus kistuch*), pink (*Oncorhynchus gorbuscha*), and sockeye salmon (*O. nerka*) (Moore et al. 2016). In recent years, concern has grown over the future viability of salmon populations in this region due to installation of fossil fuel pipelines and terminals, such as those proposed by Pacific Northwest Liquid Natural Gas, a large natural gas liquefaction and export facility (Moore et al. 2016). This would require significant alteration and, in extreme cases, complete removal of key estuarine habitat. A critical need to understand estuarine habitat use and residency by salmon species is therefore required to offer informed management of this economically and ecologically important suite of fishes (Pickard et al. 2015; Moore et al. 2016).

The study by Moore et al. (2016) constructed sulfur, carbon, and nitrogen ($\delta^{34}S$, $\delta^{13}C$, and $\delta^{15}N$) isoclocks to determine the extent of estuarine residency and assess whether residency differed both within (i.e. across different genetic populations) and across all five salmon species found within the Skeena River. To apply the isoclock to the five salmon species, the authors utilized isotopic turnover rates (λ) calculated for two tissues, liver (faster turnover) and muscle (slower turnover) of a close taxonomic relative, the steelhead (*Oncorhynchus mykiss*) (Heady and Moore 2013), estimated to be 16 ± 5 days and 39 ± 3 days, respectively. To characterize the freshwater isotopic composition of focal species (δ_0), salmon fry were sampled from freshwater rearing habitats at single locations for each species; these were assumed to adequately categorize the baseline isotopic composition of the entire freshwater watershed. Sockeye salmon fry were sampled from Babine Lake, Coho salmon fry from the Slemgeesh River, and pink, chum, and Chinook salmon fry from the Krispiox River. Estuarine isotopic baselines (δ_f) were generated by sampling resident fishes that exhibit a similar trophic role to juvenile salmon: surf smelt (*Hypomesus pretioseus*) were sampled from two regions and Pacific herring (*Clupea pallasii*) were sampled from a single region. Finally, this study used an iterative bootstrapping approach, similar to the Pacific bluefin tuna studies above, to properly account for error around parameter estimates when generating migration timing estimates.

A total of 253 juvenile salmon were sampled across the five species to estimate residency and migration timing, with sampling spanning both inner ($n = 152$) and middle ($n = 101$) regions of the estuary (Figure 5.3). Isoclock analyses showed that all species exhibit some degree of residency inside the estuary; estimates ranged from 0–18 days for sockeye, 0–43 days for Coho, 0–101 days for pink, and 7–57 days for Chinook salmon. Most Chinook salmon were estimated to spend up to 1 month in the Skeena River estuary, indicative of both rearing and growing. Individuals were estimated to have grown at a rate of 0.48 ± 0.09 mm/day while resident, illustrating that the estuary provides growth opportunities, which may be critical for individual longevity and ocean survival (Duffy and Beauchamp 2011). Estuarine residency observed for pink salmon was higher than previously assumed; juveniles were estimated to rear for up to 1 month,

whereas prior information suggested individuals moved rapidly through estuarine habitats (Levy and Northcote 1982; Weitkamp et al. 2014). Sockeye salmon were observed to utilize the estuary for the shortest timeframes; this result augmented studies from other regions (Weitkamp et al. 2014). However, significant inter-individual variability in estuarine residency and habitat use by sockeye salmon was observed, suggesting different populations spend different amounts of time inside the estuary, a result reflected in comparisons of other genetically distinct populations (e.g. Simmons et al. 2013 vs. Weitkamp et al. 2014).

Estuarine residency estimates generated for juvenile Coho salmon revealed that individuals rear in the Skeena River estuary for weeks, potentially months. Evidence was also provided for a divergent life-history strategy of Coho salmon, with longer estuarine residency by smaller individuals

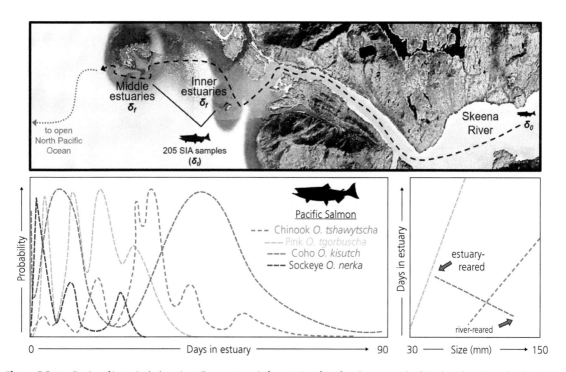

Figure 5.3 Application of isotopic clocks to juvenile estuary use in four species of Pacific salmon. Juvenile Chinook, pink, Coho, and sockeye salmon move from the Skeena River and through proximate estuaries for an unknown period of time before entering oceanic waters (top panel). Isotopic clocks using $\delta^{34}S$, $\delta^{13}C$, and $\delta^{15}N$ allowed for multi-species estimates of time spent in estuarine habitat (bottom left panel) and relationships between size and time in estuaries (bottom right panel; no significant relationship was found for sockeye salmon). Arrows in bottom right panel reveal finding of a potential divergent life-history strategy for Coho, with some fry rearing in riverine and others in estuarine habitats. Results simplified and summarized from Moore et al. (2016) with permission from Inter-Research.

versus recent estuarine entry by larger individuals (Figure 5.3). The authors concluded that some fry may enter the estuary near the beginning of life and remain for extended periods of time, while others rear in riverine habitats, a pattern observed in other watersheds (Miller and Sadro 2003; Koski 2009; Craig et al. 2014; Rebenack et al. 2015), but that had yet to be demonstrated in Skeena River populations (Gottesfeld and Rabnett 2008). Finally, isotopic data generated from chum salmon fell outside the isotopic mixing space (for $\delta^{34}S$ and $\delta^{13}C$) required to construct accurate migration histories, highlighting that a 'one-size-fits-all' isoclock approach (i.e. using the same parameters δ_0, δ_f, and λ for multiple species) may not be appropriate across all salmon species, despite ostensible taxonomic and trophic similarity. The authors attributed this discrepancy to potentially inaccurate characterization of the freshwater baseline for chum salmon, and that individuals may feed in lower regions of the Skeena River, which results in the incorporation of different freshwater baselines upon entry to the estuary.

Moore and colleagues' isoclock approach provided novel and compelling information on critical habitats used by individual salmon within the Skeena River estuary, information needed for the future management of salmonids in this system. Specifically, isoclocks revealed that estuaries should indeed be categorized as important stopover (rather than transient) habitats during the migration of juvenile salmon, as all five salmon species showed some estuarine residency in the Skeena River watershed. This has important implications for the management of this system, suggesting that development and/or alteration to estuarine habitat may have negative consequences for salmon, and may influence important processes such as growth and as a result, long-term ocean survival. More broadly, habitat connectivity studies of Pacific salmon are lacking, but deemed necessary for restoration planning and conservation (Flitcroft et al. 2019). The isoclock approach in the Skeena River estuary particularly demonstrates the utility of $\delta^{34}S$ in studies across salinity gradients (Fry and Chumchal 2011) and provides a model for other species that utilize river/estuary/marine migratory networks to ascertain critical habitat during early ontogeny. As for other species and systems, the isoclock approach

used here answered questions relating to movement, migration, and habitat use for early life-history stages of species that would be otherwise challenging to study using more conventional tracking methods.

5.5 Future isoclock applications

Early studies using SI values as chemical tracers in wild populations were limited by a lack of laboratory-derived parameters with which to accurately interpret their results (Gannes et al. 1997); this is no longer the case. Since physiological parameters for isoclock studies such as turnover rate and DTDFs can be dependent upon species, lifestage, metabolic physiology, diet, and environment, some gaps and imprecision in quantifying these parameters will always exist, especially for large predators. However, calls for laboratory-derived parameters (Gannes et al. 1997; Martínez del Rio et al. 2009) have led to enormous growth of such studies. Isotopic turnover rates, necessary for isoclocks, are now available in the marine realm for a wide range of teleost, elasmobranch, seabird, sea turtle, and marine mammal species (Table 5.1). Similarly, for species where no published values exist, DTDF estimates can be calculated by contemporary statistical approaches using closest ecological and/or taxonomic relatives (Healy et al. 2018). In contrast, the number of studies that have employed true isoclocks on dynamic tissues, utilizing lab-derived turnover rates (λ), are relatively few, restricted to tuna, salmon, gobies, and two coastal birds in the marine environment (Guelinckx et al. 2008; Moore et al. 2016; Madigan et al. 2017a; Boggie et al. 2018; Catry et al. 2018) and largely to birds in terrestrial environments (Rubenstein and Hobson 2004; Bauchinger and McWilliams 2009; Hobson 2019). In addition, two reviews of isotopic turnover rates across the animal kingdom showed allometric scaling relationships that allow half-life λ to be reasonably estimated when species-specific data are lacking (Thomas and Crowther 2015; Vander Zanden et al. 2015) as has been applied, for example, to large white sharks in the North Pacific Ocean (Carlisle et al. 2012). This implies that isoclocks could be applied to sampled tissue from almost any marine predator, provided that some knowledge of

Table 5.1 A synthesis of species with calculated stable isotope turnover rate data reported in the literature. In most cases, isotope turnover dynamics were described by single-compartment, exponential model fits, providing half-life (λ) estimates that can be applied to isotopic clock approaches.

Taxa	Species		Tissues[a]	References
Finfish	Sockeye salmon, Japanese flounder, sand goby, Atlantic salmon, croaker, mullet, California yellowtail	*Oncorhynchus nerka, Paralichthys olivaceus, Pomatoschistus minutus, Salmo salar, Micropogonias furnieri, Mugil liza, Seriola dorsalis*	Muscle	(Tominaga et al. 2003; Sakano et al. 2005; Guelinckx et al. 2007; Mont'Alverne et al. 2016; Oliveira et al. 2017; Madigan et al. in review)
	Broad whitefish, Japanese sea bass, Pacific bluefin tuna	*Coregonus nasus, Lateolabrax japonicus, Thunnus orientalis*	Muscle, liver	(Hesslein et al. 1993; Suzuki et al. 2005; Madigan et al. 2012)
	Winter flounder, red drum, Senegalese sole	*Pleuronectes americanus, Sciaenops ocellatus, Solea senegalensis*	Whole	(Herzka and Holt 2000; Bosley et al. 2002; Gamboa-Delgado et al. 2008)
	Pacific herring, European sea bass, Atlantic cod, rainbow trout, summer flounder, blue cod, tilapia, Atlantic bluefin tuna, coral trout	*Clupea pallasi, Dicentrarchus labrax, Gadus morhua, Oncorhynchus mykiss, Paralichthys dentatus, Parapercis colias, Oreochromis aureus, Thunnus thynnus, Plectropomus leopardus*	Muscle, liver, *heart, *eye, *blood, *fin, *red blood cells, *bone, *eggs	(Sweeting et al. 2005; Miller 2006; Suring and Wing 2009; Buchheister and Latour 2010; Ankjærø et al. 2012; Madigan et al. 2012; Heady and Moore 2013; Matley et al. 2016; Wang et al. 2016)
	Gag	*Mycteroperca microlepis*	Muscle, gonad	(Nelson et al. 2011)
Seabirds	Dunlin, great skua	*Calidris alpina pacifica, Catharacta skua*	Blood	(Bearhop et al. 2002; Ogden et al. 2004)
	African penguin, red knot, dunlin	*Spheniscus demersus, Calidris canutus, Calidris alpina*	Plasma, *blood, *red blood cells, *toenail	(Ogden et al. 2004; Barquete et al. 2013; van Gils and Salem 2015)
Sharks and rays	Sandbar shark, leopard shark, large-spotted dogfish	*Carcharhinus plumbeus, Triakis semifasciata, Scyliorhinus stellaris*	Muscle, *blood, *plasma, *red blood cells, *tooth, *fin	(Logan and Lutcavage 2010; Kim et al. 2012; Caut et al. 2013)
	Cownose ray	*Rhinoptera bonasus*	Skin	(St Clair 2014)
Marine mammals	Blue whale, Florida manatee	*Balaenoptera musculus, Trichechus manatus latirostris*	Skin	(Alves-Stanley and Worthy 2009; Busquets-Vass et al. 2017)
	Killer whale	*Orcinus orca*	Plasma	(Caut et al. 2011)
	Bottlenose dolphin	*Tursiops truncatus*	Blood	(Caut et al. 2011; Browning et al. 2014)
Aquatic reptiles	American alligator, broad-snouted caiman, loggerhead turtle	*Alligator mississippiensis, Caiman latirostris, Caretta caretta*	Blood, plasma, *muscle, *skin	(Reich et al. 2008; Caut 2013; Rosenblatt and Heithaus 2013)

[a] Muscle = skeletal muscle; whole = entire organism; blood = whole blood; plasma = blood plasma.

Tissues denoted with asterisk (*) indicate available turnover rates for only some species. See table references and full studies therein for information on animal size, lifestage, tissues, stable isotopes analysed, and turnover rate (λ) estimates.

potential prior habitats exists and isotopic baseline data exist for prior habitat(s).

Migration patterns of marine predators will respond to a changing ocean environment (Anderson et al. 2013; Pinsky et al. 2013; Abrahms et al. 2018), which will impact management and conservation approaches. In field studies of wild marine predators (e.g. electronic tagging), sublethal sampling of various tissues is possible and often standard during field procedures (Figure 5.4). In addition, any marine species that is recreationally and/or commercially harvested makes extensive multi-tissue sampling readily available to researchers and managers. Ongoing, standardized tissue sample

Figure 5.4 Field sampling of predator tissues for stable isotope analysis. The authors sample a (left) live tiger shark (*Galeocerdo cuvier*) for muscle tissue during deployment of electronic tags (photo: Sami Kattan and Beneath the Waves) and (right) carcass of a Pacific bluefin tuna (*Thunnus orientalis*) harvested aboard a recreational fishing vessel, from which multiple tissues (e.g. muscle, liver, bone) can be sampled and analysed for SI values, which can be used in isotopic clock studies to infer previous migration patterns.

collection, and analysis of tissues for chemical tracers, can provide insight into the response of animal migrations to environmental disturbance, habitat alteration, and/or climate change. For example, isoclocks showed that Pacific bluefin tuna migrated earlier into the East Pacific in warmer, El Niño years (Madigan et al. 2017a). Continued monitoring of such changes is key, and archived tissue samples, ongoing tissue sampling programmes, and SIA of these resources can allow for the isoclock approaches outlined here to be applied to a broad suite of marine predators. Combined with complementary technologies (e.g. biotelemetry, physiological measures of reproductive, stress, and nutritional condition), isoclocks are a powerful tool to understand how and when predators move, and how migrations will shift in space and time in a dynamically changing ocean. Isoclocks present a clear bridge between animal ecophysiology, migration dynamics, and pertinent management and conservation measures; it is the physiology of the animal that regulates tissue turnover and resource acquisition, facilitating the application of isoclocks to population-level movement dynamics that inform management and conservation. Increased application of this tool will clarify movement dynamics of migratory predators at the

population level, which will feed directly into population recovery strategies and promote conservation of sustainable predator stocks.

References

Abrahms, B., Hazen, E.L., Bograd, S.J. et al., 2018. Climate mediates the success of migration strategies in a marine predator. *Ecology Letters*, 21, 63–71.

Anderson, J.J., Gurarie, E., Bracis, C. et al., 2013. Modeling climate change impacts on phenology and population dynamics of migratory marine species. *Ecological Modelling*, 264, 83–97.

Barkley, A., Gollock, M., Samoilys, M. et al., 2019. Complex transboundary movements of marine megafauna in the Western Indian Ocean. *Animal Conservation*, 22, 420–31.

Bauchinger, U. and McWilliams, S., 2009. Carbon turnover in tissues of a passerine bird: allometry, isotopic clocks, and phenotypic flexibility in organ size. *Physiological and Biochemical Zoology*, 82, 787–97.

Baumann, H., Wells, R.J.D., Rooker, J.R. et al., 2015. Combining otolith microstructure and trace elemental analyses to infer the arrival of juvenile Pacific bluefin tuna in the California current ecosystem. *ICES Journal of Marine Science: Journal du Conseil*, 72, 2128–38.

Bayliff, W.H., 1994. A review of the biology and fisheries for northern bluefin tuna, *Thunnus thynnus*, in the Pacific Ocean. *FAO Fisheries Technical Paper*, 336, 244–95.

Bayliff, W.H., Ishizuka, Y., and Deriso, R., 1991. Growth, movement, and attrition of northern bluefin tuna, *Thunnus thynnus*, in the Pacific Ocean, as determined by tagging. *Inter-American Tropical Tuna Commission Bulletin*, 20, 3–94.

Best, P. and Schell, D., 1996. Stable isotopes in southern right whale (*Eubalaena australis*) baleen as indicators of seasonal movements, feeding and growth. *Marine Biology*, 124, 483–94.

Bird, C.S., Veríssimo, A., Magozzi, S. et al., 2018. A global perspective on the trophic geography of sharks. *Nature Ecology & Evolution*, 2, 299.

Block, B.A., Teo, S.L., Walli, A. et al., 2005. Electronic tagging and population structure of Atlantic bluefin tuna. *Nature*, 434, 1121.

Boggie, M.A., Carleton, S.A., Collins, D.P. et al., 2018. Using stable isotopes to estimate reliance on agricultural food subsidies and migration timing for a migratory bird. *Ecosphere*, 9, e02083.

Boustany, A.M., Matteson, R., Castleton, M. et al., 2010. Movements of Pacific bluefin tuna (*Thunnus orientalis*) in the Eastern North Pacific revealed with archival tags. *Progress In Oceanography*, 86, 94–104.

Carlisle, A.B., Goldman, K.J., Litvin, S.Y. et al., 2015. Stable isotope analysis of vertebrae reveals ontogenetic changes in habitat in an endothermic pelagic shark. *Proceedings of the Royal Society B*, 282, 20141446.

Carlisle, A.B., Kim, S.L., Semmens, B.X. et al., 2012. Using stable isotope analysis to understand migration and trophic ecology of northeastern Pacific white sharks (*Carcharodon carcharias*). *PLoS ONE*, 7, e30492.

Carlisle, A.B., Litvin, S.Y., Madigan, D.J. et al., 2016. Interactive effects of urea and lipid content confound stable isotope analysis in elasmobranch fishes. *Canadian Journal of Fisheries and Aquatic Sciences*, 99, 1–10.

Carter, W.A., Bauchinger, U., and McWilliams, S.R., 2019. The importance of isotopic turnover for understanding key aspects of animal ecology and nutrition. *Diversity*, 11, 84.

Catry, T., Lourenço, P.M., and Granadeiro, J.P., 2018. Quantifying population size of migrant birds at stopover sites: Combining count data with stopover length estimated from stable isotope analysis. *Methods in Ecology and Evolution*, 9, 502–12.

Caut, S., Angulo, E., and Courchamp, F., 2009. Variation in discrimination factors (Δ^{15}N and Δ^{13}C): the effect of diet isotopic values and applications for diet reconstruction. *Journal of Applied Ecology*, 46, 443–53.

Chavez, F.P., Ryan, J., Lluch-Cota, S.E., and Ñiquen C, M., 2003. From anchovies to sardines and back: multidecadal change in the Pacific Ocean. *Science*, 299, 217–21.

Christiansen, H.M., Fisk, A.T., and Hussey, N.E., 2015. Incorporating stable isotopes into a multidisciplinary framework to improve data inference and their conservation and management application. *African Journal of Marine Science*, 37(2), 189–97.

Collette, B.B., Carpenter, K.E., Polidoro, B.A. et al., 2011. High value and long life—double jeopardy for tunas and billfishes. *Science*, 333, 291–2.

Craig, B., Simenstad, C.A., and Bottom, D.L., 2014. Rearing in natural and recovering tidal wetlands enhances growth and life-history diversity of Columbia Estuary tributary Coho salmon *Oncorhynchus kisutch* population. *Journal of Fish Biology*, 85, 31–51.

Dickhut, R.M., Deshpande, A.D., Cincinelli, A. et al., 2009. Atlantic bluefin tuna (*Thunnus thynnus*) population dynamics delineated by organochlorine tracers. *Environmental Science & Technology*, 43, 8522–7.

Duffy, E.J. and Beauchamp, D.A., 2011. Rapid growth in the early marine period improves the marine survival of Chinook salmon (*Oncorhynchus tshawytscha*) in Puget Sound, Washington. *Canadian Journal of Fisheries and Aquatic Sciences*, 68, 232–40.

Flitcroft, R.L., Arismendi, I., and Santelmann, M.V., 2019. A review of habitat connectivity research for Pacific salmon in marine, estuary, and freshwater environments. *JAWRA Journal of the American Water Resources Association*, 55, 430–41.

Fry, B., 2006. *Stable Isotope Ecology*. Springer-Verlag, New York.

Fry, B. and Chumchal, M.M., 2011. Sulfur stable isotope indicators of residency in estuarine fish. *Limnology and Oceanography*, 56, 1563–76.

Fujioka, K., Fukuda, H., Tei, Y. et al., 2018. Spatial and temporal variability in the trans-Pacific migration of Pacific bluefin tuna (*Thunnus orientalis*) revealed by archival tags. *Progress in Oceanography*, 162, 52–65.

Gannes, L.Z., O'Brien, D.M., and Del Rio, C.M., 1997. Stable isotopes in animal ecology: assumptions, caveats, and a call for more laboratory experiments. *Ecology*, 78, 1271–6.

Gottesfeld, A. and Rabnett, K.A., 2008. *Skeena River Fish and Their Habitat*. OSU Press, Corvallis, OR.

Graham, B.S., Koch, P.L., Newsome, S.D. et al., 2010. Using isoscapes to trace the movements and foraging behavior of top predators in oceanic ecosystems. In J.B. West, G.J. Bowen, T.E. Dawson, and K.P. Tu, eds. *Isoscapes*, pp. 299–318. Springer, Amsterdam.

Groot, C. and Margolis, L., 1991. *Pacific Salmon Life Histories*. UBC Press, Vancouver, BC.

Guelinckx, J., Maes, J., Geysen, B., and Ollevier, F., 2008. Estuarine recruitment of a marine goby reconstructed with an isotopic clock. *Oecologia*, 157, 41–52.

Harrison, A.-L., Costa, D.P., Winship, A.J. et al., 2018. The political biogeography of migratory marine predators. *Nature Ecology & Evolution*, 2, 1571–8.

Hazen, E.L., Jorgensen, S., Rykaczewski, R.R. et al., 2013. Predicted habitat shifts of Pacific top predators in a changing climate. *Nature Climate Change, 3*, 234–8.

Hazen, E.L., Maxwell, S.M., Bailey, H. et al., 2012. Ontogeny in marine tagging and tracking science: technologies and data gaps. *Marine Ecology Progress Series, 457*, 221–40.

Heady, W.N. and Moore, J.W., 2013. Tissue turnover and stable isotope clocks to quantify resource shifts in anadromous rainbow trout. *Oecologia, 172*, 21–34.

Healey, M., 1982. Juvenile Pacific salmon in estuaries: the life support system. In V.S. Kennedy, ed. *Estuarine Comparisons*, pp. 315–41. Elsevier, Gleneden Beach, OR.

Healy, K., Guillerme, T., Kelly, S.B. et al., 2018. SIDER: an R package for predicting trophic discrimination factors of consumers based on their ecology and phylogenetic relatedness. *Ecography, 41*, 1393–400.

Hobson, K.A., 1999. Tracing origins and migration of wildlife using stable isotopes: a review. *Oecologia, 120*, 314–26.

Hobson, K.A., 2019. Application of isotopic methods to tracking animal movements. In K.A. Hobson and L.I. Wassenaar, eds. *Tracking Animal Migration with Stable Isotopes*, second edition, pp. 85–115. Academic Press, London.

Hobson, K.A. and Wassenaar, L.I., 2019. *Tracking Animal Migration with Stable Isotopes*. Academic Press, London.

Hussey, N.E., Kessel, S.T., Aarestrup, K. et al., 2015. Aquatic animal telemetry: a panoramic window into the underwater world. *Science, 348*, 1255642.

ISC, 2012. *Pacific Bluefin Stock Assessment*. International Scientific Committee for Tuna and Tuna-like Species in the North Pacific Ocean, Shimizu, Shizuoka, Japan.

Klaassen, M., Piersma, T., Korthals, H. et al., 2010. Single-point isotope measurements in blood cells and plasma to estimate the time since diet switches. *Functional Ecology, 24*, 796–804.

Koski, K.V., 2009. The fate of Coho salmon nomads: the story of an estuarine-rearing strategy promoting resilience. *Ecology and Society, 14*, article 4.

Lascelles, B., Notarbartolo Di Sciara, G. et al., 2014. Migratory marine species: their status, threats and conservation management needs. *Aquatic Conservation: Marine and Freshwater Ecosystems, 24*, 111–27.

Lennox, R.J., Chapman, J.M., Souliere, C.M. et al., 2016. Conservation physiology of animal migration. *Conservation Physiology, 4*(1), cov072.

Levy, D. and Northcote, T., 1982. Juvenile salmon residency in a marsh area of the Fraser River estuary. *Canadian Journal of Fisheries and Aquatic Sciences, 39*, 270–6.

Li, Y., Zhang, Y., Hussey, N.E., and Dai, X., 2016. Urea and lipid extraction treatment effects on $\delta^{15}N$ and $\delta^{13}C$ values in pelagic sharks. *Rapid Communications in Mass Spectrometry, 30*, 1–8.

Logan, J.M., Jardine, T.D., Miller, T.J. et al., 2008. Lipid corrections in carbon and nitrogen stable isotope analyses: comparison of chemical extraction and modelling methods. *Journal of Animal Ecology, 77*, 838–46.

MacKenzie, K., Longmore, C., Preece, C. et al., 2014. Testing the long-term stability of marine isoscapes in shelf seas using jellyfish tissues. *Biogeochemistry, 121*, 441–54.

MacKenzie, K.M., Palmer, M.R., Moore, A. et al., 2011. Locations of marine animals revealed by carbon isotopes. *Scientific Reports, 1*, 21.

Madigan, D.J., Baumann, Z., Carlisle, A.B. et al., 2014. Reconstructing trans-oceanic migration patterns of Pacific bluefin tuna using a chemical tracer toolbox. *Ecology, 95*, 1674–83.

Madigan, D.J., Baumann, Z., Carlisle, A.B. et al., 2017a. Isotopic insights into migration patterns of Pacific bluefin tuna in the eastern Pacific Ocean. *Canadian Journal of Fisheries and Aquatic Sciences, 75*, 260–70.

Madigan, D.J., Baumann, Z., Snodgrass, O.E. et al., 2013. Radiocesium in Pacific bluefin tuna *Thunnus orientalis* in 2012 validates new tracer technique. *Environmental Science & Technology, 47*, 2287–94.

Madigan, D.J., Boustany, A., and Collette, B.B., 2017b. East not least for Pacific bluefin tuna. *Science, 357*, 356–7.

Madigan, D.J., Carlisle, A.B., Dewar, H. et al., 2012a. Stable isotope analysis challenges wasp-waist food web assumptions in an upwelling pelagic food web. *Scientific Reports, 2*, e654.

Madigan, D.J., Litvin, S.Y., Popp, B.N. et al., 2012b. Tissue turnover rates and isotopic trophic discrimination factors in the endothermic teleost, Pacific bluefin tuna (*Thunnus orientalis*). *PLoS ONE, 7*, e49220.

Magozzi, S., Yool, A., Vander Zanden, H. et al., 2017. Using ocean models to predict spatial and temporal variation in marine carbon isotopes. *Ecosphere, 8*, e01763.

Martínez del Rio, C., Wolf, N., Carleton, S.A., and Gannes, L.Z., 2009. Isotopic ecology ten years after a call for more laboratory experiments. *Biological Reviews, 84*, 91–111.

Mateo, M.A., Serrano, O., Serrano, L., and Michener, R.H., 2008. Effects of sample preparation on stable isotope ratios of carbon and nitrogen in marine invertebrates: implications for food web studies using stable isotopes. *Oecologia, 157*, 105–15.

McFarlane, G., Wydoski, R.S., and Prince, E., 1990. Historical review of the development of external tags and marks. American Fisheries Society Symposium, 7, 9–29.

McMahon, K.W., Hamady, L.L., and Thorrold, S.R., 2013a. Ocean ecogeochemistry: a review. *Oceanography and Marine Biology—an Annual Review, 51*, 327–73.

McMahon, K.W., Hamady, L.L., and Thorrold, S.R., 2013b. A review of ecogeochemistry approaches to estimating movements of marine animals. *Limnology and Oceanography*, 58, 697–714.

Miller, B.A. and Sadro, S., 2003. Residence time and seasonal movements of juvenile Coho salmon in the ecotone and lower estuary of Winchester Creek, South Slough, Oregon. *Transactions of the American Fisheries Society*, 132, 546–59.

Minami, H., Minagawa, M., and Ogi, H., 1995. Changes in stable carbon and nitrogen isotope ratios in sooty and short-tailed shearwaters during their northward migration. *Condor*, 97, 565–74.

Mitani, Y., Bando, T., Takai, N., and Sakamoto, W., 2006. Patterns of stable carbon and nitrogen isotopes in the baleen of common minke whale *Balaenoptera acutorostrata* from the western North Pacific. *Fisheries Science*, 72, 69–76.

Montoya, J.P., Carpenter, E.J., and Capone, D.G., 2002. Nitrogen fixation and nitrogen isotope abundances in zooplankton of the oligotrophic North Atlantic. *Limnology and Oceanography*, 47, 1617–28.

Moore, J.W., Gordon, J., Carr-Harris, C. et al., 2016. Assessing estuaries as stopover habitats for juvenile Pacific salmon. *Marine Ecology Progress Series*, 559, 201–15.

Moore, J.W. and Semmens, B.X., 2008. Incorporating uncertainty and prior information into stable isotope mixing models. *Ecology Letters*, 11, 470–80.

Morra, K.E., Wiley, A.E., James, H.F. et al., 2018. Influence of feather selection and sampling protocol on interpretations of Hawaiian petrel (*Pterodroma sandwichensis*) nonbreeding season foraging habits from stable isotope analysis. *Waterbirds*, 41, 93–100, 108.

Nakatsuka, S., Ishida, Y., Fukuda, H., and Akita, T., 2017. A limit reference point to prevent recruitment overfishing of Pacific bluefin tuna. *Marine Policy*, 78, 107–13.

NOAA, 2016. Listing endangered or threatened species; 90-day finding on a petition to list the Pacific bluefin tuna as threatened or endangered under the Endangered Species Act. In D. Commerce, ed. Federal Register, pp. 70074–70080.

Oczkowski, A., Kreakie, B., McKinney, R.A., and Prezioso, J., 2016. Patterns in stable isotope values of nitrogen and carbon in particulate matter from the Northwest Atlantic continental shelf, from the Gulf of Maine to Cape Hatteras. *Frontiers in Marine Science*, 3, 252.

Ohshimo, S., Madigan, D.J., Kodama, T. et al., 2019. Isoscapes reveal patterns of $\delta^{13}C$ and $\delta^{15}N$ of pelagic forage fish and squid in the Northwest Pacific Ocean. *Progress in Oceanography*, 175, 124–38.

Olson, R.J., Popp, B.N., Graham, B.S. et al., 2010. Food-web inferences of stable isotope spatial patterns in copepods and yellowfin tuna in the pelagic eastern Pacific Ocean. *Progress in Oceanography*, 86, 124–38.

Orth, R.J., Carruthers, T.J., Dennison, W.C. et al., 2006. A global crisis for seagrass ecosystems. *Bioscience*, 56, 987–96.

Parnell, A.C., Inger, R., Bearhop, S., and Jackson, A.L., 2010. Source partitioning using stable isotopes: coping with too much variation. *PLoS ONE*, 5, e9672.

Pauli, J.N., Newsome, S.D., Cook, J.A. et al., 2017. Opinion: Why we need a centralized repository for isotopic data. *Proceedings of the National Academy of Sciences*, 114, 2997–3001.

Peterson, B.J. and Fry, B., 1987. Stable isotopes in ecosystem studies. *Annual Review of Ecology and Systematics*, 18, 293–320.

Pethybridge, H., Choy, C.A., Logan, J.M. et al., 2018. A global meta-analysis of marine predator nitrogen stable isotopes: relationships between trophic structure and environmental conditions. *Global Ecology and Biogeography*, 27, 1043–55.

Phillips, D. and Eldridge, P., 2006. Estimating the timing of diet shifts using stable isotopes. *Oecologia*, 147, 195–203.

Pickard, D., Porter, M., Olson, E. et al., 2015. *Skeena River Estuary Assessment: Technical Report*. Pacific Salmon Foundation, Vancouver, BC.

Pinsky, M.L., Worm, B., Fogarty, M.J. et al., 2013. Marine taxa track local climate velocities. *Science*, 341, 1239–42.

Porch, C.E., Bonhommeau, S., Diaz, G.A. et al., 2019. The journey from overfishing to sustainability for Atlantic bluefin tuna, *Thunnus thynnus*. In B.A. Block, ed. *The Future of Bluefin Tunas: Ecology, Fisheries Management, and Conservation*, pp. 3–44. Johns Hopkins University Press, Baltimore, MD.

Post, D., Layman, C., Arrington, D. et al., 2007. Getting to the fat of the matter: models, methods and assumptions for dealing with lipids in stable isotope analyses. *Oecologia*, 152, 179–89.

Quaeck-Davies, K., Bendall, V.A., MacKenzie, K.M. et al., 2018. Teleost and elasmobranch eye lenses as a target for life-history stable isotope analyses. *PeerJ*, 6, e4883.

Ramos, R. and González-Solís, J., 2012. Trace me if you can: the use of intrinsic biogeochemical markers in marine top predators. *Frontiers in Ecology and the Environment*, 10, 258–66.

Rebenack, J.J., Ricker, S., Anderson, C. et al., 2015. Early emigration of juvenile Coho salmon: implications for population monitoring. *Transactions of the American Fisheries Society*, 144, 163–72.

Rooker, J.R., Fraile, I., Liu, H. et al., 2019. Wide-ranging temporal variation in transoceanic movement and population mixing of bluefin tuna in the North Atlantic Ocean. *Frontiers in Marine Science*, 6, https://doi.org/10.3389/fmars.2019.00398

Rooker, J.R., Secor, D.H., De Metrio, G. et al., 2008. Natal homing and connectivity in Atlantic bluefin tuna populations. *Science*, 322, 742–4.

Rubenstein, D.R. and Hobson, K.A., 2004. From birds to butterflies: animal movement patterns and stable isotopes. *Trends in Ecology & Evolution*, 19, 256–63.

Runge, C.A., Martin, T.G., Possingham, H.P. et al., 2014. Conserving mobile species. *Frontiers in Ecology and the Environment*, 12, 395–402.

Shipley, O.N., Olin, J.A., Power, M. et al., 2019. Questioning assumptions of trophic behavior in a broadly ranging marine predator guild. *Ecography*, 42, 1037–49.

Simmons, R.K., Quinn, T.P., Seeb, L.W. et al, 2013. Role of estuarine rearing for sockeye salmon in Alaska (USA). *Marine Ecology Progress Series*, 481, 211–23.

Sturrock, A., Trueman, C., Darnaude, A., and Hunter, E., 2012. Can otolith elemental chemistry retrospectively track migrations in fully marine fishes? *Journal of Fish Biology*, 81, 766–95.

Sweeting, C., Polunin, N., and Jennings, S., 2006. Effects of chemical lipid extraction and arithmetic lipid correction on stable isotope ratios of fish tissues. *Rapid Communications in Mass Spectrometry*, 20, 595–601.

Tawa, A., Ishihara, T., Uematsu, Y. et al., 2017. Evidence of westward transoceanic migration of Pacific bluefin tuna in the Sea of Japan based on stable isotope analysis. *Marine Biology*, 4, 94.

Thomas, S.M. and Crowther, T.W., 2015. Predicting rates of isotopic turnover across the animal kingdom: a synthesis of existing data. *Journal of Animal Ecology*, 84, 861–70.

Thorpe, J., 1994. Salmonid fishes and the estuarine environment. *Estuaries*, 17, 76–93.

Trueman, C.N., Jackson, A.L., Chadwick, K.S. et al., 2019. Combining simulation modeling and stable isotope analyses to reconstruct the last known movements of one of nature's giants. *PeerJ*, 7, e7912.

Vander Zanden, M.J., Clayton, M.K., Moody, E.K. et al., 2015. Stable isotope turnover and half-life in animal tissues: a literature synthesis. *PLoS ONE*, 10, e0116182.

Watanabe, Y.Y., Goldman, K.J., Caselle, J.E. et al., 2015. Comparative analyses of animal-tracking data reveal ecological significance of endothermy in fishes. *Proceedings of the National Academy of Sciences*, 112, 6104–9.

Weitkamp, L.A., Goulette, G., Hawkes, J. et al., 2014. Juvenile salmon in estuaries: comparisons between North American Atlantic and Pacific salmon populations. *Reviews in Fish Biology and Fisheries*, 24, 713–36.

West, J.B., Bowen, G.J., Cerling, T.E., and Ehleringer, J.R., 2006. Stable isotopes as one of nature's ecological recorders. *Trends in Ecology & Evolution*, 21, 408–14.

Whitlock, R.E., Hazen, E.L., Walli, A. et al., 2015. Direct quantification of energy intake in an apex marine predator suggests physiology is a key driver of migrations. *Science Advances*, 1, e1400270.

Table References

Alves-Stanley, C.D. and Worthy, G.A., 2009. Carbon and nitrogen stable isotope turnover rates and diet–tissue discrimination in Florida manatees (*Trichechus manatus latirostris*). *Journal of Experimental Biology*, 212, 2349–2355.

Ankjærø, T., Christensen, J.T., and Grønkjær, P., 2012. Tissue-specific turnover rates and trophic enrichment of stable N and C isotopes in juvenile Atlantic cod *Gadus morhua* fed three different diets. *Marine Ecology Progress Series*, 461, 197–209.

Barquete, V., Strauss, V., and Ryan, P.G., 2013. Stable isotope turnover in blood and claws: a case study in captive African penguins. *Journal of Experimental Marine Biology and Ecology*, 448, 121–127.

Bearhop, S., Waldron, S., Votier, S.C., and Furness, R.W., 2002. Factors that influence assimilation rates and fractionation of nitrogen and carbon stable isotopes in avian blood and feathers. *Physiological and Biochemical Zoology*, 75, 451–458.

Bosley, K.L., Witting, D.A., Chambers, R.C., and Wainright, S.C., 2002. Estimating turnover rates of carbon and nitrogen in recently metamorphosed winter flounder *Pseudopleuronectes americanus* with stable isotopes. *Marine Ecology Progress Series*, 236, 233–240.

Browning, N.E., Dold, C., Jack, I.-F., and Worthy, G.A., 2014. Isotope turnover rates and diet–tissue discrimination in skin of ex situ bottlenose dolphins (*Tursiops truncatus*). *Journal of Experimental Biology*, 217, 214–221.

Buchheister, A., and Latour, R.J., 2010. Turnover and fractionation of carbon and nitrogen stable isotopes in tissues of a migratory coastal predator, summer flounder (*Paralichthys dentatus*). *Canadian Journal of Fisheries and Aquatic Sciences*, 67, 445–461.

Busquets-Vass, G., Newsome, S.D., Calambokidis, J. et al., 2017. Estimating blue whale skin isotopic incorporation rates and baleen growth rates: Implications for assessing diet and movement patterns in mysticetes. *PLoS ONE*, 12, e0177880.

Caut, S., 2013. Isotope incorporation in broad-snouted caimans (crocodilians). *Biology Open*, 2, 629–634.

Caut, S., Jowers, M.J., Michel, L. et al., 2013. Diet- and tissue-specific incorporation of isotopes in the shark *Scyliorhinus stellaris*, a North Sea mesopredator. *Marine Ecology Progress Series*, 492, 185–198.

Caut, S., Laran, S., Garcia-Hartmann, E., and Das, K., 2011. Stable isotopes of captive cetaceans (killer whales and bottlenose dolphins). *Journal of Experimental Biology*, 214, 538–545.

Gamboa-Delgado, J., Cañavate, J.P., Zerolo, R., and Le Vay, L., 2008. Natural carbon stable isotope ratios as indicators of the relative contribution of live and inert diets to growth in larval Senegalese sole (*Solea senegalensis*). *Aquaculture*, 280, 190–197.

Guelinckx, J., Maes, J., Van Den Driessche, P. et al., 2007. Changes in $\delta^{13}C$ and $\delta^{15}N$ in different tissues of juvenile sand goby *Pomatoschistus minutus*: a laboratory diet-switch experiment. *Marine Ecology Progress Series, 341,* 205–215.

Heady, W.N., and Moore, J.W., 2013. Tissue turnover and stable isotope clocks to quantify resource shifts in anadromous rainbow trout. *Oecologia, 172,* 21–34.

Herzka, S.Z., and Holt, G.J., 2000. Changes in isotopic composition of red drum (Sciaenops ocellatus) larvae in response to dietary shifts: potential applications to settlement studies. *Canadian Journal of Fisheries and Aquatic Sciences, 57,* 137–147.

Hesslein, R.H., Hallard, K.A., and Ramlal, P., 1993. Replacement of sulfur, carbon, and nitrogen in tissue of growing broad whitefish (*Coregonus nasus*) in response to a change in diet traced by δ^{34}S, δ^{13}C, and δ^{15}N. *Canadian Journal of Fisheries and Aquatic Sciences, 50,* 2071–2076.

Kim, S.L., Martínez del Rio, C., Casper, D., and Koch, P.L., 2012. Isotopic incorporation rates for shark tissues from a long-term captive feeding study. *The Journal of Experimental Biology,* 2495–2500.

Logan, J., and Lutcavage, M., 2010. Stable isotope dynamics in elasmobranch fishes. *Hydrobiologia, 644,* 231–244.

Madigan, D.J., Litvin, S.Y., Popp, B.N. et al., 2012. Tissue turnover rates and isotopic trophic discrimination factors in the endothermic teleost, Pacific bluefin tuna (*Thunnus orientalis*). *PLoS ONE, 7,* e49220.

Madigan, D.J., Snodgrass, O.E., Hyde, J.R., and Dewar, H., in review. Stable isotope turnover rates and fractionation in captive California yellowtail (*Seriola dorsalis*): insights for application to field studies. *Scientific Reports.*

Matley, J., Fisk, A.T., Tobin, A.J. et al., 2016. Diet-tissue discrimination factors and turnover of carbon and nitrogen stable isotopes in tissues of an adult predatory coral reef fish, *Plectropomus leopardus. Rapid Communications in Mass Spectrometry, 30,* 29–44.

Miller, T.W., 2006. Tissue-specific response of δ^{15}N in adult Pacific herring (*Clupea pallasi*) following an isotopic shift in diet. *Environmental Biology of Fishes, 76,* 177–189.

Mont'Alverne, R., Jardine, T.D., Pereyra, P.E. et al., 2016. Elemental turnover rates and isotopic discrimination in a euryhaline fish reared under different salinities: implications for movement studies. *Journal of Experimental Marine Biology and Ecology, 480,* 36–44.

Nelson, J., Chanton, J., Coleman, F., and Koenig, C., 2011. Patterns of stable carbon isotope turnover in gag, *Mycteroperca microlepis,* an economically important marine piscivore determined with a non-lethal surgical biopsy procedure. *Environmental Biology of Fishes, 90,* 243–252.

Ogden, L.J.E., Hobson, K.A., and Lank, D.B., 2004. Blood isotopic (δ13C and δ15N) turnover and diet-tissue frac-tionation factors in captive dunlin (*Calidris alpina pacifica*). *The Auk, 121,* 170–177.

Oliveira, M., Mont'Alverne, R., Sampaio, L. et al., 2017. Elemental turnover rates and trophic discrimination in juvenile Lebranche mullet *Mugil liza* under experimental conditions. *Journal of Fish Biology, 91,* 1241–1249.

Reich, K.J., Bjorndal, K.A., and Del Rio, C.M., 2008. Effects of growth and tissue type on the kinetics of ^{13}C and ^{15}N incorporation in a rapidly growing ectotherm. *Oecologia, 155,* 651–663.

Rosenblatt, A.E., and Heithaus, M.R., 2013. Slow isotope turnover rates and low discrimination values in the American alligator: implications for interpretation of ectotherm stable isotope data. *Physiological and Biochemical Zoology, 86,* 137–148.

Sakano, H., Fujiwara, E., Nohara, S., and Ueda, H., 2005. Estimation of nitrogen stable isotope turnover rate of *Oncorhynchus nerka. Environmental Biology of Fishes, 72,* 13–18.

St Clair, K.I. 2014. Stable isotope dynamics in cownose rays (*Rhinoptera bonasus*) within the Northwestern Gulf of Mexico.

Suring, E., and Wing, S.R., 2009. Isotopic turnover rate and fractionation in multiple tissues of red rock lobster (*Jasus edwardsii*) and blue cod (*Parapercis colias*): consequences for ecological studies. *Journal of Experimental Marine Biology and Ecology, 370,* 56–63.

Suzuki, K.W., Kasai, A., Nakayama, K., and Tanaka, M., 2005. Differential isotopic enrichment and half-life among tissues in Japanese temperate bass (*Lateolabrax japonicus*) juveniles: implications for analyzing migration. *Canadian Journal of Fisheries and Aquatic Sciences, 62,* 671–678.

Sweeting, C., Jennings, S., and Polunin, N., 2005. Variance in isotopic signatures as a descriptor of tissue turnover and degree of omnivory. *Functional Ecology,* 777–784.

Tominaga, O., Uno, N., and Seikai, T., 2003. Influence of diet shift from formulated feed to live mysids on the carbon and nitrogen stable isotope ratio (δ^{13}C and δ^{15}N) in dorsal muscles of juvenile Japanese flounders, *Paralichthys olivaceus. Aquaculture, 218,* 265–276.

van Gils, J.A., and Salem, M.V.A., 2015. Validating the incorporation of ^{13}C and ^{15}N in a shorebird that consumes an isotopically distinct chemosymbiotic bivalve. *PLoS ONE, 10,* e0140221.

Wang, Y., Gu, X., Zeng, Q. et al., 2016. Nitrogen stable isotope variability in tissues of juvenile tilapia *Oreochromis aureus*: empirical and modelling results. *Rapid Communications in Mass Spectrometry, 30,* 2116–2122.

PART II

Addressing Human–Wildlife Conflicts

CHAPTER 6

Using physiological tools to unlock barriers to fish passage in freshwater ecosystems

Rebecca L. Cramp, Essie M. Rodgers, Christopher Myrick, James Sakker, and Craig E. Franklin

> ⮌ **Take-home message**
>
> Artificial barriers to fish movement have contributed to worldwide freshwater fish population declines. Conventional experimental physiology approaches provide simple, cost-effective tools to assess the behavioural and physiological capacities of fish, which can inform 'fish-friendly' structure design and support remediation practices.

6.1 Introduction

Freshwater environments are widely regarded as some of the most highly threatened ecosystems on the planet. Comprising less than 1 per cent of surface waters, freshwater ecosystems support approximately half of all extant fish species (Reid et al. 2013). Anthropogenic competition for, and misuse of, aquatic resources through excessive waterway regulation, unsustainable water extraction practices, habitat destruction, invasive species, overexploitation, and pollution have led to a precipitous decline in freshwater fish numbers the world over. Indeed, freshwater fish are one of the most threatened vertebrate classes, with approximately one third of assessed species at risk of extinction (Dudgeon et al. 2006; IUCN 2019). In particular, overregulation of waterway flow is a leading cause of global freshwater fish declines. Approximately 60 per cent of the world's largest rivers are classified as highly fragmented and many no longer maintain connectivity with the sea (Reid et al. 2013; Grill et al. 2019). In fact, it is estimated that >98 per cent of waterway reaches in the continental United States are affected by human activities (Palmer et al. 2007; Grill et al. 2019). However, unimpeded waterway connectivity is essential for obligate migratory fish like salmon and eels that complete directional, long-distance movements. Further, all freshwater fish move to some degree within catchments to complete life cycles, select preferred habitats for foraging or spawning, and to maintain genetic flow between populations (Fischer and Lindenmayer 2007; Coleman et al. 2018). For this reason, artificial barriers to fish movement such as dams, weirs, culverts, altered flow regimes, water extraction practices, and pollution can lead to population fragmentation and decline even in non-migratory

Rebecca L. Cramp, Essie M. Rodgers, Christopher Myrick, James Sakker, and Craig E. Franklin, *Using physiological tools to unlock barriers to fish passage in freshwater ecosystems* In: *Conservation Physiology: Applications for Wildlife Conservation and Management.* Edited by: Christine L. Madliger, Craig E. Franklin, Oliver P. Love, and Steven J. Cooke, Oxford University Press (2021). © Oxford University Press.DOI: 10.1093/oso/9780198843610.003.0006 species (Paul and Meyer 2001; Gibson et al. 2005; Harris et al. 2017; Figure 6.1).

Structures become barriers to fish movement when they: (1) physically block fish movement up- or downstream; (2) alter the hydrodynamic conditions within a structure such that they impede and exceed fish swimming capacities, or disorient and confuse fish; (3) reduce the behavioural motivation of fish to move into the modified environment; or

(4) affect the capacity of fish to perform at an optimal level of swimming performance. Dams and weirs can physically block adult or juvenile upstream and downstream migration routes (Dugan et al. 2010) and impede the successful downstream drift of eggs and larvae (Humphries and Lake 2000; Pelicice et al. 2015). Downstream movements often require that fish (and eggs and larvae) move through hydroelectric turbines, over spillways, or via deep

(a)   (b)   (c)
(d)   (e)   (f)

Figure 6.1 Common barriers to fish passage and remediation strategies. Dams (A), causeways (B), and culverts (C) are some of the most common barriers to freshwater fish passage in waterways globally. Given the importance of fish movement for the maintenance of ecosystem health and biodiversity, remediation strategies such as vertical slot fishways (D), baffles (E), and naturalistic roughening (F) can be deployed to improve fish passage. The design and implementation of successful remediation strategies can be significantly improved through the use of physiological metrics and experimental approaches that quantify the physical, physiological, and behavioural limitations of fish using the structure. Photo credits: A, C—Rebecca Cramp; B, E, and F—Jabin Watson, and D—DPI NSW Fisheries.

water-regulating outlets, all of which can substantially reduce fish survival (Keefer et al. 2013). Dams can also affect water quality downstream, particularly when unseasonably cold water is released from deep dam outlets (thermal pollution) (Whiterod et al. 2018), or when eutrophic, algae-rich waters are released from upper outlets (Ling et al. 2016). Every place that a road crosses a small stream, whether permanent or ephemeral, culverts are used, and despite their relatively small size, road-crossing culverts can have substantial impacts on fish passage (Warren and Pardew 1998; Bouska and Paukert 2010). Culverts block fish movement when they are perched above the downstream waters or when blocked by debris or sediment. They can form hydrological impediments when stream flows are constricted by small culverts (Rodgers et al. 2014) or when they are installed on a steep grade (Baker and Boubee 2006). Culverts can also be behavioural barriers if they limit light levels inside the structure (Amtstaetter et al. 2017) or disrupt natural streambed composition (Katopodis 2005). Barriers to fish movement may also occur when agricultural runoff, sewage, or industrial waste is introduced into the migratory path of fish (Thorstad et al. 2008).

The need to remediate existing structures or to design novel structures that take into consideration the movement requirements of aquatic fauna is widely recognized (Bernhardt et al. 2005; Bernhardt and Palmer 2007). Billions of dollars every year are spent globally in an effort to restore connectivity to waterways and enhance ecosystem services (Bernhardt et al. 2005; Lapointe et al. 2013). However, designing structures that provide effective fish passage requires a thorough understanding of the ecology, behaviour, and locomotory capacities of the fish using the structures (Palmer et al. 2007; Lapointe et al. 2013). Empirical experimental fish swimming performance studies can provide important baseline data that can be used by managers to guide the design of fish passage structures, or to provide evidence-backed advice on appropriate water use practices to limit impacts on fish populations (Cooke et al. 2017). Moreover, experimental studies are uniquely positioned to provide proof of cause-and-effect relationships through careful manipulation of the factors of interest, while

controlling for the influences of other potential factors (Cooke et al. 2017). Predicted fish responses to the hydrodynamic, thermal, and physiochemical conditions generated by instream barriers can be modelled using baseline fish performance data, and then solutions tested in an iterative, controlled, and cost-effective manner in the laboratory or field environment. In the following section, we explore some of the common tools used to measure indices of fish physiological performance and how they are currently being used to inform fish passage requirements.

6.2 Physiological tools

6.2.1 Fish swimming performance

Whole-animal physiological performance metrics provide key information on the physical limitations of fish in a given environment. Because movement directly underpins key behavioural responses to predator threats, food acquisition, and reproduction, fish swimming performance is widely regarded as an important fitness metric (Hammer 1995; Brownscombe et al. 2017). Fish swimming performance tests have been used in comparative animal physiology for many decades because they provide useful information on locomotion, muscle physiology, kinematics, gas exchange, and the sensitivity of performance to external environmental factors (e.g. temperature, hypoxia, and aquatic pH). Fish utilize a range of swimming gaits including 'burst' (high swimming speeds, that can only be sustained for <30 s), 'prolonged' (moderate speeds, sustained for up to 200 min), and 'sustained' (cruising gaits and slow speeds, that can be sustained for >200 min) (Beamish 1978). Standardized performance tests can estimate the performance capacity of fish utilizing these different swimming modes.

Prolonged (or maximal sustainable) swimming speed (U_{crit}) is perhaps the most widely used and reported physiological performance metric of fish. Initially described by Brett (1964), U_{crit} tests employ a graded (ramped) velocity protocol to determine the maximum velocity that a fish can sustain, primarily aerobically, for periods of up to 200 min.

In the classic U_{crit} test (Brett 1964), individual fish are placed into a swim tunnel (Figure 6.2A, B) at a low velocity (0.1–1 body lengths [BL]/s). Water velocities are then increased stepwise (by ~0.5–1 BL/s) at set time increments (of between 5 and 60 min) until the fish fatigues. U_{crit} estimates are highly sensitive to variations in protocol (i.e. relative velocity increases, pre-test training/acclimation, velocity increment length; Brett 1967) and test equipment (Kern et al. 2018), as well as other performance variables (body mass, temperature, developmental stage, wild-caught vs. hatchery stock, etc.) meaning that U_{crit} estimates can vary enormously even within a species (Kern et al. 2018). Fixed velocity or endurance tests estimate prolonged swimming capacities by measuring the time over which fish can sustain performance at a fixed velocity (Rodgers et al. 2014). Endurance tests require fish to be swum across a range of different but fixed water velocities, which is both time-consuming and animal-intensive, making it a less commonly measured trait (Plaut, 2001). Maximal swim speed (burst speed) tests often use a ramped velocity approach similar to U_{crit} but with a much reduced time increment (i.e. 15–30 s) (Starrs et al. 2011). Termed U_{sprint} or U_{burst}, these approaches estimate the fastest swim speeds achievable over a truncated timeframe (less than a few minutes typically). Burst swimming capacity (as maximum velocity and/or acceleration) can also be estimated by startling fish to swim through a beam of photocells (Reidy et al. 2000) or by recording startle responses with a high-speed camera and analysing the velocity of the positional shifts over the movement (termed 'C-start') (Domenici and Blake 1997). Burst swimming is often exploited by fish negotiating fish passage structures such as vertical slot fish ways and baffles within culverts (Castro-Santos 2005; Chung 2006; Figure 6.1D–F).

Structure traversability estimates provide useful data on the ability of a fish to cover ground against the direction of water flow, using whichever mode of swimming they choose. Traversability or passage success indices record the capacity of fish to move a predefined distance in a swimming environment, usually against the flow of water. These tests are generally conducted in a simulated fish passage environment such as in a long flume with modified slope, roughness elements, water velocities or turbulence, a model fishway, or simulated perched culvert (Figure 6.2C, D) (e.g. Mallen-Cooper 1992; Castro-Santos 2005; Bestgen et al. 2010; Cocherell et al. 2011; Dockery et al. 2017; Watson et al. 2018). Laboratory-based traversability tests enable the hydraulic environment to be simplified and adjusted in a consistent and controllable manner in order to assess the impacts of a particular feature(s) on the capacity of fish to move through the environment. These tests are predominantly volitional so potentially provide a more ecologically realistic measure of fish performance and behaviour in a given hydrological environment. Also, because these tests require that fish physically cover ground, they provide a more accurate estimate of potential passage success unlike the stationary, non-volitional (involuntary) performance tests U_{crit} or endurance. Measures of traversability do however require substantial infrastructure, e.g. 8–12-m-long channels such that the fish is required to cover a realistic span/distance.

6.2.2 Energetic cost of locomotion

Metabolic rate is a measure of the rate at which stored energy is converted to work energy by an animal. Since energy budgets constrain performance, factors that influence the way energy is spent or acquired can have significant impacts on the scope of activities that can be undertaken by an animal (Portner and Peck 2010). For the purposes of understanding how barriers to fish passage can be addressed, the energy cost of locomotion by fish is a key physiological metric that has been used to inform remediation efforts (e.g. Reidy et al. 2000; Geist et al. 2003; Peake and Farrell 2004; Geist et al. 2005; He et al. 2013; Cai et al. 2015; Alexandre and Palstra 2017). The rate at which oxygen is used by an animal is proportional to its metabolic rate, so oxygen consumption rates are a widely used proxy for metabolic rate and estimates of the energetic cost of performance (Chabot 2016). Given that measures of oxygen consumption rate require fish to be confined within a sealed chamber, they are impossible to use to quantify energy usage costs in movement studies in free-ranging animals. However, combining fish swimming performance

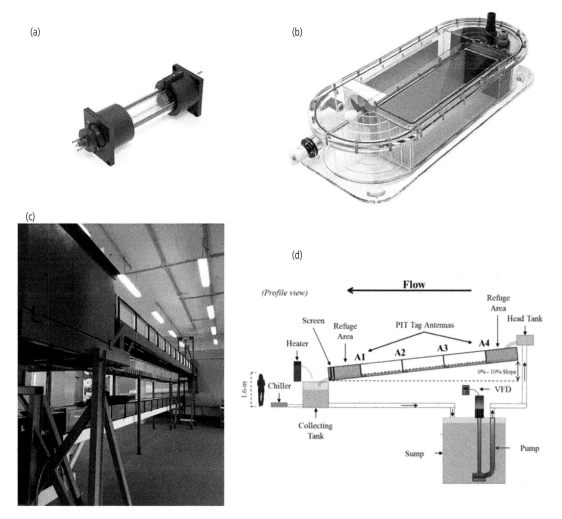

Figure 6.2 Swimming flumes for the measurement of fish swimming performance. (A) Blazka-type and (B) Steffensen-type recirculating swimming tunnels that double as respirometers for the simultaneous measurement of oxygen consumption rate (Loligo®Systems); (C) custom flume system at the University of Queensland, two gravity-fed swim channels 12 × 0.5 × 0.5 m, adjustable slope to 6°, inlet tank at the top of each flume with turbulence dampening structures, external 40 000-l header tanks; (D) custom flume system at Colorado State University.

metrics with oxygen consumption measurements in sealed swimming performance flumes can provide very valuable information on the energy cost of swimming in actively moving fish (Claireaux et al. 2006; Palstra et al. 2008; Hein and Keirsted 2012). Increasingly, the rate at which oxygen is supplied to tissues by the cardiovascular system (electrocardiogram) and used by tissues to generate electrical activity (electromyography) are being measured remotely (e.g. heartrate biologgers, accelerometer transmitters), to estimate energy use rates in

actively swimming or wild fish (e.g. Hinch et al. 1996; Geist et al. 2003; Geist et al. 2005; Eliason et al. 2013; Metcalfe et al. 2016; Brownscombe et al. 2017).

Fish swimming at sustained or prolonged swimming speeds tend to rely predominately on aerobic metabolism to fuel activities (until they approach fatigue), whereas fish moving at burst speeds rely more heavily on anaerobic metabolism (Lee et al. 2003). Quantifying anaerobic metabolism during activity is challenging since it requires measuring the by-products of anaerobic metabolism via invasive biochemical monitoring. Anaerobic metabolism is estimated by monitoring post-exercise blood lactate and pH levels or through muscle enzyme activities (e.g. lactate dehydrogenase and citrate synthase) (Reidy et al. 2000; Di Santo et al. 2017). Additionally, lactate accumulated during anaerobic activity is cleared by oxidation post-activity (i.e. exercise post oxygen consumption, EPOC) so monitoring post-exercise oxygen consumption patterns can provide some indication of the degree to which anaerobic metabolism was used during the preceding activity (Lee et al. 2003). Understanding how animals utilize anaerobic metabolism and the costs and trade-offs that result, can help inform fish passage designs. Several fish passage designs, such as vertical slot fishways and baffles in culverts, exploit fish burst swimming capacity, but also recognize the need to provide recovery points through the structure to enable fish to recover before proceeding.

6.2.3 Blood chemistry

Fish physiological performance is dependent on the ability of the cardiovascular system to supply oxygen to active tissues, so is directly proportional to blood oxygen-carrying capacity. Blood-based physiological metrics like haemoglobin levels, glucose and lactate levels, and stress markers (i.e. blood cortisol levels) have become important tools for understanding how fish performance is constrained by physiological limitations (e.g. Leonard and McCormick 1999; Cocherell et al. 2011; Hatry et al. 2014). Small blood samples can provide a relatively large amount of information that can explain/predict/reflect performance in fish. Red blood cells are responsible for carrying oxygen to the tissues and can be used to predict oxygen-limited performance capacity (Lay and Baldwin 1999; Leonard and McCormick 1999). Blood oxygen-carrying capacity can be coarsely estimated by measuring haematocrit (the proportion of the total blood sample occupied by red blood cells) and/or blood haemoglobin levels. Metabolic performance (demand) can also be estimated by examining blood glucose and/or lactate levels since glucose is the primary metabolic fuel and lactate is an anaerobic metabolism marker. Elevated levels of plasma lactate can be a negative

predictor of migration success in salmon (Young et al. 2006; Donaldson et al. 2010) while glucose levels can decline when fish perform (swim) at the upper limits of their capacity (Choi and Weber 2016).

6.2.4 Biologging

Remote monitoring of physiological or performance traits such as acceleration, muscle activity, heart rate, and body temperature enables physiological metrics to be captured from wild, free-ranging animals. Electronic tracking tools can also be used to understand how environmental conditions (depth, oxygen levels, temperature, salinity) as well as anthropogenic disturbances (e.g. dams, weirs, thermal pollution, etc.; Wilson et al. 2015) influence fish performance (Lapointe et al. 2013). Telemetry has been widely used to record free-ranging physiological function in fish moving through fish passage structures (e.g. Burnett et al. 2014a, b; Lothian et al. 2019). Telemetric tags are generally either surgically implanted into the body cavity (Jepsen et al. 2002) or attached externally to the fish (e.g. Brownscombe et al. 2018), and data are either logged directly to the tag or transmitted to a remote receiver via radio or acoustic signal or via satellite (Wilson et al. 2015). Biologging is a valuable tool in the conservation physiologist's toolbox because, unlike laboratory studies, it provides 'real-world' information on fish behaviour and performance that can be used to inform fish passage management strategies (Cooke et al. 2013).

Empirical physiological studies have been used for many decades to inform fish passage structure design and provide evidence-backed water management advice. They remain a critical component of fish passage management activities worldwide (Silva et al. 2018). In the following section, we present four cases studies that utilize empirical physiological approaches and tools to address current management issues for fish passage.

6.3 Case study 1: culverts vs. bridges in the Pacific Highway upgrade

The New South Wales (NSW) and Australian Governments are currently (2015 – present) undertaking a major upgrade of the nation's most significant road, the Pacific Highway, between Woolgoolga and Ballina on the north coast of NSW. The upgrade consisted of 120 km of dual carriageway (roadway) with some sections in undisturbed sites. Road crossings were required to traverse numerous waterways with major bridges spanning the Richmond and Clarence Rivers, smaller concrete plank bridges for creeks, and box or pipe culverts on minor drainage lines. Aquatic fauna connectivity is a key challenge to manage on a linear project of this scale. Prior to the design phase, 3 years of pre-construction fish monitoring was conducted in waterways believed to contain threatened species and a review of previous monitoring in the area was also examined. Fish species in the freshwater systems were mainly small-bodied native fish, freshwater eels, Australian bass, and one introduced pest species *Gambusia holbrooki*. One nationally significant fish species, Oxleyan pygmy perch (OPP, *Nannoperca oxleyana*) was found in 10 creeks. This species is listed federally as 'endangered' as a consequence of extensive habitat degradation and loss. Little is known about the swimming ability of OPP, but in wet years, OPP are known to move upstream into intermittent creeks in forest areas that drain into the floodplain. Five of the required road crossings for OPP were in greenfield sites that previously had no physical barrier to fish passage. Given the potential for these road crossings to impact the passage of OPP and other similarly sized native fish in the surrounding sites, information on fish swimming performance was required to guide decisions regarding the type of road crossing to be installed in significant areas.

To provide evidence-based guidelines for the provision of fish-friendly road-crossing structures along the Pacific Highway upgrade, The University of Queensland (UQ), Fisheries NSW, and Roads and Maritime Service (RMS) measured the swimming performance of over 20 Australian freshwater fish species that varied in size, shape, swimming zone (benthic or pelagic), migration requirements (meso- or macro-scale), and swimming mode (burst or sustained). Prior to this research, NSW Department of Primary Industries (DPI) fish passage guidelines recommended 0.3 m/s as the optimal water velocity through culvert structures. However, this figure was based on largely anecdotal observations of

fish swimming capacity and there was concern that this velocity overestimated the swimming capacities of many of the smallest fish species. Researchers compared U_{crit}, U_{sprint}, endurance, and traversability data across all species. As OPP are endangered, they were not used in this research; however, the closely related southern pygmy perch (*Nannoperca australis*) is a very useful analogue species and these data have been used in the design of OPP crossings. The variety of metrics examined suggested that, for the majority of fish species, the existing guidelines for water velocities through culverts were appropriate. However, for many of the weaker species, including the southern pygmy perch and several gudgeon species, the fish-friendly culvert guidelines overestimated fish swimming capacity and hence culverts were not appropriate in areas where passage for these species was required.

The research data validated the Fisheries NSW decision to insist on bridge crossings on all the greenfield sites and duplicate existing bridges on the current road alignment. Bridges cost more money to install, but they maintain existing waterway conformation, hydrology, and flows so they have minimal impact on fish passage. Velocity modelling, to ensure bridge apertures are large enough to provide fish passage through critical flood events, have since been conducted. The larger aperture also allows edge effects to slow the water velocity to the 0.1 m/s or less required by OPP to transit a structure even during peak flows. While the research was conducted with the aim of giving managers the data they need to ensure appropriate structures are used to enhance fish passage, it has also been used to design a structure to exclude OPP from accessing an intermittent creek. The creek flows into Tabbimoble Swamp Nature Reserve, which has a healthy population of OPP. Above the existing highway, the creek only flows following substantial rainfall and afterwards the remaining pools dry out, killing any stranded fish. To protect this endangered species, OPP were actively translocated below the road-crossing structure prior to construction. A pipe culvert was then installed at the crossing site with a transiting water velocity that is too fast for OPP to swim against, thus restricting their access to the ephemeral upstream sites. This structure is not only better for preventing OPP kills but has also saved taxpayers in excess of AUD$1 000 000. Prior to obtaining the swimming performance research data, and given the presence of OPP, NSW fish passage guidelines would have insisted on a bridge at this road-crossing site. These fish swimming performance data can potentially be used in the future to design structures to prevent pest fish from increasing their range upstream while still performing the required hydrological function. Having accurate performance data allows managers to select the appropriate crossing structure for a site, thus maximizing ecological benefits while minimizing cost.

6.4 Case study 2: protecting fish at water diversion pipes

Undertaking downstream migrations from freshwater nursery habitats towards seawater can be a treacherous journey for anadromous fishes. Juvenile green sturgeon (*Acipenser medirostris*), for example, must migrate past thousands of water diversion pipes that extract river water at rapid velocities to meet agricultural and municipal demands. Water diversion pipes are designed to maximize hydraulic capacity, but they essentially act as 'death traps' for young fish. Once sucked inside a pipe, a process termed 'entrainment', fish are either severely injured by pumps or left stranded in irrigation fields. All entrained fish are lost from the population—a devastating consequence for already threatened species, like the late-maturing, long-lived green sturgeon.

To quantify the susceptibility of juvenile green sturgeon to entrainment in water diversion pipes and uncover conservation solutions, the Fangue Laboratory at the University of California, Davis ran several swimming physiology and behavioural experiments (Mussen et al. 2014). Experiments were held inside a large, river-scale simulation flume containing an unscreened water diversion pipe and water extraction rates were set to match field conditions. Sixty juvenile fish (26–36 weeks old) were trialled and researchers used underwater cameras to record the number of fish swept inside the pipe. They discovered that entrainment rates were extremely high, and fish were often swept inside the pipe within 1 s of passing by. Extrapolations of

these findings estimated that approximately half of outmigrating juveniles would be killed by entrainment if they pass within 1.5 m of three diversion pipes. However, entrainment rates were dramatically reduced (78 per cent reduction) by lowering the speed at which water is extracted (from 0.57 m^3/s to 0.28 m^3/s), suggesting that limiting water extraction rates may be a successful management strategy.

Building on this, the researchers then characterized the swimming capacities of larval and juvenile green sturgeon, by measuring critical swimming speeds (U_{crit}) (Verhille et al. 2014). Absolute critical swimming speeds increased with fish size and age, but 50-cm-long (total length) juveniles were an exception to this trend. The researchers noted that juveniles of this size typically develop saltwater tolerance in preparation for outmigration, and this decline in swimming performance likely represented an energetic trade-off between physiological remodelling required for seawater acclimation, and locomotor performance, and this finding has been corroborated by others (Allen et al. 2006). The discovery of ontogenetic changes in swimming performance highlighted the need to specify limits on water diversion rates according to season and location within the river. For example, larval fish were identified as the weakest swimmers and it was recommended that water extraction speeds within critical nursery regions of the Sacramento River be restricted to ≤0.29 m/s during peak rearing times.

The Fangue Laboratory also investigated if installing deterrents or fish-protection devices at water diversion pipes could lower fish entrainment rates without reducing hydraulic efficiency (Poletto et al. 2014a, b). The efficacy of two sensory deterrents, a strobe light and acoustic vibrations, in eliciting avoidance behaviour and minimizing entrainment rates were assessed in juvenile green sturgeon. Both sensory deterrents proved ineffective, despite reports of success in other species (Maes et al. 2004; Hamel et al. 2008). The efficacy of several physical fish-protection devices in lowering entrainment rates was also assessed. Retrofit designs included: a modified upturned pipe configuration and the addition of a protective screen, a louvre box, a trash-rack box, or a perforated cylinder (Figure 6.3). All retrofit designs significantly reduced entrainment rates by 60–96 per cent. For example, the addition of a protective screen lowered entrainment rates from 44 per cent in an unmodified pipe to 13 per cent; however, fish can become impinged on screens, causing injuries and heightened levels of physiological stress (e.g. acidosis from exhaustion; Young et al. 2010). Alternatively, the upturned pipe configuration allowed water to be withdrawn from the middle of the water column rather than the bottom. This retrofit design was particularly effective at reducing entrainment rates, likely owing to the benthic behaviour of green sturgeon. The other retrofit designs (i.e. louvre box, trash-rack box, and perforated cylinder) also

reduced entrainment rates and the effectiveness of these designs was postulated to be linked to physical protection, and inflow velocities being distributed across a greater area, enabling fish to detect flow changes from a greater distance and initiate avoidance behaviour.

This series of experiments showed that fish entrainment in water diversion pipes is a significant source of early-life mortality in green sturgeon. Swimming physiology experiments provided age- and body size-specific estimates of swimming abilities, which allowed the guidance of water diversion inflow limits specific to habitat zones. Moreover,

(a)   (b)

(c)   (d)

Figure 6.3 Underwater images of water diversion pipes fitted with deterrents or modified design elements. The effectiveness of each design was tested in experimental trials by recording the number of juvenile green sturgeon (*Acipenser medirostris*) swept inside pipes. Four treatments were compared: (a) control treatment in which the pipe remained unmodified; (b) strobe light treatment, where lights were placed around the pipe inlet; (c) terminal pipe plate treatment, where a protective steel plate was fitted in front of the pipe inlet; and (d) an upturned pipe treatment, where the orientation of the water inlet was modified by adding an additional component. Figure redrawn from Poletto et al. (2014b).

these experiments identified several fish-protection device designs that significantly lowered entrainment risk, and field testing of these designs is an important next step. Currently, 98 per cent of the 3300 water diversion pipes in the Sacramento River remain unscreened (Herren and Kawasaki 2001). However, this body of research led to water diversion pipes being listed as an official, significant threat in the 2018 Recovery Plan for Green Sturgeon (National Marine Fisheries Service West Coast Region 2018), and three recovery actions were outlined so that operation and screening guidelines ensure that early-life entrainment rates are minimized.

6.5 Case study 3: making the grade—slope considerations inform the length and grade of fish passage structures in Colorado

In Colorado, USA efforts to restore the ecological connectivity of the state's heavily fragmented rivers and streams by installing fish passage structures have been underway since the early 2000s. These efforts have largely concentrated on the rivers that flow eastwards from the Rocky Mountains onto Colorado's eastern plains, where they eventually connect to the Platte River or Arkansas River systems. The fishes in these systems are generally small-bodied (30 cm in total length or less as adults) and include members of the minnow (*Cyprinidae*), darter (*Percidae*), sucker (*Catostomidae*), and North American catfish (*Ictaluridae*) families. Because these fish are not game species, little was known of their swimming performances, how those would relate to fish passage, or what types of fish passage structures or fishways would be most appropriate.

Early studies at the Colorado State University Fish Physiological Ecology Laboratory (FPEL) on four representative species of native fishes, namely Arkansas darters (*Etheostoma cragini*), brassy minnows (*Hybognathus hankinsoni*), flathead chub (*Platygobio gracilis*), and common shiners (*Luxilus cornutus*), demonstrated that while most of these species have limited jumping abilities and thus would not perform well in a pool-and-weir structure, their swimming abilities would allow them to negotiate other types of fish passage structures such as rock

ramps or vertical slot fishways. These early studies used small (<2 m long) recirculating swimming flumes and forced-swimming approaches to develop swimming performance curves relating swimming velocity to endurance. These curves provided natural resource managers in Colorado with data that were used in the design of at least one large rock ramp fishway, located at the Owens-Hall Diversion on Fountain Creek, a tributary of the Arkansas River. However, because the fish were not swimming volitionally during the creation of the curves, it remained unclear whether the results were truly representative of fish performance in actual rock ramp fishways.

To address this challenge, the FPEL designed and built a 10-m-long × 1.4-m-wide research flume and constructed a 6-m-long rock ramp fishway within the flume. The rock ramp fishway was fitted with four passive integrated transponder (PIT) tag antennas that allowed researchers to track fish movements within the structure, and the whole flume could be raised from 0 to 10 per cent slope to measure the effects of slope on the passage success of representative species of plains fishes. Unlike the small recirculating flumes used in prior studies, fish could attempt to ascend the fishway volitionally and, by tracking their movements with implanted PIT tags, researchers could determine the probability of the fish travelling different distances in the flume as a function of slope.

The first study using this system measured the effects of slopes of 2, 4, 6, 8, and 10 per cent on the passage success and distances covered of flathead chub, Arkansas darters, and stonecats (*Noturus flavus*). The results clearly showed that there are length and slope combinations that have a high probability of allowing passage of the target species (Figure 6.4). Natural resource managers can use these data to identify fishway design parameters that allow an acceptable proportion of the populations to move through the fishways. In situations where restoring the ecological connectivity of a fish community is the desired outcome (in contrast to single-species management efforts), this approach can be used to set passage guidelines that will benefit the species with the lowest passage performance. In the case study described here, Arkansas darters are the species that drive the design of structures, so structures that have an acceptable probability of

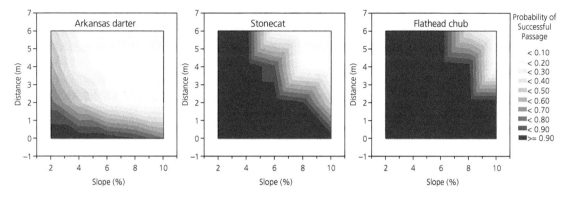

Figure 6.4 Effects of fishway slope and length on the probability of successful passage of three fish from rivers flowing onto Colorado's eastern plains. The darker areas indicate combinations of fishway slope and distance between resting areas (or absolute distance) that have the highest probability of allowing successful passage of the species in question. The Arkansas darter is the weakest swimmer, the stonecat is an intermediate swimmer, and the flathead chub is the strongest swimmer, and thus has the widest range of distance × slope combinations with a high probability of success.

allowing darter passage will also allow passage of the stronger-swimming species (i.e. if the probability is set at 0.40, then fishways should have segments no longer than 2 m at a slope of 4 per cent or less between resting areas).

As efforts continue to restore the connectivity of rivers throughout Colorado and the US Great Plains, the approach described in this case study will prove invaluable in allowing researchers to efficiently test the passage success of the species of interest in response to various fishway design parameters, including slope, placement and size of roughness elements, and substrate texture. Doing so can potentially save substantial amounts of money through the iterative and laboratory-driven test process before actual fishways are constructed on the landscape.

6.6 Case study 4: cold-water pollution—can physiological plasticity mitigate the effects of thermal pollution on Australian fish?

The Murray-Darling Basin (MDB) is the largest river system in Australia and one of the largest freshwater systems in the world, covering over 1 million km² and comprising more than 77 000 km of waterways. Waterway regulation through the installation of large instream barriers to fish movement (weirs and dams) has contributed signifi-

cantly to the massive decline of fish numbers in the MDB. Dams and weirs can also substantially affect the thermal regimes in downstream waterways. In particular, cold water released from the base of thermally stratified dams can pollute downstream waters for hundreds of kilometres with unseasonably cold water that may be as much as 15°C lower than the upstream water. There are 103 'large' dams and hundreds of smaller dams or weirs in the MDB with the potential to cause thermal pollution in downstream water reaches (Boys et al. 2009; Lugg and Copeland 2014). These water releases often occur in summer to supply agricultural requirements, but this is also a critical time for spawning of key native fish species including the critically endangered Murray cod (*Maccullochella peelii*), silver perch (*Bidyanus bidyanus*) and trout cod (*Maccullochella macquariensis*). Indeed, in some areas of the MDB, water temperatures now seldom approach levels required to trigger spawning of key native fish species (Boys et al. 2009). Presently, no guidelines exist at either the Commonwealth or state levels for ecologically safe water temperature releases from undershot dams to maintain downstream aquatic organism fitness.

The role of temperature as a key determinant of fish performance capacity is well established. As ectotherms, fish physiology and performance is directly determined by surrounding water temperatures (Angilletta 2009). For most species, performance is optimized over a relatively narrow range of environmental temperatures; at temperatures outside of these, performance declines (Guderley and Blier 1988). However, many fish species have a capacity to adjust performance in response to diel or seasonal shifts in temperature, enabling them to maintain performance across a wider range of temperatures (Guderley and Blier 1988). In highly regulated freshwater environments like the MDB, the ability of fish to sustain performance over a large thermal range may determine their capacity to persist and traverse large-scale fish passage structures (fishways or fish passes). Little is known about the thermal physiology of Australian native fish from the MDB and whether they have the capacity to adjust physiological performance following acute reductions in water temperature that occur with large dam releases.

In collaboration with NSW DPI Fisheries, the thermal sensitivities and plasticity of four MDB fish species (silver perch, golden perch [*Macquaria ambigua*], Murray rainbow fish [*Melanotaenia fluviatilis*], and spangled perch [*Leiopotherapon unicolor*]) were experimentally assessed following an acute 10°C reduction in water temperature (from 24°C to 14°C over 12 h). Researchers measured swimming performance (burst [U_{sprint}] and sustained [U_{crit}]) and energy expenditure (routine and maximal metabolic rates) at intervals over the following 10–12 weeks to track acclimation responses. A 185-l recirculating swimming flume was used to measure fish swimming performance and intermittent flow respirometry to measure oxygen consumption. All fish used were less than 10 cm in total length. For juveniles of the larger growing species (perch), growth rates were measured over the 10-week experimental period. For all species, the acute reduction in temperature had a significant effect on swimming and metabolic performance with all traits decreasing by 30–50 per cent. Importantly, performance capacity remained largely unaffected by prolonged exposure to the low temperature in all species. There was evidence of partial acclimation of both swimming and metabolic performance in some species, but none was capable of compete thermal compensation even after 10 weeks of low temperature exposure (Figure 6.5). Growth rates were suppressed in all juvenile fish that were exposed to low water temperatures. These findings

have important implications for the regulation of water flows for irrigation or environmental water, and the management of threatened fish species in the MDB.

The restoration of environmental flows in the MDB is a top national priority, critical to ensuring the protection of the 16 internationally significant (Ramsar Convention on Wetlands of International Importance) sites in the MDB. These data showed that acute water temperature drops of 10°C are too extreme for native fish to contend with. This finding is consistent with others showing that growth and survival of Murray cod and silver perch were significantly impaired by current cold-water release regimes (Astles et al. 2003; Todd et al. 2005; Whiterod et al. 2018). Remediation approaches that reduce the thermal stratification of deep dam impoundments or provide additional higher water offtake sites could reduce thermal pollution impacts on native fish populations. However, these technological solutions require vast amounts of capital outlay and debate continues as to whether they can even achieve the ecological outcomes required (Boys et al. 2009). Better management of existing water release structures underpinned by empirical studies that detail the thermal limitations for performance and reproduction in key fish species could provide a more cost-effective solution. An experimental approach could indicate whether fish have the physiological flexibility to contend with large water temperature drops if the rate of temperature change is reduced (i.e. over weeks instead of days), or if smaller temperature drops are better tolerated than 10°C drops.

6.7 Conclusions and future directions

A basic understanding of the ecology of fish, including their capacity to adjust to anthropogenic disturbances in both the short and longer terms, is important to effectively manage fish passage requirements in an increasingly fragmented landscape. The utility of experimental physiology means that the environmental boundaries within which fish operate can be determined from the level of a single stressor (e.g. water velocity) to more ecologically realistic multi-stressor studies (e.g. water movement through vertical slot fish way). Rapidly

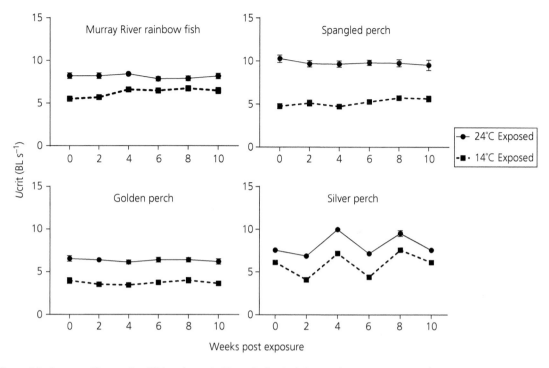

Figure 6.5 Exposure of four species of fish, native to the Murray-Darling Basin in Australia, to an acute 10°C reduction in water temperature (from 24°C to 14°C) had a marked and sustained impact on maximum sustainable fish swimming performance (U_{crit}; BL = body lengths). Over the 10-week exposure period, juvenile fish (spangled, silver, and golden perch) experienced no growth. There was evidence of partial thermal compensation of swimming performance in Murray River rainbow fish and silver perch, but none of the species was able to completely thermally acclimate performance to the cooler temperature.

advancing technologies have meant that researchers can now monitor both fish physiology and environmental characteristics remotely and, in doing so, get unprecedented access to the day-to-day lives of fish as they negotiate natural and anthropogenic barriers. By combining laboratory-based physiological studies with remote animal physiology monitoring, we are developing a greater understanding of how our water infrastructure and management practices affect fish movements, and in turn, influence fish populations. Although fish passage management in general has been relatively good at utilizing data from scientific research to inform on-the-ground actions, much remains to be done to improve the usefulness, accessibility, and incorporation of primary fish physiology research into management actions.

One of the largest impediments to incorporating laboratory-based physiological performance studies into on-the-ground management actions is their perceived irrelevance to fish movements through artificial barriers. Arguments against their utility are often made on the grounds that the tests are overly simplistic, mostly non-volitional, do not incorporate natural motivations to move (i.e. presence of predators or prey and reproductive and life cycle motivations), and are conducted under 'ideal' swimming conditions (i.e. low turbulence, constant flow, no competition) (Plaut 2001; Birnie-Gauvin et al. 2018). Several studies suggest that the commonly reported maximum sustainable swimming performance metric U_{crit}, underestimates some species' capacity to transit high flow instream structures (e.g. Mallen-Cooper 1992; Peake 2004; Tudorache et al. 2007; Bestgen et al. 2010; Dockery et al. 2017); however, burst swimming metrics (e.g. U_{sprint}) can better represent fish performance in some high-velocity environments (e.g. Nelson 2002; Starrs et al. 2011; Holmquist et al. 2018). Increasing the ecological realism in hydraulic swim flumes (e.g. making them longer and deeper and increasing bed roughening) allows fish to use multiple gaits and behaviours to negotiate higher velocities than predicted by traditional U_{crit} or burst swimming metrics (Tudorache et al. 2007). Similarly, traversability metrics conducted in near-full scale flumes (Figure 6.2C–D) allow for conditions more closely approximating field conditions (slope, turbulence, velocity, temperatures, etc.) to be 'played back' to gauge passage responses to environmental challenges. Open, large-scale flumes allow for behavioural observations to supplement physiological performance metrics to generate a more holistic view of fish movement in the environment. Despite their limitations, laboratory-based performance tests can be used to provide fish passage guidelines for a variety of structures. Moreover, compared with field tests, laboratory-based performance assessments are more controllable and repeatable, more cost-effective, and often technically less complicated to run, making them a good starting point for defining the basic parameters over which a fish passage structure should perform. Many fish passage managers recognize that some performance metrics only give conservative estimates of maximum traversable velocities, but in the absence of other information, conservatism can be acceptable when attempting to prevent inappropriate fish passage solutions.

Fish performance studies and the management issues pertaining to the impairment of fish movement have been dominated by salmonid-focused research. Although upstream adult salmon movements are of huge economic, ecological, and cultural significance, salmon are among the highest performing freshwater fish and so are not a suitable proxy species for generalizing freshwater fish movement capacity. Despite this, many fish passage structures and 'fish-friendly' culvert designs have been based on adult salmon swimming capabilities (Mallen-Cooper and Brand 2007; Baumgartner et al. 2014). Given that fish performance capacity is largely predicated by body size and body shape (Beamish 1978), remediation efforts designed to promote adult salmon movement do not perform well for small-bodied species, or juveniles of larger growing species (Mallen-Cooper and Brand 2007; Baumgartner et al. 2014). Swimming performance

data for small-bodied species and juveniles are severely underrepresented in the literature. Despite this, small-bodied species and juvenile fish make up a large proportion of the fish undertaking both local and larger-scale migrations. Future attention needs to be given to describing the variation in swimming performance capacities in small-bodied species and across life-history stages. The traditional experimental swimming performance approaches are particularly useful for small-bodied and juvenile animals that may be too small to use electronic monitoring tags.

In many cases, the need to incorporate fish passage considerations into instream structure design arises because of the presence of rare or at-risk species. Access to sufficient animals with enough physiological, genetic, and developmental diversity to capture the species' performance capacity may be limited, and the use of surrogate species may need to be considered (Cooke et al. 2017). Although other species may be more readily available (common in nature or produced at commercial scale), they remain different from the target species (Underwood et al. 2014; Cooke et al. 2017) and therefore careful consideration of their potential utility as a proxy for the target species needs to be made. Since body size and shape remain the largest predictors of swimming capacity in fish (Beamish 1978), surrogate species need to resemble the target species as closely as possible for managers to have confidence that the results approximate those of the target species. Similarly, commercially produced or captive-reared specimens are often used for performance testing because of their ease of availability. However, their suitability as indicators of the performance capacity of wild animals has been questioned. Artificial selection for traits of commercial significance, inbreeding, high-density rearing conditions, and constant 'ideal' environmental conditions can mean that large-scale, commercially produced animals are less physiologically fit than their wild counterparts (Palstra and Planas 2013). Whether they adequately represent the performance capacities of wild fish could be argued. The use of surrogate species or commercially produced fish for performance testing to inform fish passage considerations of wild fish still remains an important, albeit heavily caveated, tool in the development of fish-friendly instream structures.

Experimental physiology provides a suite of tools that together can inform the design of water management structures that take into consideration the needs of the wildlife around them. Although the challenges associated with 'scaling up' data from largely laboratory-based studies are acknowledged, traditional experimental approaches can provide important baseline information on fish performance, particularly where other data collection approaches may be limited (e.g. biologging in small-bodied animals). Moreover, experimental physiology approaches allow for both the isolation of specific factors and the capacity to explore how combinations of factors influence performance in fish passage structures. Multi-stressor studies increase the ecological realism of experimental swimming performance studies and are better placed to reveal important ecological interactions that can be of considerable interest to fish passage managers and infrastructure designers (Cooke et al. 2017). An understanding of the biological limitations that fish operate within provides a framework from which remediation efforts and the design of novel fish passage structures can be based. A strategy that incorporates a range of approaches to assess fish performance offers a more holistic picture of the physiological limits to performance and the consequences of performance at a fish's physiological limit.

References

Alexandre, C.M. and Palstra, A.P., 2017. Effect of short-term regulated temperature variations on the swimming economy of Atlantic salmon smolts. *Conservation Physiology*, 5, 9.

Allen, P.J., Hodge, B., Werner, I., and Cech, J.J., 2006. Effects of ontogeny, season, and temperature on the swimming performance of juvenile green sturgeon (*Acipenser medirostris*). *Canadian Journal of Fisheries and Aquatic Sciences*, 63, 1360–9.

Amtstaetter, F., O'Connor, J., Borg, D., Stuart, I., and Moloney, P., 2017. Remediation of upstream passage for migrating *Galaxias* (Family: Galaxiidae) through a pipe culvert. *Fisheries Management and Ecology*, 24, 186–92.

Angilletta, M., 2009. *Thermal Adaptation: A Theoretical and Empirical Synthesis*, Oxford University Press, Oxford.

Astles, K., Jh, W., Harris, J., and Gehrke, P., 2003. Experimental study of the effects of cold water pollution on native fish. *NSW Fisheries Final Report Series No. 44*. NSW Fisheries Office of Conservation, NSW Fisheries Research Institute, Orange, Australia.

Baker, C.F. and Boubee, J.A.T., 2006. Upstream passage of inanga *Galaxias maculatus* and redfin bullies *Gobiomorphus huttoni* over artificial ramps. *Journal of Fish Biology*, 69, 668–81.

Baumgartner, L.J., Zampatti, B., Jones, M.J. et al., 2014. Fish passage in the Murray-Darling Basin, Australia: not just an upstream battle. *Ecological Management & Restoration*, 15, 28–39.

Beamish, F.W.H. 1978. Swimming capacity. In W.S. Hoar and D.J. Randall, eds. *Fish Physiology*, Vol. 7, pp. 101–87. Academic Press, New York.

Bernhardt, E.S. and Palmer, M.A., 2007. Restoring streams in an urbanizing world. *Freshwater Biology*, 52, 738–51.

Bernhardt, E.S., Palmer, M.A., Allan, J.D. et al., 2005. Ecology—synthesizing US river restoration efforts. *Science*, 308, 636–7.

Bestgen, K.R., Mefford, B., Bundy, J.M. et al., 2010. Swimming performance and fishway model passage success of Rio Grande silvery minnow. *Transactions of the American Fisheries Society*, 139, 433–48.

Birnie-Gauvin, K., Franklin, P., Wilkes, M., and Aarestrup, K., 2018. Moving beyond fitting fish into equations: progressing the fish passage debate in the Anthropocene. *Aquatic Conservation: Marine and Freshwater Ecosystems*, 29, 1095–1105.

Bouska, W.W. and Paukert, C.P., 2010. Road crossing designs and their impact on fish assemblages of Great Plains streams. *Transactions of the American Fisheries Society*, 139, 214–22.

Boys, C., Miles, N., and Rayner, T., 2009. Scoping options for the ecological assessment of cold water pollution mitigation downstream of Keepit Dam, Namoi River. NSW Department of Primary Industries, Murray-Darling Basin Authority, Canberra, ACT.

Brett, J.R., 1964. The respiratory metabolism and swimming performance of young Sockeye salmon. *Journal of the Fisheries Research Board of Canada*, 21, 1183–226.

Brett, J.R., 1967. Swimming performance of sockeye salmon (*Oncorhynchus nerka*) in relation to fatigue time and temperature. *Journal of the Fisheries Research Board of Canada*, 24, 1731.

Brownscombe, J.W., Cooke, S.J., Algera, D.A. et al., 2017. Ecology of exercise in wild fish: integrating concepts of individual physiological capacity, behavior, and fitness through diverse case studies. *Integrative and Comparative Biology*, 57, 281–92.

Brownscombe, J.W., Lennox, R.J., Danylchuk, A.J., and Cooke, S.J., 2018. Estimating fish swimming metrics and metabolic rates with accelerometers: the influence of sampling frequency. *Journal of Fish Biology, 93*, 207–14.

Burnett, N.J., Hinch, S.G., Braun, D.C. et al., 2014a. Burst swimming in areas of high flow: delayed consequences of anaerobiosis in wild adult sockeye salmon. *Physiological and Biochemical Zoology, 87*, 587–98.

Burnett, N.J., Hinch, S.G., Donaldson, M.R. et al., 2014b. Alterations to dam-spill discharge influence sex-specific activity, behaviour and passage success of migrating adult sockeye salmon. *Ecohydrology, 7*, 1094–104.

Cai, L., Johnson, D., Mandal, P. et al., 2015. Effect of exhaustive exercise on the swimming capability and metabolism of juvenile Siberian sturgeon. *Transactions of the American Fisheries Society, 144*, 532–8.

Castro-Santos, T., 2005. Optimal swim speeds for traversing velocity barriers: an analysis of volitional high-speed swimming behavior of migratory fishes. *Journal of Experimental Biology, 208*, 421–32.

Chabot, D., 2016. Metabolic rate in fishes: definitions, methods and significance for conservation physiology. *Journal of Fish Biology, 88*, 1–9.

Choi, K. and Weber, J.-M., 2016. Coping with an exogenous glucose overload: glucose kinetics of rainbow trout during graded swimming. *American Journal of Physiology—Regulatory, Integrative and Comparative Physiology, 310*, R493–501.

Chung, C.-H., 2006. Forty years of ecological engineering with Spartina plantations in China. *Ecological Engineering, 27*, 49–57.

Claireaux, G., Couturier, C., and Groison, A.L., 2006. Effect of temperature on maximum swimming speed and cost of transport in juvenile European sea bass (*Dicentrarchus labrax*). *Journal of Experimental Biology, 209*, 3420–8.

Cocherell, D.E., Kawabata, A., Kratville, D.W. et al., 2011. Passage performance and physiological stress response of adult white sturgeon ascending a laboratory fishway. *Journal of Applied Ichthyology, 27*, 327–34.

Coleman, R.A., Gauffre, B., Pavlova, A. et al., 2018. Artificial barriers prevent genetic recovery of small isolated populations of a low-mobility freshwater fish. *Heredity, 120*, 520–32.

Cooke, S.J., Birnie-Gauvin, K., Lennox, R.J. et al., 2017. How experimental biology and ecology can support evidence-based decision-making in conservation: avoiding pitfalls and enabling application. *Conservation Physiology, 5*, 14.

Cooke, S.J., Midwood, J.D., Thiem, J.D. et al., 2013. Tracking animals in freshwater with electronic tags: past, present and future. *Animal Biotelemetry, 1*, 5.

Di Santo, V., Kenaley, C.P., and Lauder, G.V., 2017. High postural costs and anaerobic metabolism during swimming support the hypothesis of a U-shaped metabolism–speed curve in fishes. *Proceedings of the National Academy of Sciences, 114*, 13048–53.

Dockery, D.R., McMahon, T.E., Kappenman, K.M., and Blank, M., 2017. Evaluation of swimming performance for fish passage of longnose dace *Rhinichthys cataractae* using an experimental flume. *Journal of Fish Biology, 90*, 980–1000.

Domenici, P. and Blake, R., 1997. The kinematics and performance of fish fast-start swimming. *Journal of Experimental Biology, 200*, 1165–78.

Donaldson, M.R., Hinch, S.G., Patterson, D.A. et al., 2010. Physiological condition differentially affects the behavior and survival of two populations of sockeye salmon during their freshwater spawning migration. *Physiological and Biochemical Zoology, 83*, 446–58.

Dudgeon, D., Arthington, A.H., Gessner, M.O. et al., 2006. Freshwater biodiversity: importance, threats, status and conservation challenges. *Biological Reviews, 81*, 163.

Dugan, P.J., Barlow, C., Agostinho, A.A. et al., 2010. Fish migration, dams, and loss of ecosystem services in the Mekong Basin. *AMBIO, 39*, 344–8.

Eliason, E.J., Clark, T.D., Hinch, S.G., and Farrell, A.P., 2013. Cardiorespiratory performance and blood chemistry during swimming and recovery in three populations of elite swimmers: adult sockeye salmon. *Comparative Biochemistry and Physiology A:Molecular & Integrative Physiology, 166*, 385–97.

Fischer, J. and Lindenmayer, D.B., 2007. Landscape modification and habitat fragmentation: a synthesis. *Global Ecology and Biogeography, 16*, 265–80.

Geist, D.R., Brown, R.S., Cullinan, V.I. et al., 2005. Movement, swimming speed, and oxygen consumption of juvenile white sturgeon in response to changing flow, water temperature, and light level in the Snake River, Idaho. *Transactions of the American Fisheries Society, 134*, 803–16.

Geist, D.R., Brown, R.S., Cullinan, V.I. et al., 2003. Relationships between metabolic rate, muscle electromyograms and swim performance of adult Chinook salmon. *Journal of Fish Biology, 63*, 970–89.

Gibson, R.J., Haedrich, R.L., and Wernerheirn, C.M., 2005. Loss of fish habitat as a consequence of inappropriately constructed stream crossings. *Fisheries, 30*, 10–17.

Grill, G., Lehner, B., Thieme, M. et al., 2019. Mapping the world's free-flowing rivers. *Nature, 569*, 215–21.

Guderley, H. and Blier, P., 1988. Thermal acclimation in fish: conservative and labile properties of swimming muscle. *Canadian Journal of Zoology, 66*, 1105–15.

Hamel, M.J., Brown, M.L., and Chipps, S.R., 2008. Behavioral responses of rainbow smelt to *in situ* strobe lights. *North American Journal of Fisheries Management, 28*, 394–401.

Hammer, C., 1995. Fatigue and exercise tests with fish. *Comparative Biochemistry and Physiology A: Physiology, 112*, 1–20.

Harris, J.H., Kingsford, R.T., Peirson, W., and Baumgartner, L.J., 2017. Mitigating the effects of barriers to freshwater fish migrations: the Australian experience. *Marine and Freshwater Research*, 68, 614–28.

Hatry, C., Thiem, J.D., Binder, T.R. et al., 2014. Comparative physiology and relative swimming performance of three redhorse (*Moxostoma* spp.) species: associations with fishway passage success. *Physiological and Biochemical Zoology*, 87, 148–59.

He, W., Xia, W., Cao, Z.D., and Fu, S.J., 2013. The effect of prolonged exercise training on swimming performance and the underlying biochemical mechanisms in juvenile common carp (*Cyprinus carpio*). *Comparative Biochemistry and Physiology A: Molecular & Integrative Physiology*, 166, 308–15.

Hein, A.M. and Keirsted, K.J., 2012. The rising cost of warming waters: effects of temperature on the cost of swimming in fishes. *Biology Letters*, 8, 266–9.

Herren, J.R. and Kawasaki, S.S., 2001. Inventory of water diversions in four geographic areas in California's Central Valley. *Fish Bulletin 179: Contributions to the Biology of Central Valley Salmonids.* Available at: http://www.dfg.ca.gov/fish/Resources/Reports/Bulletin179_V2.asp (27 July 2020).

Hinch, S.G., Diewert, R.E., Lissimore, T.J. et al., 1996. Use of electromyogram telemetry to assess difficult passage areas for river-migrating adult sockeye salmon. *Transactions of the American Fisheries Society*, 125, 253–60.

Holmquist, L., Kappenman, K., Blank, M.D., and Schultz, M., 2018. Sprint swimming performance of shovelnose sturgeon in an open-channel flume. *Northwest Science*, 92, 61–71.

Humphries, P. and Lake, P.S., 2000. Fish larvae and the management of regulated rivers. *Regulated Rivers: Research & Management*, 16, 421–32.

International Union for the Conservation of Nature, 2019. The IUCN Red List of Threatened Species. Version 2019-1. Available at: www.iucnredlist.org (accessed 21 March 2019).

Jepsen, N., Koed, A., Thorstad, E.B., and Baras, E., 2002. Surgical implantation of telemetry transmitters in fish: how much have we learned? *Hydrobiologia*, 483, 239–48.

Katopodis, C., 2005. Developing a toolkit for fish passage, ecological flow management and fish habitat works *Journal of Hydraulic Research*, 43, 451–467.

Keefer, M.L., Taylor, G.A., Garletts, D.F. et al., 2013. High-head dams affect downstream fish passage timing and survival in the middle fork Willamette River. *River Research and Applications*, 29, 483–92.

Kern, P., Cramp, R.L., Gordos, M.A. et al., 2018. Measuring Ucrit and endurance: equipment choice influences estimates of fish swimming performance. *Journal of Fish Biology*, 92, 237–47.

Lapointe, N.W.R., Thiem, J.D., Doka, S.E., and Cooke, S.J., 2013. Opportunities for improving aquatic restoration science and monitoring through the use of animal electronic-tagging technology. *Bioscience*, 63, 390–6.

Lay, P.A. and Baldwin, J., 1999. What determines the size of teleost erythrocytes? Correlations with oxygen transport and nuclear volume. *Fish Physiology and Biochemistry*, 20, 31–5.

Lee, C.G., Farrell, A.P., Lotto, A., Hinch, S.G., and Healey, M.C.., 2003. Excess post-exercise oxygen consumption in adult sockeye (*Oncorhynchus nerka*) and Coho (*O. kisutch*) salmon following critical speed swimming. *Journal of Experimental Biology*, 206, 3253–60.

Leonard, J.B.K. and McCormick, S.D., 1999. Changes in haematology during upstream migration to American shad. *Journal of Fish Biology*, 54, 1218–30.

Ling, T.-Y., Soo, C.-L., Heng, T.L.-E. et al., 2016. Physicochemical characteristics of river water downstream of a large tropical hydroelectric dam. *Journal of Chemistry*, 2016, art. 7895234.

Lothian, A.J., Gardner, C.J., Hull, T. et al., 2019. Passage performance and behaviour of wild and stocked cyprinid fish at a sloping weir with a low cost baffle fishway. *Ecological Engineering*, 130, 67–79.

Lugg, A. and Copeland, C., 2014. Review of cold water pollution in the Murray–Darling Basin and the impacts on fish communities. *Ecological Management and Restoration*, 15, 71–9.

Maes, J., Turnpenny, A.W.H., Lambert, D.R. et al., 2004. Field evaluation of a sound system to reduce estuarine fish intake rates at a power plant cooling water inlet. *Journal of Fish Biology*, 64, 938–46.

Mallen-Cooper, M., 1992. Swimming ability of juvenile Australian bass, *Macquaria novemaculeata* (Steindachner), and juvenile barramundi, *Lates calcarifer* (Bloch), in an experimental vertical-slot fishway. *Marine and Freshwater Research*, 43, 823–33.

Mallen-Cooper, M. and Brand, D.A., 2007. Non-salmonids in a salmonid fishway: what do 50 years of data tell us about past and future fish passage? *Fisheries Management and Ecology*, 14, 319–32.

Metcalfe, J.D., Wright, S., Tudorache, C., and Wilson, R.P., 2016. Recent advances in telemetry for estimating the energy metabolism of wild fishes. *Journal of Fish Biology*, 88, 284–97.

Mussen, T.D., Cocherell, D., Poletto, J.B. et al., 2014. Unscreened water-diversion pipes pose an entrainment risk to the threatened green sturgeon, *Acipenser medirostris. PLoS ONE*, 9.

National Marine Fisheries Service West Coast Region, NOAA, 2018. Recovery plan for the southern distinct population segment of North American green sturgeon (*Acipenser medirostris*). Available at: https://www.westcoast.fisheries.

noaa.gov/publications/protected_species/other/green_sturgeon/noaa-sdps-green-sturgeon-recovery-plan-8-8-2018.pdf (accessed 24 June 2020).

Nelson, J.A., 2002. Beyond U(crit): matching swimming performance tests to the physiological ecology of the animal, including a new fish 'drag strip'. *Comparative Biochemistry and Physiology A: Molecular & Integrative Physiology*, 133, 289–302.

Palmer, M., Allan, J.D., Meyer, J., and Bernhardt, E.S., 2007. River restoration in the twenty-first century: Data and experiential future efforts. *Restoration Ecology*, 15, 472–81.

Palstra, A.P. and Planas, J.V. (eds), 2013. *Swimming Physiology of Fish: Towards Using Exercise to Farm a Fit Fish in Sustainable Aquaculture*. Springer, Berlin, Heidelberg.

Palstra, A., Van Ginneken, V., and Van Den Thillart, G., 2008. Cost of transport and optimal swimming speed in farmed and wild European silver eels (*Anguilla anguilla*). *Comparative Biochemistry and Physiology A: Physiology*, 151, 37–44.

Paul, M.J. and Meyer, J.L., 2001. Streams in the urban landscape. *Annual Review of Ecology and Systematics*, 32, 333–65.

Peake, S., 2004. An evaluation of the use of critical swimming speed for determination of culvert water velocity criteria for smallmouth bass. *Transactions of the American Fisheries Society*, 133, 1472–9.

Peake, S.J. and Farrell, A.P., 2004. Locomotory behaviour and post-exercise physiology in relation to swimming speed, gait transition and metabolism in free-swimming smallmouth bass (*Micropterus dolomieu*). *Journal of Experimental Biology*, 207, 1563–75.

Pelicice, F.M., Pompeu, P.S., and Agostinho, A.A., 2015. Large reservoirs as ecological barriers to downstream movements of Neotropical migratory fish. *Fish and Fisheries*, 16, 697–715.

Plaut, I., 2001. Critical swimming speed: its ecological relevance. *Comparative Biochemistry and Physiology A: Molecular and Integrative Physiology*, 131, 41–50.

Poletto, J.B., Cocherell, D.E., Ho, N. et al., 2014a. Juvenile green sturgeon (*Acipenser medirostris*) and white sturgeon (*Acipenser transmontanus*) behavior near water-diversion fish screens: experiments in a laboratory swimming flume. *Canadian Journal of Fisheries and Aquatic Sciences*, 71, 1030–8.

Poletto, J.B., Cocherell, D.E., Mussen, T.D. et al., 2014b. Efficacy of a sensory deterrent and pipe modifications in decreasing entrainment of juvenile green sturgeon (*Acipenser medirostris*) at unscreened water diversions. *Conservation Physiology*, 2, cou056.

Portner, H.O. and Peck, M.A., 2010. Climate change effects on fishes and fisheries: towards a cause-and-effect understanding. *Journal of Fish Biology*, 77, 1745–79.

Reid, G.M., Contreras Macbeath, T., and Csatádi, K., 2013. Global challenges in freshwater-fish conservation related to public aquariums and the aquarium industry. *International Zoo Yearbook*, 47, 6–45.

Reidy, S.P., Kerr, S.R., and Nelson, J.A., 2000. Aerobic and anaerobic swimming performance of individual Atlantic cod. *Journal of Experimental Biology*, 203, 347–57.

Rodgers, E.M., Cramp, R.L., Gordos, M. et al., 2014. Facilitating upstream passage of small-bodied fishes: linking the thermal dependence of swimming ability to culvert design. *Marine and Freshwater Research*, 65, 710–19.

Silva, A.T., Lucas, M.C., Castro-Santos, T. et al., 2018. The future of fish passage science, engineering, and practice. *Fish and Fisheries*, 19, 340–62.

Starrs, D., Ebner, B.C., Lintermans, M., and Fulton, C.J., 2011. Using sprint swimming performance to predict upstream passage of the endangered Macquarie perch in a highly regulated river. *Fisheries Management and Ecology*, 18, 360–74.

Thorstad, E.B., Økland, F., Aarestrup, K., and Heggberget, T.G., 2008. Factors affecting the within-river spawning migration of Atlantic salmon, with emphasis on human impacts. *Reviews in Fish Biology and Fisheries*, 18, 345–71.

Todd, C.R., Ryan, T., Nicol, S.J., and Bearlin, A.R., 2005. The impact of cold water releases on the critical period of post-spawning survival and its implications for Murray cod (*Maccullochella peelii peelii*): a case study of the Mitta Mitta River, southeastern Australia. *River Research and Applications*, 21, 1035–52.

Tudorache, C., Viaenen, P., Blust, R., and De Boeck, G., 2007. Longer flumes increase critical swimming speeds by increasing burst–glide swimming duration in carp *Cyprinus carpio*, L. *Journal of Fish Biology*, 71, 1630–8.

Underwood, Z.E., Myrick, C.A., and Compton, R.I., 2014. Comparative swimming performance of five *Catostomus* species and roundtail chub. *North American Journal of Fisheries Management*, 34, 753–63.

Verhille, C.E., Poletto, J.B., Cocherell, D.E. et al., 2014. Larval green and white sturgeon swimming performance in relation to water-diversion flows. *Conservation Physiology*, 2, cou031.

Warren, M.L. and Pardew, M.G., 1998. Road crossings as barriers to small-stream fish movement. *Transactions of the American Fisheries Society*, 127, 637–44.

Watson, J.R., Goodrich, H.R., Cramp, R.L. et al., 2018. Utilising the boundary layer to help restore the connectivity of fish habitats and populations. *Ecological Engineering*, 122, 286–94.

Whiterod, N.S., Meredith, S.N., Humphries, P. et al., 2018. Flow alteration and thermal pollution depress modelled growth rates of an iconic riverine fish, the Murray cod *Maccullochella peelii*. *Ecology of Freshwater Fish*, 27, 686–98.

Wilson, A.D. M., Wikelski, M., Wilson, R.P., and Cooke, S.J., 2015. Utility of biological sensor tags in animal conservation. *Conservation Biology*, 29, 1065–75.

Young, J.L., Hinch, S.G., Cooke, S.J. et al., 2006. Physiological and energetic correlates of en route mortality for abnormally early migrating adult sockeye salmon (*Oncorhynchus nerka*) in the Thompson River, British Columbia. *Canadian Journal of Fisheries and Aquatic Sciences*, 63, 1067–77.

Young, P.S., Swanson, C., and Cech, J.J., 2010. Close encounters with a fish screen III: behavior, performance, physiological stress responses, and recovery of adult delta smelt exposed to two-vector flows near a fish screen. *Transactions of the American Fisheries Society*, 139, 713–26.

CHAPTER 7

Transcriptome profiling in conservation physiology and ecotoxicology: mechanistic insights into organism–environment interactions to both test and generate hypotheses

Marisa L. Trego, Charles A. Brown, Benjamin Dubansky, Chelsea D. Hess, Fernando Galvez, and Andrew Whitehead

---

⊃ Take-home message

Genome-wide gene expression profiling provides detailed and unbiased data to both test hypotheses and generate hypotheses about the impacts of exposures to anthropogenic stressors on wildlife health. The mechanistic information provided can help to diagnose the causes, and predict the consequences, of environmental exposures, and thereby support effective conservation interventions.

7.1 Introduction

A key challenge in conservation biology is to identify natural populations that exist in a state of distress or compromised health and fitness, and to determine the causes of natural and human-induced environmental change (Fleishman et al. 2011). Organisms in nature are often living in suboptimal conditions, but defensive and compensatory physiologies and behaviours have evolved over long periods of time such that biological systems are generally robust to environmental perturbation. However, humans have altered the types, number, combinations, and magnitude of stressors present in the environment, and these changes can often push organisms beyond their compensatory limits, resulting in compromised health and fitness and population decline. It is within this complex matrix of natural challenges (e.g. predator–prey interactions, seasonal environmental change, diel environmental change) and anthropogenic stressors and environmental variability that we are challenged to assess the health status of organisms; we must diagnose the agents of distress prior to severe population declines, such that conservation solutions

Marisa L. Trego, Charles A. Brown, Benjamin Dubansky, Chelsea D. Hess, Fernando Galvez, and Andrew Whitehead, *Transcriptome profiling in conservation physiology and ecotoxicology: mechanistic insights into organism–environment interactions to both test and generate hypotheses* In: *Conservation Physiology: Applications for Wildlife Conservation and Management*. Edited by: Christine L. Madliger, Craig E. Franklin, Oliver P. Love, and Steven J. Cooke, Oxford University Press (2021). © Oxford University Press. DOI: 10.1093/oso/9780198843610.003.0007 may be proactively, rather than reactively, planned and enacted.

Establishing the cause–effect relationships between human-induced environmental changes and states of adverse health serves to prioritize which stressors or environmental parameters should be addressed in conservation planning. It may be that, despite exposure to a variety of stressors, there are few factors that have any appreciable impact on population health. Without a better understanding of the reasons why a population is declining, or not recovering after a decline, it will be challenging to develop a recovery plan. For example, though there are many stressors that pose threats to the persistence of killer whale populations in the US north-western Pacific coast region, research suggests that food stress causes a larger physiological challenge than stress induced by vessel traffic (Ayres et al. 2012). In this case, conservation plans could strategically prioritize management of prey populations over the mitigation of vessel traffic in a region with limited resources. Unfortunately, in most cases it is challenging to identify the relative health threat of various potential agents of stress, such as pollution, climate change, disease, invasive species, and harvest pressure. This requires tools that can distinguish health impacts caused by different agents, such that causative stressors can be identified and prioritized for mitigation along the path towards conservation.

Proper diagnosis of some health impacts does not always require sophisticated tools. For example, during crude oil spills, the external oiling of seabird carcasses is a routine, effective, and rapid diagnosis of oil-induced effects. In contrast, other phenomena such as altered sex ratios in fish from urban waterways (Jobling et al. 1998), honey-bee colony collapses (Oldroyd 2007), or elevated raptor nest failures (Ratcliffe 1963) present mysteries that require more sophisticated biological detective work to diagnose causes. Traditionally, such detective work unfolds during a lengthy process of serial elimination of candidate causes (e.g. Li et al. 2018). The recent advent of OMICS technologies has provided opportunities for accelerating this discovery process, essentially by

examining perturbation of many biological systems in a single experiment.

Transcriptomics uses second-generation sequencing technologies to characterize genome function by sequencing and quantifying the full suite of transcripts that are expressed in any tissue or species (Wang et al. 2009). This genome-wide profiling of genome function can provide detailed mechanistic insights into organism–environment interactions. Gene expression refers to the transcription of DNA into RNA. After transcription, messenger RNAs are translated into proteins that are responsible for modulating biochemical pathways that govern physiological functions within an organism. Quantifying the expression level of a gene offers an indirect estimate of the amount of proteins being generated and their associated physiological functions. Measuring the expression of all genes across the genome (i.e. the transcriptome) offers a snapshot of a full suite of biochemical pathways that are regulated in response to both internal or external signals, including those pathways associated with development (Vesterlund et al. 2011) and reproduction (Santos et al. 2007), response to perturbation by disease (Basu et al. 2012), or other biotic and abiotic environmental stressors (Kassahn et al. 2007).

Technologies for high-throughput transcriptome profiling have dramatically improved over the past decade. Microarrays were the first widely adopted technology for genome-wide gene expression profiling. One of the major limitations of this technology was that it required the prior knowledge of transcript sequences for the synthesis of microarray probes. This limited the adoption of the technology by groups studying non-traditional model species. Massively parallel RNA sequencing solved some of the limitations posed by microarrays (Zhao et al. 2014); importantly, the technology could be deployed without the need for prior knowledge of gene sequences (Wang et al. 2009). RNA sequencing (RNAseq) has largely replaced microarrays for transcriptome analyses and has improved our capacity to accurately quantify gene expression over a much broader range of expression levels. Considering RNAseq provides the genetic sequence of each transcript in addition to its expression level, it enables the detection of new genes and genetic variants that had not previously been characterized (Wang et al. 2009). This allows researchers to examine physiological changes in transcript abundance offering detailed mechanistic insight into organism–environment interactions, while at the same time providing genetic markers suitable for the types of genetics analyses traditionally used in conservation biology. Furthermore, the genome-wide nature of information on genetic variation enables detecting stressor-induced natural selection (e.g. De Wit et al. 2014) that is not feasible with the types of markers traditionally used in conservation genetics (e.g. microsatellites, amplified fragment length polymorphisms [AFLPs], single nucleotide polymorphisms [SNPs] from a few genes). As DNA- and RNA-based technologies continue to develop, so does our capacity to analyse how organisms are physiologically responding to novel environmental conditions.

Since many perturbations to physiology and health are mediated through, and manifest as, altered gene expression, transcriptome profiling provides an opportunity to rapidly provide a holistic insight into causes and consequences of environmental stress. That is, transcriptome profiles can capture molecular initiating events following perturbation thereby offering insight into the identity of agents of perturbation. These profiles provide a nuanced readout of cellular/organ status following perturbation that allows prediction of the physiological or health status of the organism. This predictive capacity can provide an early-warning signal of health impacts prior to serious population declines and improve our mechanistic understanding of physiological tipping points at a molecular level. Transcriptomic analyses allow both the testing of hypotheses about particular pathways that are suspected of perturbation, but also capture perturbations elsewhere in the transcriptome; these approaches thereby provide opportunities for the discovery of novel and often unforeseen outcomes, such that novel hypotheses about cause and effect may be generated. Scope for inference is maximized with careful and thoughtful experimental design, particularly when these molecular data are coupled with other measurements of organismal health (e.g. reproduction, physiology, behaviour) and environ-

mental metadata (e.g. weather, season, water quality, environmental chemistry, etc.).

In the following sections we offer examples of case studies of how these approaches may be deployed to predict and diagnose environmental problems, and provide information that is useful for conservation practitioners (see also Connon et al. 2018). These case studies will illustrate how transcriptomics can be applied within conservation physiology for both testing hypotheses and predictions, and for generating new hypotheses. Our focus is within the context of environmental pollution, but the paradigms and approaches are relevant to other contexts of human-induced environmental change.

7.2 Chemical pollution and wildlife physiology

Chemical pollution from human activities is one of the key markers of globalized human impact (Waters et al. 2016). Chemical contaminants can exert lethal as well as sublethal impacts on wildlife including modifying physiological performance, behaviour, immune function, or reproduction, all of which may affect fitness and therefore population dynamics. Within the context of wildlife conservation, a key goal is to detect these kinds of impacts on health prior to population decline, and link the effects to their causative agents. Establishing this cause/effect relationship is difficult for several reasons: (1) Health impacts may not manifest (or be detected) until long after an exposure event. When a cause and effect are distantly related in time it can be difficult to identify causative agents (e.g. as for cancer). For example, persistent organic pollutants (POPs) can be easily detected in marine mammal blubber. Though POPs are present in higher abundance in the blubber of diseased marine mammals than those with no signs of disease at time of death (Hall et al. 2005), this approach is unable to confirm POPs as the cause of disease or mortality. Though we know that these compounds compromise immune function (Ross et al. 1996; Desforges et al. 2017), there remain significant knowledge gaps linking reduced immune function with survival. (2) Some stressors may not be detectable or distinguishable outside of their biological impacts.

That is, some pollutants are quickly metabolized such that screening for the parent compounds themselves using analytical methods may not capture exposures, though biological impacts may be detectable. For example, diagnostic molecular and histological responses clearly indicated that killifish had been exposed to contaminating oil following an oil spill, though the analytical chemistry was unable to detect elevated body burdens of oil constituents (Whitehead et al. 2012a). (3) Multiple stressors including multiple contaminants are often encountered simultaneously, and some anthropogenic stressors are difficult to distinguish from natural stressors. For example, cetacean tissues accumulate many contaminants, some of which are structurally similar to natural compounds (Teuten and Reddy 2007). Furthermore, the chemicals that we detect include only those that we think to test for (Hoh et al. 2012; Shaul et al. 2015), potentially missing those that we do not suspect are relevant (or those that we don't know how to measure), and chemicals are often transformed by environmental and organismal processes rendering them undetectable with standard methods. In contrast, the biology of an organism integrates the effects of exposures to multiple agents across space and time. Therefore, using organisms as biosensors is appropriate and necessary for estimating impacts and risks from stressors including environmental chemicals. Especially useful would be biological measures that offer information that may aid in the diagnosis of causative agents for specific sublethal health impacts (e.g. biomarkers). Transcriptomics offers a promising way forward to link contaminant exposure to quantifiable physiological effects in the wild where animals are exposed to several anthropogenic compounds.

7.3 Case study 1: crude oil exposure on killifish in the Gulf of Mexico

Transcriptome profiling was an important component of integrative experiments testing whether fish in critical coastal marsh habitats had been exposed to the toxic fractions of oil following the Deepwater Horizon oil spill (in the Gulf of Mexico), and for generating hypotheses about physiological consequences; many of these hypotheses have since been

tested and verified. Following the Deepwater Horizon disaster in April 2010, field studies were initiated, where liver transcriptome profiles were collected for Gulf killifish (*Fundulus grandis*) inhabiting the northern Gulf of Mexico salt marshes that were heavily oiled during the spill (Whitehead et al. 2012a). Sampling included sites that were impacted by oil and sites that were not, and sites were sampled before and after oiling. This before–after control–impact (BACI) sampling design afforded much power for linking expression profiles and other physiological endpoints with the location and timing of oiling. Since these killifish have very small home ranges (Jensen et al. 2019), their health is primarily affected by their local environment. Six field sites were sampled, one of which was eventually heavily contaminated with spilled oil (Figure 7.1a). Transcriptome profiles clearly distinguished the oiled field site from the many reference sites, and profile divergence coincided with the timing of oiling (Figure 7.1b), which strongly implicated contaminating oil as the causative factor for divergence in gene expression. The gene expression profiles themselves also implicated oil as the agent of perturbation.

7.3.1 Testing hypotheses

A key question following the Deepwater Horizon oil spill was whether organisms in sensitive and crucial near-shore estuarine habitats were exposed to the toxic components of oil. Genome-wide gene expression profiles in killifish liver tissues (Figure 7.1b) showed that the aryl-hydrocarbon receptor (AHR) signalling pathway was activated in fish collected coincident with the timing and location of oil. BACI contrasts revealed divergent gene expression at the only field site (out of six) that was oiled, and this divergent response distinguished fish from the first (pre-oiling) and second (post-oiling) time points (Whitehead et al. 2012a). The transcriptome profile of fish from the oiled site also changed over time as the magnitude of oiling changed due to weathering. Oiling at field sites was confirmed with satellite imagery, visual observation, and analytical chemistry. The identity of these transcripts implicated the toxic fraction of oil as the causative agent (Figure 7.1c). Crude oil is a complex

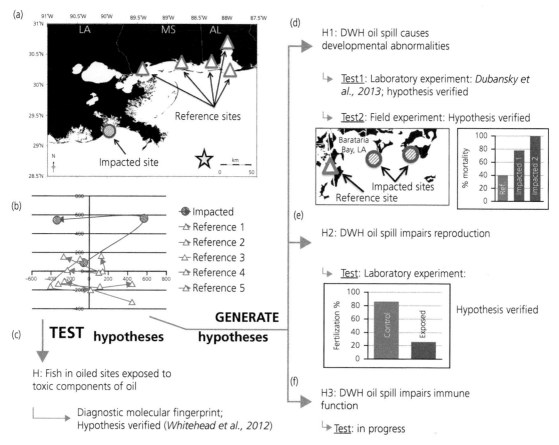

Figure 7.1 Hypotheses were both tested and generated from liver transcriptomics data from a field study using killifish collected before and after contamination from the Deepwater Horizon (DWH) oil spill. (A) The DWH explosion occurred offshore (star) and directly oiled one field site (circle) but did not oil five additional field sites (reference sites; triangles). (B) Principal components analysis (PCA) summary of the trajectory of liver transcriptome change through time from each field site (for genes with significant site-by-time interaction). Plot represents the first (*x*-axis) and second (*y*-axis) dimension of the PCA. Data points from base, middle, and tip of each arrowed line represent samples collected prior to oiling, at peak oiling, and months after peak oiling, respectively. The impacted field site showed a highly divergent transcriptome profile between the first and second sampling periods, coincident with the arrival of oil. (C) With transcriptomics data we tested the hypothesis that fish from the impacted site had been exposed to the toxic fraction of oil. This was verified since the genes known to be divergently regulated by the chemical toxicants in oil were largely responsible for the divergent response in the field study. (D–E) Transcriptome responses that were coincident with oiling (from the field study) also generated several hypotheses about biological impacts, some of which have now been tested and supported. (D) One hypothesis was that contaminants from the DWH oil spill caused developmental abnormalities. This was supported by laboratory exposures to field-collected sediments (Test1), and in field studies where developing embryos were caged at oil-impacted and clean reference sites in Barataria Bay, LA (Test2). Mortality during embryogenesis was more than two-fold higher at impacted sites than the reference site (bar graph). (E) A second hypothesis was that contaminants from the DWH oil spill impair reproduction in adult fish. This was tested by exposing adult fish to oiled water for 40 days and testing fertilization success. Fertilization was reduced by nearly 70 per cent for fish that had been exposed to oil, but recovered in fish returned to clean water for 30 days. (F) A third hypothesis is that contaminants in DWH oil cause immune dysfunction. This hypothesis is currently being tested.

mixture of chemicals, where the toxic chemicals include the polycyclic aromatic hydrocarbons (PAHs). PAHs are known to act as ligands for the AHR, and the ligand-activated AHR is known to activate transcription of a battery of genes. This response is evolutionarily conserved across vertebrates, including the killifish. Controlled laboratory

exposures to crude oil confirmed this response (Pilcher et al. 2014). Furthermore, these lab exposures were able to distinguish high-dose from low-dose liver transcriptome responses to oil. The high-dose response was more predictive of the response observed in the field than the low-dose response, suggesting that field-exposed fish had been exposed to relatively high concentrations of toxic contaminating oil.

These findings were disputed by consultants paid by BP (Jenkins et al. 2012), but the BACI design coupled with a highly diagnostic fingerprint of oil exposure enabled by transcriptomics provided for robust establishment of cause and effect (Figure 7.1b, C) (Whitehead et al. 2012a, b). Additional field sampling showed that this response persisted for at least an additional full year (Dubansky et al. 2013), indicating persistent exposure likely from crude oil that remained in the sediment in these shallow marsh habitats. Surprisingly, PAHs were almost undetectable in the tissues of exposed fish (even at peak oiling) likely because of rapid metabolism, though it was detectable at high concentrations in nearby sediments (Whitehead et al. 2012a). This was a clear demonstration that biological responses, which integrate exposures across space and time and may persist after exposures cease, can be more sensitive and diagnostic of exposures than direct chemistry measurements on their own.

7.3.2 Generating hypotheses

Inappropriate AHR activation during embryogenesis is known to impair normal organ development, particularly in heart development, leading to lifelong fitness impacts. This disruption of development can occur at very low doses of PAHs (and by related chemicals that are AHR activators) and is conserved across vertebrates. Given the AHR activation that was observed in field-collected adults, we hypothesized that these exposures would be damaging to developing animals (Figure 7.1d). Importantly, contamination of marshes by spilled oil occurred at a time of year when many species were spawning, including Gulf killifish (spring/ summer, 2010). To test this hypothesis, we exposed developing killifish embryos to sediments collected from oil-impacted field sites and clean reference sites. This exposure caused a decline in heartrate, reduced hatching success, and reduced length at hatch (Dubansky et al. 2013), thereby confirming our hypothesis. We also fertilized killifish embryos in the laboratory, submerged them (in perforated Teflon containers) into oiled and reference field sites and measured survival during embryogenesis.

Embryos (N = 23–24) were included in each container, and three containers were planted in each field site. Field sites included one reference site in Manilla Bay (29.449997°, −90.016999°) and two impacted sites at St. Mary's Point (29.447048°, −89.937137°) and West Bay Jimmy (29.451473°, −89.895121°). This field study was conducted between 24 October 2012 (embryos deployed in field) and 2 November 2012 (containers retrieved and survivors counted). Results showed that exposure to oiled field sites during development caused a significant reduction in survival (Figure 7.1d). This provided further confirmation of our hypothesis that oiled habitats were damaging to developing vertebrates.

Genome-wide gene expression profiles in the livers of field-exposed adults indicated a dramatic repression of zona pellucida, vitellogenin, and choriogenin transcripts that coincided with the timing and location of oiling (Garcia et al. 2012; Whitehead et al. 2012a). These transcripts are regulated by oestrogen signalling, and exposure to PAHs is known to interfere with oestrogen signalling, which may be mediated through AHR signalling. This knowledge contributed to a hypothesis that exposure to oil impairs reproductive success in adult killifish (Figure 7.1e). To test this hypothesis, we exposed adult male and female killifish to crude oil for 40 days. Mass spawning of treatment fish was performed by pooling manually stripped gametes followed by in vitro fertilization. We calculated the number of eggs produced per female and the percentage fertilization success of eggs by in vitro fertilization. Brood fish were then moved to clean water for 30 days and in vitro fertilization was repeated. Fertilization success was impaired by 40 days' exposure to Macondo oil (Figure 7.1e), which was consistent with our hypothesis. However, effects of the oil on fertilization success were transient. Following a 30-day recovery, fertilization success was similar between the control fish and those that had originally been exposed to oil for 40 days. Expression profiles in field-exposed animals also revealed altered regulation of immune system genes coincident with oiling, which contributed to our hypothesis that oil-exposed fish are immune-compromised and may be at elevated risk of disease (Figure 7.1f). Oil exposure-induced immune dysfunction is observed in other species (Murawski et al. 2014; Bayha et al. 2017), and ongoing experiments by our group seek to further explore this hypothesis.

7.3.3 Conservation implications

Environmental restoration activities following releases of hazardous substances by humans into the environment are often supported by a Natural Resources Damage Assessment (NRDA) conducted by federal US agencies, such as the National Oceanic and Atmospheric Administration (NOAA) or the US Fish and Wildlife Service. Through the NRDA process, these agencies partner with state and academic partners to collect data to assess the nature and extent of impacts on wildlife and other natural resources. This information is then used to pursue legal claims against responsible parties. These agencies, which are mandated with a responsibility to conserve and protect wild species and natural resources, can then use recovered funds to support restoration and conservation efforts. During the NRDA process, a key objective is to establish a cause–effect relationship between the environmental perturbation and resulting damage, such that the responsible party may be held liable. The integrative field and laboratory studies described above contributed to establishing a cause–effect relationship between the Deepwater Horizon oil spill and negative impacts in Gulf of Mexico salt marsh ecosystems (Whitehead et al. 2012a; Dubansky et al. 2013). Indeed, soon after publication of these studies, the responsible party (BP) hired consultants to write and publish articles that objected to the study's conclusions linking cause with effect (Jenkins et al. 2012; Pearson 2014) (see Whitehead et al. 2012b and Dubansky et al. 2014 for rebuttals). Expert witness testimony during the Department of Justice proceedings against BP used results from these lab and field studies to build the weight of evidence that eventually forced BP to seek a legal settlement. Furthermore, Florida Senator Bill Nelson used data from the killifish studies to argue from the floor of the US Senate for funds for science-based approaches to support conservation and restoration efforts in impacted Gulf of Mexico habitats (see Congress.gov 2011). These science-based approaches could include deployment of transcriptomics tools to track recovery of those responses that were predictive of oil impacts on reproduction and development during habitat restoration.

7.4 Case study 2: POP exposure in marine mammals in southern California

Southern California is a highly urbanized region that has a known history of high POP contamination in coastal marine habitats, most notably from dichlorodiphenyltrichloroethane (DDT) pollution (Young et al. 1976). As high-trophic-level organisms, marine mammals accumulate large quantities of POPs because the compounds are lipophilic and difficult to metabolize, and are therefore often stored in fatty tissues long-term (Goerke et al. 2004; Barón et al. 2015; Burreau et al. 2006; Peterson et al. 2015). Marine mammals in southern California specifically have some of the highest POP levels observed in the world (Blasius and Goodmanlowe 2008), but there has been little enquiry into how this could be impacting the health of those wild populations. This is, in part, because connecting lifelong POP exposure to specific physiological effects in wild animals is difficult. More generally, investigating marine mammal physiology in the wild is inherently challenging given that non-invasive sampling of tissues is difficult (Kellar et al. 2006). Biopsies composed of skin and blubber are often one of the only sample types available for assessing physiology in wild individuals (Kellar et al. 2006). Fortunately, recent developments to quantify gene expression in skin have successfully identified physiological changes related to chemical exposures (Mollenhauer et al. 2009; Buckman et al. 2011; Lunardi et al. 2016; Neely et al. 2017; Godard-Codding and Fossi, 2018).

By pairing chemical analysis of blubber POP load with the skin transcriptome it has been possible to start linking long-term POP exposure to changes in cellular pathways to enable a better mechanistic understanding of physiological impacts. Biopsies were collected from 22 bottlenose dolphins (*Tursiops truncatus*), from coastal and offshore populations, within the southern California Bight (Trego et al. 2019a). This study used transcriptomics to both test and generate hypotheses regarding the putative impacts of POP exposure in the region and whether this differs based on proximity to urban areas. A range of anthropogenic POPs were quantified in dolphin blubber; their selection for analysis was based on previous research indicating they pose a risk of endocrine disruption to other species in the area (Trego et al. 2018), including DDT-related compounds, polychlorinated biphenyl (PCBs), polybrominated diphenyl ethers (PBDEs), and chlordane-related compounds. POP levels were then compared with the skin transcriptome as well as blubber testosterone levels to test for evidence of endocrine disruption. These animals are highly mobile and their exposure to POPs has likely been lifelong, thereby precluding the use of a BACI sampling design as in the killifish studies described above. Instead, we collected considerable amounts of metadata (e.g. stock structure, temperature, gender) in an attempt to account for the many potential sources of gene expression and hormone variation in the field so as to effectively test for links between lifelong POP exposures and biological impacts. Another important aspect of this approach was to sample dolphins that occupy distinct environments (coastal and offshore) that may vary in their POP exposure and resulting physiological responses.

7.4.1 Testing hypotheses

Prior studies on marine mammal toxicology in southern California have largely focused on POP exposure without any attempt to characterize physiological consequences. As such, basic hypotheses regarding the physiological impact of POPs on marine mammals in this region remain untested, including whether POPs are activating known molecular pathways that lead to adverse health outcomes. For example, POPs are known to exert toxicity through activation of the AHR signalling pathway (Fossi et al. 2010). Coastal dolphins that had higher burdens of POPs had elevated expression of this pathway; this was consistent with our hypothesis that POP burdens in coastal dolphins were sufficiently high to be causing toxicity (Figure 7.2). In contrast, offshore dolphins with lower POP burdens had lower expression of this pathway, indicating that they had not been exposed to biologically consequential concentrations of

POPs. This aligns with our prediction that dolphins in more human-impacted coastal environments are exposed to POPs at levels that could have physiological consequences.

Since POPs are known endocrine disruptors in mammals, we further hypothesized that elevated POP burdens would affect endocrine function. Indeed, we found significant correlations between POPs and hormone-related gene expression that are consistent with endocrine disruption, particularly of oestrogen and thyroid hormones. Similar to the AHR signalling pathways, the coastal dolphins with higher POPs exhibited evidence of endocrine disruption while the same pattern was not apparent in animals located farther offshore. A similar pattern emerged when examining blubber testosterone in relation to POP levels, an approach previously used to demonstrate evidence of potential endocrine disruption in common dolphins in the same region (Figure 7.2; Trego et al. 2018). This suggests that the disruption of endocrine genes in coastal bottlenose dolphins may also translate into altered production of reproductive hormones. Integrating multiple biological endpoints, such as hormone levels, can improve insight into the link between functional genomic patterns and known markers of individual and population-level health. Thus, southern California dolphins may be at risk for POP exposure via traditional toxicological pathways, similar to other marine mammals. By looking at pathways known to respond to POP exposure, this research contributes to the weight of evidence that POP exposures across the globe cause endocrine disruption in marine mammals.

7.4.2 Generating hypotheses

Correlations between POP exposures in southern California dolphins and molecular response pathways promote hypotheses about biological impacts that can be tested in future studies, including testing for POP-related changes in immune response, hormone metabolism, and DNA repair pathways. Coastal dolphins with higher POP body burdens also exhibited lower expression of immune genes, particularly those that are important for adaptive immunity and function of T and B cells (Trego et al. 2019a). These findings contribute to several

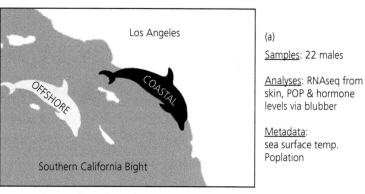

(a)

Los Angeles

OFFSHORE

COASTAL

Southern California Bight

Samples: 22 males

Analyses: RNAseq from skin, POP & hormone levels via blubber

Metadata:
sea surface temp.
Poplation

Account for variance due to sea surface femperature and population

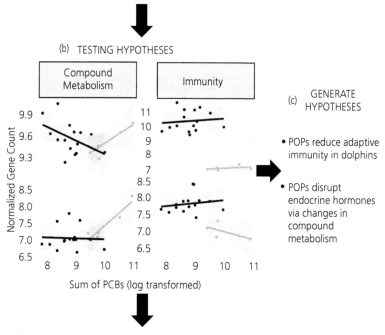

(b) TESTING HYPOTHESES

Compound Metabolism

Immunity

(c) GENERATE HYPOTHESES

• POPs reduce adaptive immunity in dolphins

• POPs disrupt endocrine hormones via changes in compound metabolism

Normalized Gene Count

Sum of PCBs (log transformed)

(d) INTEGRATE ADDITIONAL BIOMARKERS

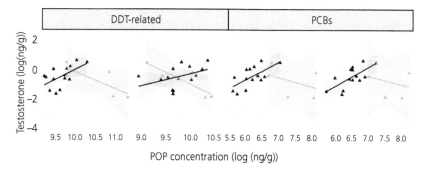

DDT-related

PCBs

Testosterone (log(ng/g))

POP concentration (log (ng/g))

Figure 7.2 The analytical workflow for investigating contaminant-associated transcriptomic patterns in southern California dolphins. (A) Twenty-two remote biopsies were collected from male bottlenose dolphins in the Southern California Bight and RNAseq, contaminant, and hormone data were collected. Variance in gene expression associated with sea surface temperature and population was first accounted for, then residual variance was examined for relationships to POP load. (B) Specific hypotheses were then examined, including whether POP load increased expression of compound metabolism (e.g. AHR pathway activation), reduced expression of immune genes, and endocrine disruption (black and grey data represent offshore and coastal samples, respectively) (see Trego et al. 2019a). (C) The full transcriptome analysis was then analysed for differential expression of individual genes and gene networks, providing a source of data with which to generate hypotheses to further explore the mechanistic relationship between POPs and health effects. (D) After confirming POP load was enough to cause gene expression changes predictive of health impacts, data were examined for POP effects on hormone levels.

hypotheses that could be addressed in future research, such as whether POPs cause reduced immune function via changes in the adaptive immune system leading to increased risk of disease. Data produced in this study also suggest that POPs may be related to the level of hormone metabolism. Though it is well understood that POPs can alter hormone levels in marine mammals, the mechanisms that lead to endocrine disruption are less clear. To determine the mechanism causing endocrine disruption after POP exposure in this species, future research could examine the hypothesis that POPs contribute to endocrine disruption through perturbation of hormone metabolism pathways. Further, a network of genes characterized by DNA repair genes significantly correlated with DDT-related compounds, indicating that high exposure to DDT-related compounds could have changed the level of DNA repair activity in dolphin skin. Testing these hypotheses could further elucidate the mechanisms of action in this region theorized to contribute to decreased reproductive success and increased disease susceptibility in wild marine mammals.

A strength of the dolphin study described here was the collection of metadata on known environmental covariates (e.g. sea surface temperature; Trego et al. 2019b) to help account for additional sources of transcriptomic variation within and among populations. Accounting for these variables in the analysis strengthened the association between POP burdens and impacts on physiological pathways. Additionally, a genome-wide approach also provided greater scope for inference of biological impacts, compared with a strictly hypothesis-driven approach: POP associations with hypothesized pathways and other pathways of potential physiological consequence were both revealed. POP exposure did not just change expression of one or two immune genes but rather altered suites of immune genes that provide a foundation for future research on POP impacts on disease susceptibility that may be of relevance for population health and sustainability.

7.4.3 Conservation implications

Successful conservation of marine mammal populations is dependent on having enough predictive

power to detect the impact of sublethal stressors before populations experience significant impacts on reproduction or survival. This is particularly critical for species with long generation times because it can take a long time for reproductive impacts to manifest, and a long time to recover from a decline. In the United States, marine mammal population management is required by the Marine Mammal Protection Act and the Endangered Species Act, both of which require an understanding of the cumulative effects of different anthropogenic stressors on marine mammals and how sublethal stressors threaten population persistence via impacts to reproduction or survival. The transcriptomic analyses included in this example provide one of few ways to mechanistically connect sublethal stressors like contaminants with reproductive and survival impacts in marine mammals before populations have passed important tipping points. At present, protective regulations are often developed well after reproductive impacts have already been apparent for several years without a clear understanding of which stressors are having the greatest impact on population fitness. These regulations can take a long time to develop and their success can be limited without substantial evidence of the source of the decline and what level of protection is needed. For example, if contaminants are the primary cause of a population decline, it is important to know the compounds having the greatest impact and the doses of these compounds that can be tolerated without significantly inhibiting physiological pathways important for survival and reproduction. This information can be used to predict how planned actions will impact marine mammal populations directly, identify how sensitive a population may be to additional exposure to stressors, and enhance our understanding of the cumulative impacts that human activities have on marine mammals.

7.5 Conclusions and future directions

Transcriptome profiling can be a useful tool for diagnosing cause–effect relationships in ecological and conservation physiology, insofar as gene transcription is a key link between stressor-specific molecular initiating events and emergent physiological responses that can alter fitness and population resilience. Since all expressed transcripts are profiled, the data can be used to both test and generate hypotheses. Transcriptomics resources can be developed and deployed for any species, at much lower cost and throughput than ever before. They are best used in combination with measures of physiological status (e.g. performance, behaviour, immune function, reproduction) to increase scope for inference and promote conservation actions. Careful experimental design, for example including enough sampling of individuals, populations, and field sites, and thorough collection of appropriate metadata (e.g. environmental conditions, health status), is crucial for strengthening inference of causative agents and delineating how they may threaten population resilience.

Transcriptomics can further illuminate the molecular mechanisms that propagate stressor exposure to states of distress, providing more detailed information about cause and effect than other more common measures of physiological distress used to monitor wildlife health (e.g. high cortisol levels). That is, these information-rich data offer an opportunity for fingerprinting agents that cause adverse physiological outcomes, and for identifying how these adverse outcomes relate to important population parameters such as reproductive success. Furthermore, knowledge of how gene networks, rather than individual genes, vary across environments can be used to develop 'early-response' diagnostics to detect physiological changes prior to population decline (Orsini et al. 2018), allowing managers to develop more efficient and effective conservation actions. As more information is generated about molecular mechanisms of action for a variety of environmental stressors, it will elevate the diagnostic power of transcriptomics and may serve as environmental forensic signatures of organism exposure and effect that could inform conservation managers on how other human perturbations may be affecting population health.

Much is to be gained from future research that establishes stronger linkages between effects at different levels of biological organization. Molecular responses that are predictive of physiological responses that affect fitness will be most useful for informing conservation strategies and allocation of management resources. The Adverse Outcome Pathway (AOP) framework was developed by the United States Environmental Protection Agency (EPA) for this purpose. AOPs provide a structured approach for linking molecular initiating events through the various levels of biological response through to population-level effects (Ankley et al. 2010) (Figure 7.3). Though developed by the EPA to estimate impacts of environmental toxicants, we think that the broader eco-physiology, conservation physiology, and climate change physiology communities could benefit from this conceptual framework; this provides a framework within which to identify and fill knowledge gaps to strengthen our abilities to diagnose perturbations and predict their impacts on population health. Genome-wide gene expression profiling provides key information to strengthen the link between molecular initiating events and physiological outcomes within the AOP framework (Brockmeier et al. 2017).

Dynamic Energy Budget (DEB) models (Figure 7.4) are a promising tool for connecting physiology with impacts on fitness and population resilience. DEBs are based on the theory surrounding how energy is allocated in an organism and how that allocation changes in response to perturbation (Kooijman, 2009). For example, when an individual is exposed to a stressor, energy is diverted towards basic maintenance and away from functions like reproduction. Lower energy investment towards reproduction can predict impacts on population dynamics. Ongoing research efforts seek to identify molecular responses that predict physiological tipping points where DEB models predict impacts at the population level; impacts that would trigger conservation interventions. Including transcriptomics within these modelling frameworks will significantly improve the predictive power of this technology and our mechanistic understanding of ecological impacts of exposure to human-induced perturbations.

Another key challenge for conservation physiology studies is to disentangle the impacts of multiple potential stressors (Crain et al. 2008) and to assign their relative responsibility for adverse impacts on population health. In scenarios where a single stressor exerts a large effect, establishing cause and effect is feasible as demonstrated by the

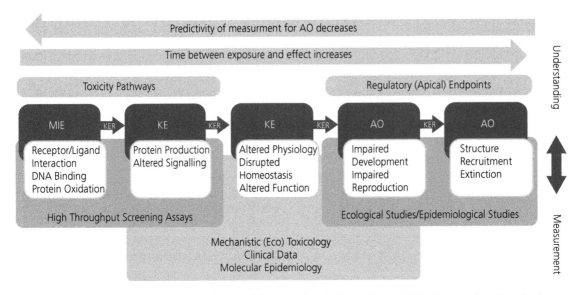

Figure 7.3 Schematic of the Adverse Outcome Pathway (AOP) framework (from Ankley and Edwards 2018). High-throughput biomarker data such as molecular responses are linked with physiology, and together are used to predict outcomes that are relevant for organismal fitness and population resilience. Well-established AOPs provide much scope for linking cause and effect, and therefore provide important information about when and how to implement conservation/mitigation activities (MIE, KE, and AO are acronyms for molecular initiating events, key events, and adverse outcomes, respectively).

examples discussed in this chapter. However, multiple stressors could interact in complex ways such that molecular signals of each individual stressor is no longer discrete or discernible (Finne et al. 2007). Given that wild species often live under suboptimal conditions and are exposed to much environmental variability, careful and thoughtful experimental design is therefore necessary. For example, the BACI field sampling design, including many field sites, was important for establishing a link between oil exposure and biological responses in killifish (as outlined in the previous section). Similarly, by including measures of sea surface temperature, Trego and colleagues (2019b) were able to more clearly discern association between POP exposure and potential immune dysfunction. The more we can disentangle the molecular mechanism of action of individual stressors from other sources of variation, the more we will expand our capacity to interpret data from experiments in the field where multi-stressor scenarios are common. Such information can refine our consideration of the relative risks posed by different stressors, and may identify particularly at-risk species or populations leading to timely conservation interventions. These approaches can

also refine our understanding of physiological limits and help predict physiological responses to future scenarios (e.g. Evans and Hofmann, 2012). Studies of multiple stressors will require better gene annotation, stronger connections between gene functions and organismal health and fitness, and dedicated comparisons between field and lab studies, when possible.

Improved annotation of gene function and characterization of molecular pathways and mechanisms of action that underlie organism–environment interactions will be important for improving the scope for inference that can be achieved with transcriptomics data. Toxicologists have been studying and revealing molecular mechanisms of action of pollutants for decades. This work provides a rich foundation for linking perturbations in gene expression with adverse health outcomes (Davis et al. 2019) that pose a threat to wildlife populations. Much less is known of molecular mechanisms of action for other human-induced stressors, such as those associated with climate change (e.g. thermal resilience, hypoxia, acidification). This will change, in part, by collecting more transcriptomics data from field and laboratory studies on diverse species

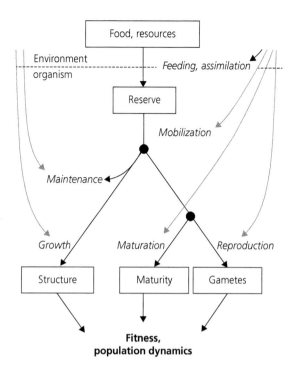

Figure 7.4 Schematic of the standard Dynamic Energy Budget (DEB) model (bold black arrows) (e.g. modified from Jager et al. 2014). Thin grey arrows represent environmental perturbations that can affect energy budget parameters. Mechanistic data, such as transcriptomics, can reveal biological mechanisms underlying responses to environmental perturbation. Responses that are indicative of exposures sufficient to impair energy budgets would be predictive of impacts on fitness and population dynamics, and would therefore be particularly informative biomarkers for conservation practitioners.

exposed to diverse conditions. As more information is generated about molecular mechanisms of action for a variety of environmental stressors, this will elevate the diagnostic power of transcriptomics to inform how other human perturbations (non-toxicants) may be affecting organismal health, and how toxicants may interact with those other stressors. However, collection of more data is not enough; these data will need to be curated in such a way as to allow for the creation of new gene ontology terms (The Gene Ontology Consortium 2019) associated with responses to stressors of physiological relevance (Pavey et al. 2012; Stanford and Rogers 2018). By improving gene annotations and pooling transcriptomic data across various ecological and evolutionary scales, we will be better able to interpret

the physiological relevance of transcriptome data to identify populations with mechanistically distinct responses that need to be prioritized for conservation.

References

Ankley, G.T., Bennett, R.S., Erickson, R.J. et al., 2010. Adverse outcome pathways: a conceptual framework to support ecotoxicology research and risk assessment. *Environmental Toxicology and Chemistry*, 29(3), 730–41. https://doi.org/10.1002/etc.34.

Ankley, G.T. and Edwards, S.W., 2018. The adverse outcome pathway: a multifaceted framework supporting 21st century toxicology. *Current Opinion in Toxicology*, 9, 1–7. https://doi.org/10.1016/j.cotox.2018.03.004

Ayres, K.L., Booth, R.K., Hempelmann, J.A. et al., 2012. Distinguishing the impacts of inadequate prey and vessel traffic on an endangered killer whale (*Orcinus orca*) population. *PLoS ONE*, 7(6), e36842. https://doi.org/10.1371/journal.pone.0036842.

Barón, E., Giménez, J., Verborgh, P. et al., 2015. Bioaccumulation and biomagnification of classical flame retardants, related halogenated natural compounds and alternative flame retardants in three delphinids from Southern European waters. *Environmental Pollution*, 203, 107–15. https://doi.org/10.1016/j.envpol.2015.03.041.

Basu, U., Almeida, L.M., Dudas, S. et al., 2012. Gene expression alterations in Rocky Mountain elk infected with chronic wasting disease. *Prion*, 6(3), 282–301. https://doi.org/10.4161/pri.19915.

Bayha, K.M., Ortell, N., Ryan, C.N. et al., 2017. Crude oil impairs immune function and increases susceptibility to pathogenic bacteria in southern flounder. *PLoS ONE*, 12(5), e0176559. https://doi.org/10.1371/journal.pone.0176559.

Blasius, M.E. and Goodmanlowe, G.D., 2008. Contaminants still high in top-level carnivores in the Southern California Bight: levels of DDT and PCBs in resident and transient pinnipeds. *Marine Pollution Bulletin*, 56(12), 1973–82. https://doi.org/10.1016/j.marpolbul.2008.08.011.

Brockmeier, E.K., Hodges, G., Hutchinson, T.H. et al., 2017. The role of omics in the application of adverse outcome pathways for chemical risk assessment. *Toxicological Sciences*, 158(2), 252–62. https://doi.org/10.1093/toxsci/kfx097.

Buckman, A.H., Veldhoen, N., Ellis, G. et al., 2011. PCB-associated changes in mRNA expression in killer whales (*Orcinus orca*) from the NE Pacific Ocean. *Environmental Science & Technology*, 45(23), 10194–202. https://doi.org/10.1021/es201541j.

Burreau, S., Zebühr, Y., Broman, D., and Ishaq, R., 2006. Biomagnification of PBDEs and PCBs in food webs from the Baltic Sea and the northern Atlantic Ocean. *Science of the Total Environment*, 366(2–3), 659–72. https://doi.org/10.1016/j.scitotenv.2006.02.005.

Congress.gov, 2011. US Senate Congressional Record for October. Available at: https://www.congress.gov/congressional-record/2011/10/11/senate-section/article/S6366-1 (accessed 25 June 2020).

Connon, R.E., Jeffries, K.M., Komoroske, L.M. et al., 2018. The utility of transcriptomics in fish conservation. *Journal of Experimental Biology*, 221(Pt 2). https://doi.org/10.1242/jeb.148833.

Crain, C.M., Kroeker, K., and Halpern, B.S., 2008. Interactive and cumulative effects of multiple human stressors in marine systems. *Ecology Letters*, 11(12), 1304–15. https://doi.org/10.1111/j.1461-0248.2008.01253.x.

Davis, A.P., Grondin, C.J., Johnson, R.J. et al., 2019. The Comparative Toxicogenomics Database: update 2019. *Nucleic Acids Research*, 47(D1), D948–54. https://doi.org/10.1093/nar/gky868.

De Wit, P., Rogers-Bennett, L., Kudela, R.M., and Palumbi, S.R., 2014. Forensic genomics as a novel tool for identifying the causes of mass mortality events. *Nature Communications*, 5(1). https://doi.org/10.1038/ncomms4652.

Desforges, J.-P., Levin, M., Jasperse, L., et al., 2017. Effects of polar bear and killer whale derived contaminant cocktails on marine mammal immunity. *Environmental Science & Technology*, 51(19), 11431–9. https://doi.org/10.1021/acs.est.7b03532.

Dubansky, B., Whitehead, A., Miller, J.T. et al., 2013. Multitissue molecular, genomic, and developmental effects of the Deepwater Horizon oil spill on resident Gulf killifish (*Fundulus grandis*). *Environmental Science & Technology*, 47(10), 5074–82.

Dubansky, B., Whitehead, A., Rice, C., Galvez, F., 2014. Response to comment on 'Multi-tissue molecular, genomic, and developmental effects of the Deepwater Horizon Oil spill on resident Gulf killifish (*Fundulus grandis*)'. *Environmental Science & Technology*, 48(13), 7679–80. https://doi.org/10.1021/es501185a.

Evans, T.G. and Hofmann, G.E., 2012. Defining the limits of physiological plasticity: how gene expression can assess and predict the consequences of ocean change. *Philosophical Transactions of the Royal Society B: Biological Sciences*, 367(1596), 1733–45. https://doi.org/10.1098/rstb.2012.0019.

Finne, E.F., Cooper, G.A., Koop, B.F. et al., 2007. Toxicogenomic responses in rainbow trout (*Oncorhynchus mykiss*) hepatocytes exposed to model chemicals and a synthetic mixture. *Aquatic Toxicology*, 81(3), 293–303. https://doi.org/10.1016/j.aquatox.2006.12.010.

Fleishman, E., Blockstein, D.E., Hall, J.A. et al., 2011. Top 40 priorities for science to inform US conservation and management policy. *BioScience*, 61(4), 290–300. https://doi.org/10.1525/bio.2011.61.4.9.

Fossi, M.C., Urban, J., Casini, S. et al., 2010. A multi-trial diagnostic tool in fin whale (*Balaenoptera physalus*) skin biopsies of the Pelagos Sanctuary (Mediterranean Sea) and the Gulf of California (Mexico). *Marine Environmental Research*, 69, S17–20. https://doi.org/10.1016/j.marenvres.2009.10.006.

Garcia, T.I., Shen, Y.J., Crawford, D. et al., 2012. RNA-Seq reveals complex genetic response to deepwater horizon oil release in Fundulus grandis. *BMC Genomics*, 13. https://doi.org/10.1186/1471-2164-13-474.

Godard-Codding, C.A.J. and Fossi, M.C., 2018. Field sampling techniques and ecotoxicologic biomarkers in cetaceans. In M.C. Fossi and C. Panti, eds. *Marine Mammal Ecotoxicology*, pp. 237–59. Academic Press, London. https://doi.org/10.1016/B978-0-12-812144-3.00009-7.

Goerke, H., Weber, K., Bornemann, H. et al., 2004. Increasing levels and biomagnification of persistent organic pollutants (POPs) in Antarctic biota. *Marine Pollution Bulletin*, 48(3–4), 295–302. https://doi.org/10.1016/j.marpolbul.2003.08.004.

Hall, A.J., McConnell, B.J., Rowles, T.K. et al., 2005. Individual-based model framework to assess population consequences of polychlorinated biphenyl exposure in bottlenose dolphins. *Environmental Health Perspectives*, 114(S-1), 60–4. https://doi.org/10.1289/ehp.8053.

Hoh, E., Dodder, N.G., Lehotay, S.J. et al., 2012. Nontargeted comprehensive two-dimensional gas chromatography/time-of-flight mass spectrometry method and software for inventorying persistent and bioaccumulative contaminants in marine environments. *Environmental Science & Technology*, 46(15), 8001–8. https://doi.org/10.1021/es301139q.

Jager, T., Barsi, A., Hamda, N.T. et al., 2014. Dynamic energy budgets in population ecotoxicology: applications and outlook. *Ecological Modelling*, 280, 140–7. https://doi.org/10.1016/j.ecolmodel.2013.06.024.

Jenkins, K.D., Branton, M.A., and Huntley, S., 2012. CYP1A expression fails to demonstrate exposure-response relationship. *Proceedings of the National Academy of Sciences of the United States of America*, 109(12), E678–8. https://doi.org/10.1073/pnas.1121372109.

Jensen, O., Martin, C., Oken, K. et al., 2019. Simultaneous estimation of dispersal and survival of the gulf killifish *Fundulus grandis* from a batch-tagging experiment. *Marine Ecology Progress Series*, 624, 183–94. https://doi.org/10.3354/meps13040.

Jobling, S., Nolan, M., Tyler, C.R. et al., 1998. Widespread sexual disruption in wild fish. *Environmental Science &*

Technology, 32(17), 2498–506. https://doi.org/10.1021/es9710870.

Kassahn, K.S., Crozier, R.H., Ward, A.C. et al., 2007. From transcriptome to biological function: environmental stress in an ectothermic vertebrate, the coral reef fish *Pomacentrus moluccensis. BMC Genomics, 8*(1), 358. https://doi.org/10.1186/1471-2164-8-358.

Kellar, N.M., Trego, M.L., Marks, C.M., and Dizon, A., 2006. Determining pregnancy from blubber in three species of delphinids. *Marine Mammal Science, 22*(1), 1–16.

Kooijman, B., 2009. *Dynamic Energy Budget Theory for Metabolic Organisation,* third edition. Cambridge University Press, Cambridge. https://doi.org/10.1017/CBO9780511805400.

Li, H., Zhang, J., and You, J., 2018. Diagnosis of complex mixture toxicity in sediments: application of toxicity identification evaluation (TIE) and effect-directed analysis (EDA). *Environmental Pollution, 237,* 944–54. https://doi.org/10.1016/j.envpol.2017.11.005.

Lunardi, D., Abelli, L., Panti, C. et al., 2016. Transcriptomic analysis of bottlenose dolphin (*Tursiops truncatus*) skin biopsies to assess the effects of emerging contaminants. *Marine Environmental Research, 114,* 74–9. https://doi.org/10.1016/j.marenvres.2016.01.002.

Mollenhauer, M.A.M., Carter, B.J., Peden-Adams, M.M. et al., 2009. Gene expression changes in bottlenose dolphin, *Tursiops truncatus,* skin cells following exposure to methylmercury (MeHg) or perfluorooctane sulfonate (PFOS). *Aquatic Toxicology, 91*(1), 10–18. https://doi.org/10.1016/j.aquatox.2008.09.013.

Murawski, S.A., Hogarth, W.T., Peebles, E.B., and Barbeiri, L., 2014. Prevalence of external skin lesions and polycyclic aromatic hydrocarbon concentrations in Gulf of Mexico fishes, post-Deepwater Horizon. *Transactions of the American Fisheries Society, 143*(4), 1084–97. https://doi.org/10.1080/00028487.2014.911205.

Neely, M.G., Morey, J.S., Anderson, P. et al., 2017. Skin transcriptomes of common bottlenose dolphins (*Tursiops truncatus*) from the northern Gulf of Mexico and southeastern U.S. Atlantic coasts. *Marine Genomics, 38,* 45–58. https://doi.org/10.1016/j.margen.2017.08.002.

Oldroyd, B.P., 2007. What's killing American honey bees? *PLoS Biology, 5*(6), e168. https://doi.org/10.1371/journal.pbio.0050168.

Orsini, L., Brown, J.B., Shams Solari, O. et al., 2018. Early transcriptional response pathways in *Daphnia magna* are coordinated in networks of crustacean-specific genes. *Molecular Ecology, 27*(4), 886–97. https://doi.org/10.1111/mec.14261.

Pavey, S.A., Bernatchez, L., Aubin-Horth, N., and Landry, C.R., 2012. What is needed for next-generation ecological and evolutionary genomics? *Trends in Ecology & Evolution, 27*(12), 673–8. https://doi.org/10.1016/j.tree.2012.07.014.

Pearson, W.H., 2014. Comment on 'Multitissue molecular, genomic, and developmental effects of the Deepwater Horizon oil spill on resident Gulf killifish (*Fundulus grandis*)'. *Environmental Science & Technology, 48*(13), 7677–8. https://doi.org/10.1021/es405220v.

Peterson, S.H., Peterson, M.G., Debier, C. et al., 2015. Deep-ocean foraging northern elephant seals bioaccumulate persistent organic pollutants. *Science of The Total Environment, 533,* 144–55. https://doi.org/10.1016/j.scitotenv.2015.06.097.

Pilcher, W., Miles, S., Tang, S. et al., 2014. Genomic and genotoxic responses to controlled weathered-oil exposures confirm and extend field studies on impacts of the Deepwater Horizon oil spill on native killifish. *PLoS ONE, 9*(9). https://doi.org/10.1371/Journal.Pone.0106351.

Ratcliffe, D.A., 1963. The status of the peregrine in Great Britain. *Bird Study, 10*(2), 56–90. https://doi.org/10.1080/00063656309476042.

Ross, P., De Swart, R., Addison, R. et al., 1996. Contaminant-induced immunotoxicity in harbour seals: wildlife at risk? *Toxicology, 112*(2), 157–69.

Santos, E.M., Workman, V.L., Paull, G.C. et al., 2007. Molecular basis of sex and reproductive status in breeding zebrafish. *Physiological Genomics, 30*(2), 111–22. https://doi.org/10.1152/physiolgenomics.00284.2006.

Shaul, N.J., Dodder, N.G., Aluwihare, L.I. et al., 2015. non-targeted biomonitoring of halogenated organic compounds in two ecotypes of bottlenose dolphins (*Tursiops truncatus*) from the Southern California Bight. *Environmental Science & Technology, 49*(3), 1328–38. https://doi.org/10.1021/es505156q.

Stanford, B.C.M. and Rogers, S.M., 2018. R(NA)-tistic expression: the art of matching unknown mRNA and proteins to environmental response in ecological genomics. *Molecular Ecology, 27*(4), 827–30. https://doi.org/10.1111/mec.14419.

Teuten, E.L. and Reddy, C.M., 2007. Halogenated organic compounds in archived whale oil: a pre-industrial record. *Environmental Pollution, 145*(3), 668–71. https://doi.org/10.1016/j.envpol.2006.08.022.

The Gene Ontology Consortium, 2019. The Gene Ontology Resource: 20 years and still GOing strong. *Nucleic Acids Research, 47*(D1), D330–8. https://doi.org/10.1093/nar/gky1055.

Trego, M.L., Hoh, E., Kellar, N.M. et al., 2018. Comprehensive screening links halogenated organic compounds with testosterone levels in male *Delphinus delphis* from the Southern California Bight. *Environmental Science & Technology, 52*(5), 3101–9. https://doi.org/10.1021/acs.est.7b04652.

Trego, M.L., Hoh, E., Whitehead, A. et al., 2019a. Contaminant exposure linked to cellular and endocrine biomarkers in southern California bottlenose dolphins. *Environmental Science & Technology, 53*(7), 3811–22. https://doi.org/10.1021/acs.est.8b06487.

Trego, M.L., Whitehead, A., Kellar, N.M. et al., 2019b. Tracking transcriptomic responses to endogenous and exogenous variation in cetaceans in the Southern California Bight. *Conservation Physiology, 7*(1), coz018. https://doi.org/10.1093/conphys/coz018.

Vesterlund, L., Jiao, H., Unneberg, P. et al., 2011. The zebrafish transcriptome during early development. *BMC Developmental Biology, 11*(1), 30. https://doi.org/10.1186/1471-213X-11-30.

Wang, Z., Gerstein, M., and Snyder, M., 2009. RNA-Seq: a revolutionary tool for transcriptomics. *Nature Reviews Genetics, 10*(1), 57.

Waters, C.N., Zalasiewicz, J., Summerhayes et al., 2016. The Anthropocene is functionally and stratigraphically distinct from the Holocene. *Science (New York, N.Y.), 351*(6269), aad2622. https://doi.org/10.1126/science.aad2622.

Whitehead, A., Dubansky, B., Bodinier, C. et al., 2012a. Genomic and physiological footprint of the Deepwater Horizon oil spill on resident marsh fishes. *Proceedings of the National Academy of Sciences of the United States of America, 109*(50), 20298–302. https://doi.org/10.1073/Pnas.1109545108.

Whitehead, A., Dubansky, B., Bodinier, C. et al., 2012b. Reply to Jenkins et al.: evidence for contaminating oil exposure is closely linked in space and time to biological effects. *Proceedings of the National Academy of Sciences, 109*(12), E679–9. https://doi.org/10.1073/pnas.1200429109.

Young, D.R., McDermott, D.J., and Heesen, T.C., 1976. Aerial fallout of DDT in southern California. *Bulletin of Environmental Contamination and Toxicology, 16*(5), 604–11. https://doi.org/10.1007/BF01685371.

Zhao, S., Fung-Leung, W.-P., Bittner, A. et al., 2014. Comparison of RNA-Seq and microarray in transcriptome profiling of activated T cells. *PLoS ONE, 9*(1), e78644. https://doi.org/10.1371/journal.pone.0078644.

CHAPTER 8

The role of conservation physiology in mitigating social-ecological traps in wildlife-provisioning tourism: a case study of feeding stingrays in the Cayman Islands

Christina A.D. Semeniuk

> ⮑ **Take-home message**
>
> Effective selection and communication of physiological metrics of animal health in wildlife-provisioning tourism can minimize problem-causing and problem-enhancing feedbacks in social-ecological systems.

8.1 Introduction

Wildlife tourism is the most rapidly expanding sector within the tourism industry, generating billions of dollars globally (Meyer et al. 2019), and contributing substantially to the gross domestic products (GDPs) of both developed and developing countries (World Travel and Tourism Council, 2015). This growing demand to interact with wildlife, with between 80–440 million estimated participants, is projected to double over the next 50 years (Moorhouse et al. 2015; Trave et al. 2017) and has given rise to a wide range of wildlife tourism activities that can be both educational and entertaining (Reynolds and Braithwaite 2001; Pratt and Suntikul 2016). When managed for sustainability, the ultimate benefit of wildlife tourism is in its potential to create a positive feedback between resource persistence and tourism demand that results in a common incentive to protect the natural environment (Wilson and Tisdell 2003). Presently, it is unclear whether wildlife tourism is succeeding in its conservation objectives, with several studies examining whether the direct and indirect negative impacts of wildlife tourism on the environment outweigh the positive (see Abrantes et al. 2018).

Coastal and marine wildlife tourism is a growing tourism subsector of wildlife tourism that has become one of the leading sources of economic earnings for countries with coastlines (Garrod and Wilson 2004). Defined as 'non-consumptive' tourism (Burgin and Hardiman 2015), it encompasses 'any tourist activity with the primary purpose of watching, studying or enjoying marine wildlife—flora and fauna that live in the coastal and maritime zone and are dependent on resources from the marine environment' (Masters 1998, p. 6). Increasingly popular is the feeding (provisioning) of marine

Christina A.D. Semeniuk, *The role of conservation physiology in mitigating social-ecological traps in wildlife-provisioning tourism: a case study of feeding stingrays in the Cayman Islands* In: *Conservation Physiology: Applications for Wildlife Conservation and Management.* Edited by: Christine L. Madliger, Craig E. Franklin, Oliver P. Love, and Steven J. Cooke, Oxford University Press (2021). © Oxford University Press. DOI: 10.1093/oso/9780198843610.003.0008 wildlife to increase the chances of viewing animals up close. This activity is frequently used by nature-based tour operators to attract visitors to marine destinations and enhance tourist satisfaction as it greatly improves the chance of sighting animals at close range (Patroni et al. 2018). Allowing visitors to feed target species can permit enhanced interactions and consequently positive visitor experiences (Brookhouse et al. 2013). Provisioning tourism therefore has the potential to significantly impact conservation directly by providing economic benefits, and indirectly through social benefits (Gallagher and Hammerschlag, 2011; Cisneros-Montemayor et al. 2013; Lowe and Tejada 2019). Nonetheless, the practice is controversial (Meekan and Lowe 2019; Ziegler et al. 2019) with concerns being raised about what long-term impacts feeding marine wildlife might have on the target animals, and is restricted or banned in some marine protected zones; however, assessment of the biological effects on marine fauna is limited, and results from different areas and taxa are frequently contradictory (Burgin and Hardiman, 2015). One reason is that the indicators used do not necessarily reflect morbidity or mortality of the target species.

8.2 Conservation physiology—a call to arms

Traditional ecological indicators used in assessing marine wildlife-provisioning effects centre on: (1) modified behaviours—foraging patterns and diel movements, habituation, aggression and other social behaviours, and parental investment; (2) morphology—injury and parasites; (3) breeding responses—reproductive success; (4) target species population dynamics; and (5) community ecology—assemblages and species composition (Bateman and Fleming 2017). With the exception of very few indicators, none outwardly reveals impacts that can be interpreted as detrimental on individual fitness (e.g. growth, reproduction, survival), or population persistence (i.e. a decline in population size can be attributable to emigration, not increased mortality rates). And yet it is these types of indicators that could serve to more definitively contribute to policy and regulation discussions. One reason why estimating provisioning

impacts is so divisive is that many studies lack the long-term monitoring that would reveal any positive or negative implications over time (Burgin and Hardiman 2015). This shortcoming is two-fold: the relative immaturity of the industry itself, and debate over what constitutes 'harmful impact' (Bateman and Fleming 2017). A category of indicators that can address both these issues can be found in physiological responses. Physiological measurements have the potential to reveal mechanistic cause-and-effect relationships and can be translated and scaled from the individual level to population levels (Illing and Rummer 2017). In wildlife provisioning, such indicators can be related to (1) stress (e.g. heartrate, circulating levels of cortisol/corticosterone and epinephrine); (2) immune response and function (e.g. humoral or cell-mediated, white blood cell parameters, parasite loads, oxidative stress); (3) metabolism/respiration rates; (4) development, growth, body condition and survival; (5) impaired responses to injuries/healing wounds (e.g. pH, partial pressure of CO_2, lactate, glucose); and (6) nutritional status (e.g. lipid profile, stable isotopes, triglyceride concentrations, essential vitamins and minerals) (Knapp et al. 2013; Burgin and Hardiman 2015; Barnett et al. 2016; Bateman and Fleming 2017). These metrics not only represent increased sensitivity to health status, but also have a greater capacity for predicting future change (Bergman et al. 2019). Physiological indicators measured at a discrete time point can thus simultaneously reflect the state of the organism and deliver critical information about the current and future state of the tourism activity.

Despite the call to action of incorporating physiology into conservation science (Wikelski and Cooke 2006; Madliger and Love 2015), very few published examples exist where physiological findings have led to improved management practices for marine wildlife in general (Cooke et al. 2017), and fewer still in regard to provisioning tourism (Madliger et al. 2016). In other words, evidence-based *ecological* results of feeding marine wildlife do not translate into adaptive management strategies for long-term sustainability of this tourism activity, despite findings being indicative of positive or negative impacts on the target species. To address this barrier, it is absolutely imperative that conservation biologists consider marine wildlife-provisioning tourism within the context of a complex socio-ecological system. It is well understood how tourism can have positive conservation outcomes by financing marine reserves and their regulation, providing alternative livelihoods, and contributing to overall socio-economic capital. But at its core, it is both wildlife and tourists that are the critical players in wildlife tourism settings, and are the ones responsible for determining the success of any tourism management plan. Too often, the visitor experience is overlooked (Patroni et al. 2018) in how it contributes to the degradation of wildlife resource, and how it can be managed so that resultant impacts on wildlife are mitigated. Human-dimensions research has found that within and across attractions, experience is influenced by ethics, values, motivations and expectations, levels of specialization, and desired wildlife interactions

(i.e. typologies). Moreover, visitors themselves are not homogeneous groups, making a 'one-size-fits-all' management plan problematic (Martin 1997; Moscardo 2000; Higham and Carr 2002; Scott and Thigpen 2003; Curtin and Wilkes 2005; Dearden et al. 2006). It is perhaps no wonder that physiological evidence of detrimental effects of wildlife feeding can have little to no impact on its management: health biomarkers are not often directly linked to impacts on visitor experience (only their actions), findings typically go uncommunicated to the visitors themselves, tourists are not informed how their actions and behaviours can impact wildlife health, and direct tourist input on alternative management plans is not always considered. Being able to translate physiological biomarkers into both animal health and visitor experience can mitigate potential negative impacts on wildlife and preserve

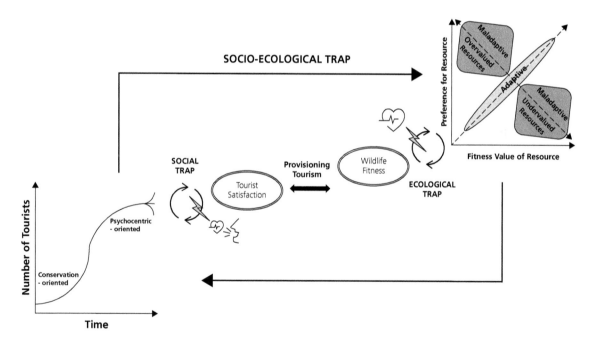

Figure 8.1 Schematic diagram of socio-ecological trap theory for marine wildlife provisioning: tourists, in their pursuit to maximize satisfaction, may engage in activities that negatively affect the tourism system, which then progresses through life-cycle characteristics that end in collapse (i.e. a social trap). Wildlife, in their quest to maximize fitness may prefer and overvalue resources with few beneficial returns, thus making maladaptive decisions that can affect their health and survival (i.e. an ecological trap). A socio-ecological trap occurs when tourist activities such as feeding wildlife directly cause an ecological trap, and as the wildlife system declines, it further exacerbates the social trap, resulting in the exhaustion of the tourist site at an accelerated rate. The careful selection of physiological indicators can uncover health impacts of fed wildlife and hence mitigate an ecological trap from occurring. Communication of physiological indicators to tourists can reveal impacts of their feeding activities, help improve the visitor experience, and prevent the progression of a social trap. Ecological trap diagram modified from Robertson et al. 2017; tourist life-cycle diagram adapted from Duffus and Dearden 1990.

or even enhance the tourist experience. Without this careful and deliberate consideration, the wildlife-tourism system may fall into a social-ecological trap (Figure 8.1). This chapter serves to argue how effective communication of physiological metrics of wildlife health can bridge the gap between science and practice and contribute to minimizing impacts of provisioning as a wildlife tourism activity. As such, I provide a framework that outlines steps that can be undertaken to achieve an enhanced understanding of wildlife feeding as a coupled human–natural system, and demonstrate with a case study on feeding stingrays how physiological indicators play a role in socio-ecological research and resultant management success.

8.3 Socio-ecological traps in wildlife-provisioning tourism

8.3.1 Ecological traps explained

Social-ecological traps (SESs) are lose–lose situations where the pressure imposed by the social system (tourist expectations, revenue) has costs for the ecological system (animal health, maladaptive behaviours, site degradation), which in turn feed back into the social system (loss of tourist satisfaction, loss of revenue), resulting in the demise of both systems (exhaustion). Social-ecological systems become embroiled in problem-causing and problem-enhancing feedbacks (Cumming and Peterson 2017), and to better understand the mechanisms driving SESs, one must look at each component as an integrated whole to help guide the selection of relevant indicators to study. In an ecological context, a trap occurs when human-driven environmental change decouples the cues that animals use to assess habitat quality from the true quality of the environment (Schlaepfer et al. 2002). When animals mistakenly rely on false or altered environmental cues to maximize their fitness, but do not immediately receive the necessary feedback of impending fitness costs, this maladaptive habitat selection leads to reduced survival or reproduction, compared with the probable outcome if the true, high-quality habitat had been selected or remained unchanged (Battin 2004). In a wildlife-tourism setting, the focal animal's natural habitat is being

altered, and these changes can create the potential for an ecological trap, as animals may be misinterpreting the cues in their environment as representing either a good-quality resource when it is not, and vice versa. Examples of ecological traps can include: birds switching to less profitable breeding sites as they believe these alternatives to be higher in quality due to human presence at traditional sites (Higham 1998); sharks attracted to feeding stations may be prone to inbreeding if they experience a reduction in dispersal (Clua et al. 2010); and dolphins with high exposure to tour vessels that choose not to leave the vicinity suffer reduced reproductive success (Higham and Bejder 2008). In provisioning tourism, marine wildlife are attracted to supplemental food sources that can result in an ecological trap should the animal's nutritional or reproductive health be affected, the food attracts predators, or if the higher congregation of competitors results in increased injury rates and/or parasite or disease transmission. In these circumstances, wildlife are making suboptimal, maladaptive decisions of where to forage, using cues that no longer maximize fitness.

8.3.2 Social traps explained

A social trap is any situation in which the short-term, local reinforcements guiding individual behaviour are inconsistent with the long-term, global best interest of the individual and society (Platt 1973; Cross and Guyer 1980). Fundamentally, the apparent short-term gains do not translate into longer-term sustainability (Costanza 1987). Moreover, while any single individual's actions may have negligible environmental consequences, the actions of many individuals damage the environment of the collective, including the individual players (Kilbourne and Pickett 2008). In wildlife tourism, a social trap arises when tourists, in their quest to maximize satisfaction with their wildlife experience, possess values, motivations, and actions that unwittingly cause impacts to the system (Higginbottom et al., 2003). Tourists do not receive the immediate feedback of the costs of their actions, and through resultant repercussions at the tourist site, a social trap occurs when tourist satisfaction becomes negatively affected, or equally, does not reach the maximum it could have under more favourable conditions (Moyle et al. 2013). Indeed, the evolution of the tourist product—a theoretically and empirically represented S-shaped growth pattern of tourist volume over time (Butler 1980, based on the economic product life-cycle concept), culminates in a fatigue of the site due to lifted restrictions to encourage greater visitation, a shift in tourist typology (e.g. conservation-oriented to psychocentric values), socio-economic factors, the appearance of other tourism competitors, and/or the reduction of environmental quality (Duffus and Dearden 1990; Catlin et al. 2011).

8.3.3 The use of physiological indicators for the prevention of social-ecological traps

What makes provisioning tourism unique to other types of wildlife tourism is that each subsystem is vulnerable to its own specific type of trap (ecological, social) independent of the other's direct influence. For example, social traps can occur when sites become overcrowded, and ecological traps can arise should the activity attract predators. However, given the importance of close proximity to and interacting with animals, marine wildlife provisioning is especially prone to both ecological and social traps inexorably linked through food. In the absence of exploring how feeding wildlife can impact both animal health and tourist satisfaction, there is a credible chance the site will become a social-ecological trap from which recovery will be a challenge (Boonstra et al. 2016). Considering provisioning tourism within the SES trap perspective has the advantage of outlining the appropriate types of questions to ask and guiding the inventory of suitable indicators which can then be translated into effective management strategies that can prevent the progression of SES traps. For instance, wildlife tourism activities can be identified that could potentially directly precipitate an ecological trap, thus providing insight into which proxies of animal fitness should be investigated. Similarly, the importance of these activities can be examined in relation to the tourist experience, satisfaction and management support, and the potential to trigger a social trap.

Physiological measures have the advantage of being informative indicators when inventorying the health status of wildlife as they can provide

crucial information on the immediate status of organisms and predict long-term consequences without the need for actual long-term monitoring of population dynamics (Cooke and O'Connor 2010). The information derived from physiological indicators on the stress, immune function, metabolism, body condition, responses to injuries, and nutritional status can then best be translated to policy makers and tourists if these indicators are further linked to tourist activities, thus playing the dual role of cause–effect–cause for both wildlife health *and* tourist satisfaction. Simply stated, tourists made aware of the impacts of their actions may be more willing to alter their activities (Moorhouse et al. 2017), and decision makers will thus be more informed about the limits of acceptable change the tourist system can undertake. While changes in tourist behaviour have been documented in other types of marine wildlife tourism (e.g. whale-watching, shark-diving, sea turtle interactions) (Trave et al. 2017), no case studies have been recorded for using physiological metrics as the driver for tourist change in marine provisioning tourism (although see the case study in Section 8.5). The ability to inform policy decisions therefore cannot be performed without an integrated, coupled model of natural and human systems (Costanza and Voinov, 2001). In a resource-management context, coupled models are used to collectively assess the impacts of policy actions from both biological and social perspectives, and can assess the importance of precaution in decision making, acceptable levels of additional risk, estimates of how long it may take for mitigating measures to take effect, whether effects are reversible, and efficient allocation of conservation resources (Thompson et al. 2000; Faust et al. 2003). Most notably, modelling can be used to build consensus among science, policy, and the public by building mutual understanding and maintaining a substantive dialogue between members of these groups (van den Belt et al. 1998).

8.4 Scenario-planning coupled systems for the management of wildlife-provisioning tourism

By coupling human- and natural systems from the outset, strategies effective in managing visitors and wildlife can be deduced much more readily, and be used to predict support for management alternatives within a scenario-planning framework (Pizzitutti et al. 2017). Scenario planning is an investigative and decision-making tool that offers managers a method for creating more resilient conservation policies by contrasting plausible scenarios to explore the uncertainty surrounding the future consequences of a decision (Peterson et al. 2003). Benefits of using scenario planning include increased understanding of key uncertainties, the incorporation of alternative perspectives and human choice into conservation planning, and greater resilience of decisions to surprise (Ceauşu et al. 2019). To successfully manage wildlife tourism, one must consider the factors that can affect both animal and tourist population persistence, and explore the differential effects alternative management scenarios can have on their respective population dynamics (Cinner et al. 2011). Within a tourism-provisioning context, scenario planning should begin, as a first step, by incorporating the tourist experience, resultant wildlife health impacts, and management actions with associated management costs (e.g. tourist fees; Burgin and Hardiman 2015) to gauge level of support and find an optimal solution for all involved (including wildlife). The crafted scenarios should be evidence-based (e.g. based on findings provided by conservation physiology research), and hence, reliant on measures indicative of population decline.

As an example, Bach and Burton (2016) employed a stated preference choice experiment—a non-market evaluation tool used to understand tourist consumer behaviour under hypothetical scenarios—to quantify preferences of visitors towards potential changes in their dolphin-feeding experience. The authors developed their survey based on the mounting evidence of negative reproductive and behavioural impacts experienced by habituated dolphins in Shark Bay, Western Australia (e.g. Mann et al. 2000). The majority of tourists felt their experience would remain the same if they were no longer able to feed the dolphins themselves, but importantly, their satisfaction would decrease should dolphin reproduction be compromised. Visitors were therefore willing to trade off management aspects should they improve dolphin welfare. What this study demonstrates is

that without physiological evidence of health impacts (if they indeed exist), the acceptance of tourist and regulators for any type of ongoing management and monitoring may be difficult to achieve, especially in the absence of noticeable, easily monitored indicators (survival- and reproductive rates) typically lacking in wildlife-tourism encounters.

8.5 Case study: feeding stingrays as a marine tourism attraction in the Cayman Islands

The case study presented here synthesizes the research conducted on the feeding of southern stingrays (*Hypanus americanus*) at 'Stingray City Sandbar' in the Cayman Islands as a wildlife-tourism attraction within the framework of SES traps. The most popular tourist site in the Cayman Islands is Stingray City Sandbar (SCS), a warm, shallow-water (1.6 m maximum depth) sandbar in the North Sound, approximately 7740 m^2 in area and located roughly 300 m inside the fringing reef. Although the southern stingray is a solitary inhabitant of all shallow bays around the Cayman Islands, only in the vicinity of SCS can stingrays be found year-round in a dense mixed-sex aggregation of individuals. This amassment results from the unregulated quantity of provisioned squid (*Illex* and *Loligo* spp.), a non-natural diet item shipped in from the North Atlantic and North Pacific (Semeniuk, pers. obs.; Gina Ebanks-Petrie, director, Cayman Islands Department of Environment, pers. comm.). The feeding routine (daily, except during the off-season, when weekends are excluded in summer) lasts from early morning until mid-afternoon as tour boats continuously deliver mainly cruise line tourists for an average 45-min visit. Due to its massive popularity, SCS supports over 50 local snorkel and dive tourism operations and hosts over 1 million visitors per year, almost half of all visitors to the island, with tourist numbers having more than doubled since 2000 (CI MoT 2002).

A day-long activity that first began in the early 1980s (Shackley 1998), by the mid-2000s, a maximum of 2500 tourists could be present at a given time at the shallow sandbar, engaged in feeding, touching, and holding of stingrays as part of their marine tourism experience. Some tour operators provided only the most rudimentary information, while others delivered an informative in-water session. The organized trip also provided photo opportunities, with some tour operators holding the ray in or out of the water or placing it on people's backs and heads while the picture is taken. Conservatively, stingray-related revenue for the local economy is estimated to be as high as US$50 million annually (Vaudo et al. 2018).

A lack of management or codes of practice since the SCS's inception in 1984 had resulted in significant tourist congestion, with stakeholders (government officials, tour operators, tourists, and locals) expressing concern about the long-term sustainability of the attraction (G. Ebanks-Petrie, C.A.D.S., pers. obs.). In 2003, Cayman Island stakeholders convened a committee to agree upon a set of detailed rules for stingray protection and crowding alleviation for SCS. While each proposed regulation considered alone could be expected to redress the known problems (e.g. limits on boat density would be likely to reduce the risk of boat-related injuries for stingrays and/or reduce congestion), the outcome of the simultaneous application of these and other regulations was uncertain, and faced opposition from some locals and tour operators fearing economic fallout as well as the inherent uncertainty in this little-studied system. As such, the potentially progressive process was in danger of being derailed by the lack of an integrative plan designed to consider both visitor satisfaction and stingray fitness. Consequently, given the tight human–animal interdependence of SCS, proposed management scenarios necessitated a true optimization of human–wildlife needs rather than simple unidirectional decisions (e.g. stingray–human interaction rules could reduce stingray injuries but also dissuade visitors from returning or promoting the site to others).

8.6 SES Trap theory and stingray tourism

8.6.1 Stingray ecological trap

At SCS, there are two human activities that have the potential to cause SES traps (i.e. affect wildlife fitness and tourist demand): feeding stingrays and 'handling' stingrays (either through direct interaction

or indirectly through collisions with boats). For an ecological trap to occur (maladaptive decision making), habitat alteration must first simultaneously alter the cue set with which the animal assesses habitat quality (i.e. increase its attractiveness), and decrease the suitability of the habitat. Next, the ability of the animal to adjust to and persist in these novel conditions must be compromised (Robertson and Hutto 2006). The ecological indicators thus chosen for study were selected to provide information on the general, physiological, and immunological health of the tourist-fed population and reveal whether these animals were being exposed to conditions of an ecological trap (Table 8.1).

Initial research into the system contrasted non-esterified fatty acid profiles between tourist- and non-tourist stingrays as a marker of diet composition, lipid requirements, and nutritional status (using non-fed stingrays as baseline). Results revealed multiple findings, with the first being non-random habitat use: stingrays at SCS were incorporating the tourist-supplemented food as the major item in their diet (Semeniuk et al. 2007), becoming habituated to the constant supply of provisioned food. This behavioural diet shift, in combination with very high yearly recapture rates (Corcoran 2006; Corcoran et al. 2013; Vaudo et al. 2018), suggested strong site fidelity, and hence, an attraction to the site. Second, this physiological indicator revealed the provisioned stingrays exhibited essential fatty acid ratios, specific to both species and habitat, comparable with those of elasmobranchs from cold-water environs, implying that the provisioned food did not provide a similar nutritional lipid composition required in tropical habitats (Semeniuk et al. 2007). The cue (provisioned food) was acting as a false attractor of a 'good-quality' foraging site. Third, an additional study demonstrated that the novel grouping behaviour that stingrays exhibited at SCS imposed significant costs in the form of increased aggression, parasite load, and injury, when compared with stingrays from control, non-tourist locations around the island (Semeniuk and Rothley 2008). What was still lacking, however, was definitive evidence of long-term costs that would reveal the occurrence of an ecological trap.

Using stingrays from non-tourist sites about Grand Cayman as a basis for comparison, findings

Table 8.1 List of ecological indicators examined between tourist-fed stingrays at Stingray City Sandbar and stingrays at three non-tourist control sites around Grand Cayman, Cayman Islands.

Wildlife fitness indicator category	Indicator	Description	Ecological condition
General health			
	Parasite load (virulence transmission)	Parasitic ectodermal isopods (*Gnathia* sp.) located in stingray spiracles	Atypical grouping
	Conspecific bite marks (aggression)	Fresh and scarred semi-circular bite marks on stingray pectoral fin margins	Atypical grouping/ non-natural diet
	Collision injuries	Fresh and scarred boat-propeller and anchor chain wounds on body	Human contact
	Predator injuries	Missing stingray tails, fresh and scarred triangular (shark-tooth) wounds on body	Atypical grouping
Physiological health			
	Serum essential fatty acids (EFAs)	Preformed long-chain fatty acids important for normal growth, development, and reproduction; relative and absolute amounts linked to metabolic demands of disease resistance and immune response	Unnatural diet
	Haematocrit	Relative amount of red blood cells in total blood volume; reflects intensity of oxygen transport. Low values indicative of bacterial or parasite infections, starvation, or scarcity of micronutrients	Unnatural diet
	Serum protein concentration	Circulating proteins in peripheral blood used as an index of total protein reserves; can be used to assess dietary inadequacies and other vital biological functions	Unnatural diet/ human contact
	Total antioxidant capacity (TAC)	Total concentration of endogenously produced enzymes, low-molecular-weight molecules, and exogenous and food-derived antioxidants (e.g. vitamin K, urea, and glutathione) used to protect against damage from reactive oxygen species	Unnatural diet/ human contact
Immunological health			
	Leukocrit	Fraction of white blood cells in total blood volume; if values are high can suggest possible pathogen infection; if low, stress-induced immunosuppression	Human contact/ atypical grouping
	White blood cells	Differential counts of lymphocytes (L), granulocytes (heterophils [H] and eosinophils [E]), monocytes (M), and thrombocytes (T). Low L, M, and T can be indicative of compromised cell-mediated immunity and antibody production; high H, E and T increase with infection, disease, and stressful conditions	Human contact/ atypical grouping
	Total oxidative status (TOS)	Reactive oxygen (and nitrogen) species as a result of cellular metabolism. Excess concentrations can damage cell structures, deplete energy, and cause early apoptosis. A low TAC:TOS ratio is sign of oxidative stress	Human contact/ atypical grouping

showed in this natural experiment that tourist-exposed stingrays exhibited haematological changes indicative of physiological costs of wildlife tourism in the form of suboptimal health and attenuation of the defence system (Semeniuk et al. 2009). More specifically, stingrays displayed lower haematocrit, total serum protein concentrations, and oxidative stress (i.e. lower total antioxidant capacity combined with higher total oxidative status).

Moreover, they showed evidence of attenuation of the defence system: for provisioned stingrays only, animals possessing both injuries and high parasite loads also exhibited lower leukocrit (packed white blood cells), serum proteins, and antioxidant potential, as well as differing proportions of differential leukocyte cell types indicative of immunosuppression (lymphocytes and heterophils) and down-regulation (eosinophils), suggesting that the

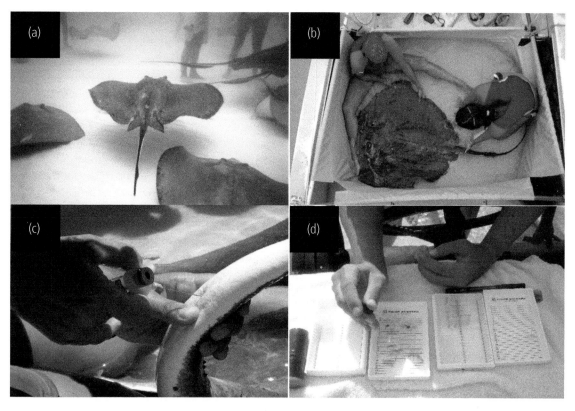

Figure 8.2 Stingray City Sandbar. (A) Stingrays and tourists. (B) Health assessment. (C) Drawing blood sample from underside of tail. (D) Performing blood smears for cell counts. Photo (A) courtesy of Matthew Potenski. Photo credit for (B), (C), and (D): Christina Semeniuk.

physiological changes of tourist stingrays were in partial response to these stressors. Although direct evidence of reduced reproductive success (e.g. measures of fecundity) and survival (e.g. observed mortality) would have been definitive evidence of an ecological cost, this was not logistically feasible. Nevertheless, the physiological indicators integrated multiple stressors: altered social behaviours, non-natural diet, and interactions with tourists and tour boats. These significant responses to provisioning tourism indicated that the long-term health and survival of tourist stingrays were being compromised, revealing the occurrence of an ecological trap (Figure 8.2).

8.6.2 Tourist social trap

To understand whether tourists were prone to a social trap—reduced satisfaction followed by tourist population declines—intercept surveys were conducted on cruise ship passengers upon their immediate return from SCS, as ship passengers comprise over 85 per cent of tourists compared with stay-over visitors at SCS (G. Ebanks-Petrie, pers. comm.). The human-dimensions indicators surveyed were chosen to reflect tourist expectations, wildlife-tourism values, and tourist preferences for management activities at SCS (via a stated-preference discrete choice experiment, similar to Bach and Burton 2016; Louviere and Timmermans 1990). These indices provided information on the conditions at SCS that would (1) have the greatest potential to add to, or detract from, tourist experiences; (2) establish whether education/interpretation would be required in promoting stewardship and awareness among tourists of the types of interactions that affect wildlife fitness; and (3) determine which management options could garner the most (or least) support by visitors. While tourists were very satisfied overall with their wildlife interaction at SCS, the majority were willing to have activities regulated to a certain extent and were willing to pay a conservation access fee to the site. However, key indicators from the data suggested that a social trap was in development: just over one half of the tourists expressed only 'mild' concern for potential negative wildlife impacts arising at the attraction and almost one quarter felt 'low' concern (Semeniuk et al. 2008). In addition, one third of the surveyed tourist population was vociferously against any management of the tourist–wildlife interaction, believing their actions caused no harm. These differences were revealed when using a decision-support tool to calculate their respective market shares of support for alternative management strategies.

The decision-support tool was created as a forecasting tool to estimate which management scenario (and its subsequent potential ecological outcome) would garner the most and least support among respondents. Despite the differences between the two tourist population typologies (i.e. 'pro-management' and 'no-management'), both exhibited a preference for the continuation of feeding and handling the stingrays (albeit at different levels of intensity). Further evidence from the decision-support tool revealed that neither tourist typology was currently experiencing maximized satisfaction from their provisioning experience—this would only be realized if crowding conditions were ameliorated and the risk of harm to stingrays was low. Indeed, tourist realization of the high risk of stingray injury would result in significant diminishment of trip experience coupled with an unwillingness to return for two thirds of the respondent population, again intimating that a social trap, in which reduced satisfaction is followed by tourist population declines, is a distinct possibility. What the collective results revealed were that visitors at SCS, in their quest to maximize their satisfaction, were in fact engaging in activities that were harmful to the health of the animals. Taken together, both ecological and social research revealed that SCS, if not adequately managed, would not be sustainable, and the use of appropriate physiological indicators was instrumental in grounding this realization.

8.7 Scenario planning for management: integrating ecological and social traps

While the survey decision-support tool was key in determining tourist support for different proposed management plans, it was not able to predict future outcomes for the makeup of the resultant tourist-population typology (i.e. proportion of tourists identified as 'pro-management' versus 'no-management'), stingray population dynamics (immigration rates, mortality), nor stingray life expectancy for individuals that choose to remain at SCS. A scenario-planning model was therefore developed using system dynamics modelling to provide illustrative results of how tourist numbers, stingray population size, and stingray life expectancy would change over time under different restrictive management plans (Figure 8.3). The development of the model was guided by the belief that sound ecological management occurs only when social values, preferences, and their resultant ecological effects are equally integrated. Essentially, plans aimed at optimizing wildlife fitness must also be acceptable to tourists. In specific, the model allowed for the ability to evaluate the impacts of alternative management plans on the sustainability of the wildlife tourism attraction by simultaneously exploring the effects of policies on both wildlife health and the tourist experience, governed by evidence-based ecological and social research at SCS. The model then outputted data on stingray population size, stingray life expectancy, and tourist visitation rates (for each tourist typology) over a future time span of 25 years. The model's main structural component was two population submodels: an ecological model of stingray population dynamics (recruitment and mortality estimates based on mark–recapture data, and subjected to sensitivity analyses) and tourist population trajectories (based on the Cayman Islands Department of Tourism's Port Authority of cruise ship tourist numbers). Each submodel was assumed to be affected by management scenarios, and data fed into the model's parameterizations were informed by previous research findings—that is, how management plans were assumed to affect (1) stingray survival and life expectancy (via research on stingray health as

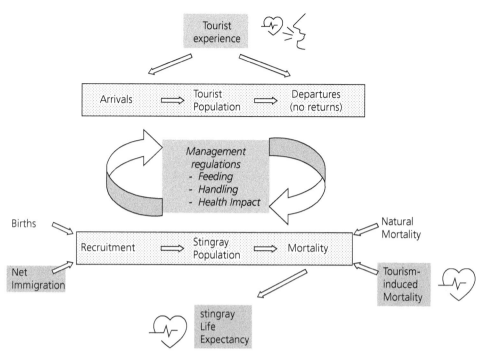

Figure 8.3 Conceptual system dynamics model of Stingray City Sandbar to investigate plausible management options within a scenario-planning framework. The aim is to find optimal plans that address both visitor experience and stingray health. The two population submodels are linked via impacts of management of tourist activities—feeding and handling stingrays—that affect both tourist satisfaction and wildlife health. These management scenarios in turn impact the tourist experience and subsequent visitation rates, stingray immigration rates due to limiting amount of provisioned food, and stingray mortality, morbidity and life expectancy (grey boxes), all of which are integral to the sustainability of the tourism system. Physiology metrics and their communication are indicated where evidence-based results were used to guide research and build realistic conservation models.

revealed by the physiological indicators), and (2) tourist visitation rates (via tourist preferences for management that was assumed to link with behavioural intentions regarding future visitation rates). Importantly, the two submodels were dynamically linked since tourist surveys indicated the extent to which the stingray population and stingray individual fitness influenced tourist experience (and hence visitation), and how tourist actions (feeding and handling) impacted stingray health (informed by the physiological indicators) and influenced stingray immigration rates (via the attraction by unnatural foods).

Scenario modelling confirmed both an ecological and a social trap in the absence of any management (business as usual option): the model predicted lower tourist numbers of both tourist typologies, a larger stingray population size (mainly through immigration), but poorer overall health—all classic hallmarks of social and ecological traps. The model also indicated that under certain management plans, tourists possessing psychocentric values and no willingness to curtail their activities with stingrays would eventually replace the more conservation value-oriented visitors. Again, the model predicted this unsustainable outcome in the long run due to the detrimental effects of these tourist-type-driven activities on stingrays. The optimal management action was one in which there was a reduction in visitor density, mild restricted stingray interactions, and an imposition of a small conservation fee. Over time, although fewer stingrays were predicted to remain at SCS, they would live longer and experience fewer stochastic disease events; the desirable tourist segment was predicted to predominate; the fee could recoup costs imposed either by on-site supervision or management actions; and model simulations were still sustainable after the allotted 25-year time span (Semeniuk et al. 2010). By understanding how management will affect tourist activities and their subsequent impacts on both wildlife health and visitor satisfaction, it was possible to explore the management alternatives that would optimize both, under the edict of preventing a social-ecological trap: a social trap (handling and feeding stingrays) triggering an ecological one (maladaptive attraction to a provisioned food resource).

8.8 Management outcomes at Stingray City Sandbar

SCS stakeholders (e.g. tour operators, general public) believed prior to this research that the tourist-fed stingrays were still foraging predominantly naturally, and that if any negative repercussions were to exist, fed stingrays—being wild and not captive—were free to leave the area. The fact that they were remaining could only be indicative of no harm. Moreover, the belief was also that access to a virtually unlimited food supply would compensate for any ill-effects of the tourism activities (e.g. via increased fecundity). The use of physiological indicators was instrumental in demonstrating these beliefs to be incorrect. Instead, provisioning stingrays was creating unsustainable conditions for the wildlife—stingrays would continue to be attracted with negative repercussions. Further, careful management was required since, if evidence of any stingray harm was unavailable, unrestricted feeding and handling stingrays would continue to influence tourist satisfaction. However, findings also revealed that, despite the importance of direct interactions with stingrays, a large majority of tourists were willing to accept some form of management should there exist a risk of injury to the rays. This management would further translate into *increased* satisfaction with the tourist experience. Again, the physiological evidence was critical during discussions with Caymanian policy makers and dissemination to the public, with research results publicized in local newspapers.

Tourist management recommendations thus emerged from the collective SCS studies to assist Caymanian resource managers charged with the responsibility of protecting the environment and providing satisfactory recreational opportunities. These are summarized as: (1) the heterogeneity of tourist types visiting SCS would require various management practices; (2) communication and education through various forms of media would play a key role in resolving behaviours or actions that prove harmful to stingray health; and (3) the wildlife tourism attraction would need to undergo marketing and promotional restructuring to implement the desirable changes, as most visual and written advertisements for SCS promote the feeding and holding of stingrays. Ecological management measures recommended were to alleviate stingray crowding conditions at SCS by limiting the number of people and boats, or by expanding the site into nearby areas to accommodate the current level. Less food provisioned to the rays would also alleviate stingray competition and subsequent aggression injuries, and ensure that the animals resumed foraging naturally and solitarily, further away from the tourist site. If food was still to be provisioned, care was to be taken to ensure that as natural a diet as possible was provided, either through locally caught food or a formulated diet that could be monetarily compensated for by the conservation access fee. Restriction of handling to the tour operator only was also recommended and safety devices on boat propellers, such as cages and guards, would also aid in reducing injuries.

Since the inception of the North Sound Committee in 2003, charged with the planning and management of SCS, new developments have transpired. Recently enacted legislation based on research findings presented here has resulted in the creation of Wildlife Interaction Zones, including the North Sound of Grand Cayman where SCS is located. This zoning act contains a regulation that no marine life may be taken out of the water, including the stingrays, and the Department of Environment will be enforcing the new regulation. Also, while feeding is allowed within these designated zones, the food must be approved by the Marine Conservation Board. Recent plans have a permanent officer for the Wildlife Interaction Zones, with a vessel purchased specifically for that role, as well as the hiring of an officer whose main responsibilities will be to patrol these areas. A campaign was also initiated to produce interpretive materials for visitors at SCS to enhance and inform their stingray experiences (Cayman Compass 2009). Lastly, continuous monitoring of SCS stingray population size and physiological health has been endorsed and participated in by the Cayman Island's Department of Environment (Corcoran et al. 2013; Vaudo et al. 2018).

8.9 Conclusions and future directions

Physiology on its own is an important tool in the conservation toolbox. In situations where impacts are revealed to be occurring at the physiological level but will still incur long-term consequences at the individual and population level, these findings must go beyond simple reporting. When translated into the human dimensions context with social science and scenario planning, it can become an exceptionally important instrument that can effect change (Kittinger et al. 2012). This change can be realized because among the diversity of situations and species involved in human–wildlife interactions, the one common thread is that the thoughts and actions of humans ultimately determine the course and resolution of any conflict (Manfredo and Dayer 2004). Effective communication of physiological metrics of wildlife health will therefore greatly contribute to minimizing problem-causing and problem-enhancing feedbacks in social-ecological systems since tourists made aware of the impacts of their actions may be more willing to alter their activities; and decision makers will thus be more informed about how much of a change a tourism system can sustainably undergo.

The socio-ecological trap theory presented here has wide applicability to other tourist–wildlife interactions, from directed ecotourism to incidental wildlife tourism, and can serve as a guiding framework for scenario-planning directives. Specifically, one should examine within a 'trap' context: (1) the motivations and behaviours of people that have the potential to simultaneously affect their own needs/livelihoods *and* wildlife fitness; and (2) the behavioural decisions of animals affected by human actions that can in turn reduce wildlife population numbers. Critically, the selection of appropriate physiological indicators reflecting this relationship between humans and wildlife can then inform management options (and explore the alternatives) to prevent both interlinked traps from occurring in wildlife tourism attractions, thus increasing the probability of a sustainable outcome for all.

References

Abrantes, K.G., Brunnschweiler, J.M., and Barnett, A., 2018. You are what you eat: examining the effects of provisioning tourism on shark diets. *Biological Conservation*, 224, 300–8.

Bach, L. and Burton, M., 2016. Proximity and animal welfare in the context of tourist interactions with habituated dolphins. *Journal of Sustainable Tourism*, 25(2), 181–97.

Barnett, A., Payne, N.L., Semmens, J.M., and Fitzpatrick, R., 2016. Ecotourism increases the field metabolic rate of whitetip reef sharks. *Biological Conservation*, 199, 132–6.

Bateman, P.W. and Fleming, P.A., 2017. Are negative effects of tourist activities on wildlife over-reported? A review of assessment methods and empirical results. *Biological Conservation*, 211, 10–19.

Battin, J., 2004. When good animals love bad habitats: ecological traps and the conservation of animal populations. *Conservation Biology*, 18, 1482–91.

Bergman, J.N., Bennett, J.R., Binley, A.D. et al., 2019. Scaling from individual physiological measures to population-level demographic change: case studies and future directions for conservation management. *Biological Conservation*, 238, 108242.

Boonstra, W. J., Björkvik, E., Haider, L.J., and Masterson, V., 2016. Human responses to social-ecological traps. *Sustainability Science*, 11, 877–89.

Brookhouse, N., Bucher, D.J., Rose, K. et al., 2013. Impacts, risks and management of fish feeding at Neds Beach, Lord Howe Island Marine Park, Australia: a case study of how a seemingly innocuous activity can become a serious problem. *Journal of Ecotourism*, 12, 165–81.

Burgin, S. and Hardiman, N., 2015. Effects of non-consumptive wildlife-oriented tourism on marine species and prospects for their sustainable management. *Journal of Environmental Management*, 151, 210–20.

Butler, R.W., 1980. The concept of a tourism area cycle of evolution: implications for management of resources. *Canadian Geographer*, 24, 5–12.

Catlin, J., Jones, R., and Jones, T., 2011. Revisiting Duffus and Dearden's wildlife tourism framework. *Biological Conservation*, 144, 1537–44.

Cayman Compass, 2009. New enforcement officer at sandbar. Available at: https://webcache.googleusercontent.com/search?q=cache:nY18v6MnPRkJ:https://www.caymancompass.com/2009/07/23/new-enforcement-officer-at-sandbar/+&cd=1&hl=en&ct=clnk&gl=ca (accessed 28 July 2020).

Cayman Islands Ministry of Tourism (CI MoT), 2002. *'Focus for the Future'—A Tourism Policy Framework for the Cayman Islands*. The Tourism Company, London.

Ceauşu, S., Graves, R.A., Killion, A.K. et al., 2019. Governing trade-offs in ecosystem services and disservices to achieve human–wildlife coexistence. *Conservation Biology*, 33, 543–53.

Cinner, J.E., Folke, C. Daw, T., and Hicks, C.C., 2011. Responding to change: using scenarios to understand how socioeconomic factors may influence amplifying or dampening exploitation feedbacks among Tanzanian fishers. *Global Environmental Change*, 21, 7–12.

Cisneros-Montemayor, A.M., Barnes-Mauthe, M., Al-Abdulrazzak, D., Navarro-Holm, E., and Sumaila, U.R., 2013. Global economic value of shark ecotourism: implications for conservation. *Oryx*, 47, 381–8.

Clua, E., Buray, N., Legendre, P. et al., 2010. Behavioural response of sicklefin lemon sharks *Negaprion acutidens* to underwater feeding for ecotourism purposes. *Marine Ecology Progress Series*, 414, 257–66.

Costanza, R., 1987. Social traps and environmental policy. *BioScience*, 37, 407–12.

Costanza, R. and Voinov, A., 2001. Modeling ecological and economic systems with STELLA: part III. *Ecological Modelling*, 143, 1–7.

Cooke, S.J., Hultine, K.R., Rummer, J.L., and Franklin, C.E., 2017. Reflections and progress in conservation physiology. *Conservation Physiology*, 5(1), cow071. https://doi.org/10.1093/conphys/cow071

Cooke, S.J. and O'Connor, C.M., 2010. Making conservation physiology relevant to policy makers and conservation practitioners. *Conservation Letters*, 3, 159–66.

Corcoran, M., 2006. The effects of supplemental feeding on the activity space and movement patterns of the southern stingray, *Dasyatis americana*, at Grand Cayman, Cayman Islands. MSc thesis, Nova Southeastern University, Fort Lauderdale, FL.

Corcoran, M.J., Wetherbee, B.M., Shivji, M.S. et al., 2013. Supplemental feeding for ecotourism reverses diel activity and alters movement patterns and spatial distribution of the southern stingray, *Dasyatis americana*. *PLoS ONE*, 8, e59235.

Cross, J.G. and Guyer, M.J. 1980. *Social Traps*. University of Michigan Press, Ann Arbor, MI.

Cumming, G.S. and Peterson, G.D., 2017. Unifying research on social–ecological resilience and collapse. *Trends in Ecology and Evolution*, 32, 695–713.

Curtin, S. and Wilkes, K., 2005. British wildlife tourism operators: current issues and typologies. *Current Issues in Tourism*, 8, 455–78.

Dearden, P., Bennett, M., and Rollins, R., 2006. Implications for coral reef conservation of diver specialization. *Environmental Conservation*, 33, 353–63.

Duffus D.A. and Dearden, P., 1990. Non-consumptive wildlife-orientated recreation: a conceptual framework. *Biological Conservation*, 5, 213–31.

Faust, L.J., Thompson, S.D., Earnhardt, J.M. et al., 2003. Using stage-based system dynamics modeling for demographic management of captive populations. *Zoo Biology*, 22, 45–64.

Gallagher, A.J. and Hammerschlag, N., 2011. Global shark currency: the distribution, frequency, and economic value of shark ecotourism. *Current Issues in Tourism*, 14, 797–812.

Garrod, B. and Wilson, J.C., 2004. Nature on the edge? Marine ecotourism in peripheral coastal areas. *Journal of Sustainable Tourism*, 12, 95–120.

Higginbottom, K., Green, R., and Northrope, C. 2003. A framework for managing the negative impacts of wildlife tourism on wildlife. *Human Dimensions of Wildlife*, 8, 1–24.

Higham, J.E.S., 1998. Tourists and albatrosses: the dynamics of tourism at the Northern Royal Albatross Colony, Taiaroa Head, New Zealand. *Tourism Management*, 19, 521–31.

Higham, J.E.S. and Bejder, L., 2008. Managing wildlife-based tourism: edging slowly towards sustainability? *Current Issues in Tourism*, 11, 75–83.

Higham, J. and Carr, A., 2002. Profiling tourists to ecotourism operations. *Annals of Tourism Research*, 29, 1171–4.

Illing, B. and Rummer, J.L., 2017. Physiology can contribute to better understanding, management, and conservation of coral reef fishes. *Conservation Physiology*, 5(1), cox005 https://doi.org/10.1093/conphys/cox005

Kilbourne, W. and Pickett, G., 2008. How materialism affects environmental beliefs, concern, and environmentally responsible behavior. *Journal of Business Research*, 61, 885–93.

Kittinger, J., Finkbeiner, E., Glazier, E., and Crowder, L., 2012. Human dimensions of coral reef social-ecological systems. *Ecology and Society*, 17(4), article 17.

Knapp, C.R., Hines, K.N., Zachariah, T.T. et al., 2013. Physiological effects of tourism and associated food provisioning in an endangered iguana. *Conservation Physiology*, 1(1), cot032.

Louviere, J. and Timmermans, H., 1990. Stated preference and choice models applied to recreation research: a review. *Leisure Sciences*, 12, 9–32.

Lowe, J. and Tejada, J.F.C., 2019. The role of livelihoods in collective engagement in sustainable integrated costal management: Oslob whale sharks. *Ocean and Coastal Management*, 170, 80–92.

Madliger, C.L., Cooke, S. J., Crespi, E. J. et al., 2016. Success stories and emerging themes in conservation physiology. *Conservation Physiology*, 4(1), cov057.

Madliger, C.L. and Love, O.P., 2015. The power of physiology in changing landscapes: considerations for the continued integration of conservation and physiology. *Integrative and Comparative Biology*, 55, 545–53.

Manfredo, M.J. and Dayer, A.A., 2004. Concepts for exploring the social aspects of human-wildlife conflict in a global context. *Human Dimensions of Wildlife*, 9, 317–28.

Mann, J., Connor, R.C., Barre, L.M., and Heithaus, M.R., 2000. Female reproductive success in bottlenose dolphins (*Tursiops* sp.): life history, habitat, provisioning, and group-size effects. *Behavioral Ecology*, 11, 210–19.

Martin, S.R., 1997. Specialization and differences in setting preferences among wildlife viewers. *Human Dimensions of Wildlife*, 2, 1–18.

Masters, D., 1998. Marine wildlife tourism: developing a quality approach in the Highlands and Islands. A report for the Tourism and Environment Initiative and Scottish Natural Heritage, May 1998. Visit Scotland, Sustainable Tourism Unit.

Meekan, M. and Lowe, J., 2019. Oslob whale sharks— preconceived ideas about provisioning? *Tourism Management*, 75, 630–1.

Meyer, L., Pethybridge, H., Beckmann, C. et al., 2019. The impact of wildlife tourism on the foraging ecology and nutritional condition of an apex predator. *Tourism Management*, 75, 206–15.

Moorhouse, T.P., Dahlsjö, C.A., Baker, S.E. et al., 2015. The customer isn't always right—conservation and animal welfare implications of the increasing demand for wildlife tourism. *PLoS ONE*, 10(10), e0138939.

Moorhouse, T., D'Cruze, N.C., and Macdonald, D.W., 2017. Unethical use of wildlife in tourism: what's the problem, who is responsible, and what can be done? *Journal of Sustainable Tourism*, 25(4), 505–16.

Moscardo, G., 2000. Understanding wildlife tourism market segments: an Australian marine study. *Human Dimensions of Wildlife*, 5, 36–53.

Moyle, B.D., Weiler, B., and Croy. G., 2013. Visitors' perceptions of tourism impacts: Bruny and Magnetic Islands, Australia. *Journal of Travel Research*, 52, 392–406.

Patroni, J., Simpson, G., and Newsome, D., 2018. Feeding wild fish for tourism—a systematic quantitative literature review of impacts and management. *International Journal of Tourism Research*, 20, 286–98.

Peterson, G.D., Cumming, G.S., and Carpenter, S.R., 2003. Scenario planning: a tool for conservation in an uncertain world. *Conservation Biology*, 17, 358–66.

Pizzitutti, F., Walsh, S.J., Rindfuss, R.R. et al., 2017. Scenario planning for tourism management: a participatory and system dynamics model applied to the Galapagos Islands of Ecuador, *Journal of Sustainable Tourism*, 25, 1117–37.

Platt, J., 1973. Social traps. *American Psychologist*, 28, 642–51.

Pratt, S. and Suntikul, W., 2016. Can marine wildlife tourism provide an 'edutaining' experience? *Journal of Travel and Tourism Marketing*, 33, 867–84.

Reynolds, P.C. and Braithwaite, D., 2001. Towards a conceptual framework for wildlife tourism. *Tourism Management*, 22, 31–42.

Robertson, B.A. and Hutto, R.L., 2006. A framework for understanding ecological traps and an evaluation of existing evidence. *Ecology*, 87, 1075–85.

Robertson, B.A., Ostfeld, R.S., and Keesing, F., 2017. Trojan females and Judas goats: evolutionary traps as tools in wildlife management. *BioScience*, 67, 983–94.

Schlaepfer M.A., Runge, M.C., and Sherman, P.W., 2002. Ecological and evolutionary traps. *Trends in Ecology and Evolution*, 17, 474–80.

Scott, D. and Thigpen, J., 2003. Understanding the birder as tourist: segmenting visitors to the Texas hummer/bird celebration. *Human Dimensions of Wildlife*, 8, 199–218.

Semeniuk, C.A.D., Bourgeon, S., Smith, S.L., and Rothley, K.D., 2009. Hematological differences between stingrays at tourist and non-visited sites suggest physiological costs of wildlife tourism. *Biological Conservation*, 142, 1818–29.

Semeniuk, C.A.D., Haider, W., Beardmore, B., and Rothley, K.D., 2008. A multi-attribute trade-off approach for advancing the management of marine wildlife tourism: a quantitative assessment of heterogeneous visitor preferences. *Aquatic Conservation: Marine and Freshwater Ecosystems*, 19, 194–208.

Semeniuk, C.A.D., Haider, W., Cooper, A., and Rothley, K.D., 2010. A linked model of animal ecology and human behaviour for the management of wildlife tourism. *Ecological Modelling*, 221, 2699–713.

Semeniuk, C.A.D. and Rothley, K.D., 2008. Costs of group-living for a normally solitary forager: effects of provisioning tourism on Southern stingrays *Dasyatis americana*. *Marine Ecology Progress Series*, 357, 271–82.

Semeniuk, C.A.D., Speers-Roesch, B., and Rothley, K.D., 2007. Using fatty acid profile analysis as an ecological indicator in the management of tourist impacts on marine wildlife: a case of stingray-feeding in the Caribbean. *Environmental Management*, 40, 665–77.

Shackley, M., 1998. 'Stingray City'—managing the impact of underwater tourism in the Cayman Islands. *Journal of Sustainable Tourism*, 6, 328–38.

Thompson, P.M., Wilson, B., Grellier, K., and Hammond, P.S., 2000. Combining power analysis and population viability analysis to compare traditional and precautionary approaches to conservation of coastal cetaceans. *Conservation Biology*, 14, 1253–63.

Trave, C., Brunnschweiler, J., Sheaves, M. et al., 2017. Are we killing them with kindness? Evaluation of sustainable marine wildlife tourism. *Biological Conservation*, *209*, 211–22.

van den Belt, M., Deutsch, L., and Jansson, Å., 1998. A consensus-based simulation model for management in the Patagonia coastal zone. *Ecological Modelling, 110*, 79–103.

Vaudo, J.J., Wetherbee, B.M., Harvey, G.C. et al., 2018. Characterisation and monitoring of one of the world's most valuable ecotourism animals, the southern stingray at Stingray City, Grand Cayman. *Marine and Freshwater Research, 69*, 144–54.

Wikelski, M. and Cooke, S.J., 2006. Conservation physiology. *Trends in Ecology & Evolution, 21*, 38–46.

Wilson, C. and Tisdell, C., 2003. Conservation and economic benefits of wildlife-based marine tourism: sea turtles and whales as case studies. *Human Dimensions of Wildlife, 8*, 49–58.

World Travel and Tourism Council (WTTC), 2015. Annual report—2015: the economic impact of travel and tourism. WTTC, London.

Ziegler, J.A., Silberg, J.N., Araujo, G. et al., 2019. Applying the precautionary principle when feeding an endangered species for marine tourism. *Tourism Management, 72*, 155–8.

PART III

Understanding Disease-Driven Population Declines

CHAPTER 9

Applying conservation physiology in response to a devastating wildlife disease, white-nose syndrome in bats

Yvonne A. Dzal and Craig K.R. Willis

⊃ **Take-home message**

A key to success in our fight against white-nose syndrome (WNS) will come from connecting physiological mechanisms with wildlife population dynamics, to understand why some bats survive WNS while others suffer devastating mortality.

9.1 Introduction

9.1.1 Emerging cutaneous fungal pathogens: a threat to wildlife diversity and conservation?

Fungal pathogens are an increasing threat to wildlife diversity and conservation due to global travel and trade, climate change, and habitat loss/fragmentation (Daszak et al. 2000; Jones et al. 2008; Fisher et al. 2012). In recent years, fungi that invade the skin of their hosts have become some of the most impactful pathogens of wildlife, causing catastrophic die-offs and extinctions of multiple animal species (Anderson et al. 2004; Skerratt et al. 2007; Frick et al. 2010; Fisher et al. 2012; Lips 2016; Casadevall 2019). For example, over the past 20 years, amphibian chytridiomycosis, a fungal skin disease caused by *Batrachochytrium dendrobatidis*, has affected over 700 species of amphibians, and has caused the rapid decline, or extinction, of dozens of species all over the world in the fastest decline of vertebrate biodiversity ever recorded (Voyles et al. 2011; Fisher et al. 2012; Lips 2016; Martel et al. 2018; The New York State Wildlife Health Program 2019). More recently (2013), salamander chytridiomycosis (caused by *Batrachochytrium salamandrivorans*, a fungal pathogen closely related to *B. dendrobatidis*), has caused widespread mortality of salamanders in Europe, although it has not yet been documented in wild salamander populations in North America (Martel et al. 2013, 2014). The prevalence and geographic range of fungal pathogens, as well as the number of hosts affected by fungi are also expanding, resulting in devastating consequences for wildlife biodiversity. In North America, specifically, the list of animals affected by cutaneous fungal outbreaks is growing at an alarming rate, as demonstrated by the recent emergence of several fatal fungal diseases, such as ophidiomycosis in snakes (caused by the fungus *Ophidiomyces ophiodiicola*) (Lorch et al. 2015; Agugliaro et al. 2020), white-nose syndrome (WNS) in bats (caused by the fungus *Pseudogymnoascus destructans*; Pd) (Blehert et al. 2009; Frick et al. 2010), and dermatophytosis in porcupines

Yvonne A. Dzal and Craig K.R. Willis, *Applying conservation physiology in response to a devastating wildlife disease, white-nose syndrome in bats*
In: *Conservation Physiology: Applications for Wildlife Conservation and Management*. Edited by: Christine L. Madliger, Craig E. Franklin, Oliver P. Love, and Steven J. Cooke, Oxford University Press (2021). © Oxford University Press. DOI: 10.1093/oso/9780198843610.003.0009

(caused by the fungi *Trichophyton mentagrophytes/Arthroderma benhamiae*) (Needle et al. 2019). What these fungal pathogens have in common is that they all preferentially invade the skin of their hosts, disrupting energy, water, and ion balance (Voyles et al. 2009; Cryan et al. 2010; Peterson et al. 2013; Verant et al. 2014; McGuire et al. 2017; Russo et al. 2018; Agugliaro et al. 2020). This homeostatic imbalance is sufficient to cause mortality in animals (such as amphibians, reptiles, and hibernating mammals) that tightly regulate the exchange of respiratory gases (O_2 and CO_2), water, and electrolytes across their skin (Voyles et al. 2009; Cryan et al. 2010; Peterson et al. 2013; Verant et al. 2014; McGuire et al. 2017; Russo et al. 2018). Furthermore, the energetic cost of coping with the fungal pathogen itself (the inflammatory response) may have negative impacts on host fitness, as indicated by decreased reproductive investment of some host species with clinical signs of fungal infection (Lind et al. 2018, 2019). Thus, host physiology is integral to understanding and predicting individual and population-level consequences of wildlife diseases and may be particularly relevant for the management of imperilled populations whose persistence is threatened by these emerging cutaneous fungal pathogens (Willis 2015).

Pathogens, such as bacteria and viruses, are not normally recognized as posing a threat of extinction to their hosts because transmission is generally thought to depend on the density of hosts so that, as host populations decline, transmission and reproduction of the pathogen also decline (Anderson 1979; McCallum and Dobson 1995; Keeling and Grenfell 1997; McCallum et al. 2001; Greer et al. 2008; Smith et al. 2009). However, so-called 'conservation pathogens' (i.e. pathogens that cause or contribute to dramatic declines in host population size; Willis 2015), often possess traits releasing them from host-density dependence such as sexual transmission or the ability to survive in the environment in the absence of hosts (McCallum et al. 2009; Willis 2015). In contrast to most bacteria and viruses, many fungal pathogens exhibit the latter characteristic, and can persist, and even grow and reproduce, within the environment (for a review, see Casadevall 2019). Another factor that makes many fungal pathogens lethal and difficult to treat is that fungi and animals

are relatively closely related (Baldauf and Palmer 1993) and, as a result, tend to share cellular and metabolic processes, which means many anti-fungal drugs that target these processes in fungi may cause severe side-effects in animal hosts (for a review, see Rohlfs and Churchill 2011; Perfect 2017). Furthermore, fungi are capable of rapid evolution (Casadevall 2019), which can lead to rapid emergence of increased virulence and anti-fungal drug resistance. These characteristics allow some fungal pathogens to cause catastrophic declines of their hosts and can make it challenging to devise treatment and management strategies for conservation of host populations. One prominent example of a lethal conservation pathogen causing alarming rates of mortality in an animal host is the cold-loving fungus, Pd (Blehert et al. 2009; Cryan et al. 2010; Willis et al. 2011; Reeder et al. 2012; Warnecke et al. 2013; Johnson et al. 2014; Verant et al. 2014; Wilcox et al. 2014; Bohn et al. 2016; McGuire et al. 2017; Moore et al. 2017; Mayberry et al. 2018).

9.1.2 White-nose syndrome: the invasive fungal skin disease decimating bat populations across North America

Since its discovery in New York State in 2007, WNS, a multi-host skin disease, has led to catastrophic declines of several species of hibernating bats across North America (Frick et al. 2010; Langwig et al. 2012; Frick et al. 2015; Langwig et al. 2015a, b, 2017; US Fish and Wildlife Service 2020a). The evidence is now clear that Pd was introduced from Eurasia (Warnecke et al. 2012; Leopardi et al. 2015; Trivedi et al. 2017) and, since then, it has been confirmed in 19 bat species, across 39 US states, and 7 Canadian provinces (Table 9.1) (US Fish and Wildlife Service 2020a, b). With the recent western and southern expansion of Pd in North America, several species of endangered bats are already at risk (Table 9.1), and some species are predicted to suffer range-wide extinction (Hoyt et al. 2016), with three formerly common species (*Myotis lucifugus*, *Myotis septentrionalis*, and *Perimyotis subflavus*) now federally endangered in Canada because of the disease (COSEWIC 2013), and *M. septentrionalis* protected as a threatened species in the United States under the Endangered Species Act (US Fish and Wildlife Service 2020c).

Table 9.1 Conservation status of North American bats affected by the fungal pathogen, *Pseudogymnoascus destructans* (Pd), responsible for the devastating disease, white-nose syndrome (WNS). Conservation status taken from the Committee on the Status of Endangered Wildlife in Canada (COSEWIC) 2013 (Canada), US Fish and Wildlife Service 2020b (United States), and International Union for Conservation of Nature (IUCN) 2020 (global).

Species	Pd status	WNS status	Canada pre-WNS listing	Canada post-WNS listing	United States pre-WNS listing	United States post-WNS listing	IUCN pre-WNS listing	IUCN post-WNS listing
Big brown bat *Eptesicus fuscus*	+	+					least concern	least concern
Cave bat *Myotis velifer*	+	+	not found in Canada	not found in Canada			least concern	least concern
Eastern red bat *Lasiurus borealis*	+	-					least concern	least concern
Eastern small-footed bat *Myotis leibii*	+	+					least concern	endangered
Fringed bat *Myotis thysanodes*	+	+	data deficient	data deficient			least concern	least concern
Gray bat *Myotis grisescens*	+	+	not found in Canada	not found in Canada	endangered	endangered	near threatened	vulnerable
Indiana bat *Myotis sodalis*	+	+	not found in Canada	not found in Canada	endangered	endangered	endangered	near threatened
Little brown bat *Myotis lucifugus*	+	+		endangered			least concern	endangered
Long-legged bat *Myotis volans*	+	+	not found in Canada	not found in Canada			least concern	least concern
Mexican free-tailed bat *Tadarida brasiliensis*	+	-	not found in Canada	not found in Canada			least concern	least concern
Northern long-eared bat *Myotis septentrionalis*	+	+		endangered		threatened	least concern	near threatened
Rafinesque's big-eared bat *Corynorhinus rafinesquii*	+	-	not found in Canada	not found in Canada			least concern	least concern
Silver-haired bat *Lasionycteris noctivagans*	+	-					least concern	least concern
Southeastern bat *Myotis austroriparius*	+	+	not found in Canada	not found in Canada			least concern	least concern
Townsend's big-eared bat *Corynorhinus townsendii*	+	-	vulnerable (in BC)	vulnerable (in BC)	endangered	endangered	least concern	least concern
Tricolored bat *Perimyotis subflavus*	+	+		endangered			least concern	vulnerable
Western long-eared bat *Myotis evotis*	+	+					least concern	least concern
Western small-footed bat *Myotis ciliolabrum*	+	-					least concern	least concern
Yuma bat *Myotis yumanensis*	+	+					least concern	least concern

Pd has been especially catastrophic to bat populations across North America because of the natural cycle of hibernation. Hibernation is an adaptive strategy that allows bats to cope with both low ambient temperatures and no access to energy sources (other than endogenous fat reserves), throughout winter (Geiser 2004). Hibernation is characterized by long and deep torpor bouts, interspersed with short periods of arousals (usually less than a day), to a normal body temperature every few weeks (Thomas et al. 1990a; Geiser 2004). During this profoundly depressed physiological state, body temperature is reduced to virtually ambient levels (2–8°C for cave-hibernating bats), metabolism and immune function are severely down-regulated, and movement is restricted (Geiser 2004). Collectively, these characteristics make hibernating bats a most accommodating host for pathogens that can tolerate low ambient temperatures, such as cold-loving Pd, the first pathogen known to cause mortality in torpid mammalian hosts.

Another reason Pd has become a highly successful emergent pathogen is that, as for many other wildlife fungal diseases, the pathogen can survive in the environment even in the absence of hosts (Hoyt et al. 2014; Langwig et al. 2015a, b). Individual bats show high fidelity to hibernacula over many years (Norquay and Willis 2014). Thus, the few individuals that make it through the winter with WNS, and any naïve individuals (e.g. young of the year) arriving at an infected hibernaculum, become reinfected or infected, with most carrying detectable levels of Pd within just a few weeks of the onset of hibernation (Langwig et al. 2015a). The ability of the fungus to persist within hibernacula increases extinction risks for bats and creates an enormous challenge for management.

Despite massive mortality across multiple bat species (e.g. Frick et al. 2010; Langwig et al. 2012), WNS does not affect all species of hibernating bats equally. Indeed, Pd has been found on some species of North American bats in the absence of any signs of disease pathology (Table 9.1) (Puechmaille et al. 2011; Warnecke et al. 2012; Frank et al. 2014; Zukal et al. 2014; Langwig et al. 2016; Davy et al. 2017; Moore et al. 2017). For example, while the pathogen and the disease are prevalent in Eurasia (Martínková et al. 2010; Wibbelt et al. 2010; Puechmaille et al.

2011; Pavlinić et al. 2015), there are striking differences in disease manifestation between North American and Eurasian bats, with no mass mortality attributed to WNS in Eurasia (Puechmaille et al. 2011). Additionally, two well-studied North American bat species, *M. lucifugus* and *Eptesicus fuscus*, are both affected by WNS, but susceptibility (as indicated by higher prevalence and severity of Pd infections), and rates of mortality are much higher for *M. lucifugus* than for *E. fuscus* (Langwig et al. 2012; Frank et al. 2014; Frick et al. 2015; Langwig et al. 2015a). These findings suggest that some bats are more resistant to (i.e. rely on mechanisms that allow hosts to reduce pathogen loads) and/or tolerant of (i.e. rely on mechanisms that limit damage caused by the pathogen enabling host survival despite high pathogen loads) Pd than others.

Even among species that experience WNS-associated mortality, population impacts of the disease vary, with mortality ranging from as low as 12 per cent to as high as 98 per cent during the first 5 years of WNS invasion (Turner et al. 2011; Langwig et al. 2012; Turner et al. 2011). Moreover, 10 years after the initial outbreak, some bat populations that have declined by more than 90 per cent appear to be stabilizing or rebounding, albeit at only 10 to 25 per cent of their pre-WNS population size (Langwig et al. 2012; Frick et al. 2015; Maslo et al. 2015; Langwig et al. 2017). While it is apparent that some bats possess mechanisms of persistence (Langwig et al. 2012; Frick et al. 2015; Maslo et al. 2015; Langwig et al. 2017), it is still unclear if this persistence reflects resistance, tolerance, or some combination of the two mechanisms.

9.1.3 Applying conservation physiology in response to white-nose syndrome

WNS is an example of a conservation problem for which integrative physiological research has been a critical part of research and management (Table 9.2). Although WNS has rarely been referred to explicitly in the context of conservation physiology (but see Willis 2015), the WNS research and management community has relied heavily on tools from the conservation physiology toolbox in an effort to understand and manage the disease (Table 9.2) (Wikelski and Cooke 2006; Cooke et al.

2013; Coristine et al. 2014; Madliger et al. 2018). Standard diagnostic techniques from veterinary pathology and microbiology have been widely used to characterize the disease. Histopathology, transillumination of bat skin with visual and ultraviolet (UV) light, and quantitative polymerase chain reaction (qPCR) are now all widely used pathogen and disease detection tools in new species and regions, tracking disease progression, and determining the efficacy of potential mitigation strategies (Table 9.2; Blehert et al. 2009; Gargas et al. 2009; Meteyer et al. 2009; Chaturvedi et al. 2010; Cryan et al. 2010; Fuller et al. 2011; Lorch et al. 2011; Muller et al. 2013; Turner et al. 2014; Field et al. 2015; Janicki et al. 2015; Davy et al. 2017; Donaldson et al. 2017). Use of infrared (IR) video in the lab and field has been crucial because of the sensitivity of hibernating bats to disturbance, and has revealed previously undocumented behaviours of hibernating bats, and important hints about the pathophysiology underlying the disease (Table 9.2; Brownlee-Bouboulis and Reeder 2013; Warnecke et al. 2014; Wilcox et al. 2014; Bohn et al. 2016). Physiological studies biologging skin temperature with temperature-sensitive transmitters and dataloggers, open- and closed-flow respirometry, isotope tracing via doubly-labelled water, and quantitative magnetic resonance (QMR) have all been effective for understanding consequences of Pd infection for hibernating bats (Table 9.2; Cryan et al. 2010; Warnecke et al. 2013; Verant et al. 2014; McGuire et al. 2017; Moore et al. 2017; Meierhofer et al. 2018). Additionally, portable blood analysers have documented perturbations in acid–base balance, electrolytes, and hydration status in WNS-affected bats, leading to the hypothesis that dehydration plays a major role in the susceptibility of bats to WNS (Table 9.2; Warnecke et al. 2013; Verant et al. 2014). Lastly, the incorporation of temperature-sensitive dataloggers and passive integrated transponder (PIT) tags has provided unprecedented data on pre- and post-WNS survival, strengthening our understanding of the links between behaviour, habitat selection, and energetic physiology (Table 9.2; Reeder et al. 2012; Warnecke et al. 2013; Britzke et al. 2014; Johnson et al. 2014; Norquay and Willis 2014; Verant et al. 2014; Turner et al. 2015; Lilley et al. 2016; Mayberry

et al. 2018; Cheng et al. 2019; Hoyt et al. 2019). These examples demonstrate that the conservation physiology toolbox continues to become more accessible from a logistical standpoint and also emphasizes the usefulness of these techniques as decision-support tools.

While the mechanisms underlying WNS-associated mortality are still not fully understood, a conservation physiology approach has allowed for rapid advances in our understanding of the disease (Table 9.2; e.g. Cryan et al. 2010; Willis et al. 2011; Warnecke et al. 2013; Verant et al. 2014; for a review, see Willis 2015). Although progress has been made, we still do not fully understand the aspects of WNS pathophysiology that best connect a simple fungal skin infection to dramatic changes in hibernation behaviour and energetics in some species and not others. Understanding the physiological causes of WNS mortality could allow a more predictive understanding of the drivers of population declines and enable scalable, physiology-based approaches to conservation. Here we review what is known about pathophysiological processes in hibernating bats with WNS, and connections between these processes and population impacts. We then outline how a conservation physiology approach has been useful for informing our understanding of resistance and tolerance mechanisms, and conclude by showing how conservation physiology could help with the design of mitigation strategies and treatments for WNS and other infectious diseases of wildlife. Identifying differences in species susceptibility could be transformative for our understanding of mechanisms underlying survival of wildlife in the face of pathogen invasion and our ability to manage populations impacted by disease.

9.2 How does a simple fungal skin infection kill hibernating bats?

9.2.1 Skin invasion with *Pseudogymnoascus destructans*

When WNS emerged in North America in 2007, biologists first documented abnormal behaviours among hibernating bats, such as movement towards the entrance of caves, and daytime flights during the winter, ultimately causing emaciation (e.g. Reichard

Table 9.2 Techniques employed from the conservation physiology toolbox to understand and manage white-nose syndrome (WNS), a fungal skin disease decimating hibernating bat populations across North America.

Technique	Information revealed relevant to conservation physiology	Potential limitations	Why is this information important to understanding and managing the disease?	References
Histopathology	• gold standard for diagnosing bats with WNS • identifies aggregates of fungal hyphae that form characteristic cupping erosions & ulceration of skin membranes	• requires a large amount of tissue from each bat, restricting analysis to dead animals or necessitates euthanasia • long turn around time & requires specialized training to analyze samples	• promotes an understanding of bat-fungus relationships • essential for devising & screening the effectiveness of biological & chemical control agents used for treating WNS • valuable for estimating the extent of disease among bat populations	• Blehert et al. 2009; Meteyer et al. 2009; Chaturvedi et al. 2010; Cryan et al. 2010; Lorch et al. 2011; Reeder et al. 2012; Field et al. 2015; Davy et al. 2017; Donaldson et al. 2017; Moore et al. 2017
Transillumination of bat skin with LED & UV light	• rapid, non-lethal field screening of bat skin to detect lesions indicative of WNS; reducing the need to euthanize bats for WNS diagnosis • LED: identifies discolouration, tears, holes, flaking, necrosis, receded wing margins, & missing bat skin tissue • UV: elicits a distinct orange–yellow fluorescence that corresponds with the fungal cupping erosions used to diagnose bats with WNS by histopathology	• gross visible signs of fungal infection in bats with WNS are often subtle & difficult to detect, especially in bats with early-stage WNS • UV: ability to discern sparse, subtle fluorescence varies by observer • wing condition alone is not a diagnostic tool to confirm WNS	• provides an accessible diagnostic tool to track disease progression & assess the presence of WNS in new species & regions • enhances ability to screen bats to determine efficacy of potential mitigation strategies • detects changes in wing condition, which is critical for assessing future WNS spread • optimizes non-lethal collection of small biopsy samples for WNS testing by histology, PCR, or culture	• Fuller et al. 2011; Turner et al. 2014; Janicki, et al. 2015; Lilley et al. 2016; McGuire et al. 2017; Moore et al. 2017
Quantitative PCR (qPCR)	• a non-invasive, sensitive, screening test that verifies the presence of WNS causing fungus in bats or environmental substrate • offers a fast, reliable, & economical alternative to histology & culture	• requires specialized training to analyze samples • not as sensitive at detecting WNS as histology • infections with a limited number of fungal foci are more likely to result in a false negative result because of the small proportion of skin/substrate sampled	• allows researchers to collect large numbers of samples to determine & track the prevalence of WNS over time • facilitates non-lethal disease monitoring among endangered bat species • provides a means to assess the efficacy of potential treatments for WNS	• Gargas et al. 2009; Chaturvedi et al. 2010; Lorch et al. 2011; Reeder et al. 2012; Muller et al. 2013; Johnson et al. 2014; Field et al. 2015; Lilley et al. 2016; Davy et al. 2017; Moore et al. 2017; Mayberry et al. 2018
Biologging skin temperature	• a temperature-sensitive datalogger is attached to the back of a bat to assess torpor-arousal patterns (e.g. arousal frequency, arousal duration, & torpor bout duration) • WNS alters hibernation patterns & reduces the length of torpor bouts in hibernating bats • primary cause of the increased mortality/disease state associated with WNS is abnormally increased arousal frequency &, consequently, fat depletion & energy expenditure • WNS causes a disruption of this torpor–arousal cycle (i.e. by increasing the frequency or duration of arousals), it could easily cause bats to metabolize fat reserves too quickly, thereby leading to starvation	• must recapture the animal to download data • limited data storage & moderate failure rate	• warming of overwintering sites (hibernacula) could reduce heat loss & the amount of energy that bats must spend during periodic arousals, thus, increasing survival of WNS-affected bats • potential of artificial heating of bat roosts & protection of high-quality natural roosts with warm microclimates as a conservation measure to increase survival & enhance recovery from WNS • artificial heating of bat roosts could be critical for WNS affected bats to maintain energy balance during hibernation, or during recovery from WNS in the spring	• Reeder et al. 2012; Johnson et al. 2014; Turner et al. 2015; Lilley et al. 2016; McGuire et al. 2017; Moore et al. 2017; Mayberry et al. 2018
Double labelled water method	• evaluates how energy use & body composition of bats with WNS changes over time • widely used to study energetics in relation to homeostasis, behavioural adaptations, & resource allocation • based on dynamic flux of hydrogen & oxygen through the body & ability to measure these flux rates over a period of time using labeled isotopes • even bats with early-stage WNS infections (prior to the onset of increased arousal frequency) spend 2x more energy than healthy bats	• inability to collect a sufficient amount of blood for isotope analysis is common in small, hibernating animals • unable to directly calculate energy use if final isotope concentrations are indistinguishable from background levels, which is common in animals with a low metabolic rate (such as hibernating bats) • relatively invasive as it requires blood sampling before & after an experiment	• links behavioural observations, such as increased arousals & increased activity of WNS bats with energy use & changes in body composition • provides a comprehensive understanding of the progression of physiological disturbances of WNS in hibernating bats • has been critical for guiding effective & properly timed management actions prior to the onset of clinical signs of the disease to minimize morbidity & mortality from WNS	• Verant et al. 2014

Table 9.2 Continued.

Technique	Information revealed relevant to conservation physiology	Potential limitations	Why is this information important to understanding and managing the disease?	References
Blood chemistry analysis (*i.e.* iSTAT: a clinical portable blood analyzer)	• offers a fast & field accessible method to determine blood chemistry perturbations (*i.e.* acid-base balance, electrolytes, & hydration status) in hibernating bats with WNS • fluid loss across damaged wing tissue reduces plasma volume (i.e., hypovolemia), & results in hypotonic dehydration & altered electrolyte balance during early-stage WNS • suggests pathologic & potentially life-threatening blood chemistry disturbances associated with early-stage WNS infections • links behavioural observations, such as increased arousals & increased activity of WNS bats with blood chemistry perturbations	• inability to collect a sufficient amount of blood for analysis is common in small, hibernating animals, often necessitating euthanasia	• provides a comprehensive understanding of the progression of blood chemistry disturbances of WNS in hibernating bats • has been critical for guiding effective & properly timed management actions prior to the onset of clinical signs of the disease to minimize morbidity & mortality from WNS	• Warnecke et al. 2013; Verant et al. 2014
Respirometry (*i.e.* whole body chamber & seperated chamber)	• whole body chamber: WNS increases energy expenditure & water loss in hibernating bats • head-body seperated chamber: disruption of passive gas exchange across the skin may be a likely mechanism of dehydration & WNS-associated mortality in bats	• inability to make measurements on free-ranging bats • bats are confined to their chamber & sometimes are stressed & do not enter torpor • head-body seperated chamber: difficult to make the seal between the head & body chamber leak proof while also making it comfortable for the bats	• understanding disease mechanisms & WNS progression is imperative for developing management interventions • highlights the extraordinary physiology of bat wings & suggests that efforts to treat WNS should consider the implications of topical anti-fungal agents for wing physiology & gas exchange across the skin	• McGuire et al. 2017; Meierhofer et al. 2018
Behavioural observation from infrared video recordings	• quantifies behavioural consequences of WNS in hibernating bats • WNS bats show a dramatic reduction in clustering behaviour & activity as infection progresses • WNS bats do not drink or groom	• bats are highly sensitive to disturbance & their overwinter sites are often remote & usually too large for effective video recording of the behaviours of individuals • the behaviour of WNS bats in captivity may differ from bats hibernating in the wild	• useful tool to complement pathophysiological studies which collectively provide insight on the effects of WNS on bats that would be impractical if not impossible to collect in the wild • influence of behavioural changes on survival & disease transmission	• Brownlee-Bouboulis and Reeder 2013; Warnecke et al. 2014; Wilcox et al. 2014; Turner et al. 2015; Bohn et al. 2016; McGuire et al. 2017; Mayberry et al. 2018
Quantitative magnetic resonance (QMR) body composition analyzer	• a significant advancement in non-destructive animal research, making it possible to calculate energy costs, fuel selection & changes in hydration in WNS bats following a simple three-minute scan	• could negatively affect free-ranging bats if it disrupts sensory systems used for geomagnetic orientation & navigation • does not detect/measure structural tissues, such as the skeleton	• useful for improving existing models of hibernation energetics which are important for predicting impacts of WNS on populations • allows for powerful longitudinal studies of the effects of WNS on individuals, & of how specific aspects of body composition influence subsequent behaviour	• Cheng et al. 2019
Passive integrated transponder (*i.e.* PIT-tag)	• small transponders with unique identifier codes that can be recorded when an animal is in range of an antenna & datalogger • collects valuable data on survival & movements of free-ranging animals • detects bats entering & exiting from overwintering sites to estimate survival • identifies bats that survive once WNS hits a certain region, & links survivors with traits that WNS selects for	• initially requires the capture & handling of bats to subcutaneously implant the PIT-tag, but once implanted, they allow for non-invasive data collection without the need to subject bats to the stress of repeatedly entering sites &/or recapturing individuals • improper implantation can lead to tag loss or failure to read the tag, or can cause infection leading to tag shedding or mortality of the individual, albeit very rare • re-detection rates are high relative to mark-recapture studies based on forearm banding but are lower than 100% as many tagged bats hibernate in unknown locations • if two tagged animals pass by the reader at the same time, the reader will not be able to read either tag • tags are relatively inexpensive, but the readers are costly	• provides unprecedented data on pre- & post-WNS survival & hibernation phenology, these data are important for determining whether a management strategy will generate positive population-level impacts • identifies traits WNS selects for, presenting an opportunity to make recommendations on appropriate mitigation strategies targeting these traits	• Britzke et al. 2014; Norquay and Willis 2014; Hoyt et al. 2019

Figure 9.1 Hibernating little brown bat (*Myotis lucifugus*) with white-nose syndrome, a multi-host skin disease caused by the fungus *Pseudogymnoascus destructans*. Note the presence of white fungal growth colonizing the muzzle and exposed skin of the ears and wing membranes. Photo by Jordi Segers from the Canadian Wildlife Health Cooperative. Used with permission.

and Kunz 2009; Blehert et al. 2009; Fuller et al. 2011). Biologists also documented dead bats on the snow outside of affected overwintering sites, and cave and mine floors littered with thousands of bat carcasses (Reichard and Kunz 2009; Blehert et al. 2009; Fuller et al. 2011). The characteristic for which WNS is named is the presence of white filamentous fungal growth colonizing the muzzle, but exposed skin of the ears, wings, and tail membranes of infected bats are also heavily colonized (Gargas et al. 2009; Meteyer et al. 2009; Blehert et al. 2009) (Figure 9.1). Aberrant from most other cutaneous fungal pathogens, Pd aggressively invades, digests, and erodes exposed skin and underlying connective tissue of hibernating bats, primarily targeting the wings and

tail membrane (Gargas et al. 2009; Meteyer et al. 2009; Reichard and Kunz 2009; Fuller et al. 2011; Lorch et al. 2011; Pikula et al. 2012; Pikula et al. 2017). Histological examination of infected skin has revealed that, once Pd invades the epidermis, it penetrates the underlying connective tissue, and causes gross epidermal erosions, and destruction of cutaneous glands (Table 9.2) (Meteyer et al. 2009; Chaturvedi et al. 2010; Cryan et al. 2010; Fuller et al. 2011). Ultimately, Pd invasion of the wing leads to necrosis of infected skin (Meteyer et al. 2009; Chaturvedi et al. 2010; Cryan et al. 2010; Fuller et al. 2011), and thus, impairment of wing structure and function (Blehert et al. 2009; Chaturvedi et al. 2010; Cryan et al. 2010). Bat wings are obviously important for flight but also play a critical role in physiological homeostasis, so damage to wing tissue by Pd appears to be a key contributor to the high mortality rates in hibernating bats (Cryan et al. 2010; Warnecke et al. 2013; Verant et al. 2014, 2018). However, the progressive disruption of physiological homeostasis that ultimately leads to mortality from this disease remains unclear.

9.2.2 White-nose syndrome pathophysiology

The precise cause of death from WNS is still not known but some basic aspects of WNS pathophysiology are understood, and infection of the skin of WNS-susceptible *M. lucifugus* with Pd is associated with several physiological consequences (Figure 9.2). Warnecke et al. (2013) confirmed that fluid loss across damaged wing tissue reduces plasma volume (i.e. hypovolaemia), and results in hypotonic dehydration and altered electrolyte balance, alongside increased arousal frequency (Willis et al. 2011) and used these data to devise a model of the pathophysiology of WNS during severe, advanced infections (Figure 9.2). Recently, McGuire et al. (2017) found support for this late-stage, dehydration model, showing that evaporative water loss and torpid metabolic rate were significantly increased for bats with advanced WNS. Pathophysiology in the early stage of infection appears quite different, however, with evidence of increased energy expenditure even before arousal patterns are disrupted, and with no evidence of increased water loss (Verant et al. 2014). These findings led Verant et al. (2014) to propose a multi-phase disease model for WNS that links skin infection with Pd with disruption of physiological homeostasis and mortality (Figure 9.2). The model suggests that the early stage of WNS infection involves the invasion, digestion, and erosion of the bat wing. This wing damage disrupts passive gas exchange and, therefore, early-stage infection is characterized by an increase in CO_2 in the blood, and severe acidosis (Verant et al. 2014).

Even with the initial colonization and invasion of the skin with Pd, prior to the onset of any clinical signs, bats exhibit a two-fold increase in energy expenditure and plasma CO_2 levels are often so high they would be considered lethal to most other mammals (Verant et al. 2014). Verant et al. (2014) suggested that, as the disease progresses, erosion of the skin and its vital structures stimulates an increase in energetically expensive arousals from hibernation, and an increase in active gas exchange across the lungs to remove elevated CO_2 from the blood and maintain acid–base homeostasis (Figure 9.2). Removing excess CO_2, and returning blood pH back to normal requires an increase in pulmonary respiration, energy expenditure, and water loss across the lungs, and will ultimately contribute to depletion of energy reserves (Thomas and Cloutier 1992; Thomas and Geiser 1997). These effects are exacerbated as WNS advances (Warnecke et al. 2012, 2013), and as Pd completely invades the skin, fluid is lost across damaged skin tissue, leading to hypovolaemia, hypotonic dehydration, and altered electrolyte balance, with arousals from torpor becoming more frequent (Warnecke et al. 2012, 2013). This cascade of physiological responses leads to an accelerated depletion of fat reserves and death (e.g. Reeder et al. 2012; Bandouchova et al. 2018; Figure 9.2). This model suggests that the physiological mechanisms bats rely on for gas exchange and acid–base balance do not appear to be stable throughout infection. More importantly, the model identifies hypotheses that need to be tested in order to develop a comprehensive understanding of the physiological effects of WNS on hibernating bats, and design strategies to minimize morbidity and mortality. It is conceivable that physiological differences among species may explain variation in WNS susceptibility among bats.

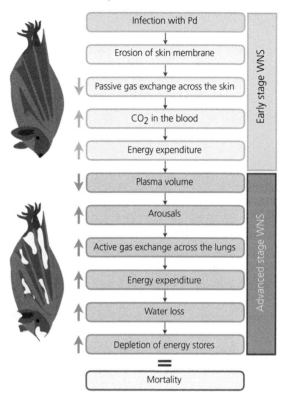

Figure 9.2 Multi-phase disease model for white-nose syndrome (WNS) that links skin infection of hibernating bats with *Pseudogymnoascus destructans* (Pd) to disruption of physiological homeostasis and mortality (modified from information in Verant et al. 2014).

9.3 Case study: why does a simple skin infection lead to dramatic changes in hibernation physiology and behaviour in some bats and not others?

9.3.1 Does disruption of circulation and gas exchange across a bat's wings play a role in white-nose syndrome pathophysiology?

Pd invasion of a bat's wings often results in the thickening and sloughing of the skin's outer layer, and consequently, functional and structural damage of the wings and their extensive vasculature (Cryan et al. 2010; Warnecke et al. 2013; Verant et al. 2014, 2018). Mammals are not typically thought to rely on their skin for gas exchange, but with its extremely thin and highly vascularized blood–gas barrier, a bat's wing is unique, and passive exchange across the skin can contribute significantly (>10 per cent) to total gas exchange (Herreid et al. 1968; Feder and Burggren 1985; Thomas et al. 1990b; Szewczak and Jackson 1992; Szewczak 1997; Makanya and Mortola 2007). Cutaneous gas exchange costs much less energy than active pulmonary gas exchange across the lungs (Herreid et al. 1968; Feder and Burggren 1985), so it may be especially important during hibernation when energy reserves are limited, and metabolism and ventilation are drastically suppressed (Thomas et al. 1990b; Szewczak and Jackson 1992; Szewczak 1997). Cryan et al. (2010) proposed that Pd invasion may restrict blood flow to the wing and impede passive gas exchange across the skin. Thus, to maintain acid–base homeostasis, Pd-infected bats may recruit energetically costly and water-intensive pulmonary gas exchange. There is now sufficient evidence to suggest that disrupted acid–base balance, owing to wing invasion by Pd, is the potential driving stimulus for the cascade of physiological changes that ultimately lead to death in bats with WNS (Cryan et al. 2010; Warnecke et al. 2013; Verant et al. 2014, 2018).

Progress has been made addressing the potential for wing damage caused by Pd to disrupt acid–base homeostasis, and increase energy expenditure and water loss (Cryan et al. 2010; Warnecke et al. 2013; Verant et al. 2014; Carey and Boyles 2015; McGuire et al. 2017; Verant et al. 2018); however, the influence of Pd invasion of the wing on gas exchange and WNS pathophysiology has received little attention. In a compelling study, Carey and Boyles (2015) tested the hypothesis that disruption of cutaneous gas exchange is a likely mechanism of WNS-associated mortality in bats. They inhibited gas exchange across bat skin by applying impermeable oil onto the wing and tail membranes of E. fuscus and then measured total energy expenditure and evaporative water loss. They found that bats did not spend more energy or lose more water when cutaneous pathways were blocked, and concluded that disruption of cutaneous gas exchange does not play a role in the dehydration and accelerated fat depletion associated with WNS pathophysiology. However, they investigated a species of bat that is resistant to WNS. Whether Pd-associated wing damage disrupts gas exchange across the wing, triggers energetically expensive pulmonary gas exchange, and contributes to mortality in a susceptible species warrants investigation. Disruption of circulation and gas exchange across a bat's wings associated with Pd invasion may be just one of many physiological mechanisms contributing to species differences in WNS susceptibility. One possibility is that WNS-susceptible M. lucifugus individuals rely more heavily on cutaneous gas exchange than WNS-resistant E. fuscus. If so, then inhibition of O_2 extraction across the skin due to fungal infection could have far more severe consequences for hibernation energetics of M. lucifugus and could help explain their susceptibility to WNS.

9.3.2 Sickness behaviour in bats with white-nose syndrome

Pd-infected bats exhibit so-called 'sickness behaviour', a coordinated response to infection mediated by inflammatory cytokines that often results in fever, and reduced behavioural activity and sociability (Hart 1988; Dantzer 1998a, b; Adelman and Martin 2009; Mayberry et al. 2018). Sickness behaviour may enhance survival by helping Pd-infected bats conserve energy and could be beneficial given that infection with Pd increases energy expenditure, arousal frequency, and fat depletion (Reeder et al. 2012; Warnecke et al. 2012, 2013; Verant et al. 2014). Several field and laboratory studies investigating behavioural responses of hibernating bats to Pd infection suggest classic symptoms of sickness behaviour in M. lucifugus (Langwig et al. 2012; Wilcox et al. 2014; Bohn et al. 2016; Mayberry et al. 2018). Healthy individuals of this species, and most species heavily impacted by WNS, tend to aggregate in large huddles or clusters during hibernation, with up to hundreds or thousands of individuals roosting together in close contact with other individual bats (McNab 1974; Clawson et al. 1980). Infection with Pd reduces this huddling behaviour, increases solitary roosting, and reduces behavioural activity in general (Langwig et al. 2012; Wilcox et al. 2014; Bohn et al. 2016; Mayberry et al. 2018; but see Brownlee-Bouboulis and Reeder 2013). This reduction in sociality could reduce transmission of the fungus to conspecifics or minimize disease severity of bats already infected with Pd by reducing additional points of contact of infection on the skin (Frick et al. 2016).

Sickness behaviour could also be energetically beneficial if roosting solitarily reduces a bat's potential to be disturbed by sick bats arousing more frequently within a huddle (Czenze and Willis 2015; Turner et al. 2015). However, reduced huddling could be detrimental for energy expenditure if it negatively impacts thermoregulatory costs. The ultimate benefit of huddling for healthy bats is not fully understood but huddling reduces a bat's exposed surface area, and thus, it may reduce thermal conductance and heat loss (Boyles et al. 2008) and it appears to reduce evaporative water loss (Boratyński et al. 2015). Furthermore, if multiple bats in a huddle warm up in synchrony it would allow them to exploit passive, social warming (Arnold 1993; Czenze and Willis 2015; Turner et al. 2015). Indeed, Czenze and Willis (2015) found that healthy *M. lucifugus* are more likely to passively rewarm and arouse in synchrony during late hibernation when their energy reserves are depleted than during early hibernation, presumably to share the costs of arousals. Thus, huddling could be especially important in bats infected with Pd, as WNS disrupts hibernation physiology and results in an increase in arousal frequency, depletion of fat stores, and dehydration (Reeder et al. 2012; Warnecke et al. 2012, 2013; Verant et al. 2014). However, the role of huddling on the winter energy and water balance of bats affected by WNS is still not fully understood.

Genes associated with the innate immune response are also known to trigger sickness behaviour and are often associated with a fever response (Cartmell et al. 2001; Fortier et al. 2004; Voss et al. 2006; Damm et al. 2012). Consistent with a fever response and sickness behaviour, Mayberry et al. (2018) showed that Pd-infected *M. lucifugus* rewarmed to higher body temperatures during arousals than healthy bats and, interestingly, genes associated with the innate immune response have been shown to be up-regulated for Pd-infected bats of this species (Rapin et al. 2014; Bure and Moore 2019), with a significant increase in pro-inflammatory cytokines associated with inflammation and regulation of immune responses to Pd infection (Rapin et al. 2014; Field et al. 2015; Lilley

et al. 2017; Davy et al. 2017; Donaldson et al. 2017; Moore et al. 2017). However, despite the up-regulation of these pro-inflammatory cytokines, they do not appear to be recruited to infection sites (Field et al. 2015; Moore et al. 2017).

Mounting an immune response during hibernation when activity, body temperature, and energy expenditure are drastically suppressed is a risky defence for Pd-infected bats, as a stimulated immune system may contribute to mortality through increased energy expenditure and premature fat depletion (Reeder et al. 2012; Lilley et al. 2016). Pd-infected *M. lucifugus* individuals also exhibit an increase in white blood cells and endogenous pyrogens (Bure and Moore 2019) and altered immune function (Moore et al. 2011, 2013), suggestive of a systemic inflammatory response. However, histopathological examination shows that Pd-infected *M. lucifugus* do not exhibit a pronounced inflammatory response during hibernation, regardless of the severity of lesions caused by the fungus (Meteyer et al. 2009; Cryan et al. 2010; Bure and Moore 2019), except for when bats emerge from hibernation the following spring (Meteyer et al. 2012). While the immune response is energetically costly and does not appear effective in WNS-susceptible species (Moore et al. 2011, 2013; Field et al. 2015; Johnson et al. 2015; Lilley et al. 2017), it may help explain the increase in energy expenditure and premature fat depletion in WNS-susceptible bats (Moore et al. 2011, 2013; Rapin et al. 2014), and could play an important role in WNS pathophysiology.

9.3.3 The potential role of huddling and protein catabolism for water conservation in white-nose syndrome-resistant/tolerant bats

One hypothesis to explain why hibernating mammals arouse from torpor during hibernation is restoration of water balance (Thomas and Cloutier 1992; Thomas and Geiser 1997; Willis et al. 2011). Hibernators lose water during torpor, albeit at dramatically reduced rates relative to normothermia, and, therefore, must arouse periodically to drink. As a result, increased water loss during Pd infection could explain why bats with WNS arouse more frequently than healthy bats. Consistent with this hypothesis, previous research has found that WNS-susceptible *M. lucifugus* individuals exhibit elevated rates of water loss and energy expenditure compared with healthy bats (McGuire et al. 2017), pointing to dehydration as a key driver of WNS-associated mortality. Interestingly, less susceptible *E. fuscus* individuals show no signs of dehydration following fungal infection (Dzal and Willis, unpublished). Yet the mechanisms that underlie this difference in water conservation among species remain unknown.

One mechanism that may play a role in water conservation is protein catabolism. The breakdown of protein yields five times more water than the breakdown of fat (Jenni and Jenni-Eiermann 1998). Thus, it is conceivable that some species do not exhibit signs of dehydration following fungal infection because they are capable of increasing the breakdown of protein to their fuel mixture, instead of solely relying on fat during hibernation. The maintenance of water balance is an immediate necessity that would reduce arousal from hibernation. To this end, the use of behavioural mechanisms, such as huddling, or physiological mechanisms, such as protein catabolism may explain how some species survive Pd invasion, whereas others do not. This information is essential to implicate the role of water loss in WNS pathophysiology and frame an understanding of what makes some bats more susceptible to WNS than others.

9.3.4 Bats persisting with white-nose syndrome accumulate larger pre-hibernation fat reserves than pre-white-nose syndrome bats from the same sites

Although WNS disrupts many aspects of physiology (Cryan et al. 2010; Warnecke et al. 2013; Verant et al. 2014, 2018; McGuire et al. 2017), it is generally thought that mortality from WNS occurs by causing bats to arouse from hibernation too frequently, and thus, WNS-affected bats prematurely deplete their fat reserves before spring (Reeder et al. 2012; Warnecke et al. 2013, Johnson et al. 2014; McGuire et al. 2017; Mayberry et al. 2018). Given these extra energetic costs, bats with larger pre-hibernation fat reserves may be more likely to survive hibernation

with WNS because they carry larger energy reserves and can tolerate more arousals. In support of this *fat-bat hypothesis*, bats persisting at some (though not all) sites following Pd invasion generally have greater fat reserves than those recorded before, or during, Pd invasion (Lacki et al. 2015; Cheng et al. 2019). An increase in pre-hibernation fat reserves after Pd invasion could result from either: (1) phenotypic plasticity in WNS survivors, such that WNS exposure leads to physiological or behavioural changes that cause WNS survivors to store more fat the following year; or (2) evolution by natural selection on fat accumulation such that only bats with large pre-hibernation fat reserves survive WNS exposure, and then pass this trait on to their offspring. Previous studies have not been able to distinguish between these two mechanisms because, to date, there are no datasets that combine measurements of pre-hibernation fat reserves prior to and following Pd invasion, with data on survival of known individuals. While recent findings support this *fat-bat hypothesis*, at least for some sites (Lacki et al. 2015; Cheng et al. 2019), mechanisms allowing bats to accumulate larger pre-hibernation fat reserves are unknown.

9.3.5 Do skin lipids or microbiome affect resistance to *Pseudogymnoascus destructans* infection?

The ability to defend the skin from structural and functional degradation caused by Pd invasion is critical to a bat's survival. Host defence mechanisms, such as the lipid composition of the epidermis (Romani 2011; Pannkuk et al. 2015; Frank et al. 2018), and the microbial community inhabiting the skin (Lemieux-Labonté et al. 2017; Ange-Stark et al. 2019, unpublished) could play a role in preventing colonization of the skin by Pd, and subsequent development of WNS. Many cutaneous lipids are known to possess anti-fungal activity (Fisher et al. 2014). These anti-fungal lipids inhibit fungal growth by inserting themselves into the cell membrane of the fungus, increasing the fluidity of the membrane, resulting in the leakage of intracellular components and, consequently, cell death (Fisher et al. 2014). Species of bats that appear to be resistant to the disease (*E. fuscus* and the European species, *Myotis*

myotis) have been found to produce unique cutaneous lipids and wax esters that prevent Pd infection (Rezanka et al. 2015; Ingala et al. 2017). Furthermore, the skin of *E. fuscus* comprises significantly more free fatty acids with potent anti-fungal properties than WNS-susceptible *M. lucifugus* (Frank et al. 2018).

In addition to epidermal lipids in the skin, variation in the microbial community inhabiting bat skin could help explain differences in WNS susceptibility among bats and variation in disease severity. The vertebrate epidermis is inhabited by microbes that protect hosts from cutaneous fungal pathogens by directly occupying sites on the skin, and producing pathogen inhibitors (Roth and James 1988; Harris et al. 2006; Woodhams et al. 2007; Grice and Segre 2011; Cho and Blaser 2012; Bataille et al. 2016). Hoyt et al. (2015) isolated a group of Pd-inhibiting bacteria (i.e. *Pseudomonas fluorescens*) that naturally inhabit the skin of WNS-resistant *E. fuscus*. When Cheng et al. (2016) applied these Pd-inhibiting bacteria to the wings of WNS-susceptible *M. lucifugus* in the laboratory, right after experimental infection with Pd, WNS severity and mortality were reduced. Interestingly, however, and highlighting the complexity of the skin microbiome and potential challenges of treatment, when Cheng et al. (2016) applied *P. fluorescens* to the wings of *M. lucifugus* several weeks prior to infection with Pd, WNS severity was augmented compared with Pd-infected controls (although there was no effect on bat survival). One explanation for this effect is that early treatment with *P. fluorescens* disrupted the natural microbiome in a way that makes it easier for Pd to grow and invade. Subsequently, Hoyt et al. (2019) tested the efficacy of *P. fluorescens* in reducing the impacts of WNS in *M. lucifugus* in the wild and found that treating free-flying *M. lucifugus* with *P. fluorescens* increased apparent overwinter survival more than five-fold (from 8.4 to 46.2 per cent), compared with untreated bats. While overall WNS-associated mortality was still high (>50 per cent), treatment effects on overwinter survival were significant, suggesting that treating WNS-affected bats with *P. fluorescens* could reduce the impact of WNS on bat populations, and prevent further declines and promote persistence.

Another probable mechanism of resistance to WNS is a shift in cutaneous microbial communities following Pd invasion. A field study by Lemieux-Labonté et al. (2017) demonstrated that the skin microbiome of bats overwintering and persisting in WNS-positive sites was enriched in bacteria with anti-fungal properties (e.g. including *Pseudomonas*), and that microbiomes of these bats had less diversity compared with bats overwintering at WNS-negative hibernacula. Their findings suggest that, as Pd arrives in a hibernaculum, bats with the highest level of anti-fungal bacteria, such as *Pseudomonas*, prior to Pd invasion, will have the highest chance of survival. This shift in cutaneous microbial communities following Pd invasion is a promising discovery for species re-establishment. However, Lemieux-Labonté et al. (2017) could not confirm that the differences in bat skin microbiome between WNS-positive and WNS-negative sites directly reflected effects of WNS because samples were not available from the same sites before and after Pd invasion. Therefore, these patterns could reflect microbiome differences that existed prior to Pd invasion due to hibernaculum-specific differences in bacterial communities (Avena et al. 2016; Lemieux-Labonté et al. 2017).

While an understanding of cutaneous microbial communities and how they change with Pd invasion is limited, Ange-Stark et al. (2019, unpublished) recently assessed the diversity of the microbes of bat skin at the same hibernacula before and after Pd invasion of three species of bats that range in WNS susceptibility, *M. lucifugus*, *P. subflavus*, and *E. fuscus*. Complementing the findings of Lemieux-Labonté et al. (2017), Ange-Stark et al. (2019, unpublished) found that all bats persisting with Pd were enriched in bacteria with anti-fungal properties. While bacterial diversity was significantly reduced in *M. lucifugus* post-WNS compared with bats pre-WNS, it was not for *P. subflavus* or *E. fuscus*, even though, similar to *M. lucifugus*, *P. subflavus* is severely impacted by WNS and experiences a high rate of mortality. These findings highlight a likely role for the microbial community inhabiting bat skin in resistance to Pd invasion and growth. Understanding interactions between Pd and the skin microbiome is important to understanding the pathophysiology and epidemiology of WNS and may be useful for developing chemical and biological treatments.

9.4 Future directions

9.4.1 Applying conservation physiology in response to white-nose syndrome: should interventions focus on treating the disease or protecting and enhancing bat habitat?

WNS is currently inducing one of the strongest natural selection events ever observed for a wildlife population in real time (e.g. Maslo and Fefferman 2015; Maslo et al. 2015). Consequently, WNS-affected bats could be subject to high selection differentials for behavioural, physiological, and immune traits that may lead to resistance to, and/or tolerance of, Pd (see Section 9.3). Since WNS emerged in North America over a decade ago, millions of dollars have been devoted to research on chemical and biological interventions to treat bats with WNS, or the environment in which they hibernate (Muller et al. 2013; Cornelison et al. 2014; Hoyt et al. 2015; O'Donoghue et al. 2015; Zhang et al. 2015; Cheng et al. 2016; Palmer et al. 2018; Rocke et al. 2019). Yet, while some promising approaches are being tested, to date, no effective treatment is ready to be deployed. Understanding differences in species susceptibility, and the physiological mechanisms underlying resistance to, and/or tolerance of Pd is critical for evaluating the safety and efficacy of chemical or biological treatments in the wild. If the skin is important for resistance and/or tolerance (see Sections 9.3.1 and 9.3.5), then any potential treatment must consider normal bat skin physiology. For example, how might chemical or biological treatments interfere with (or enhance) cutaneous or pulmonary gas exchange and water loss (Section 9.3.1), or the lipid composition and microbial community of the skin (Section 9.3.5)? What are the implications of these effects for disease severity and survival, or for healthy, pre-WNS bats? Management strategies may also differ markedly depending on whether bats acquire resistance or tolerance to Pd. Therefore, while development of chemical or biological treatments may seem highly attractive, depending on a range of factors, these treatments could have counter-productive effects.

Given limited conservation resources in areas where large aggregations of bats face WNS mortality, funding dollars may be best spent applying chemical or biological treatment for WNS once it arrives. However, if chemical or biological interventions promote the survival of any individual in the population, it is conceivable that treatment would reduce selection differentials exerted on the host by the fungus, and could ultimately slow the evolution of traits important for resistance and tolerance (Maslo and Fefferman 2015). Conversely, populations predicted to experience survivorship without intervention, due to physiology or environmental conditions, will be important for evolutionary rescue. Thus, the most urgent priorities for management of these populations should be implementing strategies that protect individuals in affected areas, and support reproduction by these survivors, rather than the application of chemical and biological treatments to kill the disease, and inadvertently decreasing a population's probability of persistence.

Understanding the potential of bats to persist, recover, and repopulate WNS-affected areas after Pd invasion, and understanding if, and how survivors differ from the bats that have died from WNS is a critical priority in light of evidence that some bats survive the winter with WNS (Langwig et al. 2012; Frick et al. 2015; Maslo et al. 2015; Langwig et al. 2017). However, information on the specific traits conferring resistance and/or tolerance to this disease are limited (for a discussion, see Sections 9.3.1 to 9.3.5). This information is needed in order to provide insight into how certain traits may translate into survival and the persistence of bats following Pd invasion, and whether natural selection has the potential to help populations recover and rebound from the disease. Infection with Pd causes a bottleneck for winter energy balance, with WNS increasing arousal frequency, fat depletion, and starvation (Reeder et al. 2012; Warnecke et al. 2012, 2013; Verant et al. 2014). However, there is also another, poorly understood bottleneck that occurs in the spring. Individuals that survive the winter with WNS emerge in spring, travel to their summer roosts, and begin to recover at a time when ambient temperature is still cold and food is scarce (Jonasson and Willis 2011; Norquay and Willis 2014; Czenze et al. 2017). These bats also immediately face a remobilized immune response associated with devastating wing damage, likely to reduce foraging ability and increase energetic costs

(Meteyer et al. 2009; Cryan et al. 2010; Meteyer et al. 2012). This spring recovery bottleneck may be especially harsh for female bats, which are capital breeders and are already energetically constrained because they depend on fat stored over the winter to initiate pregnancy before food is available in the spring (Jonasson and Willis 2011; Czenze et al. 2017). Even if females with WNS survive their winter bottleneck they may have little fat reserves left to invest in reproduction on top of the expensive process of healing and recovery from the disease. This is especially worrisome because affected species depend on social thermoregulation to maintain warm roost conditions in spring and summer (see Section 9.3.3). As bat numbers decline, fewer roostmates will be available for recovering bats to huddle with, increasing energetic costs or the need to use torpor. This, in turn, could reduce the chance of reproduction by survivors and limit potential evolution of winter survival traits. While these energetic challenges are discouraging to hopes for population recoveries, opportunities remain for proactive interventions that may mitigate impacts when WNS arrives, and their effective application will require an integrative conservation physiology approach.

Augmenting evolutionary responses of threatened populations through facilitated evolution is potentially a powerful conservation tool (Stockwell et al. 2003). For WNS survivors, management actions that promote positive energy balance during the active season could be even more important than chemical or biological interventions directly targeting effects of Pd infection on bats during winter. For example, during the autumn, bats accumulate large fat stores during a period of only a few weeks that is necessary for fuelling them for the entire hibernation period (McGuire et al. 2016). Since WNS increases energy expenditure throughout hibernation, it is likely more important that bats have access to high-quality foraging habitat close to overwintering sites for them to acquire the necessary fat stores to survive hibernation if infected with WNS. Furthermore, despite the often dramatic winter die-offs that characterize WNS invasion, physiological consequences of WNS often carry over to the active spring season. Improving energy balance of bats by protecting, and possibly enhan-

cing, high-quality roosting and foraging habitat used by bats during pre-hibernation fattening in the autumn, and immediately after emergence from hibernation in the spring is one approach that could help achieve this goal.

Bats recovering from WNS have heightened energetic costs due to the effects of wing damage on flight performance and foraging efficiency, and a reduced opportunity for social thermoregulation. This carryover effect of WNS could have pronounced implications for how we manage foraging habitats of bats. Mitigation strategies can concentrate on improving energy budgets by protecting and enhancing roosting and foraging habitat to help bats that survive WNS maintain energy balance, recover from WNS, and reproduce. However, we know next to nothing about foraging behaviour of WNS-susceptible *M. lucifugus* in the wild (but see Henry et al. 2002; Coleman et al. 2014). For example, if WNS survivors select open habitats because of reduced manoeuvrability due to WNS-associated wing damage, then wetland protection near quality roosting habitat could be especially important to provide large numbers of the emergent aquatic insects that *M. lucifugus* prefer (Clare et al. 2014), in habitat that recovering bats can reliably navigate.

Beyond protection of existing habitat known to be important to bats, habitat enhancement by providing bats with a thermal refuge (Boyles and Willis 2010; Wilcox and Willis 2016), or using insect attractants or lights to create prey patches would help identify specific roosting and habitat characteristics that could help WNS survivors maintain a positive energy balance during spring recovery. For example, WNS survivors may compensate for energetic costs of healing via selection of warm roost microclimates. At the small scale, we can provide WNS survivors with a thermal refuge, and deploy heated bat houses within summer and autumn habitat. In doing so, we can test the *roosting habitat enhancement hypothesis* to determine whether bats exploit warm roost microclimates to save energy during healing, recovery, and reproduction during the active season and whether warmer roosting habitats help WNS survivors make it through potentially harsh spring conditions, and thus, improve recovery (Wilcox and Willis 2016). Furthermore, the energy imbalance for bats resulting from WNS suggests that managing habitat to support food for bats could be especially useful in maintaining a positive energy balance. To this end, by increasing foraging efficiency of bats in autumn and spring we can improve opportunities for bats to accumulate more fat before hibernation, and improve recovery from WNS in spring, thereby increasing annual survival and aiding population recovery. We can test this *habitat enhancement hypothesis* by concentrating prey in the vicinity of known bat habitat using artificial light, as bats preferentially exploit prey patches at lights, presumably because increased prey density enhances foraging efficiency (e.g. Fenton and Morris 1976; Bell 1980; Sleep and Brigham 2003; Cravens and Boyles 2019). Recently, it has been shown that WNS selects for larger pre-hibernation fat reserves, and that larger fat reserves allow bats to tolerate Pd (Cheng et al. 2019; see Section 9.3.4). Thus, not only would concentrated prey patches reduce commuting costs from roosts to foraging areas, they would also help bats accumulate large enough overwintering fat reserves to survive Pd infection, and help WNS survivors obtain enough energy to meet the demands of survival, immune mobilization, healing, and reproduction in the spring.

At both the small scale of providing bats with a thermal refuge and artificial prey patches near overwintering sites, and at the landscape scale of protection of natural roosting and foraging habitats for bats, the research proposed here would test urgent hypotheses with great potential to inform and aid recovery of bat populations. This research would address important gaps in knowledge of ecological needs of bat species known to be susceptible to WNS, and provide evidence-based recommendations on the types of roosting and foraging habitats that should be prioritized for protection. If we can identify specific habitat types preferred by bats recovering from WNS we could dramatically improve our ability to protect survivors of WNS and prioritize management responses to best help populations of bat species at risk from WNS.

9.5 Conclusions

Since WNS was first observed over a decade ago, a conservation physiology approach has allowed for tremendous progress connecting WNS pathophysiology with population impacts, while also helping devise and inform research on possible conservation solutions. We still do not fully understand the aspects of WNS that best connect a simple skin infection with premature fat depletion and death. Determining potential effectiveness of interventions, quantifying the evolutionary potential of populations of hosts, and quantifying the importance of certain habitats for survival and reproduction, before and after WNS, have great potential to help us decide whether to focus interventions on targeting the disease with chemical or biological treatments, or alternatives, such as habitat protection to enhance reproduction by survivors. We contend that the understudied key to success in our fight against Pd will be connecting physiological mechanisms with wildlife population dynamics, to understand why some bats survive WNS while others suffer devastating mortality. Given the intrinsic value of bats to healthy ecosystems, as well as the ecological and economic benefits insectivorous bats provide, efforts to understand, predict, and appropriately manage and mitigate WNS-related mortality should be a high priority.

References

Adelman, J.S. and Martin, L.B., 2009. Vertebrate sickness behaviors: adaptive and integrated neuroendocrine immune responses. *Integrative and Comparative Biology*, 49, 202–14.

Agugliaro, J., Lind, C.M., Lorch, J.M. et al., 2019. An emerging fungal pathogen is associated with increased resting metabolic rate and total evaporative water loss rate in a winter-active snake. *Functional Ecology*, 34(2), 486–96.

Anderson, P.K., Cunningham, A.A., Patel, N.G. et al., 2004. Emerging infectious diseases of plants: pathogen pollution, climate change and agrotechnology drivers. *Trends in Ecology & Evolution*, 19(10), 535–44.

Anderson, R.M., 1979. Parasite pathogenicity and the depression of host population equilibria. *Nature*, 279, 150–2.

Ange-Stark, M., Cheng, T.L., Hoyt, J.L. et al., 2019. White-nose syndrome restructure bat skin microbiome. *bioRxiv*. https://doi.org/10.1101/614842

Arnold, W., 1993. Energetics of social hibernation. In C. Carey, G.L. Florant, B.A. Wunder, and B. Horowitz B, eds. *Life in the Cold. Ecological, Physiological, and*

Molecular Mechanisms, pp. 65–80. Westview Press, Inc., Boulder, CO.

Avena, C.V., Parfrey, L.W., Leff, J.W. et al., 2016. Deconstructing the bat skin microbiome: influences of the host and the environment. *Frontiers in Microbiology*, 7, 1753.

Baldauf, S.L. and Palmer, J.D., 1993. Animals and fungi are each other's closest relatives: congruent evidence from multiple proteins. *Proceedings of the National Academy of Science of the USA*, 90(24), 11558–11562.

Bandouchova, H., Bartonička, T., Berkova, H. et al., 2018. Alterations in the health of hibernating bats under pathogen pressure. *Scientific Reports*, 8, 6067.

Bataille, A., Lee-Cruz, L., Tripathi, B., et al., 2016. Microbiome variation across amphibian skin regions: implications for chytridiomycosis mitigation efforts. *Microbial Ecology*, 71(1), 221–32.

Bell, G.P., 1980. Habitat use and response to patches of prey by desert insectivorous bats. *Canadian Journal of Zoology*, 58(10), 1876–83.

Blehert, D.S., Hicks, A.C., Behr, M. et al., 2009. Bat white-nose syndrome: an emerging fungal pathogen? *Science*, 323(5911), 227.

Bohn, S.J., Turner, J.M., Warnecke, L. et al., 2016. Evidence of 'sickness behaviour' in bats with white-nose syndrome. *Behaviour*, 153, 981–1003.

Boratyński, J.S., Willis, C.K.R., Jefimow, M. et al., 2015. Huddling enhances survival of hibernating bats by reducing evaporative water loss. *Comparative Biochemistry and Physiology A*, 179, 125–32.

Boyles, J.G., Storm, J.J., and Brack Jr, V., 2008. Thermal benefits of clustering during hibernation-a field test of competing hypotheses on *Myotis sodalis*. *Functional Ecology*, 22, 632–6.

Boyles, J.G. and Willis, C.K.R., 2010. Could localized warm areas in cold caves reduce mortality of hibernating bats affected by white-nose syndrome? *Frontiers in Ecology and the Environment*, 8, 92–8.

Britzke, E.R., Gumbert, M.W., and Hohmann, M.G., 2014. Behavioral response of bats to passive integrated transponder tag reader arrays placed at cave entrances. *Journal of Fish and Wildlife Management*, 5(1), 146–50.

Brownlee-Bouboulis, S.A. and Reeder, D.M., 2013. White-nose syndrome affected little brown myotis (*Myotis lucifugus*) increase grooming and other active behaviors during arousals from hibernation. *Journal of Wildlife Diseases*, 49(4), 850–9.

Bure, C.M. and Moore, M.S., 2019. White-nose syndrome: a fungal disease of North American hibernating bats. In W. White, D. Culver, and T. Pipan, eds. *Encyclopedia of Caves*, pp. 1165–74. Elsevier, London.

Carey, C.S. and Boyles, J.G., 2015. Interruption to cutaneous gas exchange is not a likely mechanism of WNS-associated death in bats. *Journal of Experimental Biology*, 218(13), 1986–89.

Cartmell, T., Luheshi, G.N., Hopkins, S.J. et al., 2001. Role of endogenous interleukin-1 receptor antagonist in regulating fever induced by localised inflammation in the rat. *Journal of Physiology*, 531, 171–80.

Casadevall, A., 2019. Global catastrophic threats from the fungal kingdom: fungal catastrophic threats. *Current Topics in Microbiology and Immunology*, 424, 21–32.

Chaturvedi, V., Springer, D.J., Behr, M.J. et al., 2010. Morphological and molecular characterizations of psychrophilic fungus *Geomyces destructans* from New York bats with white nose syndrome (WNS. *PLoS ONE*, 5, e10783.

Cheng, T.L., Gerson, A., Moore, M.S. et al., 2019. Higher fat stores contribute to persistence of little brown bat populations with white-nose syndrome. *Journal of Animal Ecology*, 88, 591–600.

Cheng, T.L., Mayberry, H., McGuire, L.P. et al., 2016. Efficacy of a probiotic bacterium to treat bats affected by the disease white-nose syndrome. *Journal of Applied Ecology*, 54(3), 701–8.

Cho, I. and Blaser, M.J., 2012. The human microbiome: at the interface of health and disease. *Nature Reviews Genetics*, 13(4), 260–70.

Clare E.L., Symondson, W.O., Broders, H.G., et al., 2014. The diet of *Myotis lucifugus* across Canada: assessing foraging quality and diet variability. *Molecular Ecology*, 23, 3618–32.

Clawson, R.L., Laval, R.K., Laval, M.L. et al., 1980. Clustering behavior of hibernating *Myotis sodalis* in Missouri. *Journal of Mammalogy*, 61, 245–53.

Coleman, L.S., Ford, W.M., Dobony, C.A. et al., 2014. Comparison of radio-telemetric home-range analysis and acoustic detection for little brown bat habitat evaluation. *Northeastern Naturalist*, 21, 431–46.

Cooke, S.J., Sack, L., Franklin, C.E. et al., 2013. What is conservation physiology? Perspectives on an increasingly integrated and essential science. *Conservation Physiology*, 1, 1–23.

Coristine, L.E., Robillard, C.M., Kerr, J.T. et al., 2014. A conceptual framework for the emerging discipline of conservation physiology. *Conservation Physiology*, 2, 1–12.

Cornelison, C.T., Gabriel, K.T., Barlament, C. et al., 2014. Inhibition of *Pseudogymnoascus destructans* growth from conidia and mycelial extension by bacterially produced volatile organic compounds. *Mycopathologia*, 177, 1–10.

COSEWIC., 2013. COSEWIC assessment and status report on the little brown myotis *Myotis lucifugus*, northern myotis *Myotis septentrionalis* and tri-colored bat *Perimyotis subflavus* in Canada. Committee on the Status of Endangered Wildlife in Canada, Ottawa. (Species at Risk Public Registry website).

Cravens, Z.M. and Boyles, J.G., 2019. Illuminating the physiological implications of artificial light on an insectivorous bat community. *Oecologia*, 189, 69–77.

Cryan, P.M., Meteyer, C.U, Boyles, J.G. et al., 2010. Wing pathology of white-nose syndrome in bats suggests life-threatening disruption of physiology. *BMC Biology*, 8, 135.

Czenze, Z.C. and Willis, C.K.R., 2015. Warming up and shipping out: cues for arousal and emergence in hibernating little brown bats (*Myotis lucifugus*). *Journal of Comparative Physiology B*, 185, 575–86.

Czenze, Z.J., Jonasson, K.A., and Willis, C.K.R., 2017. Thrifty females, frisky males: winter energetics of hibernating bats from a cold climate. *Physiological and Biochemical Zoology*, 90(4), 502–11.

Damm, J., Wiegand, F., Harden, L.M. et al., 2012. Fever, sickness behavior, and expression of inflammatory genes in the hypothalamus after systemic and localized subcutaneous stimulation of rats with the toll-like receptor 7 agonist imiquimod. *Neuroscience*, 201, 166–83.

Dantzer, R., Bluth, R., Gheusi, G. et al., 1998a. Molecular basis of sickness behavior. *Annals of the New York Academy of Sciences*, 856, 132–8.

Dantzer, R., Bluth, R., Laye, S. et al., 1998b. Cytokines and sickness behavior. *Annals of the New York Academy of Sciences*, 840, 586–90.

Daszak, P., Cunningham, A.A., and Hyatt, A.D., 2000. Emerging infectious diseases of wildlife: threats to biodiversity and human health. *Science*, 287, 443–9.

Davy, C.M., Donaldson, M.E., Willis, C.K.R. et al., 2017. The other white-nose syndrome transcriptome: Tolerant and susceptible hosts respond differently to the pathogen *Pseudogymnoascus destructans*. *Ecology and Evolution*, 7, 7161–70.

Donaldson, M.E., Davy, C.M., Willis, C.K.R. et al., 2017. Profiling the immunome of little brown myotis provides a yardstick for measuring the genetic response to white-nose syndrome. *Evolutionary Applications*, 10, 1076–90.

Feder, M.E. and Burggren, W.W., 1985. Cutaneous gas exchange in vertebrates: design, patterns, control and implications. *Biological Reviews*, 60, 1–45.

Fenton, M.B., and Morris, G.K., 1976. Opportunistic feeding by desert bats (*Myotis* spp.. *Canadian Journal of Zoology*, 54(4), 526–30.

Field, K.A., Johnson, J.S., Lilley, T.M. et al., 2015. The white-nose syndrome transcriptome: activation of antifungal host responses in wing tissue of hibernating little brown myotis. *PLoS Pathogens*, 11, e1005168.

Fisher, C.L., Blanchette, D.R., Brogden, K.A. et al., 2014. The roles of cutaneous lipids in host defense. *Biochimica et Biophysica Acta*, 1841, 319–22.

Fisher, M.C., Henk, D.A., Briggs, C.J. et al., 2012. Emerging fungal threats to animal, plant and ecosystem health. *Nature*, 484, 186–94.

Fortier, M.E., Kent, S., Ashdown, H. et al., 2004. The viral mimic, polyinosinic: polycytidylic acid, induces fever in rats via an interleukin-1 dependent mechanism. *American Journal of Physiology—Regulatory, Integrative and Comparative Physiology*, 287, 759–66.

Frank, C.L., Michalski, A., McDonough, A.A. et al., 2014. The resistance of a North American bat species (*Eptesicus fuscus*) to white-nose syndrome (WNS. *PLoS ONE*, 9, e113958.

Frank, C.L., Sitler-Elbel, K.G., Hudson, A.J. et al., 2018. The antifungal properties of epidermal fatty acid esters: insights from white-nose syndrome (WNS) in bats. *Molecules*, 23, 1986.

Frick, W.F., Pollock, J.F., Hicks, A.C. et al., 2010. An emerging disease causes regional population collapse of a common North American bat species. *Science*, 329(5992), 679–82.

Frick, W.F., Puechmaille, S.J., Hoyt, J.R. et al., 2015. Disease alters macroecological patterns of North American bats. *Global Ecology and Biogeography*, 24(7), 741–9.

Frick, W.F., Puechmaille, S.J., and Willis, C.K.R., 2016. White-nose syndrome in bats. In C.C. Voigt and T. Kingston, eds. *Bats in the Anthropocene: Conservation of Bats in a Changing World*, pp.245–62. Springer, Cham, Switzerland.

Fuller, N.W., Reichard, J.D., Nabhan, M.L. et al., 2011. Free-ranging little brown myotis (*Myotis lucifugus*) heal from wing damage associated with white-nose syndrome. *EcoHealth*, 8, 154–62.

Gargas, A., Trest, M.T., Christensen, M. et al., 2009. *Geomyces destructans* sp. nov. associated with bat white-nose syndrome. *Mycotaxon*, 108, 147–54.

Geiser, F., 2004. Metabolic rate and body temperature reduction during hibernation and daily torpor. *Annual Review of Physiology*, 66, 239–74.

Greer, A.L., Briggs, C.J. and Collins, J.P., 2008. Testing a key assumption of host pathogen theory: density and disease transmission. *Oikos*, 117(11), 1667–73.

Grice, E.A. and Segre, J.A., 2011. The skin microbiome. *Nature Reviews Microbiology*, 9(4), 244–53.

Harris, R.N., James, T.Y., Lauer, A. et al., 2006. Amphibian pathogen *Batrachochytrium dendrobatidis* is inhibited by the cutaneous bacteria of amphibian species. *EcoHealth*, 3(1), 53.

Hart, B.L., 1988. Biological basis of the behaviour of sick animals. *Neuroscience and Biobehavioral Reviews*, 12, 123–37.

Henry, M., Thomas, D.W., Vaudry, R. et al., 2002. Foraging distances and home ranges of pregnant and lactating little brown bats (*Myotis lucifugus*). *Journal of Mammalogy*, 83(3), 767–74.

Herreid, C.F., Bretz, W.L., and Schmidt-Nielsen, K., 1968. Cutaneous gas exchange in bats. *American Journal of Physiology*, 215, 506–8.

Hoyt, J.R., Cheng, T.L., Langwig, K.E. et al., 2015. Bacteria isolated from bats inhibit the growth of *Pseudogymnoascus destructans*, the causative agent of white-nose syndrome. *PLoS ONE, 10*(4), e0121329.

Hoyt, J.R., Langwig, K.E., Okoniewski, J. et al., 2014. Long-term persistence of *Pseudogymnoascus destructans*, the causative agent of white-nose syndrome, in the absence of bats. *EcoHealth, 12*(2), 330–3.

Hoyt, J.R., Langwig, K.E., Sun, K. et al., 2016. Host persistence or extinction from emerging infectious disease: insights from white-nose syndrome in endemic and invading regions. *Proceedings of the Royal Society B: Biological Sciences, 283*(1826), 20152861.

Hoyt, J.R., Langwig, K.E., White, J.P. et al., 2019. Field trial of a probiotic bacteria to protect bats from white-nose syndrome. *Scientific Reports, 9*(1), 9158.

Ingala, M.R., Ravenelle, R.E., Monro, J.J. et al., 2017. The effects of epidermal fatty acid profiles, 1-oleoglycerol, and triacylglycerols on the susceptibility of hibernating bats to *Pseudogymnoascus destructans*. *PLoS ONE, 12*, e0187195.

International Union for the Conservation of Nature (IUCN), 2020. The IUCN Red List of Threatened Species. January 2020. Available at: www.iucnredlist.org/ (accessed 26 June 2020).

Janicki, A.F., Frick, W.F., Kilpatrick, A.M. et al., 2015. Efficacy of visual surveys for white-nose syndrome at bat hibernacula. *PLoS ONE, 10*(7), e0133390.

Jenni, L. and Jenni-Eiermann, S., 1998. Fuel supply and metabolic constraints in migrating birds. *Journal of Avian Biology, 29*, 521–8.

Johnson, J.S., Reeder, D.M., Lilley, T.M. et al., 2015. Antibodies to *Pseudogymnoascus destructans* are not sufficient for protection against white-nose syndrome. *Ecology and Evolution, 5*, 2203–14.

Johnson, J.S., Reeder, D.M., McMichael, J.W. et al., 2014. Host, pathogen and environmental characteristics predict white-nose syndrome mortality in captive little brown bats (*Myotis lucifugus. PLoS ONE, 9*, e112502.

Jonasson, K.A. and Willis, C.K.R., 2011. Changes in body condition of hibernating bats support the thrifty female hypothesis and predict consequences for populations with white-nose syndrome. *PLoS ONE, 6*(6), e21061.

Jones, K.E., Patel, N.G., Levy, M.A. et al., 2008. Global trends in emerging infectious diseases. *Nature, 451*, 990–3.

Keeling, M.J. and Grenfell, B.T., 1997. Disease extinction and community size: modeling the persistence of measles. *Science, 275*, 65–7.

Lacki, M.J., Dodd, L.E., Toomey, R.S. et al., 2015. Temporal changes in body mass and body condition of cave-hibernating bats during staging and swarming. *Journal of Fish and Wildlife Management, 6*(2), 360–70.

Langwig, K.E., Frick, W.F., Bried, J.T. et al., 2012. Sociality, density-dependence and microclimates determine the persistence of populations suffering from a novel fungal disease, white-nose syndrome. *Ecology Letters, 15*(9), 1050–7.

Langwig, K.E., Frick, W.F., Hoyt, J.R. et al., 2016. Drivers of variation in species impacts for a multi-host fungal disease of bats. *Philosophical Transactions of the Royal Society of London. Series B, Biological Sciences, 371*, 20150456.

Langwig, K.E., Frick, W.F., Reynolds, R. et al., 2015a. Host and pathogen ecology drive the seasonal dynamics of a fungal disease, white-nose syndrome. *Proceedings of the Royal Society. Biological Sciences, 282*(1799), 20142335.

Langwig, K.E., Hoyt, J.R., Parise, K.L. et al., 2015b. Invasion dynamics of white-nose syndrome fungus, Midwestern United States, 2012–2014. *Emerging Infectious Diseases, 21*(6), 1023–6.

Langwig, K.E., Hoyt, J.R., Parise, K.L. et al., 2017. Resistance in persisting bat populations after white-nose syndrome invasion. *Philosophical Transactions of the Royal Society of London. Series B, Biological Sciences, 372*(1712), 20160044.

Lemieux-Labonté, V., Simard, A., Willis, C.K.R. et al., 2017. Enrichment of beneficial bacteria in the skin microbiota of bats persisting with white-nose syndrome. *Microbiome, 5*, 115.

Leopardi, S., Blake, D., and Puechmaille, S.J., 2015. White-nose syndrome fungus introduced from Europe to North America. *Current Biology, 25*, R217–19.

Lilley, T.M., Johnson, J.S., Ruokolainen, L., et al., 2016. White-nose syndrome survivors do not exhibit frequent arousals associated with *Pseudogymnoascus destructans* infection. *Frontiers in Zoology, 13*, 12.

Lilley, T.M., Prokkola, J.M., Johnson, J.S. et al., 2017. Immune responses in hibernating little brown myotis (*Myotis lucifugus*) with white-nose syndrome. *Proceedings of the Royal Society B, 284*, 20162232.

Lind, C., Moore, I.T., Akçay, Ç., et al., 2018. Patterns of circulating corticosterone in a population of rattlesnakes afflicted with snake fungal disease: stress hormones as a potential mediator of seasonal cycles in disease severity and outcomes. *Physiological and Biochemical Zoology, 91*, 765–75.

Lind, C.M., Lorch, J.M., Moore, I.T. et al., 2019. Seasonal sex steroids indicate reproductive costs associated with snake fungal disease. *Journal of Zoology, 307*, 104–10.

Lips, K.R., 2016. Overview of chytrid emergence and impacts on amphibians. *Philosophical Transactions of the Royal Society of London. Series B, Biological Sciences, 371*(1709), 20150465.

Lorch, J.M., Lankton, J., Werner, K. et al., 2015. Experimental infection of snakes with *Ophidiomyces ophiodiicola* causes pathological changes that typify snake fungal disease. *mBio*, 6, 1–9.

Lorch, J.M., Meteyer, C.U., Behr, M.J. et al., 2011. Experimental infection of bats with *Geomyces destructans* causes white-nose syndrome. *Nature*, 480, 376–8.

Madliger, C.L., Love, O.P., Hultine, K.R. et al., 2018. The conservation physiology toolbox: status and opportunities. *Conservation Physiology*, 6(1), coy029.

Makanya, A.N. and Mortola, J.P., 2007. The structural design of the bat wing web and its possible role in gas exchange. *Journal of Anatomy*, 211, 687–97.

Martel, A., Blooi, M., Adriaensen, C. et al., 2014. Recent introduction of a chytrid fungus endangers Western Palearctic salamanders. *Science*, 346(6209), 630–1.

Martel, A., Pasmans, F., Fisher, M.C. et al., 2018. Chytridiomycosis. In S. Seyedmousavi, G. de Hoog, J. Guillot, and P. Verweij, eds. *Emerging and Epizootic Fungal Infections in Animals*, pp. 309–35. Springer, Cham, Switzerland.

Martel, A., Spitzen-van der Sluijs, A., Blooi, M. et al., 2013. *Batrachochytrium salamandrivorans* sp. nov. causes lethal chytridiomycosis in amphibians. *Proceedings of the National Academy of Sciences*, 110(38), 15325–9.

Martínková, N., Bačkor, P., Bartonička, T. et al., 2010. Increasing incidence of *Geomyces destructans* fungus in bats from the Czech Republic and Slovakia. *PLoS ONE*, 5, e13853.

Maslo, B. and Fefferman, N.H., 2015. A case study of bats and white-nose syndrome demonstrating how to model population viability with evolutionary effects. *Conservation Biology*, 29(4), 1176–85.

Maslo, B., Valent, M., Gumbs, J.F. et al., 2015. Conservation implications of ameliorating survival of little brown bats with white-nose syndrome. *Ecological Applications*, 25, 1832–40.

Mayberry, H.W., McGuire, L. P., and Willis, C.K.R., 2018. Body temperatures of hibernating little brown bats reveal pronounced behavioural activity during deep torpor and suggest a fever response during white-nose syndrome. *Journal of Comparative Physiology B*, 188, 333–43.

McCallum, H., Barlow, N., and Hone, J., 2001. How should pathogen transmission be modelled? *Trends in Ecology and Evolution*, 16, 295–300.

McCallum, H. and Dobson, A., 1995. Detecting disease and parasite threats to endangered species and ecosystems. *Trends in Ecology and Evolution*, 10, 190–4.

McCallum, H., Jones, M., Hawkins, C. et al., 2009. Transmission dynamics of Tasmanian devil facial tumor disease may lead to disease-induced extinction. *Ecology*, 90(12), 3379–92.

McGuire, L.P., Mayberry, H.W., and Willis, C.K.R., 2017. White-nose syndrome increases torpid metabolic rate and evaporative water loss in hibernating bats. *American Journal of Physiology: Regulatory, Integrative, and Comparative Physiology*, 313, R680–6.

McGuire, L.P., Muise, K.A., Shrivastav, A. et al., 2016. No evidence of hyperphagia during pre-hibernation fattening in a northern population of little brown bats (*Myotis lucifugus*. *Canadian Journal of Zoology*, 64, 821–7.

McNab, B.K., 1974. The behavior of temperate cave bats in a subtropical environment. *Ecology*, 55, 943–58.

Meierhofer, M.B., Johnson, J.S., Field, K.A. et al., 2018. Bats recovering from white-nose syndrome elevate metabolic rate during wing healing in spring. *Journal of Wildlife Diseases*, 54(3), 480–90.

Meteyer, C.U., Barber, D., and Mandl, J.N., 2012. Pathology in euthermic bats with white nose syndrome suggests a natural manifestation of immune reconstitution inflammatory syndrome. *Virulence*, 3, 583–8.

Meteyer, C.U., Buckles, E.L., Blehert, D.S. et al., 2009. Histopathologic criteria to confirm white-nose syndrome in bats. *Journal of Veterinary Diagnostic Investigation*, 21, 411–14.

Moore, M.S., Field, K.A., Behr, M.J. et al., 2017. Energy conserving thermoregulatory patterns and lower disease severity in a bat resistant to the impacts of white-nose syndrome. *Journal of Comparative Physiology B*, 188, 164–76.

Moore, M.S., Reichard, J.D., Murtha, T.D. et al., 2011. Specific alterations in complement protein activity of little brown myotis (*Myotis lucifugus*) hibernating in white-nose syndrome affected sites. *PLoS ONE*, 6, e27430.

Moore, M.S., Reichard, J.D., Murtha, T.D. et al., 2013. Hibernating little brown myotis (*Myotis lucifugus*) show variable immunological responses to white-nose syndrome. *PLoS ONE*, 8, e58976.

Muller, L.K., Lorch, J.M., Lindner, D.L. et al., 2013. Bat white-nose syndrome: a real-time TaqMan polymerase chain reaction test targeting the intergenic spacer region of *Geomyces destructans*. *Mycologia*, 105, 253–9.

Needle, D.B., Gibson, R., Hollingshead, N.A. et al., 2019. Atypical dermatophytosis in 12 North American porcupines (*Erethizon dorsatum*) from the northeastern United States 2010–2017. *Pathogens*, 8(4), 171.

Norquay, K.J.O. and Willis, C.K.R., 2014. Hibernation phenology of *Myotis lucifugus*. *Journal of Zoology*, 294(2), 85–92.

O'Donoghue, A.J., Knudsen, G.M., Beekman, C. et al., 2015. Destructin-1 is a collagen-degrading endopeptidase secreted by *Pseudogymnoascus destructans*, the causative agent of white-nose syndrome. *Proceedings of the National Academy of Science of the USA*, 112(24), 7478–83.

Palmer, J.M., Drees, K.P., Foster, J.T. et al., 2018. Extreme sensitivity to ultraviolet light in the fungal pathogen causing white-nose syndrome of bats. *Nature Communications, 9*, 35.

Pannkuk, E.L., McGuire, L.P., Warnecke, L. et al., 2015. Glycerophospholipid profiles of bats with white nose syndrome. *Physiological and Biochemical Zoology, 88*, 425–32.

Pavlinić, I., Ðaković, M., and Lojkić, I., 2015. *Pseudogymnoascus destructans* in Croatia confirmed. *European Journal of Wildlife Research, 61*, 325–8.

Perfect, J.R., 2017. The antifungal pipeline: a reality check. *Nature Reviews Drug Discovery, 16*(9), 603–16.

Peterson, J.D., Steffen, J.E., Reinert, L.K. et al., 2013. Host stress response is important for the pathogenesis of the deadly amphibian disease, chytridiomycosis, in *Litoria caerulea*. *PLoS ONE, 8*, e62146.

Pikula, J., Amelon, S.K., Bandouchova, H. et al., 2017. White-nose syndrome pathology grading in Nearctic and Palearctic bats. *PLoS ONE, 12*, e0180435.

Pikula, J., Bandouchova, H., Novotnỳ, L. et al., 2012. Histopathology confirms white-nose syndrome in bats in Europe. *Journal of Wildlife Diseases, 48*, 207–11.

Puechmaille, S.J., Wibbelt, G., Korn, V. et al., 2011. Pan-European distribution of white-nose syndrome fungus (*Geomyces destructans*) not associated with mass mortality. *PLoS ONE, 6*, e19167.

Rapin, N., Johns, K., Martin, L. et al., 2014. Activation of innate-response genes in little brown bats (*Myotis lucifugus*) infected with the fungus *Pseudogymnoascus destructans*. *PLoS ONE, 9*, e112285.

Reeder, D.M., Frank, C.L., Turner, G.G. et al., 2012. Frequent arousal from hibernation linked to severity of infection and mortality in bats with white-nose syndrome. *PLoS ONE, 7*, e38920.

Reichard, J.D. and Kunz, T.H., 2009. White-nose syndrome inflicts lasting injuries to the wings of little brown myotis (*Myotis lucifugus*). *Acta Chiropterologica, 11*, 457–64.

Rezanka, T., Viden, I., Nováková, A. et al., 2015. Wax ester analysis of bats suffering from white nose syndrome in Europe. *Lipids, 50*, 633–45.

Rocke, T.E., Kingstad-Bakke, B., Wüthrich, M. et al., 2019. Virally-vectored vaccine candidates against white-nose syndrome induce anti-fungal immune response in little brown bats (*Myotis lucifugus. Scientific Reports, 9*, 6788.

Rohlfs, M. and Churchill, A.C.L., 2011. Fungal secondary metabolites as modulators of interactions with insects and other arthropods. *Fungal Genetics and Biology, 48*(1), 23–34.

Romani, L., 2011. Immunity to fungal infection. *Nature Reviews Immunology, 11*, 275–88.

Roth, R.R. and James, W.D., 1988. Microbial ecology of the skin. *Annual Review of Microbiology, 42*, 441–64.

Russo, C.J.M., Ohmer, M.E.B., Cramp, R.L. et al., 2018. A pathogenic skin fungus and sloughing exacerbate cuta-neous water loss in amphibians. *Journal of Experimental Biology, 221*, jeb167445.

Skerratt, L.F., Berger, L., Speare, R. et al., 2007. Spread of chytridiomycosis has caused the rapid global decline and extinction of frogs. *EcoHealth, 4*, 125–34.

Sleep, D.J.H. and Brigham, R.M., 2003. An experimental test of clutter tolerance in bats. *Journal of Mammalogy, 84*(1), 216–24.

Smith, M.J., Telfer, S., Kallio, E.R. et al., 2009. Host–pathogen time series data in wildlife support a transmission function between density and frequency dependence. *Proceedings of the National Academy of Sciences, 106*(19), 7905–9.

Stockwell, C.A., Hendry, A.P. and Kinnison, M.T., 2003. Contemporary evolution meets conservation biology. *Trends in Ecology and Evolution, 18*(2), 94–101.

Szewczak, J.M., 1997. Matching gas exchange in the bat from flight to torpor. *American Zoologist, 37*, 92–100.

Szewczak, J.M. and Jackson, D.C., 1992. Apneic oxygen uptake in the torpid bat, *Eptesicus fuscus*. *Journal of Experimental Biology, 173*, 217–27.

The New York State Wildlife Health Program, 2019. Chytridiomycosis. September 2019. Available at: https://cwhl.vet.cornell.edu/disease/chytridiomycosis (accessed 26 June 2020).

Thomas, D.W. and Cloutier, D., 1992. Evaporative water loss by hibernating little brown bats, *Myotis lucifugus*. *Physiological Zoology, 65*(2), 443–56.

Thomas, D.W., Cloutier, D., and Gagné, D., 1990b. Arrhythmic breathing, apnea and non-steady-state oxygen uptake in hibernating little brown bats. *Journal of Experimental Biology, 149*, 395–406.

Thomas, D.W., Dorais, M., and Bergeron, J.M., 1990a. Winter energy budgets and cost of arousals for hibernating little brown bats, *Myotis lucifugus*. *Journal of Mammalogy, 71*, 475–9.

Thomas, D.W. and Geiser, F., 1997. Periodic arousals in hibernating mammals: is evaporative water loss involved? *Functional Ecology, 11*, 585–91.

Trivedi, J., Lachapelle, J., Vanderwolf, K.J. et al., 2017. Fungus causing white-nose syndrome in bats accumulates genetic variability in North America with no sign of recombination. *mSphere, e2*, e00271-17.

Turner, G.G., Meteyer, C.U., Barton, H. et al., 2014. Nonlethal screening of bat-wing skin with the use of ultraviolet fluorescence to detect lesions indicative of white-nose syndrome. *Journal of Wildlife Diseases, 50*(3), 566–73.

Turner, G.G., Reeder, D.M., and Coleman, J.T.H., 2011. A five-year assessment of mortality and geographic spread of white-nose syndrome in North American bats and a look to the future. *Bat Research News, 52*, 13–27.

Turner, J.M., Warnecke, L., Wilcox, A. et al., 2015. Conspecific disturbance contributes to altered hibernation patterns in bats with whitenose syndrome. *Physiology and Behavior*, 140, 71–8.

US Fish and Wildlife Service, 2020a. Where is WNS now? July 2020. Available at: https://www.whitenosesyndrome.org/where-is-wns (accessed 8 July 2020).

US Fish and Wildlife Service, 2020b. Bats affected by WNS. July 2020. Available at: https://www.whitenosesyndrome.org/static-page/bats-affected-by-wns (accessed 8 July 2020).

US Fish and Wildlife Service, 2020c. Endangered species. January 2020. https://www.fws.gov/endangered/ (accessed 26 June 2020).

Verant, M.L., Bouhuski, E.A., Richgels, K.L.D. et al., 2018. Determinants of *Pseudogymnoascus destructans* within bat hibernacula: Implications for surveillance and management of white-nose syndrome. *Journal of Applied Ecology*, 55, 820–9.

Verant, M.L., Meteyer, C.U., Speakman, J.R. et al., 2014. White-nose syndrome initiates a cascade of physiologic disturbances in the hibernating bat host. *BMC Physiology*, 14, 10.

Voss, T., Rummel, C., Gerstberger, R. et al., 2006. Fever and circulating cytokines induced by double stranded RNA in guinea pigs: dependence on the route of administration and effects of repeated injections. *Acta Physiologica*, 187, 379–89.

Voyles, J., Rosenblum, E.B., and Berger, L., 2011. Interactions between *Batrachochytrium dendrobatidis* and its amphibian hosts: a review of pathogenesis and immunity. *Microbes and Infection*, 13, 25–32.

Voyles, J., Young, S., Berger, L. et al., 2009. Pathogenesis of chytridiomycosis, a cause of catastrophic amphibian declines. *Science*, 326(5952), 582–85.

Warnecke, L., Turner, J.M., Bollinger, T.K. et al., 2012. Inoculation of bats with European *Geomyces destructans* supports the novel pathogen hypothesis for the origin of white-nose syndrome. *Proceedings of the National Academy of Science of the USA*, 109, 6999–7003.

Warnecke, L., Turner, J.M., Bollinger, T.K. et al., 2013. Pathophysiology of white-nose syndrome in bats: a mechanistic model linking wing damage to mortality. *Biology Letters*, 9(4), 20130177.

Warnecke, L., Turner, J.M., Craig, K.R. et al., 2014. Behaviour of hibernating little brown bats experimentally inoculated with the pathogen that causes white-nose syndrome. *Animal Behaviour*, 88, 157–64.

Wibbelt, G., Kurth, A., Hellmann, D. et al., 2010. White-nose syndrome fungus (*Geomyces destructans*) in bats, Europe. *Emerging Infectious Diseases*, 16, 1237–43.

Wikelski, M. and Cooke, S.J., 2006. Conservation physiology. *Trends in Ecology and Evolution*, 21, 38–46.

Wilcox, A., Warnecke, L., Turner, J.M. et al., 2014. Behaviour of hibernating little brown bats experimentally inoculated with the pathogen that causes white-nose syndrome. *Animal Behaviour*, 88, 157–64.

Wilcox, A. and Willis, C.K.R., 2016. Energetic benefits of enhanced summer roosting habitat for little brown bats (*Myotis lucifugus*) recovering from white-nose syndrome. *Conservation Physiology*, 4(1), cov070.

Willis, C.K.R., 2015. Conservation physiology for conservation pathogens: white-nose syndrome and integrative biology for host-pathogen systems. *Integrative and Comparative Biology*, 55, 631–41.

Willis, C.K.R., Menzies, A.K., Boyles, J.G. et al., 2011. Cutaneous water loss is a plausible explanation for mortality of bats from white-nose syndrome. *Integrative and Comparative Biology*, 51, 364–73.

Woodhams, D.C., Ardipradja, K., Alford, R.A. et al., 2007. Resistance to chytridiomycosis varies among amphibian species and is correlated with skin peptide defenses. *Animal Conservation*, 10(4), 409–17.

Zhang, T., Chaturvedi, V., and Chaturvedi, S., 2015. Novel *Trichoderma polysporum* strain for the biocontrol of *Pseudogymnoascus destructans*, the fungal etiologic agent of bat white nose syndrome. *PLoS ONE*, 10(10), e0141316.

Zukal, J., Bandouchova, H., Bartonicka, T. et al., 2014. White-nose syndrome fungus: a generalist pathogen of hibernating bats. *PLoS ONE*, 9, e97224.

CHAPTER 10

Physiology provides a window into how the multi-stressor environment contributes to amphibian declines

Michel E.B. Ohmer, Lesley A. Alton, and Rebecca L. Cramp

> ⮌ **Take-home message**
>
> Disease and a changing climate are wreaking havoc on amphibian populations worldwide in complex and multifaceted ways. By examining this problem from a physiological perspective, we can determine the mechanisms driving disease susceptibility, and better design strategies for conservation and management.

10.1 Introduction

Rapid environmental change associated with the onset of the Anthropocene has resulted in catastrophic biodiversity loss that many believe to be the beginning of the sixth mass extinction (Barnosky et al. 2011; Dirzo et al. 2014). Amphibians have become the poster child of this biodiversity crisis not only because proportionally they are the most threatened vertebrate taxon (International Union for Conservation of Nature 2019), but because hundreds of species have declined rapidly and within pristine environments all over the world (Stuart et al. 2004). The main driver of these rapid declines was unknown until the discovery of the fungal pathogen *Batrachochytrium dendrobatidis* (*Bd*) in 1998 (Berger et al. 1998) and its sister species *B. salamandrivorans* (*Bsal*) in 2013 (Martel et al. 2013). These pathogenic chytrids infect the skin of post-metamorphic amphibians and the mouthparts of larvae and cause the deadly disease chytridiomycosis.

Chytridiomycosis is now generally accepted as the proximate cause of many of these 'enigmatic' mass die-offs, population declines, and species extinctions that were primarily observed in Australia, the Americas, and Europe (Stuart et al. 2004; Skerratt et al. 2007). The chytridiomycosis panzootic is responsible for what is considered the greatest loss of biodiversity attributable to a disease in recorded history (Scheele et al. 2019).

Why have amphibians only recently succumbed to the devastating effects of the chytrid fungus? There are two non-mutually exclusive hypotheses that explain the emergence of this infectious disease: (1) the novel pathogen hypothesis, where the pathogen has encountered naïve hosts because it has either recently evolved or spread into new geographic areas; and (2) the endemic pathogen hypothesis, where recent environmental change has caused an existing pathogen to become more prevalent either by enhancing its growth/virulence or by compromising the immune response of hosts

Michel E.B. Ohmer, Lesley A. Alton, and Rebecca L. Cramp, *Physiology provides a window into how the multi-stressor environment contributes to amphibian declines* In: *Conservation Physiology: Applications for Wildlife Conservation and Management.* Edited by: Christine L. Madliger, Craig E. Franklin, Oliver P. Love, and Steven J. Cooke, Oxford University Press (2021). © Oxford University Press. DOI: 10.1093/oso/9780198843610.003.0010

(Rachowicz et al. 2005). There is mounting evidence for both hypotheses. Genetic analyses have found that *Bd* radiated out of Asia into several genetically distinct lineages, including the hypervirulent Global Panzootic Lineage (*Bd*GPL), which has been implicated in declines worldwide. *Bd*GPL likely emerged in the early 20th century and spread via international trade, resulting in the introduction of this novel pathogen to naïve populations worldwide (O'Hanlon et al. 2018). Additionally, susceptibility to *Bd* is influenced by environmental factors, which could have contributed to amphibian declines under past environmental change, and may increase susceptibility under future predicted change. Increasing global air temperatures and a more variable climate (Cohen et al. 2019), encroachment of human modified landscapes on amphibian habitat (Becker et al. 2017), and rising impacts from pollutants (Hua et al. 2017) can all affect the immune response of the host.

Variation in susceptibility to *Bd* is the combined result of complex interactions among the host, pathogen, and environment (Fisher et al. 2009). Both the host and the pathogen have physiological constraints and environmental conditions in which they perform optimally, and abiotic and biotic factors can influence the host–pathogen relationship further. For example, climate and seasonality (Rohr and Raffel 2010), population size and community structure (Searle et al. 2011; Becker et al. 2014), host behaviour (Richards-Zawacki, 2010), and innate and acquired immunity (Ramsey et al. 2010; McMahon et al. 2014) have all been demonstrated to modulate *Bd* infection dynamics (Daszak et al. 2003; James et al. 2015). Thus, a better understanding of host physiological responses to environmental change, and how this may vary by species and with ecology and evolutionary history, will improve our understanding of the mechanisms behind amphibian declines and extinctions, and inform management decisions (Lips 2016).

This chapter will focus on how environmental change may influence amphibian disease susceptibility via impacts on immune function. In the following sections we provide a brief overview of the amphibian immune system, the chytrid fungal pathogens, and some of the physiological tools available to assess immune function. We then illustrate how the environment can influence immune function via several case studies that use physiological metrics to understand disease processes in amphibian hosts. Overall, this chapter will illustrate the power of physiology to uncover mechanisms driving declines, which can inform on-the-ground conservation action.

10.1.1 The amphibian immune system

Immune function is a critical physiological performance trait that can be stimulated or suppressed by exposure to different environmental drivers (Dhabhar 2009). It is also a key fitness determinant because it underlies the capacity of animals to resist or tolerate potential infections, although the fitness costs of mounting an immune defence also need to be considered (Hawley and Altizer 2011). In all vertebrates, the immune system is a complex and multifaceted assortment of non-specific (innate) and specific (acquired) mechanisms that essentially discriminate 'self' from 'non-self' (Schulenburg et al. 2009). Innate immune defences serve as a first line of defence providing rapid, non-specific protection against a variety of potential threats. The innate immune system includes physical structures like the skin, which restricts the movement of microbes into the body; mucus, which has antimicrobial chemical components; cell-mediated defences in the form of white blood cells; and chemical cascades like complement proteins and lysozyme. The acquired or adaptive immune system is a set of immune responses highly specific to individual pathogens and includes the production of antibodies against foreign antigens and the retention of immunological memory (Cramp and Franklin 2018).

The amphibian immune system is similar to that of most other vertebrates (Carey et al. 1999). In amphibians, early-life immune defences (embryonic and larval) are less well developed than those of adult frogs. While some embryonic immune defences are established very early (Rollins-Smith 1998), the primary lymphoid tissues, the spleen and thymus, only begin to develop approximately 2 weeks post-fertilization in *Xenopus laevis* larvae (Robert and Ohta 2009). Larval amphibians typically have fewer circulating lymphocytes and reduced antibody diversity compared with adults (Rollins-Smith 2017). During metamorphosis, larvae experience a sharp reduction in larval lymphocyte abundance (i.e. immunosuppression) to prevent the larval immune system from attacking the newly formed adult tissues (Rollins-Smith 1998; Robert and Ohta 2009). Following metamorphosis, immune function gradually increases; however, adult-type immune defences may not be achieved for up to a year post-metamorphosis (Rollins-Smith 1998, 2017).

10.1.2 Amphibian chytrid fungal pathogens

As a generalist pathogen, *Bd* infects a wide taxonomic range of amphibian hosts and has likely contributed to the decline of at least 500 amphibian species worldwide (Scheele et al. 2019). *Bd* infection is limited to the keratin and prekeratin-containing layers of an amphibian's skin, and the keratin-containing mouthparts of larvae (Berger et al. 2005b). The two described species of *Batrachochytrium* pathogens are the only chytridiomycete taxa known to parasitize vertebrates (Berger et al. 1998; Pessier et al. 1999; Longcore et al. 1999). *Bd* has a two-part

life cycle that consists of an infective and reproductive stage: a motile flagellated zoospore, and a thallus that develops into the reproductive zoosporangium (Berger et al. 2005a). First, aquatic zoospores migrate towards a suitable substrate, after which they encyst and develop a germ tube to penetrate host cell membranes (Van Rooij et al. 2012). Then, zoospore cellular contents are transferred through this germ tube to allow zoosporangia to develop intracellularly, and likely evade detection by the immune system (Berger et al. 2005a; Grogan et al. 2018). Finally, zoosporangia produce additional zoospores, which leave via a discharge papilla that may be directed externally or internally, resulting in dissemination of zoospores or reinfection of the host (Berger et al. 2005a).

Bd infection can lead to significant morbidity in amphibian hosts. Infection causes localized thinning and thickening (hypo- and hyperkeratosis) of the stratum corneum and stratum granulosum, and in extreme cases can lead to ulceration and erosions of the skin (Berger et al. 1998, 2005b; Pessier et al. 1999; Figure 10.1). *Bd* infection also increases skin sloughing or shedding rates (Ohmer et al. 2015), particularly in highly infected individuals, which

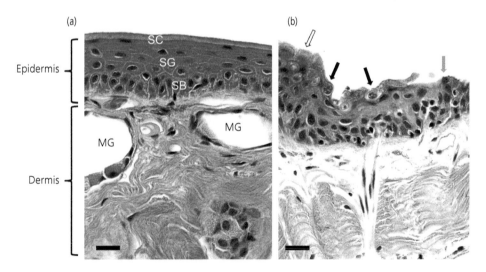

Figure 10.1 Epidermal and dermal layers of *Litoria caerulea* skin: (A) Healthy skin and (B) skin clinically infected with *Batrachochytrium dendrobatidis* (Bd). Skin sectioned at 5 μm and stained with haematoxylin and eosin. SC = stratum corneum, SG = stratum granulosum or stratum spinosum, SB = stratum basale or stratum germinativum, MG = mucous gland, GG = granular gland, black arrows = Bd zoosporangia growing intracellularly, grey arrow = erosion, white arrow = hyperkeratosis. Reproduced from Michel Ohmer's PhD thesis.

reduces *Bd* infection loads and can assist in infection clearance in some species (Ohmer et al. 2017). However, physical damage to the skin, in combination with increased sloughing rates, results in a systemic loss of electrolytes and a reduction in cutaneous sodium uptake (Wu et al. 2018b). These physiological changes can cause reduced cardiac electrical activity, ultimately leading to cardiac arrest (Voyles et al. 2009; Wu et al. 2018b). Furthermore, increased haematocrit levels concurrent with a decrease in mass indicate dehydration in infected frogs in the wild (Voyles et al. 2012), and *Bd*-infected frogs have higher cutaneous water loss rates that are exacerbated by an increase in sloughing (Russo et al. 2018). Thus, environmental change that significantly alters the habitats that amphibians occupy may further impact physiological homeostasis.

10.2 Tools for exploring immune responses to environmental factors and *Bd* infection in amphibians

Ecoimmunological studies commonly measure aspects of the immune system as a proxy for individual fitness, because immune function can predict the survival outcome of a pathogen challenge (Downs and Stewart 2014). Many immunological assays, developed largely in laboratory model systems (i.e. mice), are employed to measure aspects of immune function in non-model organisms. However, for most non-model organisms, the usefulness of the particular metric, its temporal or inter-individual variability, or its environmental sensitivity may be unknown (Demas et al. 2011). In this section, we explore some of the commonly used tools for measuring immune function metrics in amphibians, as well as other physiological processes that can affect the capacity of the immune system to respond to an antigenic challenge.

10.2.1 Antimicrobial peptides

Skin secretions form an important component of the innate immune system of adult amphibians. Secretions are primarily produced by granular glands within the dermal layer of the skin and form both a physical and chemical barrier to microbes,

containing a variety of proteins and peptides with antimicrobial activities (Woodhams et al. 2016). Antimicrobial peptides (AMP) are thought to be important immune defences against *Bd*, as differences in AMP diversity and abundance appear to contribute to interspecific and inter-population differences in susceptibility to *Bd* (Rollins-Smith 2009; Woodhams et al. 2016). AMPs are also thought to be a critical component of a newly metamorphosed frog's immune system since adaptive immune functions are not fully developed at this life stage (Groner et al. 2013; Woodhams et al. 2016). The production, composition, and activity of several amphibian AMPs can be influenced by a range of anthropogenic or natural stressors. For example, prolonged exposure to high levels of traffic noise is correlated with impaired AMP production in metamorphic frogs (Tennessen et al. 2018). In addition, prolonged treatment with synthetic glucocorticoid (stress) hormones, as well as freezing, dehydration, and anoxia, can increase the production of AMP gene transcripts in skin secretions (Katzenback et al. 2014; Tatiersky et al. 2015). The benefits of AMPs as a measure of amphibian immune function include their relative ease of collection, that sampling is repeatable, non-terminal, and can be conducted in the field or in the laboratory, and that the pathogen inhibition assay is relatively straightforward to undertake. Additionally, they can be used to assess a metric of immune function that is present across life stages since larvae of several frog species also possess AMPs (Woodhams et al. 2016). AMPs are also highly relevant to host fitness because they can directly influence pathogen survival.

10.2.2 Haematological measures

The composition of blood (red and white cell counts, white cell discrimination, and haematocrit) has been widely used in ecoimmunological studies to assess both baseline and stimulated immune system function. For many animals, haematological tests are quick to perform, are minimally invasive, repeatable, and provide a method for coarsely assessing aspects of both the innate and adaptive immune systems simultaneously. White blood cells (WBC) are the circulating cells of the immune system (also called leukocytes or splenocytes). Basic WBC counts (differentials or complete blood count) and changes in the relative abundance of particular leukocytes (i.e. neutrophils-to-lymphocytes, N:L, ratio) are commonly employed immune function metrics. The relative proportion of neutrophils to lymphocytes (N:L ratio) is widely used in avian ecoimmunology as a measure of stress, and is increasingly being used in amphibian ecoimmunology (Davis and Maerz, 2008; Davis and Durso 2009; Shutler and Marcogliese 2011; Young et al. 2014). In response to an increase in stress hormones (glucocorticoids), lymphocytes are sequestered from the circulating blood to tissues such as the spleen and skin, accompanied by an ingress of neutrophils into the blood (Davis et al. 2008). The resulting shift in the relative abundance of lymphocytes and neutrophils is proportional to glucocorticoid levels (Davis et al. 2008), thus N:L ratios have been used to assess responses to infection with *Bd* in amphibians and tend to reflect pathogen-induced physiological stress (e.g. Gervasi et al. 2014; Young et al. 2014; Savage et al. 2016).

10.2.3 Bacterial killing assays

Bacterial killing assays are a measure of the capacity of a host's blood to kill microorganisms. This is a relevant and important metric of immune function because the effectiveness of the process is highly advantageous to the host (Beck et al. 2017). Bacterial killing assays can be performed with whole blood, serum, or plasma. The cell free assay exploits complement, lysozyme, and/or natural antibody-mediated bacterial killing capacities (Beck et al. 2017), while the whole blood assay also incorporates cell-mediated bacterial killing processes. To perform the assay, plasma or whole blood is added to an aliquot of actively growing bacteria and allowed to incubate for 12–24 h after which time its absorbance is measured in a spectrophotometer (Liebl and Martin II 2009). The absorbance of the solution gives a measure of bacterial concentration, and the difference in absorbance between cultures with and without blood provides an indication of the efficiency of an organism's blood to kill bacteria *in vitro*. This assay has been used to assess the impacts of nutritional status on immune function in tadpoles (Venesky et al.

2012), the effects of infection with *Bd* on immune capacity (Hopkins and DuRant 2011; Savage et al. 2016), and the impact of environmental temperatures on immune function in frogs and salamanders (Maniero and Carey 1997; Terrell et al. 2013).

10.2.4 Delayed-type hypersensitivity assay

Tissue and metabolic responses to challenge by a non-pathogenic but foreign peptide can provide useful information on both innate and adaptive immune responses, without the risks associated with a pathogen challenge. The delayed-type hypersensitivity (DTH) assay has been widely used in avian and mammalian ecoimmunology studies, and has recently been validated for use in amphibians (Clulow et al. 2015). In this assay, a lectin, often phytohaemagglutinin (PHA) or keyhole limpet haemocyanin (KLH), is injected under the skin and the magnitude of the resulting inflammatory response quantified. These antigens trigger a localized response in which leukocytes (particularly eosinophils, macrophages, and neutrophils) infiltrate the affected area, causing swelling and inflammation (Fites et al. 2014). By measuring skin thickness before and after the administration of the lectin, the magnitude of the response can be positively correlated with immune responsiveness (Kennedy and Nager 2006; Clulow et al. 2015). The assay has been used to demonstrate that adverse environmental conditions can suppress immune system responses in amphibians. For example, tadpoles and metamorphic frogs have weaker PHA-induced immune responses to pond drying stress (Gervasi and Foufopoulos 2008; Brannelly et al. 2019) and following rearing on low-protein diets (Venesky et al. 2012). In addition, *Limnodynastes peronii* metamorphs exposed to UV-B radiation during larval development had weaker PHA responses than those that were not exposed (Ceccato et al. 2016), while simultaneous exposure to both moderate UV levels and a herbicide reduced cellular immune responses in tadpoles of *Ambystoma maculatum* (Levis and Johnson, 2015). Finally, juvenile *Litoria aurea* infected with *Bd* had reduced PHA responses compared with healthy animals (Abu Bakar et al. 2016).

10.2.5 Physiological stress markers

Physiological stress is widely understood to influence the performance of the immune system and can result from exposure to both natural and anthropogenic factors such as elevated temperatures, crowding, food deprivation, pollutants, handling, and habitat loss (Kiesecker 2011; Rollins-Smith 2017). Exposure to such stressors results in an acute glucocorticoid (GC) response, which is mediated by the main glucocorticoid in amphibians, corticosterone (Walls and Gabor 2019). Short-term elevations in GCs via the acute stress response can promote beneficial responses to environmental stressors by suppressing non-essential functions, mobilizing energy reserves, and restoring homeostasis (Sapolsky et al. 2000). However, chronically elevated GC levels can lead to deleterious effects on immune function, and can influence other fitness-related traits (Sapolsky et al. 2000). Therefore, the timing, magnitude, and duration of GC responses to stressors can provide an indication of the likely physiological costs of the response. Glucocorticoid levels are increasingly used as a metric to assess physiological responses to a *Bd* challenge or infection (e.g. Kindermann et al. 2012; Peterson et al. 2013; Crespi et al. 2015; Gabor et al. 2015; Fonner et al. 2017) or to inform predictions about how environmental conditions may influence the subsequent health and susceptibility of amphibians to *Bd* infection (Warne et al. 2011; Burraco and Gomez-Mestre 2016; Gabor et al. 2018). While instantaneous GC levels can be measured directly using blood and tissue samples, recent technological advancements have seen the development of somewhat less invasive sampling protocols that can utilize faeces, urine, saliva, skin swabs, or water bath samples to measure excreted GC metabolites that have accumulated over a defined period of time (Narayan et al. 2011; Sheriff et al. 2011; Hammond et al. 2018; Santymire et al. 2018; Scheun et al. 2019). In most cases, these samples are then purified and quantitated using antibody-based immunoassays. However, since GC levels can rise and fall quickly in response to handling or disturbance stress, other metrics of chronic physiological stress such as neutrophil-to-leukocyte ratios (see Section 10.2.2), oxidative stress biomarkers (Burraco et al. 2013), GC-binding globulin levels, and telomere lengths (Burraco et al. 2017) may be integrated with GC measures to provide a more holistic picture of short- and long-term physiological stress responses in amphibians.

10.2.6 Other measures of physiological performance

The environment can exert its influence on immune function through its capacity to impact other physiological processes and endocrine functions. The rate of energy metabolism is a fundamental physiological trait that reflects the energetic cost of living (McNab 2002). Animals fuel their metabolism by taking up energy from the environment and allocate that energy to the competing functions of maintenance, activity, growth, storage, and reproduction (Zera and Harshman 2001). In theory, when an organism is exposed to stressful environmental conditions that cause physiological damage or disrupt homeostasis, more energy is required for maintenance, leaving less energy available for other pathways. Immune responses and resistance to pathogens and parasites are physiologically demanding (Demas et al. 1997; Martin et al. 2003; Hawley and Altizer 2011). Energetic costs associated with exposure to environmental stressors may result in trade-offs, reducing an animal's capacity to mount an effective immune response. To understand how different environmental factors affect rates of energy expenditure in amphibians, researchers can measure the rate of oxygen consumption and/or carbon dioxide production by an individual under different conditions to calculate metabolic rate (e.g. Alton et al. 2012). Measuring rates of gas exchange in amphibians using standard respirometry techniques is conceptually simple but technically demanding. Researchers are encouraged to consult Lighton (2019) as an instructional guide to measuring metabolic rates, as well as other methodological papers (e.g. Harter et al. 2017; Winwood-Smith and White 2018) to ensure best practice and to take advantage of the technological advancements that have been made that allow for accurate and precise estimates of gas exchange in small organisms in both aerial and aquatic phases. Researchers should also familiarize themselves with the definitions of different metabolic states, including standard, resting, routine, and maximum metabolic rate, as the scientific question being asked will dictate the most appropriate metabolic state and associated experimental protocol (Careau and Garland 2012). Ideally, researchers should also simultaneously quantify behaviour (e.g. activity) to account for such sources of variation in their measures of metabolic rate (see Alton et al. 2012). Finally, in accounting for variation in metabolic rate associated with body mass, researchers should include body mass as a covariate in their statistical analyses (Lighton 2019). Measures of metabolic rate under varying environmental conditions and disease states can improve our understanding of the energetic costs associated with mounting an immune defence (Bonneaud et al. 2016, Agugliaro et al. 2019).

10.3 Links between environmental factors and immune function

10.3.1 Environmental temperature drives host immune performance

Environmental temperature is one of the most important drivers of biological systems owing to the thermal dependence of physiochemical reactions. Recent rapid changes in environmental temperatures, linked to anthropogenic activities, have increased focus on the role of temperature in underpinning ecosystem functioning and species relationships. While predictions indicate a general increase in global mean temperatures, the greatest impacts of climate warming on amphibians will likely arise from increased thermal variability (including increased incidence of heat waves and cold snaps) and altered precipitation patterns (Raffel et al. 2013; Rollins-Smith 2017). For most ectotherms, physiological performance is often maintained across a relatively narrow range of 'optimal' environmental temperatures with performance declining at temperatures either side of optimum. As ectotherms, amphibians are likely to be disproportionally vulnerable to climate warming because their body temperatures are dependent on environmental temperature (Huey et al. 2012).

Immune function, like most physiological processes in amphibians, is thermally sensitive. There

is a great deal of literature to show that immune function can vary in response to seasonal shifts in temperature. For example, in the frog *Rana temporaria*, the thymus, which produces the majority of lymphocytic cells in amphibians, shrinks during winter when animals are dormant and recovers its size during summer when the animals are active (Miodoński et al. 1996; Rollins-Smith and Woodhams 2012). Similarly, numbers of circulating lymphocytes and eosinophils are positively correlated with seasonal environmental temperature in adult *Notophthalmus viridescens* (Raffel et al. 2006) and prolonged exposure to low temperatures can reduce antimicrobial peptide secretion in *R. sylvatica* (Matutte et al. 2000). For the majority of published studies, immune function metrics tend to decline in response to cold temperatures. Some amphibians can also combat infections by displaying behavioural fever (i.e. seeking out warm environments to improve their immune system performance and reduce pathogen performance; Lefcort and Blaustein 1995). However, individual thermal preferences, rather than behavioural fever per se, may more strongly drive infection outcomes (Sauer et al. 2018).

As a small ectothermic organism, *Bd* is highly sensitive to temperature, growing best between 17 and 25°C in culture, with a thermal minima close to 4°C and a thermal maxima between 26 and 28°C (Piotrowski et al. 2004; Voyles et al. 2017). However, evidence suggests that optimal growth on the amphibian host is the combined outcome of pathogen reproductive rate and host immune function at a given temperature (Cohen et al. 2017; Sonn et al. 2017). Termed the *thermal mismatch hypothesis*, relatively cold-adapted species are predicted to be more susceptible to *Bd* at warmer temperatures, while warm-adapted species are predicted to be more susceptible to *Bd* at cooler temperatures (Cohen et al. 2017). Based on an analysis of over 32 000 *Bd* swabs from live amphibians globally, Cohen et al (2019) demonstrated that climate variability is correlated with higher susceptibility of cold-adapted amphibians to *Bd*, particularly at low latitudes and high elevations. Future work investigating the temperature sensitivity and acclimation capacity of amphibian immune function, and how these interplay with seasonality, will help elucidate the mechanisms underlying this general trend.

10.3.2 Impacts of hydroperiod on the development of the immune system

One of the many impacts of increasing average air temperatures is elevated rates of evapotranspiration, which has already led to alterations in the global water cycle (Milly et al. 2005). Increases in the length and occurrence of droughts will shift wetland hydroperiod, or the duration of water availability, and this can have profound impacts on amphibians at multiple life stages (McMenamin et al. 2008). Drying wetlands can directly impact amphibian survival and fitness by reducing the availability of freshwater for spawning and larval development. In addition, a reduction in water volume can impose environmental stressors on developing larvae, resulting in a sub-optimal rearing environment with higher larval densities, reduced foraging space, and potentially reduced water quality (Walls et al. 2013). While pond drying may induce developmental plasticity in larvae, in which development is sped up to minimize the risk of death by desiccation (Richter-Boix et al. 2011; Edge et al. 2016), this can result in trade-offs in growth, immune function, and body condition or size (Wilbur and Collins 1973; Gervasi and Foufopoulos, 2008), which can impact future fecundity and survival (Semlitsch et al. 1988; Altwegg and Reyer 2003).

The hypothalamic–pituitary–interrenal (HPI) axis mediates an amphibian's response to its environment, including the timing of metamorphosis (Rollins-Smith 2017). GCs, the stress hormones of vertebrates, also play a role in metamorphic timing, and both thyroid hormone and corticosterone peak at the moment of front limb emergence (Rollins-Smith 1998). Corticosterone produced as a result of external environmental stimuli, such as reduced hydroperiods, may synergize with thyroid hormones to result in accelerated development and early metamorphosis (Denver 1997, 2009). Thus, since the immune system and stress axis are coupled, amphibians experiencing accelerated development may have short- or long-term alterations to their immune function (Gervasi and Foufopoulos 2008; Kohli et al. 2019). However, because the stress response is dose-dependent, responses to environmental change may depend on severity.

From laboratory and semi-natural mesocosm experiments, there is evidence that hydroperiod-related stressors during development can have lasting impacts on amphibian phenotype, physiology, and immune function. An increased rate of development in response to pond drying is often associated with a reduction in body size at metamorphosis, which may have important fitness consequences in terms of size at maturity and fecundity later in life (Richter-Boix et al. 2015; Edge et al. 2016). For example, reduced hydroperiods resulted in increased development rates in *Rana sylvatica*, lower cell-mediated immune responses to PHA, and reduced leukocyte counts in juvenile frogs (Gervasi and Foufopoulos 2008). In addition, shorter larval periods are associated with a reduction in bacterial killing ability of whole blood, and a small size at metamorphosis is correlated with reduced total antibody production (Brannelly et al. 2019). Overall, reduced immune function following accelerated metamorphosis may result in increased susceptibility to pathogens, such as *Bd*, post-metamorphosis (Rollins-Smith et al. 2011).

10.4 Case study 1: using sloughing physiology to understand variation in susceptibility to chytridiomycosis

The disease chytridiomycosis is so striking because the pathogen, *Bd*, is confined to the skin of adult amphibians, yet can cause severe morbidity and mortality (Berger et al. 1998). Thus, understanding the structure and function of amphibian skin may provide insight into why some species are more susceptible to the pathogen than others, and under what environmental conditions amphibians will be most at risk of disease. Amphibians undergo a regular skin renewal process called sloughing or shedding, which involves the physical removal of the stratum corneum, or outermost layer of skin, via a series of limb and body movements (Larsen 1976; Ohmer et al. 2017; Figure 10.2). Skin sloughing has been shown to remove cutaneous microbes in amphibians (Cramp et al. 2014), increase with infection loads (Ohmer et al. 2015), and was hypothesized to play a role in regulating pathogen growth,

(a)

(b)

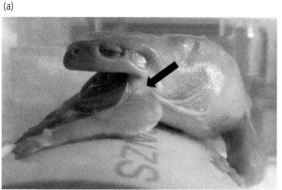

Figure 10.2 Adult *Litoria caerulea* exhibiting classic sloughing behaviour, in which the back is arched, and the mouth opens (A) and closes (B) with the rhythmic movement of the limbs to pull the shed skin towards the corners of the mouth (see black arrow), where it is ingested. Photo credit: Michel Ohmer.

particularly *Bd* (Ohmer et al. 2015). Thus, variation in how often species shed their skin, and their ability to increase sloughing rates during a pathogen challenge, was posited to play a role in amphibian susceptibility to *Bd*.

Using a multi-species comparison, Ohmer et al. (2017) found that sloughing indeed removed *Bd* from the skin of amphibians, helping to regulate pathogen loads in less susceptible species, and even resulting in infection clearance. However, in more susceptible species, the temporary reduction in *Bd* load after sloughing did not help clear infections, and those animals often still succumbed to the pathogen (Ohmer et al. 2017). Thus, it would appear that an increase in skin sloughing in frogs infected with *Bd* is a double-edged sword: while it can help remove pathogens and parasites from the skin surface, it also results in increased ion loss due to the increased permeability of the skin during sloughing (Wu et al. 2018b). In addition, both chytridiomycosis and skin sloughing increase the rate of cutaneous water loss, potentially making infected amphibians more physiologically vulnerable to changing environmental conditions (Russo et al. 2018). Furthermore, smaller frogs are more susceptible to greater ion and water loss when infected with *Bd*, indicating why chytridiomycosis may be more severe in juvenile frogs (Russo et al. 2018; Wu et al. 2018a).

But how does the rate of routine sloughing impact amphibian susceptibility to chytridiomycosis? Amphibians slough their skin anywhere from every day (Castanho and de Luca 2001) to every other week (Budtz and Larsen 1973; Meyer et al. 2012), thus innate differences in skin shedding rates across species may predict how susceptible animals are to *Bd* infection and subsequent disease. Using a phylogenetic framework, Ohmer et al. (2019) compared the sloughing rates of 21 frog species across eight families, as well as structural aspects of the ventral skin, in order to determine the role of structural and functional skin traits in amphibian susceptibility to chytridiomycosis. Overall, they found that sloughing rate demonstrated a strong phylogenetic signal, with more closely related species sloughing at more similar rates than distantly related species. Evidence for *Bd*-related declines in the literature was not related to skin sloughing rates in the amphibians studied, but for some particularly vulnerable species, such as those in the genus *Atelopus*, slow sloughing rates in combination with other aspects of their physiology such as their thermal ecology (Cohen et al. 2017) may shed light on their acute susceptibility to *Bd* infection. Furthermore, as sloughing can regulate cutaneous microbial loads (Cramp et al. 2014), as well as pathogens such as *Bd* (Ohmer et al. 2017), an understanding of the role of routine skin maintenance processes may improve microbial *Bd* mitigation strategies. Future work investigating the role of skin sloughing on cutaneous microbial community structure and function, particularly at different ambient temperatures, will improve models of infection progression across amphibian species.

10.5 Case study 2: using stress physiology to aid in the conservation of imperilled amphibians

Acute or chronic exposure to environmental stressors can induce a physiological stress response, mediated in part by GC hormones like corticosterone. Short-term, unpredictable stresses can result in increases in circulating GC levels that have beneficial effects on immune system functioning including the initiation of an acute inflammatory response and the stimulation of subsequent adaptive immune responses (Martin 2009). However, protracted exposure to elevated GC levels can cause suppression of these same responses (Martin 2009), which could contribute to stress-associated population declines (Dantzer et al. 2014). For example, chronic exposure to some agrochemicals can affect immune function by influencing physiological stress pathways (Hayes et al. 2010). Understanding how temporal patterns in GC hormone levels correlate with exposure to acute or chronic stressors is vital to understand whether environmental stressors, anthropogenic or otherwise, actually pose a risk to exposed individuals or can be tolerated/managed. They are also potential indicators of the impacts of cumulative environmental changes on physiological functioning. For example, stress induced by predator cues caused an increase in tadpole mortality following exposure to a pesticide (Relyea and Mills 2001). Hormones can now be routinely monitored to quantify and provide a mechanistic basis for how animals respond to short- and long-term environmental change, and as a proxy to assess the physiological robustness of populations, which could provide early warning of 'at-risk' species or populations (Narayan 2019).

Exposure of amphibians to negative stimuli such as pathogens and infections can initiate a GC stress response, which can be important for managing acute responses to these stressors (Peterson et al. 2013); although the exact direction of the response (i.e. increase or decrease in GC levels) can be species- and/or pathogen-specific (Warne et al. 2011; Gabor et al. 2013, 2018; DuRant et al. 2015; Koprivnikar et al. 2019). However, prolonged exposure of amphibian larvae to chronically artificially elevated GC levels (for 10 days) resulted in increased trematode infection rates in two species, and rapid infection clearance following cessation of GC exposures (Belden and Kiesecker 2005; LaFonte and Johnson 2013), demonstrating that elevated GC levels can lead to chronic immunosuppression and increased pathogen infection rates in amphibians. Elevated GC levels have also been widely used as a marker of (generally negative) physiological impacts of natural and anthropogenic environmental stressors in amphibians (Graham et al. 2013; Burraco and Gomez-Mestre 2016). Although individual and interspecific GC responses to some stressors can vary widely, and are not always correlated with fitness outcomes, there is general agreement that GC levels can provide an indication of individual or population health (McMahon et al. 2011; Hammond et al. 2018; Van Meter et al. 2019). Therefore, monitoring the GC response to environmental stressors can give a broad estimation of the degree to which general homeostasis is being disrupted by exposure to the stressor. Moreover, the recent development of non-invasive techniques for monitoring of GC hormone levels in captive and wild amphibian populations (e.g. Gabor et al. 2013; Graham et al. 2013) provides an important tool to assess the physiological status of vulnerable species where more invasive physiological monitoring would be precluded on ethical grounds (Wikelski and Cooke 2006; Hammond et al. 2018).

One of the biggest obstacles faced in establishing successful captive breeding programmes for imperilled amphibians is ensuring the health and physiological fitness of captive animals. Captivity (Narayan et al. 2011), handling (Bliley and Woodley 2012), communal housing (Forsburg et al. 2019), abiotic environmental conditions (Narayan and Hero 2014; de Bruijn and Romero 2018' Novarro et al. 2018), and food availability/quality (Crespi and Warne 2013) can all alter GC levels in amphibians. In addition to their negative effects on immune

Figure 10.3 Non-invasive stress hormone (glucocorticoid, GC) monitoring from faecal, urine, or skin secretion samples could become an important tool in the management of healthy amphibian insurance populations, such as here at Currumbin Wildlife Sanctuary in Queensland, Australia. Monitoring of GC levels in captive amphibians can be used to establish robust GC baselines, ensure captive environments are optimized to monitor physiological health, and, when coupled with reproductive hormone monitoring, can help inform the management of reproductive activities. Photo credit: Ed Meyer.

function, elevated GC levels can also inhibit reproductive behaviours in some species (Moore and Miller 1984). Non-invasive monitoring of GC levels in captive amphibians can be used to establish robust GC baselines, ensure captive environments are optimized, monitor physiological health, and, when coupled with reproductive hormone monitoring, can help inform the management of reproductive activities (Figure 10.3).

10.6 Conclusions and future directions

Examining the link between environmental change and disease-driven declines in amphibians is a challenge due to the complex drivers of susceptibility to chytridiomycosis. Examining this problem from a physiological prospective provides a toolbox for testing hypotheses regarding the impacts of stressors, and their interactions, on amphibian disease susceptibility, growth, survival, and fitness. By linking experimental laboratory work with replicated *in situ* (e.g. mesocosm, field enclosures) and field studies, we can elucidate the role of multiple stressors in driving population impacts. Furthermore, by focusing on the physiology and ecoimmunology of populations that are rebounding after epidemics of *Bd* or *Bsal*, we will better understand the limits and extent of host resilience and evolutionary potential (Voyles et al. 2018).

A major obstacle to better understanding the drivers of amphibian disease susceptibility continues to be our relatively rudimentary understanding of the amphibian immune response to *Bd* infection, and the mechanisms by which hosts resist or tolerate infection (Grogan et al. 2018). Understanding which defences are most important in combating or tolerating *Bd* infection will help pinpoint the best assays of host immune function relevant to this pathogen. Studies that employ single immune assays, like WBC counts or lectin-induced swelling assays, are likely overly simplistic or overinterpreted due to the challenge of linking immune measures to pathogen susceptibility, resistance, or recovery (Hawley and Altizer 2011). Moreover, there is often a lack of correlation among immune assays within single individuals or species

(Hawley and Altizer 2011). Consequently, it is important that ecoimmunological studies examine a suite of traits that reflect the diverse ways that the immune system may respond to a challenge. The use of transcriptomics is already revolutionizing the field, allowing for the comparison of immune gene expression in response to *Bd* infection under different environmental conditions, such as varying temperatures (Ellison et al. 2019). For example, Ellison et al. (2019) found that transcripts associated with inflammation are increased at cooler temperatures, while those associated with adaptive immune genes are increased at warmer temperatures, demonstrating that immune function strategies may vary with temperature. This has implications for understanding how environmental change can impact disease susceptibility, and future work should investigate the effects of other biotic and abiotic stressors on the immune transcriptome. However, a key challenge will be identifying the mechanisms underlying variation in immune gene up-regulation under different environmental conditions for non-model amphibian species.

Combining immunological tests with physiological measurements, including metabolic rate, physical performance, and thermal tolerances, will help to clarify the trade-offs associated with investment in immune function under different environmental conditions. Immune defences are costly and require significant energetic resources (Lochmiller and Deerenberg 2000; Demas and Nelson 2012). Thus, measuring amphibian physiology both in the field and laboratory under different infection loads can help identify the impact of infection. For example, recent work has investigated the baseline and stress-induced corticosterone response in free-ranging amphibians infected with *Bd* (Hammond et al. 2020), the metabolic demands of infection with *Bd* and routine skin sloughing (Wu et al. 2018a), and the impact of higher infection loads and sloughing on evaporative water loss rates (Russo et al. 2018). Future work investigating the causes and consequences of infection for physiological functioning under free-ranging conditions, as well as under realistic scenarios such as high densities, fluctuating temperatures, or in response to land-use change, will provide a more complete picture of the impact of infection in the absence of disease. While tracking amphibians in the field can be difficult, the use of passive integrated transponder (PIT) tags, as well as the ever-decreasing size of surgically implantable radio transmitters and temperature loggers, may pave the way for a better understanding of the physiology and immune function of free-ranging animals infected with *Bd* or *Bsal*. Furthermore, investigating aspects of amphibian physiology across populations with varying levels of disease impact will help determine how animals are coping with infection through life-history or metabolic trade-offs, or plasticity and/or local adaptation in relevant physiological traits.

Although immune system performance at constant temperatures predicts optimal temperatures reasonably well in ectotherms, environmental thermal variability means that organisms rarely experience constant temperatures in nature. The non-linearity of natural temperatures means that thermal variability, particularly as a result of climate change, may be a strong modulator of amphibian immune function (Rohr and Raffel 2010; Rohr et al. 2013). Exposure to realistic changes in environmental temperature that mimic daily patterns in terrestrial environments have been shown to both improve (Terrell et al. 2013) and worsen disease outcomes and/or immune function metrics (Raffel et al. 2015). This variation may be due to the thermal physiology of the host, for example, whether they are cool- or warm-adapted, or if they are typically thermal conformers or thermoregulators in the environment. Future work employing biophysical equations and mechanistic models, including aspects of dynamic energy budget theory that are based on measurements of organismal physiology, may improve predictions of the effects of temperature variation on immune function (Bovo et al. 2018; Kearney and Porter 2019).

In conclusion, physiology provides the tools to better understand the lethal and sublethal effects of chytrid fungal pathogens on amphibians, and how this may impact individual survival and population persistence. Physiological and ecoimmunological assays can pinpoint the effects of individual and combinations of biotic and abiotic stressors on amphibians, which can inform conservation management decisions. Putting questions in an ecological and evolutionary context, by comparing aspects of physiology across life stages, species, and populations will improve our understanding of the effects of current and future threats on amphibians, and their potential to resist extinction.

References

Abu Bakar, A., Bower, D.S., Stockwell, M.P. et al., 2016. Susceptibility to disease varies with ontogeny and immunocompetence in a threatened amphibian. *Oecologia*, 181, 997–1009.

Agugliaro, J., Lind, C.M., Lorch, J.M., and Farrell, T.M., 2019. An emerging fungal pathogen is associated with increased resting metabolic rate and total evaporative water loss rate in a winter-active snake. *Functional Ecology*, 34, 486–96.

Alton, L.A., White, C.R., Wilson, R.S., and Franklin, C.E., 2012. The energetic cost of exposure to UV radiation for tadpoles is greater when they live with predators. Functional Ecology, 26, 94–103.

Altwegg, R. and Reyer, H.U., 2003. Patterns of natural selection on size at metamorphosis in water frogs. *Evolution*, 57, 872–82.

Barnosky, A.D., Matzke, N., Tomiya, S. et al., 2011. Has the Earth's sixth mass extinction already arrived? *Nature*, 471, 51–7.

Beck, M.L., Thompson, M., and Hopkins, W.A., 2017. Repeatability and sources of variation of the bacteria-killing assay in the common snapping turtle. *Journal of Experimental Zoology Part A: Ecological and Integrative Physiology*, 327, 293–301.

Becker, C.G., Longo, A.V., Haddad, C.F. B., and Zamudio, K.R., 2017. Land cover and forest connectivity alter the interactions among host, pathogen and skin microbiome. *Proceedings of the Royal Society B: Biological Sciences*, 284, 20170582.

Becker, C.G., Rodriguez, D., Toledo, L.F., Longo, A.V. et al., 2014. Partitioning the net effect of host diversity on an emerging amphibian pathogen. *Proceedings of the Royal Society of London B: Biological Sciences*, 281, 20141796.

Belden, L. and Kiesecker, J., 2005. Glucocorticosteroid hormone treatment of larval treefrogs increases infection by *Alaria* sp. trematode cercariae. *Journal of Parasitology*, 91, 686–9.

Berger, L., Hyatt, A.D., Speare, R., and Longcore, J.E., 2005a. Life cycle stages of the amphibian chytrid *Batrachochytrium dendrobatidis*. *Diseases of Aquatic Organisms*, 68, 51–63.

Berger, L., Speare, R., Daszak, P. et al., 1998. Chytridio-mycosis causes amphibian mortality associated with population declines in the rain forests of Australia and Central America. *Proceedings of the National Academy of Sciences*, 95, 9031–6.

Berger, L., Speare, R., and Skerratt, L.F., 2005b. Distribution of *Batrachochytrium dendrobatidis* and pathology in the skin of green tree frogs *Litoria caerulea* with severe chytridiomycosis. *Diseases of Aquatic Organisms*, 68, 65–70.

Bliley, J.M. and Woodley, S.K., 2012. The effects of repeated handling and corticosterone treatment on behavior in an amphibian (Ocoee salamander: *Desmognathus ocoee*). *Physiology & Behavior*, 105, 1132–9.

Bonneaud, C., Wilson, R.S.., and Seebacher, F. 2016. Immune-challenged fish up-regulate their metabolic scope to support locomotion. *PLoS ONE*, 11, e0166028.

Bovo, R., Navas, C., Tejedo, M. et al., 2018. Ecophysiology of amphibians: information for best mechanistic models. *Diversity*, 10, 118.

Brannelly, L.A., Ohmer, M.E. B., Saenz, V., and Richards-Zawacki, C.L., 2019. Effects of hydroperiod on growth, development, survival and immune defences in a temperate amphibian. *Functional Ecology*, 33, 1952–61.

Budtz, P.E. and Larsen, L.O., 1973. Structure of the toad epidermis during the moulting cycle. *Zeitschrift für Zellforschung und Mikroskopische Anatomie*, 144, 353–68.

Burraco, P., Díaz-Paniagua, C., and Gomez-Mestre, I., 2017. Different effects of accelerated development and enhanced growth on oxidative stress and telomere shortening in amphibian larvae. *Scientific Reports*, 7, 7494.

Burraco, P., Duarte, L.J., and Gomez-Mestre, I., 2013. Predator-induced physiological responses in tadpoles challenged with herbicide pollution. *Current Zoology*, 59, 475–84.

Burraco, P. and Gomez-Mestre, I., 2016. Physiological stress responses in amphibian larvae to multiple stressors reveal marked anthropogenic effects even below lethal levels. *Physiological and Biochemical Zoology*, 89, 462–72.

Careau, V. and Garland, T., Jr, 2012. Performance, personality, and energetics: correlation, causation, and mechanism. Physiological and Biochemical Zoology, 85, 543–71.

Carey, C., Cohen, N., and Rollins-Smith, L., 1999. Amphibian declines: an immunological perspective. *Developmental & Comparative Immunology*, 23, 459–72.

Castanho, L.M. and De Luca, I.M.S., 2001. Moulting behavior in leaf-frogs of the genus *Phyllomedusa* (Anura: Hylidae). *Zoologischer Anzeiger—A Journal of Comparative Zoology*, 240, 3–6.

Ceccato, E., Cramp, R.L., Seebacher, F., and Franklin, C.E., 2016. Early exposure to ultraviolet-B radiation decreases immune function later in life. *Conservation Physiology*, 4, (1), cow037. https://doi.org/10.1093/conphys/cow037

Clulow, S., Harris, M., and Mahony, M.J., 2015. Optimization, validation and efficacy of the phytohaemagglutinin inflammation assay for use in ecoimmunological studies of amphibians. *Conservation Physiology*, 3, cov042.

Cohen, J.M., McMahon, T.A., Ramsay, C. et al., 2019. Impacts of thermal mismatches on chytrid fungus *Batrachochytrium dendrobatidis* prevalence are moderated by life stage, body size, elevation and latitude. *Ecology Letters*, 22, 817–25.

Cohen, J.M., Venesky, M.D., Sauer, E.L. et al., 2017. The thermal mismatch hypothesis explains host susceptibility to an emerging infectious disease. *Ecology Letters*, 20, 184–93.

Cramp, R.L. and Franklin, C.E., 2018. Exploring the link between ultraviolet B radiation and immune function in amphibians: implications for emerging infectious diseases. Conservation Physiology, 6(1), coy035. https://doi.org/10.1093/conphys/coy035

Cramp, R.L., McPhee, R.K., Meyer, E.A. et al., 2014. First line of defence: the role of sloughing in the regulation of cutaneous microbes in frogs. *Conservation Physiology*, 2, cou012.

Crespi, E.J., Rissler, L.J., Mattheus, N.M. et al., 2015. Geophysiology of wood frogs: landscape patterns of prevalence of disease and circulating hormone concentrations across the eastern range. *Integrative and Comparative Biology*, 55, 602–17.

Crespi, E.J. and Warne, R.W., 2013. Environmental conditions experienced during the tadpole stage alter postmetamorphic glucocorticoid response to stress in an amphibian. *Integrative and Comparative Biology*, 53, 989–1001.

Dantzer, B., Fletcher, Q.E., Boonstra, R., and Sheriff, M.J., 2014. Measures of physiological stress: a transparent or opaque window into the status, management and conservation of species? *Conservation Physiology*, 2, cou023.

Daszak, P., Cunningham, A.A., and Hyatt, A.D., 2003. Infectious disease and amphibian population declines. *Diversity and Distributions*, 9, 141–50.

Davis, A.K. and Durso, A.M., 2009. White blood cell differentials of northern cricket frogs (*Acris c. crepitans*) with a compilation of published values from other amphibians. *Herpetologica*, 65, 260–7.

Davis, A.K. and Maerz, J.C., 2008. Comparison of hematological stress indicators in recently captured and captive paedomorphic mole salamanders, *Ambystoma talpoideum*. *Copeia*, 2008, 613–17.

Davis, A.K., Maney, D.L., and Maerz, J.C., 2008. The use of leukocyte profiles to measure stress in vertebrates: a review for ecologists. *Functional Ecology*, 22, 760–72.

De Bruijn, R. and Romero, L.M., 2018. The role of glucocorticoids in the vertebrate response to weather. *General and Comparative Endocrinology*, 269, 11–32.

Demas, G.E., Chefer, V., Talan, M.I., and Nelson, R.J., 1997. Metabolic costs of mounting an antigen-stimulated immune response in adult and aged C57BL/6J mice. *American Journal of Physiology—Regulatory, Integrative and Comparative Physiology*, 273, R1631–7.

Demas, G. and Nelson, R., 2012. *Ecoimmunology*. Oxford University Press, Oxford.

Demas, G.E., Zysling, D.A., Beechler, B.R. et al., 2011. Beyond phytohaemagglutinin: assessing vertebrate immune function across ecological contexts. *Journal of Animal Ecology*, 80, 710–30.

Denver, R.J., 1997. Proximate mechanisms of phenotypic plasticity in amphibian metamorphosis. *American Zoologist*, 37, 172–84.

Denver, R.J., 2009. Stress hormones mediate environment-genotype interactions during amphibian development. *General and Comparative Endocrinology*, 164, 20–31.

Dhabhar, F.S., 2009. Enhancing versus suppressive effects of stress on immune function: implications for immunoprotection and immunopathology. *Neuroimmunomodulation*, 16, 300–17.

Dirzo, R., Young, H.S., Galetti, M. et al., 2014. Defaunation in the Anthropocene. *Science*, 345, 401–6.

Downs, C.J. and Stewart, K.M., 2014. A primer in ecoimmunology and immunology for wildlife research and management. *California Fish and Game*, 100, 371–95.

Durant, S.E., Hopkins, W.A., Davis, A.K., and Romero, L.M., 2015. Evidence of ectoparasite-induced endocrine disruption in an imperiled giant salamander, the eastern hellbender (*Cryptobranchus alleganiensis*). *Journal of Experimental Biology*, 218, 2297–304.

Edge, C.B., Houlahan, J.E., Jackson, D.A., and Fortin, M.J., 2016. The response of amphibian larvae to environmental change is both consistent and variable. *Oikos*, 125, 1700–11.

Ellison, A., Zamudio, K., Lips, K., and Muletz-Wolz, C., 2019. Temperature-mediated shifts in salamander transcriptomic responses to the amphibian-killing fungus. *Molecular Ecology*, 29, 325–43.

Fisher, M., Garner, T., and Walker, S., 2009. Global emergence of *Batrachochytrium dendrobatidis* and amphibian chytridiomycosis in space, time, and host. *Microbiology*, 63, 291–310.

Fites, J.S., Reinert, L.K., Chappell, T.M., and Rollins-Smith, L.A., 2014. Inhibition of local immune responses by the frog-killing fungus *Batrachochytrium dendrobatidis*. *Infection and Immunity*, 82, 4698–706.

Fonner, C.W., Patel, S.A., Boord, S.M. et al., 2017. Effects of corticosterone on infection and disease in salamanders exposed to the amphibian fungal pathogen *Batrachochytrium dendrobatidis*. *Diseases of Aquatic Organisms*, 123, 159–71.

Forsburg, Z.R., Goff, C.B., Perkins, H.R. et al., 2019. Validation of water-borne cortisol and corticosterone in tadpoles: recovery rate from an acute stressor, repeatability, and evaluating rearing methods. *General and Comparative Endocrinology*, 281, 145–52.

Gabor, C.R., Fisher, M.C., and Bosch, J., 2013. A non-invasive stress assay shows that tadpole populations infected with *Batrachochytrium dendrobatidis* have elevated corticosterone levels. *PLoS ONE, 8*, e56054.

Gabor, C.R., Fisher, M.C., and Bosch, J., 2015. Elevated corticosterone levels and changes in amphibian behavior are associated with *Batrachochytrium dendrobatidis (Bd)* infection and Bd lineage. *PLoS ONE, 10*, e0122685.

Gabor, C.R., Knutie, S.A., Roznik, E.A., and Rohr, J.R., 2018. Are the adverse effects of stressors on amphibians mediated by their effects on stress hormones? *Oecologia, 186*, 393–404.

Gervasi, S.S. and Foufopoulos, J., 2008. Costs of plasticity: responses to desiccation decrease post-metamorphic immune function in a pond-breeding amphibian. *Functional Ecology, 22*, 100–8.

Gervasi, S.S., Hunt, E.G., Lowry, M., and Blaustein, A.R., 2014. Temporal patterns in immunity, infection load and disease susceptibility: understanding the drivers of host responses in the amphibian-chytrid fungus system. *Functional Ecology, 28*, 569–78.

Graham, C.M., Narayan, E.J., McCallum, H., and Hero, J.-M., 2013. Non-invasive monitoring of glucocorticoid physiology within highland and lowland populations of native Australian great barred frog (*Mixophyes fasciolatus*). *General and Comparative Endocrinology, 191*, 24–30.

Grogan, L.F., Robert, J., Berger, L. et al., 2018. Review of the amphibian immune response to chytridiomycosis, and future directions. *Frontiers in Immunology, 9*, 2536.

Groner, M.L., Buck, J.C., Gervasi, S. et al., 2013. Larval exposure to predator cues alters immune function and response to a fungal pathogen in post-metamorphic wood frogs. *Ecological Applications, 23*, 1443–54.

Hammond, T.T., Au, Z.A., Hartman, A.C., and Richards-Zawacki, C.L., 2018. Assay validation and interspecific comparison of salivary glucocorticoids in three amphibian species. *Conservation Physiology, 6*, coy055.

Hammond, T.T., Blackwood, P.E., Shablin, S.A., and Richards-Zawacki, C.L., 2020. Relationships between glucocorticoids and infection with *Batrachochytrium dendrobatidis* in three amphibian species. *General and Comparative Endocrinology, 285*, 113269.

Harter, T.S., Brauner, C.J., and Matthews, P.G.D., 2017. A novel technique for the precise measurement of CO2 production in small aquatic organisms as validated on aeshnid dragonfly nymphs. Journal of Experimental Biology, 220, 964–68.

Hawley, D.M. and Altizer, S.M., 2011. Disease ecology meets ecological immunology: understanding the links between organismal immunity and infection dynamics in natural populations. *Functional Ecology, 25*, 48–60.

Hayes, T., Falso, P., Gallipeau, S., and Stice, M., 2010. The cause of global amphibian declines: a developmental endocrinologist's perspective. *Journal of Experimental Biology, 213*, 921–33.

Hopkins, W.A. and Durant, S.E., 2011. Innate immunity and stress physiology of eastern hellbenders (*Cryptobranchus alleganiensis*) from two stream reaches with differing habitat quality. *General and Comparative Endocrinology, 174*, 107–15.

Hua, J., Wuerthner, V.P., Jones, D.K. et al., 2017. Evolved pesticide tolerance influences susceptibility to parasites in amphibians. *Evolutionary Applications, 10*, 802–12.

Huey, R.B., Kearney, M.R., Krockenberger, A. et al., 2012. Predicting organismal vulnerability to climate warming: roles of behaviour, physiology and adaptation. *Philosophical Transactions of the Royal Society B, 367*, 1665–79.

International Union for Conservation of Nature, 2019. The IUCN Red List of Threatened Species. Version 2019-1. Available: www.iucnredlist.org (accessed 25 March 2019).

James, T.Y., Toledo, L.F., Rödder, D., 2015. Disentangling host, pathogen, and environmental determinants of a recently emerged wildlife disease: lessons from the first 15 years of amphibian chytridiomycosis research. *Ecology and Evolution, 5*, 4079–97.

Katzenback, B.A., Holden, H.A., Falardeau, J. et al., 2014. Regulation of the *Rana sylvatica* brevinin-1SY antimicrobial peptide during development and in dorsal and ventral skin in response to freezing, anoxia and dehydration. *Journal of Experimental Biology, 217*, 1392–401.

Kearney, M.R. and Porter, W.P., 2019. NicheMapR—an R package for biophysical modelling: the ectotherm and Dynamic Energy Budget models. *Ecography, 43*, 85–96.

Kennedy, M.W. and Nager, R.G., 2006. The perils and prospects of using phytohaemagglutinin in evolutionary ecology. *Trends in Ecology & Evolution, 21*, 653–5.

Kiesecker, J.M., 2011. Global stressors and the global decline of amphibians: tipping the stress immunocompetency axis. *Ecological Research, 26*, 897–908.

Kindermann, C., Narayan, E.J., and Hero, J.-M., 2012. Urinary corticosterone metabolites and chytridiomycosis disease prevalence in a free-living population of male Stony Creek frogs (*Litoria wilcoxii*). *Comparative Biochemistry and Physiology Part A: Molecular & Integrative Physiology, 162*, 171–6.

Kohli, A., Brannelly, L., Ohmer, M. et al., 2019. Disease and the drying pond: examining possible links between drought, immune function, and disease development in amphibians. Physiological and Biochem*ical* Zoology, 92, 339–48.

Koprivnikar, J., Hoye, B.J., Urichuk, T.M.Y., and Johnson, P.T.J., 2019. Endocrine and immune responses of larval amphibians to trematode exposure. *Parasitology Research, 118*, 275–88.

Lafonte, B.E. and Johnson, P.T., 2013. Experimental infection dynamics: using immunosuppression and *in vivo* parasite tracking to understand host resistance in an amphibian–trematode system. *Journal of Experimental Biology*, 216, 3700–8.

Larsen, L.O., 1976. Physiology of molting. In J.A. Moore and B. Lofts, B., eds. *Physiology of the Amphibia*, pp. 54–100. Academic Press, New York.

Lefcort, H. and Blaustein, A.R., 1995. Disease, predator avoidance, and vulnerability to predation in tadpoles. *Oikos*, 74, 469–74.

Levis, N.A. and Johnson, J.R., 2015. Level of UV-B radiation influences the effects of glyphosate-based herbicide on the spotted salamander. *Ecotoxicology*, 24, 1073–86.

Liebl, A.L. and Martin II, L.B., 2009. Simple quantification of blood and plasma antimicrobial capacity using spectrophotometry. *Functional Ecology*, 23, 1091–6.

Lighton, J.R.B., 2019. Measuring Metabolic Rates: A Manual for Scientists, second edition. Oxford University Press, Oxford.

Lips, K.R., 2016. Overview of chytrid emergence and impacts on amphibians. *Philosophical Transactions of the Royal Society B: Biological Sciences*, 371, 20150465.

Lochmiller, R.L. and Deerenberg, C., 2000. Trade-offs in evolutionary immunology: just what is the cost of immunity? *Oikos*, 88, 87–98.

Longcore, J.E., Pessier, A.P., and Nichols, D.K., 1999. *Batrachochytrium dendrobatidis* gen. et sp. nov., a chytrid pathogenic to amphibians. Mycologia, 91, 219–27.

Maniero, G.D. and Carey, C., 1997. Changes in selected aspects of immune function in the leopard frog, *Rana pipiens*, associated with exposure to cold. *Journal of Comparative Physiology B*, 167, 256–63.

Martel, A., Spitzen-Van Der Sluijs, A., Blooi, M. et al., 2013. *Batrachochytrium salamandrivorans* sp. nov. causes lethal chytridiomycosis in amphibians. *Proceedings of the National Academy of Sciences of the USA*, 110, 15325–9.

Martin, L.B., 2009. Stress and immunity in wild vertebrates: timing is everything. *General and Comparative Endocrinology*, 163, 70–6.

Martin, L.B., Scheuerlein, A., and Wikelski, M., 2003. Immune activity elevates energy expenditure of house sparrows: a link between direct and indirect costs? *Proceedings of the Royal Society of London. Series B: Biological Sciences*, 270, 153–8.

Matutte, B., Storey, K.B., Knoop, F.C., and Conlon, J.M., 2000. Induction of synthesis of an antimicrobial peptide in the skin of the freeze-tolerant frog, *Rana sylvatica*, in response to environmental stimuli. *Febs Letters*, 483, 135–8.

McMahon, T.A., Halstead, N.T., Johnson, S. et al., 2011. The fungicide chlorothalonil is nonlinearly associated with corticosterone levels, immunity, and mortality in amphibians. *Environmental Health Perspectives*, 119, 1098–103.

McMahon, T.A., Sears, B.F., Venesky, M.D. et al., 2014. Amphibians acquire resistance to live and dead fungus overcoming fungal immunosuppression. *Nature*, 511, 224–7.

McMenamin, S.K., Hadly, E.A., and Wright, C.K., 2008. Climatic change and wetland desiccation cause amphibian decline in Yellowstone National Park. *Proceedings of the National Academy of Sciences*, 105, 16988–93.

McNab, B.K., 2002. *The Physiological Ecology of Vertebrates: a View from Energetics*. Cornell University Press, Ithaca, NY.

Meyer, E.A., Cramp, R.L., Bernal, M.H., and Franklin, C.E., 2012. Changes in cutaneous microbial abundance with sloughing: possible implications for infection and disease in amphibians. *Diseases of Aquatic Organisms*, 101, 235.

Milly, P.C., Dunne, K.A., and Vecchia, A.V., 2005. Global pattern of trends in streamflow and water availability in a changing climate. *Nature*, 438, 347.

Miodoński, A.J., Bigaj, J., Mika, J., and Płytycz, B., 1996. Season-specific thymic architecture in the frog, *Rana temporaria*: SEM studies. *Developmental & Comparative Immunology*, 20, 129–37.

Moore, F.L. and Miller, L.J., 1984. Stress-induced inhibition of sexual behavior: corticosterone inhibits courtship behaviors of a male amphibian (*Taricha granulosa*). *Hormones and Behavior*, 18, 400–10.

Narayan, E.J., 2019. Introductory chapter: applications of stress endocrinology in wildlife conservation and livestock science. *Comparative Endocrinology of Animals*, 1–8. http://dx.doi.org/10.5772/intechopen.86523

Narayan, E.J. and Hero, J.-M., 2014. Acute thermal stressor increases glucocorticoid response but minimizes testosterone and locomotor performance in the cane toad (*Rhinella marina*). *PLoS ONE*, 9, e92090.

Narayan, E.J., Molinia, F.C., Kindermann, C. et al., 2011. Urinary corticosterone responses to capture and toe-clipping in the cane toad (*Rhinella marina*) indicate that toe-clipping is a stressor for amphibians. *General and Comparative Endocrinology*, 174, 238–45.

Novarro, A.J., Gabor, C.R., Goff, C.B. et al., 2018. Physiological responses to elevated temperature across the geographic range of a terrestrial salamander. *Journal of Experimental Biology*, 221, jeb178236.

O'Hanlon, S.J., Rieux, A., Farrer, R.A., 2018. Recent Asian origin of chytrid fungi causing global amphibian declines. *Science*, 360, 621.

Ohmer, M.E.B., Cramp, R.L., Russo, C.J.M., et al., 2017. Skin sloughing in susceptible and resistant amphibians regulates infection with a fungal pathogen. *Scientific Reports*, 7, 3529.

Ohmer, M.E.B., Cramp, R.L., White, C.R., and Franklin, C.E., 2015. Skin sloughing rate increases with chytrid fungus infection load in a susceptible amphibian. *Functional Ecology, 29,* 674–82.

Ohmer, M.E.B., Cramp, R.L., White, C.R. et al., 2019. Phylogenetic investigation of skin sloughing rates in frogs: relationships with skin characteristics and disease-driven declines. *Proceedings of the Royal Society B: Biological Sciences, 286,* 20182378.

Pessier, A.P., Nichols, D.K., Longcore, J.E., and Fuller, M.S. 1999. *Batrachochytrium dendrobatidis* gen. et sp. nov., a chytrid pathogenic to amphibians. *Mycologia, 91,* 219–27.

Peterson, J.D., Steffen, J.E., Reinert, L.K. et al., 2013. Host stress response is important for the pathogenesis of the deadly amphibian disease, chytridiomycosis, in *Litoria caerulea. PLoS ONE, 8,* e62146.

Piotrowski, J.S., Annis, S.L., and Longcore, J.E., 2004. Physiology of *Batrachochytrium dendrobatidis*, a chytrid pathogen of amphibians. *Mycologia, 96,* 9–15.

Rachowicz, L.J., Hero, J.-M., Alford, R.A. et al., 2005. The novel and endemic pathogen hypotheses: competing explanations for the origin of emerging infectious diseases of wildlife. *Conservation Biology, 19,* 1441–8.

Raffel, T., Rohr, J., Kiesecker, J., and Hudson, P., 2006. Negative effects of changing temperature on amphibian immunity under field conditions. *Functional Ecology, 20,* 819–28.

Raffel, T.R., Halstead, N.T., McMahon, T.A. et al., 2015. Temperature variability and moisture synergistically interact to exacerbate an epizootic disease. *Proceedings of the Royal Society of London B: Biological Sciences, 282,* 20142039.

Raffel, T.R., Romansic, J.M., Halstead, N.T. et al., 2013. Disease and thermal acclimation in a more variable and unpredictable climate. *Nature Climate Change, 3,* 146–51.

Ramsey, J.P., Reinert, L.K., Harper, L.K. et al., 2010. Immune defenses against *Batrachochytrium dendrobatidis*, a fungus linked to global amphibian declines, in the South African clawed frog, *Xenopus laevis. Infection and Immunity, 78,* 3981–92.

Relyea, R.A., and Mills, N., 2001. Predator-induced stress makes the pesticide carbaryl more deadly to gray treefrog tadpoles (*Hyla versicolor*). *Proceedings of the National Academy of Sciences, 98,* 2491–6.

Richards-Zawacki, C.L., 2010. Thermoregulatory behaviour affects prevalence of chytrid fungal infection in a wild population of Panamanian golden frogs. *Proceedings of the Royal Society B—Biological Sciences, 277,* 519–28.

Richter-Boix, A., Katzenberger, M., Duarte, H. et al., 2015. Local divergence of thermal reaction norms among amphibian populations is affected by pond temperature variation. *Evolution, 69,* 2210–26.

Richter-Boix, A., Tejedo, M., and Rezende, E.L., 2011. Evolution and plasticity of anuran larval development in response to desiccation. A comparative analysis. *Ecology and Evolution, 1,* 15–25.

Robert, J. and Ohta, Y., 2009. Comparative and developmental study of the immune system in *Xenopus. Developmental Dynamics, 238,* 1249–70.

Rohr, J.R. and Raffel, T.R., 2010. Linking global climate and temperature variability to widespread amphibian declines putatively caused by disease. *Proceedings of the National Academy of Sciences, 107,* 8269–74.

Rollins-Smith, L.A., 1998. Metamorphosis and the amphibian immune system. *Immunological Reviews, 166,* 221–30.

Rollins-Smith, L.A., 2009. The role of amphibian anti-microbial peptides in protection of amphibians from pathogens linked to global amphibian declines. *Biochimica et Biophysica Acta, 1788,* 1593–9.

Rollins-Smith, L.A., 2017. Amphibian immunity—stress, disease, and climate change. *Developmental & Comparative Immunology, 66,* 111–19.

Rollins-Smith, L.A., Pask, J.D., Ramsey, J.P. et al., 2011. Amphibian immune defenses against chytridiomycosis: impacts of changing environments. *Integrative and Comparative Biology, 51,* 552–62.

Rollins-Smith, L.A. and Woodhams, D.C., 2012. *Amphibian Immunity.* Oxford University Press, New York.

Russo, C., Ohmer, M., Cramp, R., and Franklin, C., 2018. A pathogenic skin fungus and sloughing exacerbate cutaneous water loss in amphibians. *Journal of Experimental Biology, 221,* jeb167445.

Santymire, R., Manjerovic, M., and Sacerdote-Velat, A., 2018. A novel method for the measurement of glucocorticoids in dermal secretions of amphibians. *Conservation Physiology, 6,* coy008.

Sapolsky, R.M., Romero, L.M., and Munck, A.U., 2000. How do glucocorticoids influence stress responses? Integrating permissive, suppressive, stimulatory, and preparative actions. *Endocrine Reviews, 21,* 55–89.

Sauer, E.L., Fuller, R.C., Richards-Zawacki, C.L. et al., 2018. Variation in individual temperature preferences, not behavioural fever, affects susceptibility to chytridiomycosis in amphibians. *Proceedings of the Royal Society B: Biological Sciences, 285,* 20181111.

Savage, A.E., Terrell, K.A., Gratwicke, B. et al., 2016. Reduced immune function predicts disease susceptibility in frogs infected with a deadly fungal pathogen. *Conservation Physiology, 4,* cow011.

Scheele, B.C., Pasmans, F., Skerratt, L.F. et al., 2019. Amphibian fungal panzootic causes catastrophic and ongoing loss of biodiversity. *Science, 363,* 1459–63.

Scheun, J., Greeff, D., Medger, K., and Ganswindt, A., 2019. Validating the use of dermal secretion as a matrix for monitoring glucocorticoid concentrations in African amphibian species. *Conservation Physiology*, 7, coz022.

Schulenburg, H., Kurtz, J., Moret, Y., and Siva-Jothy, M.T., 2009. Introduction. Ecological immunology. *Philosophical Transactions of the Royal Society B: Biological Sciences*, 364, 3–14.

Searle, C.L., Biga, L.M., Spatafora, J.W., and Blaustein, A.R., 2011. A dilution effect in the emerging amphibian pathogen *Batrachochytrium dendrobatidis*. *Proceedings of the National Academy of Sciences*, 108, 16322–6.

Semlitsch, R.D., Scott, D.E., and Pechmann, J.H., 1988. Time and size at metamorphosis related to adult fitness in *Ambystoma talpoideum*. *Ecology*, 69, 184–92.

Sheriff, M.J., Dantzer, B., Delehanty, B. et al., 2011. Measuring stress in wildlife: techniques for quantifying glucocorticoids. *Oecologia*, 166, 869–87.

Shutler, D. and Marcogliese, D.J., 2011. Leukocyte profiles of northern leopard frogs, *Lithobates pipiens*, exposed to pesticides and hematozoa in agricultural wetlands. *Copeia*, 2011, 301–7.

Skerratt, L.F., Berger, L., Speare, R. et al., 2007. Spread of chytridiomycosis has caused the rapid global decline and extinction of frogs. *EcoHealth*, 4, 125–34.

Sonn, J.M., Berman, S., and Richards-Zawacki, C.L., 2017. The influence of temperature on chytridiomycosis in vivo. *EcoHealth*, 14, 762–70.

Stuart, S.N., Chanson, J.S., Cox, N.A. et al., 2004. Status and trends of amphibian declines and extinctions worldwide. *Science*, 306, 1783–6.

Tatiersky, L., Rollins-Smith, L.A., Lu, R. et al., 2015. Effect of glucocorticoids on expression of cutaneous antimicrobial peptides in northern leopard frogs (*Lithobates pipiens*). *BMC Veterinary Research*, 11, 191–1.

Tennessen, J.B., Parks, S.E., Swierk, L. et al., 2018. Frogs adapt to physiologically costly anthropogenic noise. *Proceedings of the Royal Society B—Biological Sciences*, 285, 20182194.

Terrell, K.A., Quintero, R.P., Murray, S. et al., 2013. Cryptic impacts of temperature variability on amphibian immune function. *Journal of Experimental Biology*, 216, 4204–11.

Van Meter, R.J., Adelizzi, R., Glinski, D.A., and Henderson, W.M., 2019. Agrochemical mixtures and amphibians: the combined effects of pesticides and fertilizer on stress, acetylcholinesterase activity, and bioaccumulation in a terrestrial environment. *Environmental Toxicology and Chemistry*, 38, 1052–61.

Van Rooij, P., Martel, A., D'Herde, K. et al., 2012. Germ tube mediated invasion of *Batrachochytrium dendrobatidis* in amphibian skin is host dependent. *PLoS ONE*, 7, e41481.

Venesky, M.D., Wilcoxen, T.E., Rensel, M.A. et al., 2012. Dietary protein restriction impairs growth, immunity, and disease resistance in southern leopard frog tadpoles. *Oecologia*, 169, 23–31.

Voyles, J., Johnson, L.R., Rohr, J. et al., 2017. Diversity in growth patterns among strains of the lethal fungal pathogen *Batrachochytrium dendrobatidis* across extended thermal optima. *Oecologia*, 184, 363–73.

Voyles, J., Vredenburg, V.T., Tunstall, T.S. et al., 2012. Pathophysiology in mountain yellow-legged frogs (*Rana muscosa*) during a chytridiomycosis outbreak. *PLoS ONE*, 7, e35374.

Voyles, J., Woodhams, D.C., Saenz, V. et al., 2018. Shifts in disease dynamics in a tropical amphibian assemblage are not due to pathogen attenuation. *Science*, 359, 1517–19.

Voyles, J., Young, S., Berger, L. et al., 2009. Pathogenesis of chytridiomycosis, a cause of catastrophic amphibian declines. *Science*, 326, 582–5.

Walls, S., Barichivich, W., and Brown, M., 2013. Drought, deluge and declines: the impact of precipitation extremes on amphibians in a changing climate. *Biology*, 2, 399–418.

Walls, S.C. and Gabor, C.R., 2019. Integrating behavior and physiology into strategies for amphibian conservation. *Frontiers in Ecology and Evolution*, 7, 234.

Warne, R.W., Crespi, E.J., and Brunner, J.L., 2011. Escape from the pond: stress and developmental responses to ranavirus infection in wood frog tadpoles. *Functional Ecology*, 25, 139–46.

Wikelski, M. and Cooke, S.J., 2006. Physiological assessment of environmental stressors. *Trends in Ecology & Evolution*, 1, 38–46.

Wilbur, H.M. and Collins, J.P., 1973. Ecological aspects of amphibian metamorphosis: nonnormal distributions of competitive ability reflect selection for facultative metamorphosis. *Science*, 182, 1305–14.

Winwood-Smith, H.S. and White, C.R., 2018. Short-duration respirometry underestimates metabolic rate for discontinuous breathers. *Journal of Experimental Biology*, 221, jeb175752.

Woodhams, D.C., Bell, S.C., Bigler, L. et al., 2016. Life history linked to immune investment in developing amphibians. *Conservation Physiology*, 4, cow025.

Wu, N.C., Cramp, R.L., and Franklin, C.E., 2018a. Body size influences energetic and osmoregulatory costs in frogs infected with *Batrachochytrium dendrobatidis*. *Scientific Reports*, 8, 3739.

Wu, N.C., Cramp, R.L., Ohmer, M.E.B., and Franklin, C.E., 2018b. Epidermal epidemic: unravelling the pathogenesis of chytridiomycosis. *The Journal of Experimental Biology*, 222, jeb191817.

Young, S., Whitehorn, P., Berger, L. et al., 2014. Defects in host immune function in tree frogs with chronic chytridiomycosis. *PLoS ONE*, 9, e107284.

Zera, A.J. and Harshman, L.G., 2001. The physiology of life history trade-offs in animals. *Annual Review of Ecology and Systematics*, 32, 95–126.

PART IV

Determining the Multi-Scale Effects of Human-Induced Environmental Change

Improving 'shark park' protections under threat from climate change using the conservation physiology toolbox

Ian A. Bouyoucos and Jodie L. Rummer

⮎ Take-home message

Combining ecophysiology techniques with community outreach and education are valuable steps towards achieving improved protection for shark populations predicted to be vulnerable to climate change.

11.1 Introduction: why protect sharks and rays?

Chondrichthyan fishes (sharks, rays, and chimaeras; hereafter, sharks and rays) are among the most threatened aquatic vertebrate taxa. Many species are thought to be vulnerable to anthropogenic disturbance owing to life-history traits that are characteristic of *K*-selected organisms: slow growth, late maturity, and small litters (Cortés 2000). In general, large-bodied species that occupy shallow habitats with a large geographic range are most at risk of exposure to anthropogenic impacts and categorization under a threatened status (Dulvy et al. 2014; Stein et al. 2018). To date, approximately one quarter of all shark and ray species are threatened with extinction according to the International Union for Conservation of Nature (IUCN) criteria (Dulvy et al. 2014). Targeted fishing—mostly for consumptive purposes—and bycatch (i.e. landings that are unused or without sustainable management) are the predominant threats driving global population declines in sharks and rays (Dulvy et al. 2014; Oliver et al. 2015). Declining shark and ray abundance is well documented, as demonstrated by declining global catch from 2000 to 2011 (Worm et al. 2013; Davidson, Krawchuk and Dulvy 2016). Following exploitation in fisheries, global declines in populations of sharks and rays are driven by habitat loss and degradation (e.g. development, pollution), species control measures (e.g. culling campaigns), and climate change (ocean warming, acidification, and deoxygenation) (Dulvy et al. 2014). Given the sheer number of species and geographic range occupied by sharks and rays across jurisdictional boundaries, their economic value (i.e. fisheries and tourism), and significance as sources of protein, management and conservation strategies for sharks and rays are, at best, complicated (Dulvy et al. 2017; Simpfendorfer and Dulvy 2017).

Ian A. Bouyoucos and Jodie L. Rummer, *Improving 'shark park' protections under threat from climate change using the conservation physiology toolbox*
In: *Conservation Physiology: Applications for Wildlife Conservation and Management.* Edited by: Christine L. Madliger, Craig E. Franklin, Oliver P. Love, and Steven J. Cooke, Oxford University Press (2021). © Oxford University Press. DOI: 10.1093/oso/9780198843610.003.0011

Marine protected areas (MPAs) have the potential to halt or reverse declines in shark and ray populations. There is evidence that well-managed and enforced MPAs can contribute to population recovery among reef sharks (Bond et al. 2012; Speed et al. 2018). For some populations of species threatened with extinction, however, MPA coverage can be inadequate such that species' ranges are not well represented in MPAs (Chin et al. 2017; Davidson and Dulvy 2017; White et al. 2017). As conservation tools, MPAs are perceived as moderately effective at achieving shark and ray conservation goals, and MPAs specifically for sharks and rays (e.g. 'shark sanctuaries') are marginally more effective than general MPAs (MacKeracher et al. 2019). MPAs can be effective conservation tools, but not without additional conservation measures (Ward-Paige and Worm 2017; MacKeracher et al. 2019). So-called shark sanctuaries (MPAs specifically for sharks that span a country's entire exclusive economic zone; EEZ) have potential to reduce fishing pressure and population declines for sharks, but a number of unmanaged threats within sanctuaries (e.g. bycatch and ghost fishing, wildlife tourism, climate change) may limit their effectiveness (Ward-Paige 2017; Ward-Paige and Worm 2017). Climate change is a newly recognized threat to sharks and rays globally that will first require evaluation before adequate management and conservation planning can be implemented within MPAs (Chin et al. 2010; Ward-Paige and Worm 2017).

Climate change was only recognized as a threat to sharks and rays in the first decade of the 21st century. Following this, studies started considering the sensitivities of various species to anthropogenic climate change for conservation and management planning (Chin et al. 2010; Fuentes et al. 2016). To this end, research efforts have been dedicated to tracking animals to model future distributions (Hazen et al. 2013; Sequeira et al. 2014; Payne et al. 2018), and experimental studies have been designed to generate a mechanistic understanding of animals' responses to predicted climate change conditions (Rosa et al. 2017; Bouyoucos et al. 2019). Climate change is predicted to alter the physiology and behaviour of sharks and rays through myriad global change phenomena: ocean warming (Rosa et al. 2014; Pistevos et al. 2015; Di Santo 2016; Gervais et al. 2018), increases in storm frequency (Morash et al. 2016; Tunnah et al. 2016), increases in environmental carbon dioxide (i.e. ocean acidification; Rosa et al. 2017), and deoxygenation of ocean habitats (Di Santo et al. 2016; Crear et al. 2019; Schwieterman et al. 2019b). Ocean acidification was first considered a threat in 2014 after targeted studies provided evidence of physiological and behavioural impairments to sharks (Rosa et al. 2017) and skates (Di Santo 2015), despite the previously held notion that this group would be resilient to high environmental carbon dioxide, given their evolutionary history (Rummer and Munday 2017). These global change stressors are predicted to reduce organismal fitness through sublethal impairments to physiological performance, such that adaptive responses include acclimation or redistribution (Sunday et al. 2012; Habary et al. 2017; Donelson et al. 2019). At the time this chapter was written, however, only one species of shark (New Caledonia catshark, *Aulohalaelurus kanakorum*) had been identified as threatened by climate change (Dulvy et al. 2014). Yet, this mostly reflects the scarcity of studies rather than the robustness of these taxa. Species' capacity to acclimate to multiple stressors experienced within an MPA and their potential for redistribution under climate change have clear implications for MPA planning.

Protecting and managing shark and ray populations that are threatened by climate change is a conservation challenge that can be addressed using the conservation physiology toolbox (Madliger et al. 2018). As mentioned above, climate change is predicted to alter the physiology of marine ectotherms (Lefevre 2016). Thus, the conservation physiology toolbox can help improve our understanding of these physiological responses and the mechanisms underpinning changes in performance and fitness (Horodysky et al. 2016; McKenzie et al. 2016). This chapter will discuss how the conservation physiology toolbox has been applied to predict the vulnerability of populations of tropical reef sharks in Australia and French Polynesia to climate change. The overarching goal is to better understand the extent of this threat and suggest conservation strategies. We also discuss the importance of implementing community outreach, education, and social media initiatives, through the collective branding of these projects as the Physioshark project.

11.2 Physioshark: baby sharks and climate change

How can we better protect and manage sharks and rays threatened by climate change if we lack a basic understanding as to the magnitude of the threat to these taxa? To address this key question, the Physioshark project was initiated in 2013 with its purpose being to understand how human-induced stressors, primarily climate change, will affect tropical sharks and rays at all life-history stages, and how that will play into the health and viability of populations. Tropical species represent a major knowledge gap in our understanding of the effects of climate change on sharks and rays (Rosa et al. 2017). Yet, there is an inherent urgency in defining responses of tropical fishes because these species are thought to live close to their upper boundaries for thermal performance and tolerance, such that 1–3°C increases in water temperature may reduce organismal performance with consequences for fitness (Nilsson et al. 2009; Rummer et al. 2014; Comte and Olden 2017). Climate change is not the only threat to sharks being evaluated by Physioshark: bycatch is a considerable problem for sharks and rays within some MPAs and especially within shark sanctuaries (Ward-Paige and Worm 2017). To achieve our goal of generating a mechanistic understanding of the stress tolerance and vulnerability of tropical sharks and rays to anthropogenic stress, Physioshark focuses on species across a range of ecological niches: a small, benthic mesopredator, a medium-sized reef-associated mesopredator, and a large apex predator. In working towards identifying the mechanisms behind tropical species' vulnerabilities to anthropogenic stressors, these data have significance towards managing climate change risk

to tropical sharks globally in addition to managing and protecting local populations.

11.3 Case study 1: walking sharks down under

In this first case study we present a series of studies aimed at understanding the climate change vulnerability of a reef-associated shark population, forming the basis of the Physioshark programme. At the time, the work was novel in producing some of the first evidence of physiological and behavioural tolerance of a shark to three major global change stressors (warming, acidification, and deoxygenation), and providing one of the only accounts of the responses of a tropical elasmobranch fish (Rosa et al. 2017; Lopes et al. 2018; Pegado et al. 2018). Here, we describe the comprehensive model approach that initiated this research programme and is now being applied to ongoing projects in French Polynesia.

The Great Barrier Reef (GBR) is home to 133 species of sharks and rays and is part of an MPA that is larger than New Zealand. This World Heritage Area stretches 2300 km along Australia's tropical eastern coastline, includes a series of no-take and no-entry zones, and offers protection to nine species of sharks, five mobulid rays, and all species of sawfishes (GBRMPA 2019). Reef-associated sharks and rays on the GBR are mostly mesopredators with functionally redundant ecosystem roles similar to that of large teleost fishes (Heupel et al. 2014; Roff et al. 2016); regardless, the loss of millions of years of evolutionary history through local extirpations is not an acceptable conservation outcome (Stein et al. 2018). The health of the GBR ecosystem is under considerable threat from climate change, with recent recurrent mass coral bleaching events being responsible for the loss of coral cover and reduced recruitment (Hughes et al. 2017, 2019). Other serious sources of stress include declining water quality and sound pollution (Pratchett et al. 2019), all of which can be investigated with the conservation physiological toolbox (Hess et al. 2017; Jain-Schlaepfer et al. 2018). Sharks on the GBR seem to have benefited from no-entry zones more than no-take areas (Roff et al. 2016; Cinner et al. 2018; Frisch and Rizzari 2019); yet, these protections do not directly buffer against the effects of climate change (Chin et al. 2010). Environmental conditions are important drivers of distribution for some shark species on the GBR, such that some species may move to locate more suitable habitats (Yates et al. 2015; Schlaff et al. 2017). On the other hand, benthic species that are much less capable of redistribution must be able to tolerate local conditions.

The purpose of Physioshark's first investigation into sharks on the GBR was to understand the capacity of a benthic shark with limited dispersal potential to tolerate extreme environmental conditions predicted to occur with climate change. The study species was the epaulette shark (*Hemiscyllium ocellatum*), a species renowned for its remarkable hypoxia and anoxia tolerance (Wise et al. 1998; Routley et al. 2002; Nilsson and Renshaw 2004; Devaux et al. 2019) and ability to 'walk' between tide pools across exposed reef flat using its paired fins (Goto et al. 1999). The epaulette shark is truly a product of its environment. Epaulette sharks on the GBR are found on shallow reef platforms with locally extreme conditions. For instance, diurnal variation in carbon dioxide partial pressures (pCO_2) on a reef flat at Lady Elliot Island on the southern GBR can exceed 1000 µatm (Shaw et al. 2012), the predicted open ocean pCO_2 for the year 2100 under business-as-usual projections (Meinshausen et al. 2011). Reef flats around Heron Island can experience daily summer high temperatures exceeding 35°C (Gervais et al. 2018) and can become hypoxic during low nocturnal tides (Routley et al. 2002). In addition to living a benthic lifestyle, epaulette sharks lay eggs; therefore, developing embryos must also be able to survive such extreme local conditions. Epaulette sharks are listed as a 'least-concern species' by the IUCN Red List, but the report acknowledges a need to research possible threats to epaulette shark populations (Bennet et al. 2015). Therefore, our objectives were to test experimentally, the effects of simulated ocean acidification and warming conditions on physiological performance metrics of epaulette sharks during early ontogeny.

A key tool that is heavily relied upon by Physioshark is respirometry (Madliger et al. 2018). Specifically, intermittent-flow respirometry is a widely used technique to measure oxygen uptake

rates ($\dot{M}O_2$) that serve as proxies of whole-organism metabolic rates (Svendsen et al. 2016). Respirometry systems can be used in the field and laboratory (Farrell et al. 2003; Mochnacz et al. 2017). Moreover, systems come in a variety of designs so that $\dot{M}O_2$ can be measured at a range of activity levels to elicit an individual's full range of $\dot{M}O_2$ or aerobic scope, which represents an organism's capacity to take up oxygen to meet metabolic demands (Rummer et al. 2016). Oxygen uptake rates are also widely tested in climate change studies to define thermal performance (via thermal performance of aerobic scope) and hypoxia tolerance (via the critical saturation minimum, P_{crit}), for instance (Bouyoucos et al. 2019). It can be difficult to conduct respirometry on large shark species that must be continuously active (Payne et al. 2015); fortuitously, small, benthic sharks (e.g. epaulette sharks) and juveniles of species that are capable of stationary respiration are well suited to respirometry.

The first series of experiments tested the effects of ocean acidification on sharks developing *in ovo* and juveniles. Across these experiments, sharks were exposed from 1–3 months to a range of static pCO_2 approximating current (~400 µatm), mid-century (~600 µatm), and end-of-century (~900 µatm) values predicted for business-as-usual climate change scenarios (Heinrich et al. 2014, 2016; Johnson et al. 2016). For developing sharks, growth rates, yolk consumption, ventilation rates (i.e. gill movement and tail oscillations), and survival upon hatching were measured. There was no effect of elevated pCO_2 (400 and 900 µatm, ~80 days of exposure) on any trait (Johnson et al. 2016). Next, we tested the effect of elevated pCO_2 (400, 600, and 900 µatm, ~90 days of exposure) on oxygen uptake (resting $\dot{M}O_2$, as a proxy of basic maintenance costs), hypoxia tolerance (P_{crit}), and a suite of blood- and tissue-based metrics in juvenile epaulette sharks (Heinrich et al. 2014). Sharks had similar P_{crit} and $\dot{M}O_2$ among pCO_2 treatments that were associated with: (1) elevated plasma bicarbonate ion concentrations to buffer against acidosis; (2) increased whole blood and mean corpuscular haemoglobin concentrations to maintain oxygen uptake; and (3) no change in haematocrit, spleen-somatic index, plasma electrolyte concentrations, and tissue citrate synthase activity (Heinrich et al. 2014). Finally, behavioural endpoints were investigated because of evidence suggesting that behavioural impairments stem from interference with the GABA-A inhibitory neurotransmitter receptor or olfactory epithelium (Heuer et al. 2016; Tresguerres and Hamilton 2017; Porteus et al. 2018). Juveniles exposed to elevated pCO_2 (400, 600, and 900 µatm, ~30 days of exposure) did not exhibit differences in foraging behaviours or shelter-seeking behaviours between treatment groups (Heinrich et al. 2016). Overall, epaulette sharks could tolerate static mid- and end-of-century pCO_2 for months with little to no change across a suite of physiological and behavioural metrics.

A second series of experiments tested the effects of elevated temperature on physiological performance and tolerance of epaulette sharks from embryos to adults. The first study tested for differences in survival of sharks *in ovo* reared at an average summer temperature (28°C) and a predicted end-of-century temperature (32°C) (Gervais et al. 2016). The percentage of embryos that survived to hatch was halved at 32°C, and neonates reared under end-of-century temperatures even lacked pigmentation in their distinctive epaulette patterning (Gervais et al. 2016). Moving forward, juveniles reared at 28°C were tested for differences in growth rates, food consumption rates, temperature preference, and survival following ~140 days of acclimation to either 28 or 32°C (Gervais et al. 2018). Juveniles held at 32°C maintained similar food consumption rates to juveniles held at 28°C but had reduced growth rates and 100 per cent mortality by 80 days into acclimation (Gervais et al. 2018). Sharks maintained at 32°C generally preferred higher temperatures, but a time-of-day effect was also apparent (Gervais et al. 2018). Finally, adult epaulette sharks were collected to characterize seasonal differences in thermal tolerance and ventilation rates (Gervais et al. 2018). Thermal tolerance was quantified using critical thermal maximum (CT_{Max}), which is a measure of the highest temperature an animal can tolerate, beyond which the animal cannot serve a functional ecological role and will eventually die (Dabruzzi et al. 2013). Briefly, sharks are acutely heated until a repeatable, non-lethal behavioural endpoint is achieved, thereby defining an individual's upper thermal tolerance limit. This study found that epaulettes have a thermal safety margin to tolerate some warming; CT_{Max} exceeded maximum

winter (~24°C) and summer (~35°C) temperatures (Gervais et al. 2018). Ongoing research by this extended team of researchers into the effects of temperature on physiological performance of epaulette sharks on the GBR aims to investigate development, survival, and acclimatization of embryos, and thermal preference and performance *in situ*.

According to these data, epaulette sharks on the GBR appear well adapted to tolerate two of the three physicochemical challenges of climate change: ocean acidification and deoxygenation. Data suggest that epaulette sharks can tolerate sustained end-of-century pCO_2 conditions without compromising survival (Johnson et al. 2016), predator–prey behaviours (Heinrich et al. 2016), or hypoxia tolerance across this species' early ontogeny (Heinrich et al. 2014). In other words, it's 'business as usual' for epaulette sharks under acidification and hypoxia conditions; although, additional work is needed to characterize sharks' responses to hypoxia and acidification under more natural fluctuating conditions and in combination with other stressors, like warming. Regarding temperature, 4°C of warming drastically reduced embryo, neonate, and juvenile survival (Gervais et al. 2016; Gervais et al. 2018). Furthermore, adult epaulette sharks' upper thermal tolerance limit (~39°C) when held at predicted end-of-century conditions was within several degrees of currently observed summer maximum temperatures (~35°C), although juveniles did exhibit some capacity for using behaviour to select a preferred temperature under summer conditions (Gervais et al. 2018). It remains to be tested whether epaulette sharks are capable of transgenerational acclimation to improve thermal tolerance (Donelson et al. 2018) and whether populations on the GBR are capable of redistribution. Therefore, more work is necessary to determine the extent to which epaulette sharks can maintain their ecological function as mesopredators on the GBR under ocean warming.

In conclusion, the conservation physiology toolbox was employed to identify environmental tolerance limits of epaulette sharks—small-bodied, benthic, tropical mesopredators—under the threat of climate change. The tools used include: bioenergetics and nutritional physiology (growth, body condition, and temperature), cardiorespiratory physiology (oxygen uptake, haematocrit and haemoglobin concentration, muscle enzymes, respiratory rate), and stress physiology (plasma ion concentration, resistance, thermal tolerance) (Madliger et al. 2018). Our data suggest that ocean warming may be a threat to epaulette shark populations on the GBR and can be used in risk assessments to inform species-specific management priorities, especially for this species' northernmost populations (Chin et al. 2010; Gallagher et al. 2012; Fuentes et al. 2016).

11.4 Case study 2: blacktip reef and sicklefin lemon sharks

In this second case study we review an ongoing series of experiments at the core of Physioshark. Here, we discuss the steps taken to establish a research programme focusing on understanding the climate change vulnerability of reef shark populations. In so doing, we describe work done to establish baseline environmental data and quantify performance *in situ* and to experimentally test species' responses to predicted climate change conditions. The research described herein is at the forefront of reef shark science and profits from the success of the epaulette shark case study.

French Polynesia protects all shark species within its 4.7 million km^2 jurisdiction, making it the largest so-called 'shark sanctuary' in the world *c.* 2012 (Andréfouët and Adjeroud 2019). There is apparent high compliance with enforcing regulations and protecting sharks in French Polynesia (Ward-Paige 2017; Ward-Paige and Worm 2017). Among other shark sanctuaries, French Polynesia boasts the largest shark abundance and above-average species richness (Ward-Paige and Worm 2017), including the densest documented aggregation of grey reef sharks (*Carcharhinus amblyrhynchos*) in the world (Mourier et al. 2016). Shark populations are generally considered to be in good health and not in need of further conservation (Ward-Paige and Worm 2017; MacKeracher et al. 2019), yet anthropogenic disturbances related to bycatch (Mourier, Brown and Planes 2017; Bouyoucos et al. 2018b), tourism (Clua et al. 2010; Brena et al. 2015), and marine debris are still considered important threats to sharks in French Polynesia (Ward-Paige and Worm 2017). Climate change is predicted to be a

threat to pearl oysters (*Pinctada margaritifera*), giant clams (*Tridacna maxima*), and corals (only under unabated climate change), but ocean warming and acidification are considered threats of least concern to sharks in French Polynesia (Ward-Paige and Worm 2017; Andréfouët and Adjeroud 2019). However, recent evidence documents reductions in performance and survival of tropical sharks, including epaulette sharks, brown-banded bamboo sharks (*Chiloscyllium punctatum*) (Rosa et al. 2014; Rosa et al. 2016a; Rosa et al. 2016b), and white-spotted bamboo sharks (*Chiloscyllium plagiosum*) (Lopes et al. 2018; Pegado et al. 2018). While these species do not occur in French Polynesia, this information was the only baseline for tropical sharks on which to gauge how those of French Polynesia might respond and catalysed much of the research that commenced. The vulnerabilities of shark populations to climate change are slowly becoming apparent, but a considerable knowledge gap remains regarding responses of tropical shark species, and there are no data for high trophic level species (Rosa et al. 2017; Heupel et al. 2019). The best-protected MPAs are not immune to climate change, and French Polynesia's healthy shark populations offer an excellent system for understanding the isolated effects of climate change stressors on reef sharks.

The purpose of our investigation in French Polynesia has been to understand the capacity for high trophic level predators to support fitness-related processes under climate change stress. A secondary goal has been to use this case study as a model for using the conservation physiology toolbox to improve MPA protections for sharks threatened by climate change. Our study species were the 'near threatened' blacktip reef shark (*Carcharhinus melanopterus*), a large-bodied mesopredator, and the 'vulnerable' sicklefin lemon shark (*Negaprion acutidens*), a large-bodied apex predator (Pillans 2003; Heupel 2009). Unlike the small-bodied, benthic species that have previously been investigated (e.g. epaulette and tropical bamboo sharks), large-bodied, high trophic level species can be highly mobile and migratory, have higher energy requirements, and can influence lower trophic level populations through top-down consumptive and fear effects (Heupel et al. 2014; Roff et al. 2016). Climate change-driven impairments to the performance of species like blacktip reef sharks or sicklefin lemon sharks could have population- and ecosystem-level outcomes (Nagelkerken and Munday 2016; Rosa et al. 2017). Similar to previously studied tropical species, however, blacktip reef and sicklefin lemon sharks rely on shallow, nearshore habitats within lagoons as parturition grounds, and neonates appear site-attached to these—potentially shark nursery—habitats (Mourier, Buray et al. 2013; Mourier, Mills et al. 2013). Blacktip reef and sicklefin lemon sharks are also site-attached for reproduction (Mourier and Planes 2013; Mourier, Buray et al. 2013), such that it is unclear whether sharks can select new parturition grounds if conditions reduce survival of neonates and juveniles. Therefore, our objectives were to investigate anthropogenic and ecological sources of stress for neonatal and juvenile blacktip reef and sicklefin lemon sharks and to identify key stressors for targeted management.

Populations of neonatal and juvenile reef sharks have been monitored around Moorea since 2007 by collaborators at the Centre de Recherches Insulaires et Observatoire de l'Environnement (CRIOBE) (Mourier, Mills et al. 2013). Through research efforts by the CRIOBE and Physioshark, a series of ecological and anthropogenic stressors were identified. The breeding populations of blacktip reef and sicklefin lemon sharks are small enough that inbreeding could affect the survival of offspring (Mourier and Planes 2013; Mourier, Buray et al. 2013). Low interannual recapture rates and low juvenile abundance relative to neonate and adult populations suggest that survival of neonates within their first year of life is intrinsically low (Mourier, Mills et al. 2013). Evidence of infrequent foraging success and scarring/injuries suggest that starvation and predation are common sources of mortality (Chin, Mourier and Rummer, 2015; Weideli et al. 2019). Fishing pressure within the lagoon has affected reef fish communities in the same habitats where neonatal sharks occur; efforts to regulate fishing pressure in these habitats with protected areas have not benefited non-target species, though data for sharks are lacking (Thiault et al. 2017, 2019). Bycatch does occur (Mourier, Brown and Planes, 2017; Bouyoucos et al. 2018b), as indicated by retained fishing hooks, although rates have not been properly quantified and sharks quickly expel hooks (I. Bouyoucos, pers. obs.).

The other issue confounding predictions as to the fate of newborn sharks around Moorea is that these sharks are born during austral summer months (Porcher, 2005; Mourier et al. 2013) when water temperatures average ~30°C (Bouyoucos et al. 2018b). Continuous temperature monitoring and opportunistic dissolved oxygen and pH monitoring at shark capture sites have revealed daily summer trends in the abiotic environment (Figure 11.1a–c). During these monitoring periods, temperatures ranged from 26–35°C with an average daily range of 4°C and extreme range of 8°C in shallow, nearshore water in 2016. Oxygen content was found to decline to ~50% O_2 saturation (3.2 mg O_2/l at ~28°C) during the night and early morning (0100–0500); yet, shallow waters experienced oxygen supersaturation (~120% O_2) during the afternoon and evening owing to high primary productivity. Seawater pH was found to remain relatively high (pH = 7.92–8.47), reflecting oxygen supersaturation and a range of low pCO_2 values (~280–550 µatm) for Moorea's lagoon (Comeau et al. 2014; Edmunds and Burgess, 2016). Given this information, we sought to address the current threat of bycatch, and to quantify physiological performance under warming, low oxygen, and high pCO_2/low pH conditions to understand sharks' climate change vulnerability.

The first series of experiments consulted the conservation physiology toolbox's stress physiology 'kit' to investigate sharks' responses to accidental capture (i.e. bycatch) within the lagoon. As mentioned earlier, it was not uncommon to catch a neonate with fresh or healing bite scars or an embedded fishing hook, suggesting that these animals do have some capacity for surviving and outperforming a predation attempt or fishing encounter (Figure 11.2a–b). We first investigated neonates' capacity for wound healing using the closure of sharks' umbilicus as a proxy for closure of a naturally occurring wound (Chin et al. 2015). These data suggest that blacktip reef sharks have a high capacity for wound healing: umbilicus surface area decreased by ~70 per cent within a week, and these data are corroborated by rapid healing observed in adults with considerable injuries (Chin et al. 2015). Next, sharks faced a fishing stressor *in situ* (gillnet capture with air exposure) to evaluate the

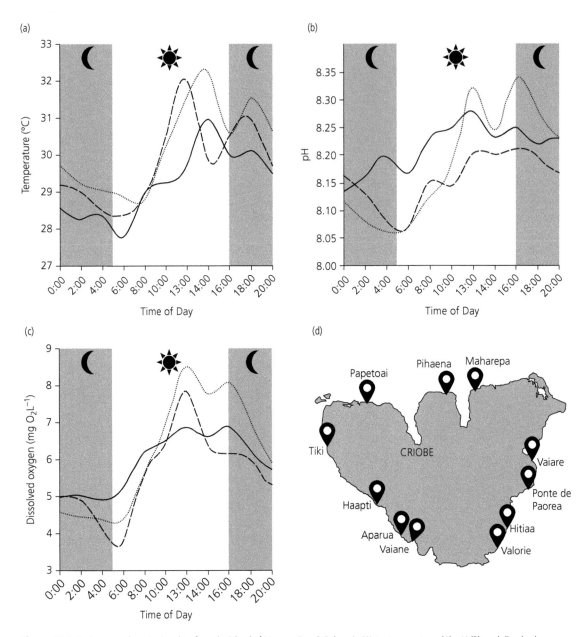

Figure 11.1 Environmental monitoring data from the island of Moorea, French Polynesia. Water temperatures (A), pH (B), and dissolved oxygen concentrations (C) were monitored at 11 locations where newborn reef sharks are found starting in November 2016 through February 2017 (i.e. parturition months). Data were collected at water depths of approximately 50 cm within 10 m of the shoreline, which is where newborn sharks are collected. Values are presented as means in 2-h bins over 24 h for three capture sites (D; Apaura, Pihaena, and Valorie); some of these sites may be defined as shark nursery areas.

physiological status and survival of neonates under stress. Blacktip reef and sicklefin lemon sharks both had increased blood lactate loads and decreased blood pH, relative to minimally stressed reference values, suggesting a reliance on anaerobic metabolism

during capture (Bouyoucos et al. 2018b). Other metrics (i.e. blood glucose, haematocrit, and haemoglobin concentration) were unaffected by stress, but glucose and haemoglobin concentrations increased with temperature in blacktip reef and sicklefin lemon

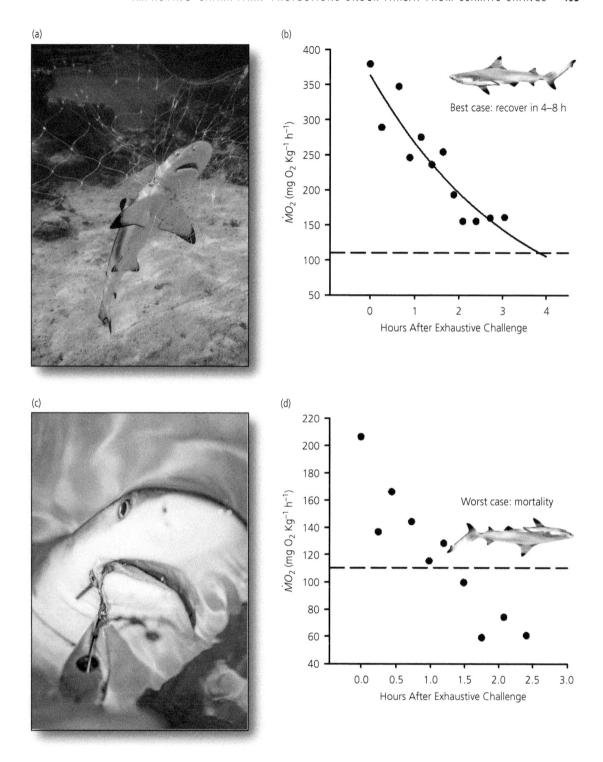

Figure 11.2 Shark bycatch and recovery scenarios for newborn blacktip reef sharks (*Carcharhinus melanopterus*). Around the island of Moorea, French Polynesia, protected reef sharks can be caught accidentally (i.e. bycatch) in nets (A) or on hook-and-line (B). Bycatch interactions initiate a physiological stress response that we assessed using intermittent-flow respirometry *in situ* immediately following capture in a net. We quantified recovery of oxygen uptake rates ($\dot{M}O_2$, a proxy of whole-organism metabolic rate) to a baseline value that was determined in the lab (dashed line in panels C and D). A best-case scenario (C) is full physiological recovery within 4–8 h following net capture. Conversely, a worst-case scenario (D) is delayed mortality. Data are redrawn from Bouyoucos et al. (2018b). Photo credit: Tom Vierus.

sharks, respectively (Bouyoucos et al. 2018b). Both species exhibited delayed mortality; however, blacktip reef sharks had lower (6 per cent) mortality than sicklefin lemon sharks (25 per cent) (Bouyoucos et al. 2018b). Finally, we used respirometry in the field to generate estimates of recovery time for $\dot{M}O_2$ for blacktip reef sharks (Figure 11.2c–d), and estimated that recovery from exhaustive exercise *in situ* could last up to 8.5 h (Bouyoucos et al. 2018b). Taken together, these results suggest that blacktip reef shark neonates are more robust to stress than sicklefin lemon sharks; however, this claim only holds for sharks under stress under the study's 'ideal' temperature conditions (~28–31°C) (Bouyoucos et al. 2018b). Follow-up studies aim to characterize sharks' abilities to cope with stress across a range of environmental conditions.

Examining the overall physiological performance of sharks under variable environmental conditions is an ongoing objective for Physioshark. A similar approach to investigate the effects of ocean acidification in epaulette sharks (Heinrich et al. 2014, 2016) was applied to blacktip reef sharks. Respirometry and blood sampling were used to quantify $\dot{M}O_2$ and underlying physiological status under ambient and end-of-century pCO_2 (~600 and 1000 µatm, respectively) at ambient temperatures (~29°C). Preliminary data suggest that aerobic scope is similar between treatment groups following 30 days of exposure, which is possibly explained by similarities in blood pH, lactate, haematocrit, and haemoglobin concentration between treatment groups (Figure 11.3; Rummer, et al. 2020).

Behavioural responses were also investigated; sharks' behavioural response when exposed to an odour cue (i.e. increased time spent near the odour source and activity level) was the same between treatment groups after 20 days of exposure (Figure 11.4; J. Rummer, unpublished data). Finally, preliminary P_{crit} data suggest that blacktip reef sharks have a higher P_{crit} (~2.6 mg O_2L^{-1} at ~29°C) and are, therefore, less hypoxia-tolerant than epaulette sharks (Heinrich et al. 2014). At a glance, there are similarities among responses of tropical sharks to ocean acidification. Limited physiological responses and no behavioural changes have been documented; although, it remains to be seen how

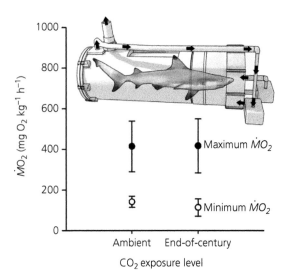

Figure 11.3 Effects of elevated CO_2 exposure level on oxygen uptake rates ($\dot{M}O_2$) of newborn blacktip reef sharks (*Carcharhinus melanopterus*) quantified via intermittent-flow respirometry. Sharks were exposed to ambient (~600 µatm) or end-of-century (~1000 µatm) CO_2 partial pressures (pCO_2) for 30 days. Intermittent-flow respirometry—the respirometry chamber is embedded in the top of the graph—was used to measure sharks' maximum (i.e. upper limit to oxygen uptake) and minimum (i.e. oxygen demand to sustain life) $\dot{M}O_2$. There was no effect of CO_2 exposure. Preliminary data are presented from sharks (ambient n = 8, end-of-century n = 4) tested at 29°C during 2014 (Rummer et al. 2020).

temperature affects reef sharks and how it interacts with ocean acidification conditions (Rosa et al. 2017).

In conclusion, the conservation physiology toolbox was used to characterize sharks' stress tolerance *in situ* and test sharks' responses to climate change stressors. The tools used include: bioenergetics and nutritional physiology (energy expenditure, metabolic rate, plasma glucose, plasma lactate), cardiorespiratory physiology (aerobic scope, haematocrit and haemoglobin concentration, respiratory rate), and stress physiology (blood pH, resistance) (Madliger et al. 2018). These data suggest that exercise stress (e.g. bycatch, during and following predator evasion) is a considerable stressor for blacktip reef and sicklefin lemon sharks, although blacktip reef sharks at least appear to have a

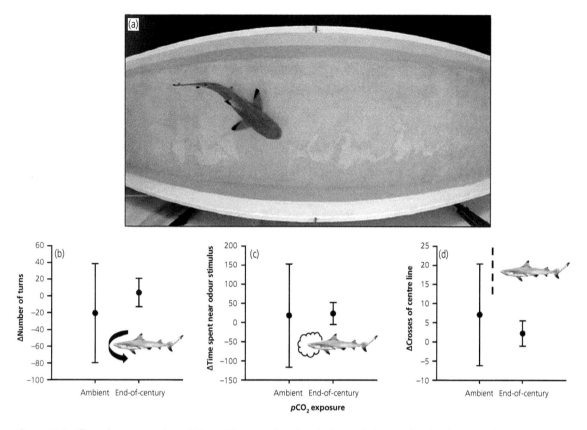

Figure 11.4 Effects of exposure to elevated CO_2 partial pressures (pCO_2) on the foraging behaviour of newborn blacktip reef sharks (*Carcharhinus melanopterus*). Sharks were exposed to ambient (~600 μatm) or end-of-century (~1000 μatm) pCO_2 for 20 days. A raceway tank was used to quantify sharks' behavioural responses to exposure to a prey odour cue (A). The number of times sharks turned (B), the time spent near the odour cue (C), and the number of times sharks crossed the centre line (D) were scored. The differences between behaviours scored during 10 min before and after introducing the odour cue were compared within treatments. There was no effect of CO_2 exposure. Preliminary data are presented from sharks (ambient $n = 8$, end-of-century $n = 4$) tested at 29°C during 2014 (J. Rummer, unpublished data).

remarkable capacity for recovery (Chin et al. 2015; Bouyoucos et al. 2018b). Preliminary data suggest that sharks' stress response was temperature-sensitive and that ocean acidification conditions had little effect on physiological and behavioural traits of blacktip reef sharks. Furthermore, nearshore habitats currently do not become too hypoxic for blacktip reef sharks. Ongoing research by Physioshark into compounding stressors will draw conclusions on these populations' vulnerability to climate change to better prepare French Polynesia to protect biodiversity and its natural resources. These data will be the first of their kind for these Indo-Pacific species that occur in many MPAs across their range. Furthermore, these data are directly relevant to addressing the most recent evaluation of French Polynesia's shark sanctuary that claims ocean warming and acidifica-

tion are considered relatively non-important threats (Ward-Paige and Worm 2017).

11.5 Collaboration towards a global approach

Physioshark is only one of a series of research programmes dedicated to generating conservation outcomes for elasmobranch fishes by applying the conservation physiology toolbox. Other research teams around the world have taken a physiological approach to understanding conservation problems for sharks, including understanding shark declines owing to bycatch and the impact of climate change. For instance, a team based in Australia has done tremendous work characterizing physiological responses in a diversity of shark, ray, and chimaera species to a diversity of gear types (Frick et al. 2010; Heard et al. 2014; Dapp et al. 2017; Martins et al. 2018), validating new physiological and behavioural metrics to measure stress (Van Rijn and Reina 2010; Guida et al. 2016; Guida et al. 2017a), and predicting mortality (Dapp et al. 2016a; Dapp et al. 2016b; Dapp et al. 2016c), and were the first to document sublethal reproductive consequences of capture for sharks and rays (Guida et al. 2017b). In addition, collaboration between groups based in the United States, UK, and The Bahamas has contributed to an understanding of sharks' vulnerabilities to longline (Brooks et al. 2012; Bouyoucos et al. 2017; Bouyoucos et al. 2018a) and drumline capture (Gallagher et al. 2014, 2017; Jerome et al. 2018), suggested alternative methods and techniques to reduce stress on sharks used in research (Brooks et al. 2011a; Brooks et al. 2011b; Sloman et al. 2019), and provided some of the first data on the physiological responses of deep-sea sharks—a group that has recently been exploited by fisheries—to capture (Brooks et al. 2015; Talwar et al. 2017b). Researchers have even contributed to validating veterinary and food industry pH meters and a haemoglobin point-of-care device for use in elasmobranch stress physiology (Harter et al. 2015; Talwar et al. 2017a; Schwieterman et al. 2019a).

The way in which shark stress physiology research has contributed to conservation physiology is well established, yet climate change-driven research has recently begun to emerge *c.* 2014. Much of this work has been predominantly driven by a research group from Portugal (Rosa et al. 2017). This group has contributed to an understanding of the effects of ocean warming and acidification on the embryonic development and performance in juveniles for two tropical benthic sharks (Rosa et al. 2017). Specifically, this group has identified responses to climate change in the form of oxidative damage (Rosa et al. 2016a; Lopes et al. 2018), digestive impairment (Rosa et al. 2016b), and overall reductions in fitness and survival (Rosa et al. 2014; Pegado et al. 2018). Other significant contributions to the growing field come from a team based in the United States that took the first steps towards characterizing physiological responses of embryonic and juvenile rays to ocean acidification (Lauder and Di Santo 2015; Di Santo 2019), warming (Di

Santo 2015, 2016), and deoxygenation (Di Santo et al. 2016). Physioshark's focus on bycatch and climate change as conservation issues is not meant to undervalue the considerable research effort of others who have applied the conservation physiology toolbox to sharks and rays (Madliger et al. 2016, 2018; Illing and Rummer 2017). Indeed, Physioshark would not have been possible without collaboration with research teams in the United States, Australia, and France. Thus, it is our hope that continued collaborative efforts of projects like Physioshark ultimately contribute significantly to shark and ray conservation.

11.6 Education and outreach

Public outreach and education about the conservation issues facing sharks is an important component of Physioshark. On the ground in French Polynesia, our team visits primary and secondary schools and receives school groups at the CRIOBE laboratory on Moorea to give presentations on the value of sharks in French Polynesia, the threat of climate change, and the significance of our work to achieving conservation outcomes. Physioshark has also prioritized opportunities to present research at seminar series hosted by the CRIOBE that are open to the public, including France's *Fête de la Science* (Science Celebration) and the International Year of the Reef. Cooperation with respectful media outlets has resulted in numerous print publications and several televised specials broadcast on local and international television networks (e.g. ABC, Polynésie Première), and the ability to interact with media companies (e.g. Disney Nature, Discovery Channel, and National Geographic) for consultation on sharks in French Polynesia. We employ a similar approach in Australia, prioritizing press releases for key papers, appearing on local and national radio, and presenting to diverse audiences at public events.

Physioshark's reach has dramatically expanded, however, largely through use of social media to disseminate information on the programme's key concepts, publications, presentations, media, and successes. Physioshark began on social media using Facebook and Instagram and via the search term #physioshark. During the first two seasons operating on social media, Physioshark's following (56 per cent female, 42 per cent male, 2 per cent other, predominantly ages 25–34, English-speaking, with a bit of French) on Facebook and Instagram steadily increased (Figure 11.5). During our first year using social media, total Facebook impressions exceeded 53 000, with an average of nearly 400 impressions per day (Figure 11.5). This metric represents the number of times our page entered a person's screen. On Instagram, we uploaded 95 posts during this time period, which attracted nearly 12 000 clicks, averaging 125 clicks per post (Figure 11.5). Our top video during our first year using social media, with over 1600 views on Facebook and nearly 500 views on Instagram, was regarding laboratory temperature tolerance tests with juvenile blacktip reef sharks. Our top photo on Instagram was regarding sharks' umbilical scars (i.e. their belly buttons). During our second season employing social media (October 2018 through March 2019), our top video on Facebook reached nearly 19 000 people, resulted

in nearly 27 000 impressions, and was regarding our 8-min baby shark documentary that debuted in New York City at the Wildlife Conservation Film Festival and then again at several other international film festivals (Figure 11.5). Clearly, presence on social media has extended the reach of our research and made the questions and answers—our findings—more accessible to the general public. We have continued to upload content to social media while research is ongoing; the 'reach' of social media posts is dramatically higher when content (photographs or videos) is current and of high quality, but also during the off season to keep our following engaged.

Indeed, storytelling through digital media is an important means for attracting attention to research, and an attempt at exposing the viewer to content that can achieve an emotional connection and, in many instances, can result in support and even action. For Physioshark, posting photos and videos

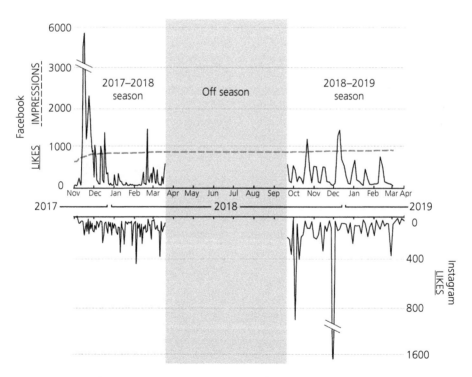

Figure 11.5 Graphical depiction of social media engagement via Facebook (left *y*-axis) and Instagram (right *y*-axis) over two field seasons (2017–2018 and 2018–2019), including an intermediate off-season (April–September 2018) on the *x*-axis. For Facebook, page likes (solid line) and post impressions (the number of times a post appeared on someone's screen; hashed line) are displayed. For Instagram, individual post likes/clicks (solid line) are displayed.

on social media, and engaging in conversations around the topics that are being highlighted as well as the creation of several short documentaries, have also attracted the attention of larger video projects, including a feature-length documentary and television special. Finally, we would be remiss if we did not mention the importance of influential members of the local community, both in French Polynesia and Australia, and their support in promoting Physioshark's research through their channels.

11.7 Conclusions and future directions

In this chapter, we discussed two case studies investigating tropical reef sharks' physiological tolerances to stress and climate change with the conservation physiology toolbox. Two of the study species, the epaulette shark and blacktip reef shark, appear to possess physiological mechanisms to tolerate local hypercapnia and hypoxia without experiencing reductions in physiological performance (Heinrich et al. 2014, 2016; Johnson et al. 2016). Extreme high temperatures appear to reduce survival of epaulette sharks, and ocean warming is predicted to be a threat, at least for this species. Our investigation has also revealed that bycatch interactions can affect the survival of neonatal blacktip reef and sicklefin lemon sharks under current environmental conditions (Chin et al. 2015; Bouyoucos et al. 2018b). Identifying these stressors in our study species and systems has been essential to addressing a knowledge gap for many shark and ray species: is climate change a conservation problem for sharks and rays (Rosa et al. 2017)? Physioshark's research in French Polynesia is in its sixth year at the time that this work is being written (*c.* 2019), and research on epaulette sharks in Australia and with collaborators in the United States are ongoing to achieve conservation outcomes for reef sharks.

The conservation physiology toolbox has much to offer in the way of helping Physioshark achieve its desired conservation outcomes (Madliger et al. 2018). Overall, Physioshark relies on the conservation physiology toolbox primarily for bioenergetics and nutritional physiology, cardiorespiratory physiology, and stress physiology. However, a clear extension of this work is incorporating physiological genomics for a deeper mechanistic under-

standing of sharks' responses to climate change. Neurophysiology and sensory biology tools are relevant to identifying mechanisms underlying susceptibility to behavioural change in response to ocean acidification. In general, relating ecophysiology to behaviour would be relevant to predicting habitat suitability, an important consideration when studying species with documented site fidelity that may exhibit natal philopatry (Mourier and Planes 2013; Mourier et al. 2013). For our work in Moorea, utilizing tools to characterize reproductive physiology would be particularly relevant for identifying environmental drivers of population abundance and successful lineages with 'climate change-resistant' phenotypes. Immunology has been used for characterizing fisheries stress in elasmobranch fishes but, to our knowledge, has not yet been applied to sharks and rays in an environmental change context. Finally, toxicology tools can help identify cryptic stressors that shark and ray MPAs may not manage. In addition to the conservation physiology toolbox, Physioshark aims to support research through education and outreach. Indeed, we aim to conduct research with deliverables in mind that are palatable to conservation practitioners (Cooke and O'Connor 2010). It is paramount to the success of Physioshark that the conservation problems we address are understood by the general public.

Acknowledgements

The authors thank Erin Walsh for help with the illustrations used for figures. Funding to both I.A.B. and J.L.R. is from the Australian Research Council Centre of Excellence for Coral Reef Studies.

References

Andréfouët, S. and Adjeroud, M., 2019. French Polynesia. In C. Sheppard, ed. *World Seas: An Environmental Evaluation*, second edition, *Volume Two: The Indian Ocean to the Pacific*, pp. 827–54. Academic Press, London.

Bennet, M.B., Kyne, P.M., and Heupel, M.R., 2015. *Hemiscyllium ocellatum*, epaulette shark, the IUCN Red List of Threatened Species. Available at: http://dx.doi.org/10.2305/IUCN.UK.2015-4.RLTS.T41818A68625284.en

Bond, M. E., Babcock, E., Pikitch, E. et al., 2012. Reef sharks exhibit site-fidelity and higher relative abundance in marine reserves on the Mesoamerican Barrier Reef. *PLoS ONE*, 7(3), e32983. doi:10.1371/journal.pone.0032983

Bouyoucos, I.A., Simpfendorfer, C.A., and Rummer, J.L., 2019. Estimating oxygen uptake rates to understand stress in sharks and rays. *Reviews in Fish Biology and Fisheries*, 29(2), 297–311. doi:10.1007/s11160-019-09553-3

Bouyoucos, I.A., Suski, C., Mandelman, J. et al., 2017. The energetic, physiological, and behavioral response of lemon sharks (*Negaprion brevirostris*) to simulated longline capture. *Comparative Biochemistry and Physiology Part A: Molecular & Integrative Physiology*, 207, pp. 65–72. doi: 10.1016/j.cbpa.2017.02.023

Bouyoucos, I.A., Talwar, B.S., Brooks, E. et al., 2018a. Exercise intensity while hooked is associated with physiological status of longline-captured sharks. *Conservation Physiology*, 6(1), coy074. doi:10.1093/conphys/coy074

Bouyoucos, I.A., Weideli, O.C., Planes, S. et al., 2018b. Dead tired: evaluating the physiological status and survival of neonatal reef sharks under stress. *Conservation Physiology*, 6(1), coy053. doi:10.1093/conphys/coy053

Brena, P.F., Mourier, J., Planes, S., and Clua, E., 2015. Shark and ray provisioning functional insights into behavioral, ecological and physiological responses across multiple scales. *Marine Ecology Progress Series*, 538, 273–83. doi:10.3354/meps11492

Brooks, E.J., Sloman, K.A., Liss, S. et al., 2011a. The stress physiology of extended duration tonic immobility in the juvenile lemon shark, *Negaprion brevirostris* (Poey 1868). *Journal of Experimental Marine Biology and Ecology*, 409(1), 351–60. doi:10.1016/j.jembe.2011.09.017.

Brooks, E.J., Sloman, K.A., Sims, D.W. et al., 2011b. Validating the use of baited remote underwater video surveys for assessing the diversity, distribution and abundance of sharks in the Bahamas. *Endangered Species Research*, 13(3), 231–43. doi:10.3354/esr00331

Brooks, E.J., Mandelman, J., Sloman, K. et al., 2012. The physiological response of the Caribbean reef shark (*Carcharhinus perezi*) to longline capture. *Comparative Biochemistry and Physiology Part A: Molecular & Integrative Physiology*, 162(2), 94–100. doi:10.1016/j.cbpa.2011.04.012

Brooks, E.J., Brooks, A., Williams, S. et al., 2015. First description of deep-water elasmobranch assemblages in the Exuma Sound, The Bahamas. *Deep Sea Research Part II: Topical Studies in Oceanography*, 115, 81–91. doi:10.1016/j.dsr2.2015.01.015

Chin, A., Kyne, P., Walker, T., McAuley, R., 2010. An integrated risk assessment for climate change: analysing the vulnerability of sharks and rays on Australia's Great Barrier Reef. *Global Change Biology*, 16(7), 1936–53. doi:10.1111/j.1365-2486.2009.02128.x

Chin, A., Mourier, J., and Rummer, J.L., 2015. Blacktip reef sharks (*Carcharhinus melanopterus*) show high capacity for wound healing and recovery following injury. *Conservation Physiology*, 3(1), cov062. doi:10.1093/conphys/cov062

Chin, A., Simfendorfer, C., White, W. et al., 2017. Crossing lines: a multidisciplinary framework for assessing connectivity of hammerhead sharks across jurisdictional boundaries. *Scientific Reports*, 7(March), 46061. doi:10.1038/srep46061

Cinner, J.E., Maire, E., Huchery, C. et al., 2018. Gravity of human impacts mediates coral reef conservation gains. *Proceedings of the National Academy of Sciences*, 115(27), E6116–25. doi:10.1073/pnas.1708001115

Clua, E., Buray, N., Legendre, P. et al., 2010. Behavioural response of sicklefin lemon sharks *Negaprion acutidens* to underwater feeding for ecotourism purposes. *Marine Ecology Progress Series*, 414, 257–66. doi:10.3354/meps08746

Comeau, S., Edmunds, P., Spindel, N. et al., 2014. Diel pCO_2 oscillations modulate the response of the coral *Acropora hyacinthus* to ocean acidification. *Marine Ecology Progress Series*, 501, 99–111. doi:10.3354/meps10690

Comte, L. and Olden, J.D., 2017. Climatic vulnerability of the world's freshwater and marine fishes. *Nature Climate Change*, 7(10), 718–22. doi:10.1038/nclimate3382

Cooke, S.J. and O'Connor, C.M., 2010. Making conservation physiology relevant to policy makers and conservation practitioners. *Conservation Letters*, 3(3), 159–66. doi:10.1111/j.1755-263X.2010.00109.x

Cortés, E., 2000. Life history patterns and correlations in sharks, *Reviews in Fisheries Science*, 8(4), 299–344. doi:10.1080/10408340308951115

Crear, D.P., Brill, R., Bushnell, P. et al., 2019. The impacts of warming and hypoxia on the performance of an obligate ram ventilator. *Conservation Physiology*, 7, coz026. doi:10.1093/conphys/coz026

Dabruzzi, T.F., Bennett, W., Rummer, J., and Fangue, N., 2013. Juvenile ribbontail stingray, *Taeniura lymma* (Forsskål, 1775) (Chondrichthyes, Dasyatidae), demonstrate a unique suite of physiological adaptations to survive hyperthermic nursery conditions. *Hydrobiologia*, 701(1), 37–49. doi:10.1007/s10750-012-1249-z

Dapp, D.R., Huveneers, C., Walker, T.I., Drew, M. et al., 2016a. Moving from measuring to predicting bycatch mortality: predicting the capture condition of a longline-caught pelagic shark. *Frontiers in Marine Science*, 2, 126. doi:10.3389/fmars.2015.00126

Dapp, D.R., Huveneers, C., Walker, T.I., Mandelman, J. et al., 2016b. Using logbook data to determine the immediate mortality of blue sharks (*Prionace glauca*) and tiger sharks (*Galeocerdo cuvier*) caught in the commercial U.S. pelagic longline fishery. *Fishery Bulletin*, 115(1), 27–41. doi:10.7755/FB.115.1.3

Dapp, D.R., Walker, T.I., Huveneers, C., Reina, R., 2016c. Respiratory mode and gear type are important determinants of elasmobranch immediate and post-release mortality. *Fish and Fisheries*, 17(2), 507–24. doi:10.1111/faf.12124

Dapp, D.R., Huveneers, C., Walker, T., and Reina, R., 2017. Physiological response and immediate mortality of gill-net-caught blacktip reef sharks (*Carcharhinus melanopterus*). *Marine and Freshwater Research*, 68(9), 1734–40. doi:10.1071/MF16132

Davidson, L.N.K. and Dulvy, N.K., 2017. Global marine protected areas to prevent extinctions. *Nature Ecology & Evolution*, 1(2), 0040. doi:10.1038/s41559-016-0040

Davidson, L.N.K., Krawchuk, M.A., and Dulvy, N.K., 2016. Why have global shark and ray landings declined: Improved management or overfishing? *Fish and Fisheries*, 17(2), 438–58. doi:10.1111/faf.12119

Devaux, J.B.L., Hickey, A.J.R., and Renshaw, G.M.C., 2019. Mitochondrial plasticity in the cerebellum of two anoxia-tolerant sharks: contrasting responses to anoxia/re-oxygenation. *The Journal of Experimental Biology*, 222(6), jeb191353. doi:10.1242/jeb.191353

Di Santo, V., 2015. Ocean acidification exacerbates the impacts of global warming on embryonic little skate, *Leucoraja erinacea* (Mitchill). *Journal of Experimental Marine Biology and Ecology*, 463, 72–8. doi:10.1016/j.jembe.2014.11.006

Di Santo, V., 2016. Intraspecific variation in physiological performance of a benthic elasmobranch challenged by ocean acidification and warming. *The Journal of Experimental Biology*, 219(11), 1725–33. doi:10.1242/jeb.139204

Di Santo, V., 2019. Ocean acidification and warming affect skeletal mineralization in a marine fish. *Proceedings of the Royal Society B: Biological Sciences*, 286(1894), 20182187. doi:10.1098/rspb.2018.2187

Di Santo, V., Tran, A.H., and Svendsen, J.C., 2016. Progressive hypoxia decouples activity and aerobic performance of skate embryos. *Conservation Physiology*, 4(1), cov067. doi:10.1093/conphys/cov067

Donelson, J.M. Salinas, S., Munday, P., and Shama, L., 2018. Transgenerational plasticity and climate change experiments: where do we go from here? *Global Change Biology*, 24(1), 13–34. doi:10.1111/gcb.13903

Donelson, J.M., Sunday, J., Figueira, W. et al., 2019. Understanding interactions between plasticity, adaptation and range shifts in response to marine environmental change. *Philosophical Transactions of the Royal Society B: Biological Sciences*, 374(1768), 20180186. doi:10.1098/rstb.2018.0186

Dulvy, N.K., Fowler, S., Musick, J. et al., 2014. Extinction risk and conservation of the world's sharks and rays. *eLife*, 3, 1–35. doi:10.7554/eLife.00590

Dulvy, N.K., Simfendorfer, C., Davidson, L. et al., 2017. Challenges and priorities in shark and ray conservation. *Current Biology*, 27(11), R565–72. doi:10.1016/j.cub.2017.04.038

Edmunds, P.J. and Burgess, S.C., 2016. Size-dependent physiological responses of the branching coral Pocillopora verrucosa to elevated temperature and P CO_2. *Journal of Experimental Biology*, 219(24), 3896–906. doi:10.1242/jeb.146381

Farrell, A.P., Lee, C., Tierney, K. et al., 2003. Field-based measurements of oxygen uptake and swimming performance with adult Pacific salmon using a mobile respirometer swim tunnel. *Journal of Fish Biology*, 62, 64–84. doi:10.1046/j.0022-1112.2003.00010.x

Frick, L.H., Reina, R.D., and Walker, T.I., 2010. Stress related physiological changes and post-release survival of Port Jackson sharks (*Heterodontus portusjacksoni*) and gummy sharks (*Mustelus antarcticus*) following gill-net and longline capture in captivity. *Journal of Experimental Marine Biology and Ecology*, 385(1–2), 29–37. doi:10.1016/j.jembe.2010.01.013

Frisch, A.J. and Rizzari, J.R., 2019. Parks for sharks: human exclusion areas outperform no-take marine reserves. *Frontiers in Ecology and the Environment*, 17(3), 145–50. doi:10.1002/fee.2003

Fuentes, M., Chambers, L., Chin, A. et al., 2016. Adaptive management of marine mega-fauna in a changing climate. *Mitigation and Adaptation Strategies for Global Change*, 21(2), 209–24. doi:10.1007/s11027-014-9590-3

Gallagher, A.J., Kyne, P.M., and Hammerschlag, N., 2012. Ecological risk assessment and its application to elasmobranch conservation and management. *Journal of Fish Biology*, 80(5), 1727–48. doi:10.1111/j.1095-8649.2012.03235.x

Gallagher, A.J., Serafy, J., Cooke, S., and Hammerschlag, N., 2014. Physiological stress response, reflex impairment, and survival of five sympatric shark species following experimental capture and release. *Marine Ecology Progress Series*, 496, 207–18. doi:10.3354/meps10490

Gallagher, A.J., Staaterman, E., Cooke, S., and Hammerschlag, N. 2017. Behavioural responses to fisheries capture among sharks caught using experimental fishery gear. *Canadian Journal of Fisheries and Aquatic Sciences*, 74(1), 1–7. doi:10.1139/cjfas-2016-0165

Gervais, C., Mourier, J., and Rummer, J.L., 2016. Developing in warm water: irregular colouration and patterns of a neonate elasmobranch. *Marine Biodiversity*, 46(4), 743–4. doi:10.1007/s12526-015-0429-2

Gervais, C.R., Nay, T., Renshaw, G. et al., 2018. Too hot to handle? Using movement to alleviate effects of elevated temperatures in a benthic elasmobranch, *Hemiscyllium ocellatum*. *Marine Biology*, 165(11), 162. doi:10.1007/s00227-018-3427-7

Goto, T., Nishida, K., and Nakaya, K., 1999. Internal morphology and function of paired fins in the epaulette shark, *Hemiscyllium ocellatum*. *Ichthyological Research*, 46(3), 281–7. doi:10.1007/BF02678514

Great Barrier Reef Marine Park Authority, 2019. *Great Barrier Reef Marine Park Regulations*. GBRMPA, Townsville, Australia.

Guida, L., Awruch, C., Walker, T., and Reina, R., 2017a. Prenatal stress from trawl capture affects mothers and neonates: a case study using the southern fiddler ray (*Trygonorrhina dumerilii*). *Scientific Reports*, 7(1), 46300. doi:10.1038/srep46300

Guida, L., Dapp, D.R., Huveneers, C. et al., 2017b. Evaluating time-depth recorders as a tool to measure the behaviour of sharks captured on longlines. *Journal of Experimental Marine Biology and Ecology*, 497, 120–6. doi:10.1016/j.jembe.2017.09.011

Guida, L., Walker, T.I., and Reina, R.D., 2016. The adenylate energy charge as a new and useful indicator of capture stress in chondrichthyans. *Journal of Comparative Physiology B*, 186(2), 193–204. doi:10.1007/s00360-015-0948-y

Habary, A., Johansen, J.L., Nay, T.J. et al., 2017. Adapt, move or die—how will tropical coral reef fishes cope with ocean warming? *Global Change Biology*, 23(2), 566–77. doi:10.1111/gcb.13488

Harter, T.S., Morrison, P.R., Mandelman, J.W. et al., 2015. Validation of the i-STAT system for the analysis of blood gases and acid–base status in juvenile sandbar shark (*Carcharhinus plumbeus*). *Conservation Physiology*, 3(1), cov002. doi:10.1093/conphys/cov002

Hazen, E.L., Jorgensen, S., Rykaczewski, R.R. et al., 2013. Predicted habitat shifts of Pacific top predators in a changing climate. *Nature Climate Change*, 3(3), 234–8. doi:10.1038/nclimate1686

Heard, M., Van Rijn, J.A., Reina, R.D., and Huveneers, C., 2014. Impacts of crowding, trawl duration and air exposure on the physiology of stingarees (family: Urolophidae). *Conservation Physiology*, 2, cou040. doi:10.1093/conphys/cou040.Introduction

Heinrich, D.D.U., Rummer, J.L., Morash, A.J. et al., 2014. A product of its environment: the epaulette shark (*Hemiscyllium ocellatum*) exhibits physiological tolerance to elevated environmental CO_2. *Conservation Physiology*, 2(1), cou047. doi:10.1093/conphys/cou047

Heinrich, D.D.U., Watson, S.A., Rummer, J.L. et al., 2016. Foraging behaviour of the epaulette shark *Hemiscyllium ocellatum* is not affected by elevated CO_2. *ICES Journal of Marine Science: Journal du Conseil*, 73(3), 633–40. doi:10.1093/icesjms/fsv085

Hess, S., Prescott, L.J., Hoey, A.S. et al., 2017. Species-specific impacts of suspended sediments on gill structure and function in coral reef fishes. *Proceedings of the Royal Society B: Biological Sciences*, 284(1866), 20171279. doi:10.1098/rspb.2017.1279

Heuer, R.M., Welch, M.J., Rummer, J.L. et al., 2016. Altered brain ion gradients following compensation for elevated CO_2 are linked to behavioural alterations in a coral reef fish. *Scientific Reports*, 6, 33216. doi:10.1038/srep33216

Heupel, M., 2009. *Carcharhinus melanopterus*, blacktip reef shark. The IUCN Red List of Threatened Species, e. T39375A10219032. Available at: http://dx.doi.org/10.2305/IUCN.UK.2009-2.RLTS.T39375A10219032.en (accessed 6 July 2020).

Heupel, M.R., Knip, D.M., Simpfendorfer, C.A., and Dulvy, N.K., 2014. Sizing up the ecological role of sharks as predators, *Marine Ecology Progress Series*, 495, 291–8. doi:10.3354/meps10597

Heupel, M.R., Papastamatiou, Y.P., Espinoza, M. et al., 2019. Reef shark science – key questions and future directions. *Frontiers in Marine Science*, 6, 1–14. doi:10.3389/fmars.2019.00012

Horodysky, A.Z., Cooke, S.J., Graves, J.E., and Brill, R.W., 2016. Fisheries conservation on the high seas: linking conservation physiology and fisheries ecology for the management of large pelagic fishes. *Conservation Physiology*, 4(1), cov059. doi:10.1093/conphys/cov059

Hughes, T.P., Kerry, J.T., Álvarez-Noriega, M. et al., 2017. Global warming and recurrent mass bleaching of corals. *Nature*, 543(7645), 373–7. doi:10.1038/nature21707

Hughes, T.P., Kerry, J.T., Baird, A.H. et al., 2019. Global warming impairs stock–recruitment dynamics of corals. *Nature*, 568(7752), 387–90. doi:10.1038/s41586-019-1081-y

Illing, B. and Rummer, J.L., 2017. Physiology can contribute to better understanding, management, and conservation of coral reef fishes. *Conservation Physiology*, 5(1), cox005. doi:10.1093/conphys/cox005

Jain-Schlaepfer, S., Fakan, E., Rummer, J.L. et al., 2018. Impact of motorboats on fish embryos depends on engine type. *Conservation Physiology*, 6(1), 1410–093. doi:10.1093/conphys/coy014

Jerome, J.M., Gallagher, A.J., Cooke, S.J., and Hammerschlag, N., 2018. Integrating reflexes with physiological measures to evaluate coastal shark stress response to capture. *ICES Journal of Marine Science*, 75(2), 796–804. doi:10.1093/icesjms/fsx191

Johnson, M.S., Kraver, D.W., Renshaw, G.M.C., and Rummer, J.L., 2016. Will ocean acidification affect the early ontogeny of a tropical oviparous elasmobranch (*Hemiscyllium ocellatum*)? *Conservation Physiology*, 4, 1–11. doi:10.1093/conphys/cow003.Introduction

Lauder, G.V. and Di Santo, V., 2015. Swimming mechanics and energetics of elasmobranch fishes. In R.E. Shadwick, A.P. Farrell, and C.J. Brauner, eds. *Physiology of Elasmobranch Fishes: Structure and Interaction with*

Environment, pp. 219–53. Elsevier, New York. doi:10.1016/B978-0-12-801289-5.00006-7

Lefevre, S., 2016. Are global warming and ocean acidification conspiring against marine ectotherms? A meta-analysis of the respiratory effects of elevated temperature, high CO₂ and their interaction. *Conservation Physiology*, 4(1), cow009. doi:10.1093/conphys/cow009

Lopes, A.R., Sampaio, E., Santos, C. et al., 2018. Absence of cellular damage in tropical newly hatched sharks (*Chiloscyllium plagiosum*) under ocean acidification conditions. *Cell Stress and Chaperones*, 23(5), 837–46. doi:10.1007/s12192-018-0892-3

MacKeracher, T., Diedrich, A., and Simpfendorfer, C.A., 2019. Sharks, rays and marine protected areas: a critical evaluation of current perspectives. *Fish and Fisheries*, 20(2), 255–67. doi:10.1111/faf.12337

Madliger, C.L., Cooke, S.J., Crespi, E.J. et al., 2016. Success stories and emerging themes in conservation physiology. *Conservation Physiology*, 4(1), cov057. doi:10.1093/conphys/cov057

Madliger, C.L., Love, O.P., Hultine, K.R., and Cooke, S.J., 2018. The conservation physiology toolbox: status and opportunities. *Conservation Physiology*, 6(1), coy029. doi:10.1093/conphys/coy029

Martins, C.L., Walker, T.I., and Reina, R.D., 2018. Stress-related physiological changes and post-release survival of elephant fish (*Callorhinchus milii*) after longlining, gillnetting, angling and handling in a controlled setting. *Fisheries Research*, 204, 116–24. doi:10.1016/j.fishres.2018.01.016

McKenzie, D.J., Axelsson, M., Chabot, D. et al., 2016. Conservation physiology of marine fishes: state of the art and prospects for policy. *Conservation Physiology*, 4(1), cow046. doi:10.1093/conphys/cow046

Meinshausen, M., Smith, S.J., Calvin, K. et al., 2011. The RCP greenhouse gas concentrations and their extensions from 1765 to 2300. *Climatic Change*, 109(1), 213–41. doi:10.1007/s10584-011-0156-z

Mochnacz, N.J., Kissinger, B.C., Deslauriers, D. et al., 2017. Development and testing of a simple field-based intermittent-flow respirometry system for riverine fishes. *Conservation Physiology*, 5(1), cox048. doi:10.1093/conphys/cox048

Morash, A.J., Mackellar, S.R.C., Tunnah, L. et al., 2016. Pass the salt: physiological consequences of ecologically relevant hyposmotic exposure in juvenile gummy sharks (*Mustelus antarcticus*) and school sharks (*Galeorhinus galeus*). *Conservation Physiology*, 4(1), cow036. doi:10.1093/conphys/cow036

Mourier, J., Buray, N., Schultz, J. et al., 2013. Genetic network and breeding patterns of a sicklefin lemon shark (*Negaprion acutidens*) population in the Society Islands, French Polynesia. *PLoS ONE*, 8(8), e73899. doi:10.1371/journal.pone.0073899

Mourier, J., Brown, C., and Planes, S., 2017. Learning and robustness to catch-and-release fishing in a shark social network. *Biology Letters*, 13(3), 20160824. doi:10.1098/rsbl.2016.0824

Mourier, J., Maynard, J., Parravicini, V. et al., 2016. Extreme inverted trophic pyramid of reef sharks supported by spawning groupers. *Current Biology*, 26(15), 2011–16. doi:10.1016/j.cub.2016.05.058

Mourier, J., Mills, S.C., and Planes, S., 2013. Population structure, spatial distribution and life-history traits of blacktip reef sharks *Carcharhinus melanopterus*. *Journal of Fish Biology*, 82(3), 979–93. doi:10.1111/jfb.12039

Mourier, J. and Planes, S., 2013. Direct genetic evidence for reproductive philopatry and associated fine-scale migrations in female blacktip reef sharks (*Carcharhinus melanopterus*) in French Polynesia. *Molecular Ecology*, 22(1), pp. 201–14. doi:10.1111/mec.12103

Nagelkerken, I. and Munday, P.L., 2016. Animal behaviour shapes the ecological effects of ocean acidification and warming: moving from individual to community-level responses. *Global Change Biology*, 22(3), 974–89. doi:10.1111/gcb.13167

Nilsson, G.E., Crawley, N., Lunde, I.G., and Munday, P.L., 2009. Elevated temperature reduces the respiratory scope of coral reef fishes. *Global Change Biology*, 15(6), 1405–12. doi:10.1111/j.1365–2486.2008.01767.x

Nilsson, G.E. and Renshaw, G.M.C., 2004. Hypoxic survival strategies in two fishes: extreme anoxia tolerance in the North European crucian carp and natural hypoxic preconditioning in a coral-reef shark. *Journal of Experimental Biology*, 207(18), 3131–9. doi:10.1242/jeb.00979

Oliver, S., Braccini, M., Newman, S.J., and Harvey, E.S., 2015. Global patterns in the bycatch of sharks and rays. *Marine Policy*, 54, 86–97. doi:10.1016/j.marpol.2014.12.017

Payne, N.L., Snelling, E.P., Fitzpatrick, R. et al., 2015. A new method for resolving uncertainty of energy requirements in large water breathers: the 'mega-flume' seagoing swim-tunnel respirometer. *Methods in Ecology and Evolution*, 6(6), 668–77. doi:10.1111/2041-210X.12358

Payne, N.L., Meyer, C.G., Smith, J.A. et al., 2018. Combining abundance and performance data reveals how temperature regulates coastal occurrences and activity of a roaming apex predator. *Global Change Biology*, 24(5), 1884–93. doi:10.1111/gcb.14088

Pegado, M.R., Santos, C., Couto, A. et al., 2018. Reduced impact of ocean acidification on growth and swimming performance of newly hatched tropical sharks (*Chiloscyllium plagiosum*). *Marine and Freshwater Behaviour and Physiology*, 51(6), 347–57. doi:10.1080/10236244.2019.1590120

Pillans, R., 2003. *Negaprion acutidens*, sharptooth lemon shark. The IUCN Red List of Threatened Species, e. T41836A10576957. Available at: http://dx.doi.org/

10.2305/IUCN.UK.2003.RLTS.T41836A10576957.en (accessed 6 July 2020).

Pistevos, J.C.A., Nagelkerken, I., Rossi, T. et al., 2015. Ocean acidification and global warming impair shark hunting behaviour and growth. *Scientific Reports, 5,* 16293. doi:10.1038/srep16293

Porcher, I.F., 2005. On the gestation period of the blackfin reef shark, *Carcharhinus melanopterus,* in waters off Moorea, French Polynesia. *Marine Biology, 146*(6), 1207–11. doi:10.1007/s00227-004-1518-0

Porteus, C.S., Hubbard, P.C., Uren Webster, T.M. et al., 2018. Near-future CO_2 levels impair the olfactory system of a marine fish, *Nature Climate Change, 8*(8), 737–43. doi:10.1038/s41558-018-0224-8

Pratchett, M.S., Bridge, T.C.L., Brodie, J. et al., 2019. Australia's Great Barrier Reef. In C. Sheppard, ed. *World Seas: An Environmental Evaluation,* second edition, *Volume Two: The Indian Ocean to the Pacific,* pp. 333–62. Academic Press, London.

Roff, G., Doropoulos, C., Rogers, A. et al., 2016. The ecological role of sharks on coral reefs. *Trends in Ecology and Evolution, 31*(5), 395–407. doi:10.1016/j.tree.2016.02.014

Rosa, R., Baptista, M., Lopes, V.M. et al., 2014. Early-life exposure to climate change impairs tropical shark survival. *Proceedings of the Royal Society B: Biological Sciences, 281*(1793), 20141738. doi:10.1098/rspb.2014.1738

Rosa, R., Pimentel, M., Galan, J.G. et al., 2016a. Deficit in digestive capabilities of bamboo shark early stages under climate change, *Marine Biology, 163*(3), 60. doi:10.1007/s00227-016-2840-z

Rosa, R., Ricardo Paula, J., Sampaio, E. et al., 2016b. Neuro-oxidative damage and aerobic potential loss of sharks under elevated CO_2 and warming. *Marine Biology, 163*(5), 119. doi:10.1007/s00227-016-2898-7

Rosa, R., Rummer, J.L., and Munday, P.L., 2017. Biological responses of sharks to ocean acidification. *Biology Letters, 13*(3), 20160796. doi:10.1098/rsbl.2016.0796

Routley, M.H., Nilsson, G.E., and Renshaw, G.M.C., 2002. Exposure to hypoxia primes the respiratory and metabolic responses of the epaulette shark to progressive hypoxia. *Comparative Biochemistry and Physiology Part A: Molecular & Integrative Physiology, 131*(2), 313–21. doi:10.1016/S1095-6433(01)00484-6

Rummer, J.L., Couturier, C.S., Stecyk, J.A.W. et al., 2014. Life on the edge: thermal optima for aerobic scope of equatorial reef fishes are close to current day temperatures. *Global Change Biology, 20*(4), 1055–66. doi:10.1111/gcb.12455

Rummer, J.L., Binning, S.A., Roche, D.G., and Johansen, J.L., 2016. Methods matter: considering locomotory mode and respirometry technique when estimating metabolic rates of fishes. *Conservation Physiology, 4*(1), cow008. doi:10.1093/conphys/cow008

Rummer, J.L. and Munday, P.L., 2017. Climate change and the evolution of reef fishes: past and future, *Fish and Fisheries, 18*(1), 22–39. doi:10.1111/faf.12164

Rummer, J.L., Bouyoucos, I.A., Mourier, J., Nakamura, N., Planes, S. (2020) Responses of a coral reef shark acutely exposed to ocean acidification conditions. *Coral Reefs* 39,1215-1220. doi: 10.1007/s00338-020-01972-0

Schlaff, A.M., Heupel, M.R., Udyawer, V., and Simpfendorfer, C.A., 2017. Biological and environmental effects on activity space of a common reef shark on an inshore reef. *Marine Ecology Progress Series, 571,* 169–81. doi:10.3354/meps12107

Schwieterman, G.D., Bouyoucos, I.A., Potgieter, K. et al., 2019a. Analyzing tropical elasmobranch blood samples in the field: blood stability during storage and validation of the HemoCue® haemoglobin analyzer. *Conservation Physiology, 7*(1), coz081.

Schwieterman, G.D., Crear, D.P., Anderson, B.N. et al., 2019b. Combined effects of acute temperature change and elevated pCO_2 on the metabolic rates and hypoxia tolerances of clearnose skate (*Rostaraja eglanteria*), summer flounder (*Paralichthys dentatus*), and thorny skate (*Amblyraja radiata*). *Biology, 8*(3), 56. doi:10.3390/biology8030056

Sequeira, A.M.M., Mellin, C., Fordham, D.A. et al., 2014. Predicting current and future global distributions of whale sharks, *Global Change Biology, 20*(3), 778–89. doi:10.1111/gcb.12343

Shaw, E.C., McNeil, B.I., and Tilbrook, B., 2012. Impacts of ocean acidification in naturally variable coral reef flat ecosystems. *Journal of Geophysical Research: Oceans, 117*(3), 3038. doi:10.1029/2011JC007655

Simpfendorfer, C.A. and Dulvy, N.K., 2017. Bright spots of sustainable shark fishing, *Current Biology, 27*(3), R97–8. doi:10.1016/j.cub.2016.12.017

Sloman, K.A., Bouyoucos, I.A., Brooks, E.J., and Sneddon, L.U., 2019. Ethical considerations in fish research, *Journal of Fish Biology. 94,* 61–79. doi:10.1111/jfb.13946

Speed, C.W., Cappo, M., and Meekan, M.G., 2018. Evidence for rapid recovery of shark populations within a coral reef marine protected area. *Biological Conservation, 220,* 308–19. doi:10.1016/j.biocon.2018.01.010

Stein, R.W., Mull, C.G., Kuhn, T.S. et al., 2018. Global priorities for conserving the evolutionary history of sharks, rays and chimaeras. *Nature Ecology & Evolution, 2*(2), 288–98. doi:10.1038/s41559-017-0448-4

Sunday, J.M., Bates, A.E., and Dulvy, N.K., 2012. Thermal tolerance and the global redistribution of animals. *Nature Climate Change, 2*(9), 686–90. doi:10.1038/nclimate1539

Svendsen, M.B.S., Bushnell, P.G., and Steffensen, J.F., 2016. Design and setup of intermittent-flow respirometry system for aquatic organisms. *Journal of Fish Biology*, *88*(1), 26–50. doi:10.1111/jfb.12797

Talwar, B., Bouyoucos, I.A., Shipley, O. et al., 2017a. Validation of a portable, waterproof blood pH analyser for elasmobranchs. *Conservation Physiology*, *5*(1), cox012. doi:10.1093/conphys/cox012

Talwar, B., Brooks, E.J., Mandelman, J.W., and Grubbs, R.D., 2017b. Stress, post-release mortality, and recovery of commonly discarded deep-sea sharks caught on longlines. *Marine Ecology Progress Series*, *582*, 147–61. doi:10.3354/meps12334

Thiault, L., Collin, A., Chlous, F. et al., 2017. Combining participatory and socioeconomic approaches to map fishing effort in small-scale fisheries. *PLoS ONE*, *12*(5), e0176862. doi:10.1371/journal.pone.0176862

Thiault, L., Kernaléguen, L., Osenberg, C.W. et al., 2019. Ecological evaluation of a marine protected area network: a progressive-change BACIPS approach. *Ecosphere*, *10*(2), e02576. doi:10.1002/ecs2.2576

Tresguerres, M. and Hamilton, T.J., 2017. Acid–base physiology, neurobiology and behaviour in relation to CO_2-induced ocean acidification. *Journal of Experimental Biology*, *220*(12), 2136–48. doi:10.1242/jeb.144113

Tunnah, L., MacKellar, S.R.C., Barnett, D.A. et al., 2016. Physiological responses to hypersalinity correspond to nursery ground usage in two inshore shark species (*Mustelus antarcticus* and *Galeorhinus galeus*). *The Journal of Experimental Biology*, *219*(13), 2028–38. doi:10.1242/jeb.139964

Van Rijn, J.A. and Reina, R.D., 2010. Distribution of leukocytes as indicators of stress in the Australian swellshark, *Cephaloscyllium laticeps*. *Fish & Shellfish Immunology*, *29*(3), 534–8. doi:10.1016/j.fsi.2010.04.016

Ward-Paige, C.A., 2017. A global overview of shark sanctuary regulations and their impact on shark fisheries. *Marine Policy*, *82*, 87–97. doi:10.1016/j.marpol.2017.05.004

Ward-Paige, C.A. and Worm, B., 2017. Global evaluation of shark sanctuaries. *Global Environmental Change*, *47*, 174–89. doi:10.1016/j.gloenvcha.2017.09.005

Weideli, O.C., Bouyoucos, I.A., Papastamatiou, Y.P. et al., 2019. Same species, different prerequisites: investigating body condition and foraging success in young reef sharks between an atoll and an island system. *Scientific Reports*, *9*(1), 13447. doi:10.1038/s41598-019-49761-2

White, T.D., Carlisle, A.B., Kroodsma, D.A. et al., 2017. Assessing the effectiveness of a large marine protected area for reef shark conservation. *Biological Conservation*, *207*, 64–71. doi:10.1016/j.biocon.2017.01.009

Wise, G., Mulvey, J.M., and Renshaw, G.M.C., 1998. Hypoxia tolerance in the epaulette shark (*Hemiscyllium ocellatum*). *Journal of Experimental Zoology*, *281*(June), 1–5. doi:10.1002/(SICI)1097-010X(19980501)281:1<1::AID-JEZ1>3.0.CO;2-S

Worm, B., Davis, B., Kettemer, L. et al., 2013. Global catches, exploitation rates, and rebuilding options for sharks. *Marine Policy*, *40*, 194–204. doi:10.1016/j.marpol.2012.12.034

Yates, P.M., Heupel, M.R., Tobin, A.J., and Simpfendorfer, C.A., 2015. Ecological drivers of shark distributions along a tropical coastline. *PLoS ONE*, *10*(4), e0121346. doi:10.1371/journal.pone.0121346

A tale of two whales: putting physiological tools to work for North Atlantic and southern right whales

Kathleen E. Hunt, Alejandro Fernández Ajó, Carley Lowe, Elizabeth A. Burgess, and C. Loren Buck

---

⊃ **Take-home message**

Creative use of non-traditional physiological tools can help discriminate acute from chronic stress, identify causes of mortality, and track changes in reproduction, even in taxa that are challenging to study in the wild.

12.1 Introduction

Ocean environments are increasingly impacted by anthropogenic activities (McCauley et al. 2015; Schipper et al. 2008; Avila et al. 2018) and mitigating these threats requires understanding adverse effects on individuals and their populations (Cooke et al. 2013). Large whales, for example, face a mix of anthropogenic impacts that can include collisions with vessels, entanglement in fishing gear, acoustic exposure (from seismic exploration, commercial ship traffic, military sonar, etc.), exposure to a variety of toxins and pollutants, and increasing effects of global climate change (Clapham 2016; Thomas et al. 2016; Tulloch et al. 2019). Conservation-relevant impacts of these threats on whale stress, reproduction, health, and survival have been difficult to assess, partly because traditional physiological tools are largely impractical or impossible to employ on large cetaceans. These huge animals are difficult to follow, range over vast geographic areas, cannot be safely captured alive for sampling, and have protracted reproductive cycles and long lifespans that may exceed those of the researchers. Blood samples cannot presently be collected from free-swimming whales, and even affixing a tag to a specific individual remains challenging (and for some populations is not legally permitted) (Szesciorka et al. 2016). Compounding the problem, many whale populations are presently subjected to multiple and complex anthropogenic and environmental stressors that overlap in time and space, with impacts unfolding gradually over months or years (Schipper et al. 2008; McCauley et al. 2015; Avila et al. 2018). These numerous and diverse stressors (of fishing gear entanglement, wounds from vessel strikes, etc.) could each have gradually accruing detrimental effects on individual whales, with the physiological toll not necessarily apparent until years or even decades after the initial impact(s). Overall, assessing physiological effects of anthropogenic impacts on the large whales is a challenging endeavour.

Kathleen E. Hunt, Alejandro Fernández Ajó, Carley Lowe, Elizabeth A. Burgess, and C. Loren Buck, *A tale of two whales: putting physiological tools to work for North Atlantic and southern right whales* In: *Conservation Physiology: Applications for Wildlife Conservation and Management.* Edited by: Christine L. Madliger, Craig E. Franklin, Oliver P. Love, and Steven J. Cooke, Oxford University Press (2021). © Oxford University Press. DOI: 10.1093/oso/9780198843610.003.0012

Fortunately, a growing array of innovative physiological methods can deliver vital data for whale conservation. As we outline below, physiological tools may help discriminate acute from chronic stress prior to mortality, can help identify specific stressors, and could identify and track early changes in reproductive rate. In some cases, physiological data may detect such changes before birth or death rates are notably affected, identifying sublethal impacts that could precede population consequences by months or years. Nonetheless, the effort to adapt physiological tools for whale conservation has required substantial investment in developing creative and novel approaches. Considerable ingenuity has been used to elucidate physiological patterns in the absence of any physical sample from the animal, using, for example, visualization of the animal's body condition (visual health assessment, photogrammetry, ultrasound measurement of blubber thickness), behavioural data (sensory physiology, diving physiology), and bioenergetics (modelling drag and buoyancy, estimates of lactational transfer of energy from mother to young). Additionally, several physiological analyses, particularly endocrine assays, have been adapted to quantify analytes of interest in 'alternative' sample types such as faeces, respiratory vapour, blubber samples, baleen, and earplugs (i.e. sample types other than blood). To date, these conservation physiology tools for large cetaceans have been most thoroughly employed in right whales (*Eubalaena* spp.), with two well-studied species providing an instructive comparison. With these species as models, we aim to provide examples of how perseverance with evolving techniques can make headway in gaining conservation-relevant physiological knowledge of a particularly challenging taxon.

12.2 A brief introduction to the right whales

Right whales comprise three species within the genus *Eubalaena*: the North Atlantic right whale (*Eubalaena glacialis*, 'NARW'), the southern right whale (*Eubalaena australis*, 'SRW'), and the poorly understood North Pacific right whale (*Eubalaena japonica*, 'NPRW') (Rosenbaum et al. 2000). As their names suggest, the three species have non-overlapping distributions, with the NARW and NPRW restricted to their respective northern ocean basins, whereas the SRW occurs only in the southern hemisphere (Figure 12.1; Rosenbaum et al. 2000; Harcourt et al. 2019). All species of right whales differ genetically but are morphologically similar, and share several life-history traits such as long gestation lengths (at least 12–13 months; Best 1994), long lifespans (estimated at 100 years; Hamilton et al. 1998), delayed age of sexual maturity (age at first parturition is 8–9 years; Hamilton et al. 1998; Best et al. 2001), a typical reproductive interval of 3 years between calves ('inter-calving interval'; Burnell 2001; Kraus et al. 2001; International Whaling Commission 2012), and annual long-distance migrations from summer high-latitude feeding grounds to winter lower-latitude calving grounds (Knowlton et al. 1992; Best et al. 1993; Zerbini et al. 2016, 2018; Kenney 2018; Harcourt et al. 2019). These life-history traits make the right whales especially vulnerable to extrinsic mortality and rapid environmental change (Fowler 1981; Crespo and Hall 2002).

Historically, the common name 'right whale' has been sadly appropriate, referring to the fact that these whales were the 'right' ones to hunt. For centuries, right whales were a preferred target for whaling, as they are a relatively slow-moving, coastal whale that floated after being killed and yielded long baleen plates as well as high-quality whale oil. In the North Atlantic, whaling started as early as 1000 CE with 'shore whaling' of NARW by Basque whalers (Reeves and Smith 2006). Once the calving populations of near-shore waters off Europe were depleted, offshore whaling began in earnest. Whalers of multiple nations (British, French, Norwegian, Portuguese, and Dutch, among others), targeted all three species of right whales around the world, along with the related bowhead whale (*Balaena mysticetus*) (Reeves and Smith 2006). By the early 20th century, all populations of right whales had been reduced to less than 1 per cent of their historical numbers (Reeves and

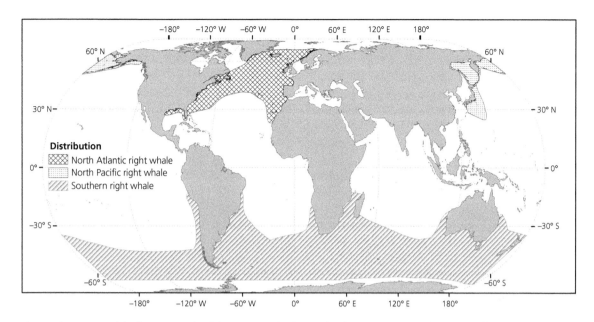

Figure 12.1 Ranges of the three extant right whale species: North Atlantic right whale (*Eubalaena glacialis*), North Pacific right whale (*Eubalaena japonica*), and southern right whale (*Eubalaena australis*). Ranges shown here are based on historic patterns that may be changing. Map courtesy of Dr Logan Berner, Northern Arizona University; data used with permission of the IUCN Red List of Threatened Species, 2019.

Smith 2006; Corkeron et al. 2018; Harcourt et al. 2019). Pre-whaling abundance of NARW is a subject of debate, but recent studies estimate a historic population size of between 9000 and 22 000 individuals in the North Atlantic (Monsarrat et al. 2015). By 1900, probably only a few dozen NARW remained (Katona and Kraus 1999; Harcourt et al. 2019). The SRW, similarly, was reduced from an initial population size of 60 000–100 000 to ~400 individuals worldwide by 1920 (Harcourt et al. 2019). Little data exist on the NPRW, which will not be discussed further here, but it is thought to have suffered a similarly severe population bottleneck (Harcourt et al. 2019).

Commercial hunting of all right whale species ceased in 1935 with the ratification of the Convention for the Regulation of Whaling, which specifically protected right whales (Gambell 1993). Though some illegal or 'pirate' whaling of right whales did occur past 1935 (especially by the Soviet Union in the 1960s and 1970s, impacting the Argentina/Brazil population of SRW in particular; Tormosov et al. 1998), many right whale populations began to recover. However, today SRW and NARW populations offer an intriguing contrast. The global SRW population size was last estimated at 12 000–15 000 in 2012 (International Whaling Commission 2012; Cooke and Zerbini 2018). Though still well below the SRW's historic population size, this represents a considerable recovery from the 1920 population nadir for this species. The NARW population, however, currently comprises fewer than 500 individuals (Pace et al. 2017). The SRW is thus often described as the 'success story' of the right whales, whereas the NARW, in contrast, remains on the edge of extinction and is classed as critically endangered (Cooke 2018; Harcourt et al. 2019). These two right whale species are illustrative examples of similar species on different population trajectories due to diverse anthropogenic and environmental pressures. This chapter addresses whether physiological tools have been useful for conservation of either species with an emphasis on how resulting data can inform conservation, welfare, and policy.

12.3 Case study 1: North Atlantic right whales—a species on the brink

12.3.1 Background and current status

A long-term, collaborative effort among multiple US and Canadian research teams has supported a sustained research focus on the NARW since the 1970s (Kraus and Rolland 2007). Intensive annual surveys of the single known calving ground (southeastern United States), migration route, and spring and summer feeding areas (e.g. Bay of Fundy, Cape Cod Bay) have resulted in detailed information on reproductive rate, survival, mortality (inferred by disappearance of individuals for multiple years), and habitat use (Brown et al. 2007; Kraus and Rolland 2007). These studies have revealed that the NARW population growth rate has been erratic in recent decades. After a period of slow, but steady, population recovery in the 1970s–1980s, NARW population growth stalled in the mid-1990s, recovered in the 2000s, and most recently has gone into decline again during the 2010s (Meyer-Gutbrod and Greene 2017; Harcourt et al. 2019).

Population models indicate that a primary factor constraining the recovery of the NARW population is anthropogenic mortality (Corkeron et al. 2018). Efforts to necropsy all available carcasses in both the United States and Canada have identified two predominant causes of mortality: entanglement in fishing gear and collisions with large vessels ('ship strikes'; Moore et al. 2007; Sharp et al. 2019; Figure 12.2). Of the two, entanglement is now the greatest source of mortality, responsible for 85 per cent of diagnosed mortalities from 2010–2015 (Knowlton et al. 2012; Pettis and Hamilton 2015; Hayes et al. 2017; Corkeron et al. 2018; Sharp et al. 2019). As of 2017, 86 per cent of NARW individuals bore the characteristic linear white scars indicating at least one entanglement, 88 per cent of females have been entangled more than once, and each year 27 per cent of individuals acquire scars indicating new entanglement(s) (Knowlton et al. 2017).

Reproductive rate is low for NARW, which further compromises the recovery of this species (Reeves et al. 2000; Kraus et al. 2016; Pace et al. 2017;

Figure 12.2 Examples of common sources of stress and mortality in right whales. Top, fishing gear entanglement visible on carcass of an 11-year-old female North Atlantic right whale, Eg#3603 'Starboard'; this whale's name references a prior propeller injury that severed the right tip of the flukes, visible at top of photo (carcass is ventral side up) (photo © Peter Duley/NOAA Northeast Fisheries Science Center, NOAA permit #17355). Middle, propeller wounds from a lethal ship strike event on a 2-year-old male North Atlantic right whale, Eg#3508 (photo © Florida Fish and Wildlife Conservation Commission, NOAA permit #932-1905-MA-009526). Bottom, kelp gull (*Larus dominicanus*) parasitizing, harassing, and wounding a southern right whale, with inset showing close-up of a gull-inflicted wound; this interaction is unique to the southern right whale population off Argentina (photo © Diego Taboada/Instituto de Conservación de Ballenas de Argentina; inset, © Paula Faiferman/Instituto de Conservación de Ballenas de Argentina).

Pettis and Hamilton 2018). Healthy females of all *Eubalaena* spp. are thought to follow a 3- to 4-year reproductive cycle or 'inter-calving interval', assuming sufficient nutritional reserves: pregnancy lasts 1 year, followed by lactation for most of the second year, and a 'resting' phase in the third year (and sometimes, in NARW, a fourth year) during which the female replenishes blubber stores (Knowlton et al. 1994; Kraus et al. 2007; Kenney 2018). In the late 1990s, however, mean NARW inter-calving interval shifted from 3.67 to 5.3 years (Kraus et al. 2001). Extensive discussion of possible causes identified food limitation and/or infectious disease as potential factors, among others (Knowlton et al. 1994, 2012; Kraus et al. 2001). Calving rates recovered in the 2000s, but in the late 2010s the population again showed low calf counts and indications of prolonged inter-calving interval (i.e. depressed reproductive rate), again for unclear reasons (Kraus et al. 2016; Pettis and Hamilton 2017, 2018). This modern period of poor reproduction coincided with partial abandonment of traditional summer feeding grounds in the Bay of Fundy (Davis et al. 2017; Davies et al. 2019).

Thus, the recent NARW population decline appears largely a result of high rates of entanglement and ship strikes, combined with low and erratic rates of reproduction. However, we still know little regarding why exactly the whales encounter fishing gear and vessels so often, nor why reproductive rate is so variable. Here physiological information could be of great utility. Reproductive physiology studies could help clarify the exact nature of reproductive failure, which could be occurring at various points during the reproductive cycle—lack of conception, loss of the foetus, high calf mortality—each of which could imply problems in very different months and locations, potentially requiring quite different management solutions. Sensory physiology studies, as well, could help clarify whether NARW are able to see fishing gear or detect approaching vessels. Finally, even if a whale survives a given anthropogenic impact (entanglement, vessel strike), it is not clear what the potential long-term sublethal effects may be. If a whale survives an entanglement, what are the physiological costs? Many entanglements are not fatal and are eventually shed. Some surviving NARW may be entangled for months or years, towing the heavy fishing gear north and south along their migration route, often with severe wounding as the lines lacerate flippers, flukes, or blowholes (Moore and Van der Hoop 2012; Knowlton et al. 2015). In many cases, the fishing line extends

through the baleen and presumably affects foraging; in some cases, the mouth may be essentially tied shut. Even if a whale eventually sheds such an entanglement, the associated energetic burden and physical toll of the injuries could cause long-term physiological effects such as impaired reproduction, immunity, and health (Moore and van der Hoop 2012; van der Hoop et al. 2017; Lysiak et al. 2018). Similarly, ship-struck individuals can sustain large wounds from propeller strikes, and may even lose parts of their flukes or flippers (Sharp et al. 2019). Such injuries surely require extended healing times and may impact swimming efficiency afterwards, but potential long-term physiological effects remain unclear. Assessing variability in stress and energetic physiology may offer insight into these conservation-relevant questions.

To try to answer these questions, physiology research on NARW has proceeded on multiple fronts. Since the late 1970s, methods for stress assessment, health assessment (viewed here as a component of physiology), energetic physiology, sensory physiology, and reproductive physiology have all been under active investigation, with multiple approaches pursued simultaneously by several different research teams (Kraus and Rolland 2007). In the late 1990s, an effort towards adapting physiological tools for right whales received renewed attention due to the sharp decline in NARW calving (Reeves et al. 2000), and formed part of a multi-pronged, inter-disciplinary effort to elucidate not only the precise causes, but also possible long-term consequences of the shifting trends in mortality, non-lethal injury, and reduced reproductive output. Here we give a brief overview of some of these approaches. We do not intend this as an exhaustive review of NARW physiological research (some major fields of inquiry are not addressed here—e.g. diving physiology, bioacoustics, ecotoxicology), but only as an introduction to some of the many creative approaches that have been devised to study physiology in right whales.

12.3.2 Photo-identification

Any discussion of physiological research on right whales must mention, at the outset, the immense value of an extensive long-term photo-identification catalogue for NARW, the North Atlantic Right Whale Identification and Sightings Databases (Hamilton et al. 2007; North Atlantic Right Whale Consortium 2019). This catalogue has underpinned NARW physiological research to such an extent that we regard it here as an essential part of the whale physiology toolkit. In brief, right whales are individually identifiable via unique patterns of callosities on the head (and, sometimes, via scars from entanglements and ship strikes; Payne et al. 1983; Hamilton et al. 2007). Considerable survey effort is expended each year to attempt to photograph every NARW possible, whether on calving grounds (south-eastern United States along Georgia and northern Florida coasts), summer foraging grounds (e.g. Bay of Fundy, Roseway Basin, and recently, parts of the Gulf of St. Lawrence), or along migration routes and feeding areas in between (e.g. Cape Cod Bay in spring months). Through a cooperative data-sharing arrangement (North Atlantic Right Whale Consortium, www.narwc.org), photographs from different research teams are unified in a single database, matched to known individuals, and ultimately compiled into individual sightings histories that are made available to all participating researchers (Hamilton et al. 2007; North Atlantic Right Whale Consortium 2019). The NARW catalogue began in 1980 and now totals over 1 000 000 whale images from over 75 000 sighting events (Hamilton et al. 2007; Pettis and Hamilton 2018). As a result, the large majority of NARW are known individually, with detailed sightings records that span decades in some cases, documenting calving history, lactation, appearance of scars indicative of injury or entanglement, minimum age (or, for whales first sighted as calves, exact age), sex, habitat-use patterns, etc. (Hamilton et al. 2007). Such a photo-identification catalogue has proven invaluable, not just to elucidate patterns in life history and changing rates of reproduction, but also for validation of physiological methods using whales of known physiological state. For example, many endocrine assays presently in use for whales use non-traditional sample types including faeces, respiratory vapour, and baleen. Most of these assays were first tested in NARW by using the catalogue of well-known individuals to verify that measured hormones

reflect patterns expected for various physiological states, for example, comparing endocrine data of confirmed pregnant females, known mature males, known-entangled whales, etc. (e.g. Rolland et al. 2005; Hunt et al. 2006, 2016, 2018; Burgess et al. 2017, 2018; Corkeron et al. 2017; Lysiak et al. 2018).

12.3.3 Assessment of body condition

Photographs can also provide information on the physiological state of the whale, especially estimates of body condition and fuel stores (blubber). In cetaceans, changes in the thickness of the blubber layer cause a visible alteration in the contour of the animal. Even a partial view of the animal's blowhole region, captured when the whale surfaces to breathe, contains useful information. A 'visual health assessment' (VHA) method has been devised for right whales that ranks the contour of fat stores just posterior to the blowhole, as well as occurrence of striations ('rake marks') in the skin of emaciated animals, and skin colour (related to presence of certain cyamid parasites) (Pettis et al. 2004). Such VHA scores have been shown to correlate with reproductive state and exposure to anthropogenic stressors (Rolland et al. 2016; Pettis et al. 2017). This method has the advantage of enabling consistent body condition assessment via the same ship-board photography that is already being used for photo-identification; that is, photographs taken to identify an individual whale may also be used to assess some aspects of physiological state. Overhead views, however, when possible, offer additional information. Aerial views of the entire whale can be used to quantify fat stores using 'photogrammetry' techniques to measure girth:length ratios at various defined points along the body. This technique, originally developed for use from aircraft (Best 1994), is now increasingly used via cameras mounted on unmanned aerial systems (UAS) or drones (e.g. Dawson et al. 2017). Photogrammetry girth:length ratios can be used to estimate blubber thickness and hence body fat reserves and can also diagnose pregnancy (Pettis et al. 2004). Finally, some direct assessment of body condition is also possible. Necropsies of stranded carcasses can produce direct data on blubber thickness, whereas pole-mounted ultrasound devices have been used to briefly contact free-swimming whales for direct measurement of blubber thickness of living individuals (Moore et al. 2001; Miller et al. 2011; van der Hoop et al. 2017; Figure 12.3).

These diverse methods of assessing body condition have shown convincingly that right whales are capital breeders (i.e. they accumulate and store blubber reserves in advance of initiating breeding; Miller et al. 2011). Females build fat stores during productive summer feeding in the year before pregnancy (the 'resting' year), and then exhibit rapid thinning of blubber stores—and associated declines in VHA scores—during pregnancy and especially lactation, during which NARW calves grow at an extremely rapid rate (Fortune et al. 2012; Miller et al. 2012; Pettis et al. 2017). These VHA scores have also been shown to decline on a population-wide basis during periods when mean calving rate declines, such as during the mid- to late 1990s, suggesting a potential causal relationship between body condition and subsequent calving rate (Rolland et al. 2016). The inventive use of ultrasound on free-swimming whales has shown that blubber thickness is highest in females in the year immediately prior to pregnancy, suggesting that females may only initiate a pregnancy if they have sufficient nutritional reserves (Miller et al. 2012; Figure 12.3, top). Similarly, calving females have higher VHA scores than non-calving females (Rolland et al. 2016). Importantly, entanglement in fishing gear causes a reduction in these nutritional reserves, apparent both from blubber thickness measurements (Figure 12.3, bottom; van der Hoop et al. 2017) and VHA scores (Pettis et al. 2017). Modelling studies further indicate that entanglement exacts an extremely costly energetic burden similar to that of lactation (Figure 12.3, bottom; van der Hoop et al. 2017). Thus, even if a female whale survives an entanglement, blubber reserves may be so depleted that the whale may require a year or more to replenish fuel stores before it can initiate another pregnancy. Studies also indicate that the periodic declines in NARW calving parallel declines in its principal copepod prey, *Calanus finmarchicus*, on summer feeding grounds (which in turn reflects oceanographic and climatic factors, such as the North Atlantic Oscillation Index and long-term effects of global climate change; Meyer-Gutbrod and Green 2014). All of these sources of physio-

logical data point to nutritional state and blubber reserves as a key factor for successful reproduction in subsequent years.

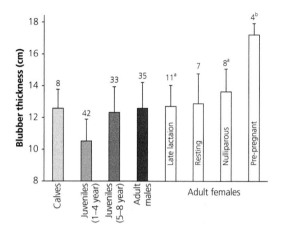

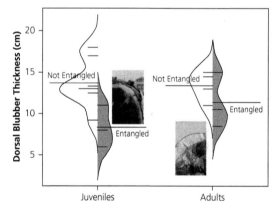

Figure 12.3 Blubber thickness studies on live whales and stranded carcasses can reveal effects of reproductive state and entanglement on nutritional reserves. Top, ultrasound measurements of blubber thickness on free-swimming North Atlantic right whales, divided by age class and reproductive state; females in the year before pregnancy ('pre-pregnant') have greater blubber thickness than other demographic groups (sample sizes shown above bars, significant differences marked with different letters; used with permission from Miller et al. 2011, © Inter-Research 2011). Bottom, dorsal axillary blubber thickness measured directly during necropsies of stranded North Atlantic right whale carcasses, divided by age class; within each age class, non-entanglement cases shown on left and entanglement cases shown on right, with inset photos illustrating typical examples for entangled and non-entangled. Curved vertical lines indicate population distribution, short horizontal lines indicate individual samples, and widest horizontal lines indicate population means; entangled and non-entangled population means are significantly different (used with permission from van der Hoop et al. 2017, © Wiley Online 2017).

12.3.4 Endocrine analyses

12.3.4.1 Faecal hormone analysis

Endocrine studies of NARW began in 1999 with the first attempts to adapt faecal hormone analytic techniques—originally developed for samples from terrestrial wildlife—for floating whale faeces. A series of validation studies (Rolland et al. 2005; Hunt et al. 2006; Corkeron et al. 2017) confirmed that hormone content of right whale faeces does reflect the whales' physiological state: faecal progestins, which are usually associated with successful pregnancy in mammals, are significantly higher in pregnant females (confirmed from calf sightings) compared with non-pregnant females, while faecal androgens (e.g. testosterone) are significantly higher in mature males than in immature males, and so forth. Of particular relevance to conservation studies, faecal glucocorticoids (i.e. 'stress' hormones) are elevated not only during certain reproductive states (especially pregnancy) but are significantly, and often dramatically, elevated during periods of chronic stress such as entanglement (e.g. Hunt et al. 2006; Rolland et al. 2017). Faecal glucocorticoid analyses have been employed in studies of the physiological impact of both acute and chronic stressors. For example, a reduction in vessel traffic, and associated ocean noise, immediately after the terrorist attacks in the United States on September 11th, 2001, was followed by a significant decline in faecal glucocorticoids in NARW in September of that year, a pattern that did not occur in any other year (Rolland et al. 2012). This finding suggests that faecal glucocorticoids in NARW may be chronically elevated due to nearly continuous exposure to human-caused ocean noise. Faecal glucocorticoids can also be useful in post-mortem investigations by providing insight into acute vs. chronic contributors to mortality. For instance, faeces collected from NARW carcasses have dramatically higher concentrations of glucocorticoids in cases of sustained chronic stress (e.g. fishing gear entanglement) as opposed to an acute event (e.g. ship strike) (Rolland et al. 2017). Thus, chronic stress (chronic ocean noise, entanglement) affects whales not just behaviourally, but also physiologically.

More recently, Burgess et al. (2017) have expanded NARW faecal analyses to include the adrenal hormone aldosterone (a steroid hormone involved in osmotic regulation as well as stress), which has proven to be a useful complement to glucocorticoid analysis. Faecal glucocorticoid data of all mammals, including NARW, tend to show high variation due not only to extrinsic factors of interest (e.g. anthropogenic impacts) but also due to intrinsic confounding factors (sex, life-history stage, individual, etc.; Goymann 2012) and even methodological effects (Palme 2019). Analyses of additional adrenal hormones such as aldosterone can help mitigate these confounding factors, thus allowing for improved interpretation of the causes and mechanisms underlying glucocorticoid elevation in whales. For example, an elevation in glucocorticoids that is accompanied by a parallel elevation in aldosterone is likely to be 'real', representing a period of true adrenal activation above baseline. Additionally, thyroid hormones can also be quantified in cetacean faeces (e.g. Ayres et al. 2012; Hunt et al. 2019) and, given that thyroid hormones are important regulators of metabolic rate in mammals, measurement of this hormone may help clarify the suspected impact of shifts in nutritional state, again helping to interpret the potential cause of an elevation in glucocorticoids. For example, elevations in glucocorticoids that co-occur with declines in thyroid hormones are more likely to be related to prolonged nutritional stress specifically, as opposed to a non-nutritional stressor or an acute stressor. Conversely, elevations in glucocorticoids accompanied by a parallel elevation in thyroid hormones may represent stressors that entail high energetic burden, such as entanglement, prolonged exercise, or cold stress (reviewed in Behringer et al. 2018). In this way, combining glucocorticoid measurements with other hormone measures (aldosterone, thyroid hormones, and the reproductive hormones) improves the ability to discriminate among specific anthropogenic stressors.

12.3.4.2 Respiratory vapour analysis

Exhaled droplets of respiratory vapour ('blow') can be collected from large whales by using a small aerial drone or extended pole to position a collection device above the blowholes at the moment a whale exhales. Though the resulting sample of condensed aerosol droplets is often tiny—nothing more than a few microlitres—several teams have used this method to study microbial ecology of the cetacean respiratory tract (Apprill et al. 2017) or endocrine physiology. The use of exhaled breath for exploring the respiratory microbiome is a significant advancement for health and disease monitoring of large whales, and will undoubtedly have important future applications for studying right whales (e.g. potential effects of stress on immunity and health). Hormone measurements of respiratory vapour from large whales were first trialled on NARW and humpback whales (*Megaptera novaeangliae*) by Hogg et al. (2009), who demonstrated presence and detectability of progesterone and testosterone. Follow-up studies on NARW found that blow samples contain all major steroid and thyroid hormones, and that concentrations of these hormones could be quantified using enzyme immunoassays (Hunt et al. 2014a; Burgess et al. 2016). A field trial by Burgess et al. (2018) provided the first evidence that blow hormone data, when corrected for water content, may indeed reflect individual state of the whale (e.g. high blow progesterone in a known-pregnant female, matching high faecal progesterone in a concurrently collected faecal sample, confirmed with a calf sighting the following year). Although blow hormone analysis has not yet been implemented in studies to address questions of conservation relevance, this approach holds great promise, since sampling is non-invasive and could allow targeted sampling of substantial numbers of individuals immediately before and after a given stressor.

12.3.4.3 Baleen analysis

Baleen is the filter-feeding apparatus of mysticete whales, consisting of slender, long, flexible pieces ('plates') of keratinized tissue that are suspended from the upper jaw in parallel. A given baleen plate grows slowly and continuously from a highly vascularized root region in the upper palate, incorporating stable isotopes (from prey) and hormones as it grows. A full-length baleen plate thus represents a chronological record of the whale's physiological state during the period of baleen growth, which spans 9–10 years for adult right whales (Lysiak 2008; Hunt et al. 2016). In NARW, stable isotope (SI) ratios often vary cyclically along the length of a baleen plate, due to seasonal prey-switching between isotopically distinct feeding grounds along the annual

migratory route (Lysiak 2008). Endocrine data can then be combined with the timeline provided by SI data, enabling reconstruction of nearly a decade of individual endocrine history (Hunt et al. 2016, 2017, 2018; Hunt et al. 2014b). For example, patterns in baleen progesterone can be used to reconstruct the calving intervals of females. Studies of two NARW females of known reproductive history confirm that the spacing of areas of high baleen progesterone content (i.e. pregnancies) in their baleen plates match the timing of subsequent calf sightings recorded in the NARW Catalog (Hunt et al. 2016). Baleen data from these two females also shed some light on the proximate causes of the unusually long inter-calving intervals in the 1990s, a period when reproductive rate declined in the entire NARW population. Both females' baleen plates included extended inter-calving intervals of 6 or 7 years, during which neither female was seen on the calving grounds and neither was ever observed with a calf. Baleen progesterone remained extremely low during these extended inter-calving intervals, that is, with no indications of pregnancy or luteal phase, which both typically involve prolonged high progesterone (Hunt et al. 2016). Thus, in these two cases it appears that the potential cause of the reproductive decline of the late 1990s may be a cessation of ovulation rather than mortality of foetuses or calves. Considering the observed poor body condition in many NARW during the same time period (Rolland et al. 2016), a hypothesis emerges in which poor body condition may inhibit ovarian activity, such that 'thin' females cease reproductive cycling.

Baleen analysis has also provided evidence of the physiological effects of prolonged entanglement. Three baleen glucocorticoid profiles are now available from entangled NARW, and all three cases (two adult females and one adult male) exhibit pronounced elevations in baleen glucocorticoids during the period of entanglement (Hunt et al. 2017, 2018; Lysiak et al. 2018). In one case the whale died due to the entanglement; this baleen specimen also exhibited a spike in thyroid hormones (regulators of metabolic rate) as the whale neared death, possibly reflecting the energetic burden of dragging the fishing gear (van der Hoop et al. 2017; Lysiak et al. 2018). In the two other cases the whales shed the entanglements, yet even so, these individuals appeared to reduce their investment in reproduction in subsequent years, the female prolonging her inter-calving cycle while the male had marked reduction in testosterone in the subsequent year (both these cases may also be related to possible disease) (Hunt et al. 2017, 2018). While these three cases are anecdotal, such endocrine analyses suggest that sublethal stress might depress future reproduction in both sexes.

12.3.4.4 Additional endocrine sample types

Though faeces, respiratory vapour, and baleen have been the primary sources of endocrine information in NARW, other sample types exist that could also be informative of energetic or reproductive state (Hunt et al. 2013; de Mello and de Oliveira 2015). Earplugs, retrieved from the ear canal during necropsy of other large whales, contain annually deposited layers of cerumen (earwax) that capture the whale's entire lifespan (Trumble et al. 2018). Though this method does not have sufficient temporal resolution to reveal acute stressors or seasonal changes within a year, such data may be ideal for studying chronic multi-year stressors and lifetime reproductive success (Trumble et al. 2018). Blubber samples recovered from biopsy darts also may contain valuable physiological information such as reproductive state and stress responses (e.g. Kellar et al. 2009, 2013; Champagne et al. 2018). For right whales in particular, blubber biopsy samples have primarily been collected for genetic analyses and some toxicology studies (e.g. Malik et al. 2006; Frasier et al. 2007; Weisbrod et al. 2009), but are also now being investigated for endocrine research in tandem with these other approaches (K. Graham, pers. comm.). Several other potential endocrine sample types (sloughed skin, bone; Bechshoft et al. 2015; Charapata et al. 2018) remain to be investigated in large whales.

12.3.5 Inferring physiology from behaviour

Animal behaviour has often been a close partner to physiology (Cooke et al. 2014), and behavioural research tools can sometimes directly inform physiological questions (Madliger et al. 2018). In the case of cetaceans, behavioural observations are usually limited to brief periods at the surface but have nonetheless clarified aspects of swimming and dive physiology, digestive physiology, and even sensory physiology. One sensory experiment used a behavioural test to address the conservation-relevant question of whether NARW can visually detect fishing lines. By placing 'fake ropes' of different colours in front of free-swimming whales and observing whether the whales changed course and at what distance, Kraus et al. (2014) demonstrated that NARW may have specific difficulty detecting green objects. This study immediately suggested a possible gear change—altering the colour of the lines in fixed fishing gear to red or orange, which NARW seem able to detect at greater distance (Kraus and Hagbloom 2016; Howle et al. 2019).

Underwater behaviour, and hence some aspects of underwater physiology, can also be inferred from dive data derived from tags. Tagging studies have only rarely been performed with right whales, however, partly because NARW permitting restrictions currently disallow application of certain types of tags due to concerns about wounding. However, advances in suction-cup attachment methods for digital acoustic recording tags (DTAGs) have allowed determination of swimming speed, drag, baleen filtering rate, and other aspects of swimming and foraging physiology in right whales (e.g. Nowacek et al., 2001; Baumgartner and Mate 2003; Argüelles et al., 2016; van der Hoop et al., 2019). For example, van der Hoop et al. (2019) used suction-cup tags to document a high-drag foraging strategy used by NARW. This strategy works well in high-density prey patches, but may actually induce an energetic deficit for any NARW that must forage in low-density prey patches, a situation believed to be occurring more frequently in the western North Atlantic with advancing climate change (Sorochan et al. 2019; Tulloch et al. 2019). A full discussion of tag-based approaches to whale physiological research is beyond the scope of this chapter, but continued improvements in tagging technology make it likely that future research on both NARW and SRW will make greater use of DTAGs and similar devices.

In sum, body condition and food limitation have emerged as key factors that may be driving the decline of NARW reproduction. Sublethal impacts have also been confirmed to have physiological effects on NARW, including likely impacts on later reproduction. Thus, even if whales survive a given anthropogenic impact, they may experience marked adrenal stress responses as well as direct energetic impacts, and both sexes may then subsequently reduce reproductive investment in the following year(s).

12.4 Case study 2: southern right whales—better trajectory, different problems

12.4.1 Background and current status

While both northern hemisphere right whale species remain in a perilous state, the SRW is faring better. This species has a wider range than either of the northern species, with a circumpolar distribution that includes the entire Southern Ocean (Figure 12.1). From summer feeding grounds in offshore and sub-Antarctic waters (Ohsumi and Kasamatsu 1986; Bannister et al. 2016; Nijs and Rowntree 2016; Zerbini et al. 2016, 2018), SRW individuals undertake migrations to specific calving areas in shallow coastal waters around several major southern land masses—effectively subdividing the species into distinct calving populations. Several major SRW calving populations have been recognized, including populations off Argentina/ Brazil, South Africa, Australia, and New Zealand (Cooke and Zerbini 2018; Harcourt et al. 2019). An additional, very small, calving population off Chile/Peru remains critically endangered (Reilly et al. 2008). In consequence of the remoteness and inaccessibility of SRW summer feeding grounds, most information on SRW biology has been collected from winter calving grounds at lower latitudes.

After the SRW gained international legal protection in 1935, the major SRW calving populations began recovering. The first estimates of SRW population growth in the 1970s documented a healthy population growth rate of approximately 7 per cent /year, sustained over the following decades (Tormosov et al. 1998; Cooke and Zerbini 2018; Harcourt et al. 2019). However, since 2000, the growth rate of the SRW population of Argentina/ Brazil that calves off Península Valdés ('PV'), Argentina, started to decelerate markedly (Crespo et al. 2019). There has been no consensus regarding the cause of this decline in growth rate, with possibilities including reduced recruitment of adult females, redistribution of individuals along the shoreline, and/or recolonization of pre-whaling calving grounds (Marón et al. 2015b; Arias et al. 2018; Crespo et al. 2019). Perhaps not coincidentally, since 2000 the PV population has also experienced recurrent high calf mortalities (Rowntree et al. 2013). Prior to 2000, annual calf mortality at PV tracked the population growth rate (Rowntree et al. 2013). However, between 2003 and 2013, at least 672 whales died at PV, of which 91 per cent were calves (Sironi et al. 2014). Average total whale deaths per year increased tenfold, from 8.2 in 1993–2002 to 80 in 2007–2013. These calf mortality rates are the highest ever documented for any population of large whale, and it has been suggested that this may be the cause of the change in population growth rate, that is, with mortality of female calves causing reduced recruitment of adult females to the breeding population (Marón et al. 2015b). The cause of these recurrent calf dieoffs at PV is unknown. In the remainder of this section, we focus largely on the PV population of SRW as a case study for conservation physiology.

12.4.2 Photo-identification

All populations of SRW have photo-identification catalogues that are each associated with a calving ground. The PV population has perhaps the most extensive catalogue, since this population of SRW has been monitored continuously from 1971 to the present (Frasier et al. 2009). The PV photo-identification catalogue currently includes thousands of individual whales with known life histories, in many cases comprising complete family histories with up to five generations of whales. As has been discussed earlier for NARW, this SRW catalogue information is of immense value for physiological research, allowing inferences about many population parameters and physiological states, including calving intervals, age of sexual maturity, and gestation length (Payne et al. 1983; Best and Rüther 1992; Best et al. 2001). Catalogue records indicate that

SRW females typically give birth to their first calf at around 8 years of age (Cooke et al. 2015), and normally calve once every 3 years (Whitehead and Payne 1981; Burnell 2001; Kraus et al. 2007; International Whaling Commission 2012; Marón et al. 2015b). Though in NARW 4-year inter-calving intervals are not uncommon, in SRW, any deviations from the normal 3-year calving interval are interpreted as resulting from calving failures (Knowlton et al. 1994; Burnell 2001; Marón et al. 2015b). Individual SRW females have been traced for decades in the PV population (Payne 1986; Payne et al. 1990), allowing detection of changes in reproductive rate that can be related to environmental factors such as prey availability (Leaper et al. 2006).

12.4.3 Assessment of body condition

VHA methodology has been applied to several SRW populations, with photographs derived from cliff-based observation platforms overlooking calving grounds, from boat-based and aerial photographs, and more recently using underwater viewing platforms (Sironi et al. 2019). As in NARW, body condition and nutritional state can be inferred from body girth and length using photogrammetry techniques (reviewed in Christiansen et al. 2018). For example, in a study of Australian SRW mother–calf pairs, repeated photogrammetry measurements via UAS or drones have documented shifts in maternal body condition that correspond to calf growth rate (in volume and length), allowing inferences regarding the high energy investment of lactation for SRW mothers (Christiansen et al. 2018).

12.4.4 Endocrine analyses

12.4.4.1 Hormone analysis in faeces and respiratory vapour

Faecal hormone analysis, a mainstay in NARW physiological research, has been difficult to apply to SRW due to limited sample availability on the calving grounds when the population is most accessible to researchers. All species of right whales exhibit seasonal fasting while at their calving grounds, resulting in few observed defecations. As is typical of capital breeders, female SRW rely almost exclusively on fat reserves during the time they spend at the calving grounds (Miller et al. 2011). However, females have been observed to feed occasionally towards the end of the calving season when food becomes available; therefore, some limited faecal-based studies may be possible (D'Agostino et al. 2016). Some of these late-season faecal samples have been collected opportunistically for analysis of prey selection and exposure to biotoxins (D'Agostino et al. 2016), but the use of such samples for endocrine studies of SRW has not yet been rigorously explored.

Respiratory vapour collection from SRW has been tested in Patagonia, Argentina, using a small UAS equipped with a collection arm holding a sterile Petri dish (I. Kerr, pers. comm.). This technique appears to be a promising tool for studying endocrine physiology of SRW, particularly given that faecal collection is problematic for this species. Additional strengths of this technique are that it enables repeated sampling of targeted individual whales, throughout seasonal movements and across all life-history stages, and could improve the efficiency of data collection (Burgess et al. 2018). Respiratory vapour analysis using UAS can potentially be combined with photogrammetric measurement to enable linkages between body condition, behaviour, reproductive stage, and physiological state.

12.4.4.2 Baleen analysis

Baleen hormone analysis has proven useful for informing SRW conservation, allowing retrospective assessment of patterns of stress and reproductive physiology from samples obtained at necropsy. This technique, initially developed for NARW and bowheads (Hunt et al. 2014b; Hunt et al. 2016), has been successfully transferred to SRW, with the specific aim of distinguishing acute from chronic stress in recurrent calf die-offs of SRW at PV. The application of this method has allowed investigation of the potential involvement of kelp gulls (*Larus dominicanus*) in SRW calf mortalities. At PV, kelp gulls have been observed to perform a unique parasitic behaviour focused on SRW calves, in which the gulls land on the backs of resting or surfacing young SRW calves and peck at their backs in sustained attacks (Figure 12.2, bottom). Harassment of an individual whale can last for hours and causes cutaneous wounds that accumulate in number and severity over weeks (Rowntree et al. 1998; Marón et al. 2015a). Harassment by kelp gulls is unique to the PV SRW calving population, and it has been hypothesized to be the cause, or a contributing cause, of the extremely high calf mortality in this area (Thomas et al. 2013; Sironi et al. 2018). Endocrine analysis of baleen of deceased SRW calves provides insight into how gull-wounding affects calves physiologically. Fernández Ajó et al. (2018) demonstrated that baleen glucocorticoid

concentrations were linked to the severity of cutaneous wounds observed on calf carcasses. Results showed a positive correlation between baleen glucocorticoids and the number of gull-inflicted lesions, with highly wounded calves exhibiting a progressive elevation of baleen glucocorticoids prior to death (two representative cases illustrated in Figure 12.4; Fernández Ajó et al. 2018). Thus, the physiological record preserved in baleen has provided evidence that kelp gull harassment is a significant environmental stressor for SRW calves, and hence may be a contributor to calf deaths in this SRW population. Additionally, as an unexpected

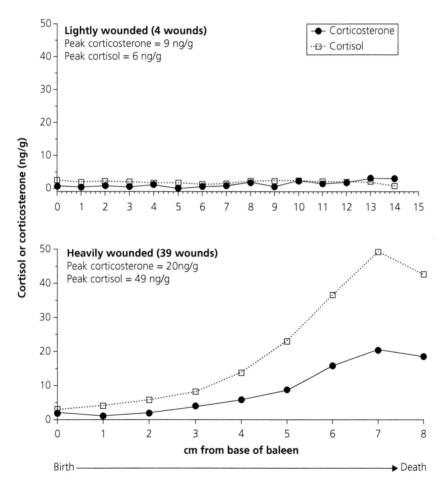

Figure 12.4 Baleen endocrine analyses from samples collected at necropsy may enable retrospective assessment of chronic stress in the weeks before death. Two illustrative lifetime baleen glucocorticoid profiles shown of cortisol (dashed lines) and corticosterone (solid lines) in baleen of two southern right whale calves, one with only mild wounding by kelp gulls (top, whale 102905PV-Ea28) vs. a calf with severe chronic wounding (bottom, whale 091208PV-Ea49). Note different scales of x-axes; the severely wounded calf had shorter baleen (indicating shorter lifespan) as compared with the mildly wounded calf. Adapted from Fernández Ajó et al. (2018).

outcome of this research, it was found that pre-natally grown baleen at the tip of a calf's baleen plate appears to reflect changes in endocrine state *in utero*, which may enable investigation of prenatal physiological condition of calves as well as mothers (Fernández Ajó et al. 2018). Thus, physiological research using a novel matrix has helped address conservation-relevant questions such as distinguishing acute from chronic stress and, potentially, determining cause of death.

12.4.5 Inferring physiology from behaviour

Mother–calf pairs at PV show changes in behaviour that could be associated with increased energy expenditure. In capital breeders that may be operating at the limit of their capacities, any additional energetic burden has the potential to negatively impact breeding success, as discussed previously for entangled NARW. At PV, gull attacks impact SRW calves not just physically but also behaviourally. Gull harassment often results in rapid-motion postural adjustments (Thomas 1988). Whales under attack spend more time underwater and engage in more energetically expensive activities, for example, travelling faster than non-attacked animals (Rowntree et al. 1998). Adult SRW in the PV region adjust their normal surface posture, adopting an arced position that limits the area of the back that is exposed above water (Thomas 1988; Sironi et al. 2009). Similarly, calves in the PV region have adopted an avoidance behaviour termed 'oblique breathing' in which they surface to breathe with only their heads above the surface, thus keeping their backs underwater (Fazio et al. 2015). These postural adjustments are presumably a learned response to avoid discomfort of gull attacks and might impose extra energy costs. All these observations suggest that gull attacks could represent an energetic drain on both mothers and calves during this sensitive part of their reproductive cycle. Lactation is known to be the most energy-intensive period of an adult female's life and is also the period of most rapid calf growth; it is exactly this period when kelp gull harassment occurs, suggesting that important energetic costs could be incurred from these alterations in behaviour (though see also Marón et al. 2018). Aiming to answer this question,

current research (Fernández Ajó et al., 2020) is evaluating the physiological tradeoffs between two endocrine axes, the hypothalamic–pituitary–adrenal (HPA) and the hypothalamic–pituitary–thyroid (HPT) axes, in response to kelp gull parasitism by analysing the lifetime concentrations of tri-iodothyronine (the active form of thyroid hormone) and the adrenal glucocorticoids (cortisol and corticosterone) in baleen recovered from dead SRW calves.

12.5 Similarities, differences, and lessons learned

The NARW and SRW are facing different conservation challenges—high calf mortality of unknown cause for the PV population of SRW vs. the NARW's more complex and challenging situation of low calving rate, frequent fishing gear entanglements, frequent vessel strikes, and decline in body condition. However, both species have very similar life histories and present similar logistical challenges for researchers. Both are highly migratory species with only part of the annual range amenable to study, but near-coast calving grounds in both species, as well as near-coast feeding grounds for some NARW, have enabled comprehensive surveys of mothers and calves. Photo-identification is possible in both species, but with the NARW's much lower numbers allowing more comprehensive coverage.

In both species, discrimination of acute from chronic stress has been a pressing concern for discerning cause of mortality (SRW calves), impact of sublethal entanglement (NARW) or cause of failure of reproduction (NARW). Even when a cause of mortality is known, it has been invaluable from a conservation, welfare, and policy perspective to be able to demonstrate whether such deaths entail a prolonged period of physiological stress before death (e.g. Rolland et al. 2017; Fernández Ajó et al. 2018; Lysiak et al. 2018). Confirmation of sublethal impacts in NARW, even in surviving whales, may clarify whether entanglements 'take' more whales than has been historically realized (i.e. not just killing NARW outright, but also depressing the reproductive fitness of survivors). It remains to be seen whether gull-attacked SRW suffer similar long-term impacts. In other mammals, prolonged stress experienced by juveniles affects stress responses and reproductive performance years later (Lindström 1999; Metcalfe and Monaghan 2001; Marcil-Ferland et al. 2013; Douhard et al. 2014), but whether this phenomenon occurs in SRW cannot yet be confirmed. Further exploration of these questions can benefit from the creative conservation physiology toolkit described here.

Notably, physiological tools developed in one species are often used successfully in another. Data-management and data-sharing techniques first devised for managing the NARW's immense photo-identification catalogue are now being used in several SRW populations. VHA scoring techniques, too, have been ported from NARW to SRW. Photogrammetry was trialled first on SRW and then employed on NARW; estimates of gestation duration were derived first from SRW foetal length data and then adapted for NARW; baleen hormones were validated first using NARW individuals of known history, then applied to SRW calves, and the SRW calf study has resulted in improvements of the technique (e.g. optimization of hormone extraction, minimization of sample mass) that will further enable the method's use in future NARW studies. Respiratory vapour and faecal hormone analysis techniques can also be ported from NARW to SRW largely unchanged. Thus, the pairing of a critically endangered population with a near complete census of well-known individuals (NARW) and a much larger population of largely unknown individuals (SRW) has proven a powerful combination, with techniques repeatedly traded back and forth between teams of researchers. This approach has benefited from periodic meetings between NARW and SRW research groups (e.g. in workshops at major international meetings), and has been enhanced by a notable attitude of cooperation across institutional and international boundaries to share field and lab methods, photo-catalogue management methods, statistical techniques, and even, in some cases, personnel.

12.6 Conclusions

Ultimately, as Madliger et al. (2018) states, conservation physiological tools are 'decision-support tools'. That is, any physiological tool can become a 'conservation physiology' tool if it is employed in a conservation context with the goal of enabling a policy or management decision. Conservation physiology, as a field, was at first rather narrowly focused on quantification of glucocorticoids, and several researchers have called for colleagues to broaden 'beyond just cort', and indeed beyond just hormones (Wikelski and Cooke 2006; Cooke et al. 2013; Madliger et al. 2018). In the large whales, however, the observational tools (VHA, photogrammetry, etc.) have historically been the more dominant technique, with various methods of externally assessing blubber stores and body condition still the most common physiological tool. The use of endocrine tools in cetacean conservation has trailed behind the terrestrial literature by approximately two decades, with researchers having to adapt terrestrially oriented field methodologies to a challenging marine environment. However, faecal assays are now well-proven for whales, and a multitude of alternative (non-blood) endocrine sample types have been successfully validated (baleen, earwax, blubber, respiratory vapour). Endocrine physiology is at last coming to the fore as a major tool for the whale conservation biologist. We can hope that tandem advances in other technologies (e.g. suction-cup tags and UAS/drone sampling techniques) will continue to result in new physiological tools for use on free-swimming large whales.

Have these physiological tools truly added anything useful to the effort to conserve the right whales? Our answer, to date, is a qualified yes. In some cases, physiological tools have clarified a particular question: SRW baleen data now strongly support the kelp gull hypothesis of mortality; NARW faecal hormone data show that whales are indeed affected by chronic ocean noise. Some novel discoveries have emerged—it had not before been clear that entangled male whales might 'skip a testosterone cycle' the following year, or that they even had annual testosterone cycles that could be skipped. In other cases, physiological data have been less surprising, reinforcing hypotheses that were already well-supported—for example, it is not unexpected that a severely entangled whale should have elevated glucocorticoids. Yet even in these latter cases, the physiological data can add a powerful punch to the story that can be told to policy makers as well as to the public. It is one thing to hypothesize that a whale is 'probably stressed'; it is another to be able to present solid data showing orders-of-magnitude increases in adrenal glucocorticoids as a badly entangled whale nears death. Finally, modelling studies are beginning to draw these diverse sources of information together, combining physiological information with ecological and oceanographic data, ultimately providing convincing evidence of important linkages between initial impact (entanglement, disease, poor food), subsequent physiological state (health or body condition), and long-term effects on reproduction, mortality, and resultant population trends.

Though the relevant physiological data have accrued slowly, and validations have often required years, we may at last be reaching the point where good physiological data are indeed improving our understanding of the problems faced by the right whales. The sustained, decades-long research effort to adapt physiological tools to right whales will, we think, ultimately help identify practical and viable policy solutions, hopefully in time to help not only the NARW and its southern cousins, but also the NPRW and indeed all the large cetaceans.

Acknowledgements

The writing of this manuscript was supported by (1) the Smithsonian-Mason School of Conservation and George Mason University [KH]; and (2) the Fulbright program and the Ministerio de Educación y Deportes de la Nación Argentina (AFA). We would like to thank all members, researchers, volunteers, and collaborators of the Southern Right Whale Health Monitoring Program (SRWHMP), the Instituto de Conservación de Ballenas de Argentina, the North Atlantic Right Whale Consortium, the US marine mammal stranding networks, and all others who have assisted in the ongoing effort to advance understanding and conservation of right whales.

References

Apprill, A., Miller, C.A., Moore, M. et al., 2017. Extensive core microbiome in drone-captured whale blow supports a framework for health monitoring. *mSystems*, 2, e00119-17. doi:10.1128/mSystems.00119-17

Argüelles, M.B., Fazio, A., Fiorito, C. et al., 2016. Diving behavior of southern right whales (*Eubalaena australis*) in a maritime traffic area in Patagonia, Argentina. *Aquatic Mammals*, 42, 104–8.

Arias, M., Coscarella, M., Romero, M. et al., 2018. Southern right whale *Eubalaena australis* in Golfo San Matías (Patagonia, Argentina): evidence of recolonisation. *PLoS ONE*, 13, e0207524. doi:10.1371/journal.pone.0207524

Avila, I., Kaschner, K., and Dormann, C., 2018. Current global risks to marine mammals: taking stock of the threats. *Biological Conservation*, 221, 44–58.

Ayres, K., Booth, R., Hempelmann, J. et al., 2012. Distinguishing the impacts of inadequate prey and vessel traffic on an endangered killer whale (*Orcinus orca*) population. *PLoS ONE*, 7(6), e36842 doi:10.1371/journal.pone.0036842

Bannister, J., Hammond, P., and Double, M., 2016. Population trends in right whales off southern Australia 1993–2015. Report SC/66b/BRG09 presented to the International Whaling Commission Scientific Committee, Bled, Slovenia. Available at: https://iwc.int/hom (accessed 6 July 2020).

Baumgartner, M.R. and Mate, B.R., 2003. Summertime foraging ecology of North Atlantic right whales. *Marine Ecology Progress Series*, 264, 123–35.

Bechshoft, T., Wright, A., Weisser, J. et al., 2015. Developing a new research tool for use in free-ranging cetaceans: recovering cortisol from harbour porpoise skin. *Conservation Physiology*, 3, cov016. doi:10.1093/conphys/cov016

Behringer, V., Deimel, C., Hohmann, G. et al., 2018. Applications for non-invasive thyroid hormone measurements in mammalian ecology, growth, and maintenance. *Hormones and Behavior*, 105, 66–85.

Best, P., Brandao, A., and Butterworth, D., 2001. Demographic parameters of southern right whales off South Africa. *Report of the International Whaling Commission (Special Issue)*, 2, 161–9.

Best, P. and Rüther, H., 1992. Aerial photogrammetry of southern right whales, *Eubalaena australis*. *Journal of Zoology*, 228, 595–614. doi:10.1111/j.1469-7998.1992.tb04458.x

Best, P.B., 1994. Seasonality of reproduction and the length of gestation in southern right whales *Eubalaena australis*. *Journal of Zoology*, 232, 175–89.

Best, P.B., Payne, R. Rowntree, V. et al., 1993. Long-range movements of south Atlantic right whales *Eubalaena australis*. *Marine Mammal Science*, 9, 227–34.

Brown, M., Kraus, S., Slay, C., and Garrison, L., 2007. Surveying for discovery, science and management. In S.D. Kraus and R.M. Rolland, eds. *The Urban Whale: North Atlantic Right Whales at the Crossroads*, pp. 105–137. Harvard University Press, Cambridge, MA.

Burgess, E., Hunt, K.E., Kraus, S.D., and Rolland, R.M., 2016. Get the most out of blow hormones: validation of sampling materials, field storage and extraction techniques for whale respiratory vapor samples. *Conservation Physiology*, 4, cow024.

Burgess, E.A., Hunt, K.E., Kraus, S.D., and Rolland, R.M., 2018. Quantifying hormones in exhaled breath for physiological assessment of large whales at sea. *Scientific Reports*, 8, 10031.

Burgess, E.A., Hunt, K.E., Rolland, R.M., and Kraus, S.D., 2017. Adrenal responses of large whales: integrating fecal aldosterone as a complementary biomarker. *General and Comparative Endocrinology*, 252, 103–10.

Burnell, S., 2001. Aspects of the reproductive biology and behavioral ecology of right whales off Australia. *Journal of Cetacean Research and Management (Special Issue)*, 2, 89–102.

Champagne, C., Kellar, N., Trego, M. et al., 2018. Comprehensive endocrine response to acute stress in the bottlenose dolphin from serum, blubber and feces. *General and Comparative Endocrinology*, 266, 178–93.

Charapata, P., Horstmann, L., Jannasch, A., and Misarti, N., 2018. A novel method to measure steroid hormone concentrations in walrus bone from archeological, historical, and modern time periods using liquid chromatography/tandem mass spectrometry. *Rapid Communications in Mass Spectrometry*, 32, 9999–2023. doi:10.1002/rcm.8272

Christiansen, F., Vivier, F., Charlton, C. et al., 2018. Maternal body size and condition determine calf growth rates in southern right whales. *Marine Ecology Progress Series*, 592, 267–82. doi:10.3354/meps12522

Clapham, P.J., 2016. Managing leviathan: conservation challenges for the great whales in a post-whaling world. *Oceanography*, 29, 214–25.

Cooke, J., Rowntree, V., and Sironi, M., 2015. Southwest Atlantic right whales: interim updated population assessment from photo-id collected at Península Valdés, Argentina. SC/66a/BRG/23 report presented to the International Whaling Commission Scientific Committee, San Diego, USA. Available at: https://iwc.int/home (accessed 6 July 2020).

Cooke, J.G., 2018. *Eubalaena glacialis*. The IUCN Red List of Threatened Species 2018: e.T41712A50380891. Available at: 10.2305/IUCN.UK.2018-1.RLTS.T41712A50380891.en (accessed 6 July 2020).

Cooke, J.G. and Zerbini, A., 2018. *Eubalaena australis*. The IUCN Red List of Threatened Species 2018: e.T8153A50354147. Available at: 10.2305/IUCN.UK.2018-1.RLTS.T8153A50354147.en (accessed 6 July 2020).

Cooke, S., Sack, L., Franklin, C. et al., 2013. What is conservation physiology? Perspectives on an increasingly integrated and essential science. *Conservation Physiology*, 1, cot001.

Cooke, S.J., Blumstein, D.T., Buchholz, R. et al., 2014. Physiology, behavior, and conservation. *Physiologcal and Biochemical Zoology*, 87, 1–14.

Corkeron, P., Hamilton, P., Bannister, J. et al., 2018. The recovery of North Atlantic right whales, *Eubalaena glacialis*, has been constrained by human-caused mortality. *Royal Society Open Science*, 5, 180892. doi:10.1098/rsos.180892

Corkeron, P.J., Rolland, R.M., Hunt, K.E., and Kraus, S.D., 2017. A right whale PooTree: fecal hormones and classification trees identify reproductive states in North Atlantic right whales (*Eubalaena glacialis*). *Conservation Physiology*, 5, cox006. doi:10.1093/conphys/cox006

Crespo, E., Pedraza, S., Dans, S. et al., 2019. The southwestern Atlantic southern right whale, *Eubalaena australis*, population is growing but at a decelerated rate. *Marine Mammal Science*, 35, 93–107. doi:10.1111/mms.12526

Crespo, E.A. and Hall, M.A., 2002. Interactions between aquatic mammals and humans in the context of ecosystem management. In P.G.H. Evans and J.A. Raga, eds. *Marine Mammals*, pp. 463–90. Springer, Boston, MA.

D'Agostino, V., Hoffmeyer, M., and Degrati, M., 2016. Faecal analysis of southern right whales (*Eubalaena australis*) in Península Valdés calving ground, Argentina: *Calanus australis*, a key prey species. *Journal of the Marine Biological Association of the United Kingdom*, 96 (Special Issue 4, Marine Mammals), 859–68. doi:10.1017/S0025315415001897

Davies, K.T.A., Brown, M., Hamilton, P. et al., 2019. Variation in North Atlantic right whale *Eubalaena glacialis* occurrence in the Bay of Fundy, Canada, over three decades. *Endangered Species Research*, 39, 159–71. doi:10.3354/esr00951

Davis, G.E., Baumgartner, M.F., Bonnell, J.M., et al., 2017. Long-term passive acoustic recordings track the changing distribution of North Atlantic right whales (*Eubalaena glacialis*) from 2004 to 2014. *Scientific Reports*, 7, 13460.

Dawson, S.M., Bowman, M.H., Leunissen, E., and Sirguey, P., 2017. Inexpensive aerial photogrammetry for studies of whales and large marine animals. *Frontiers in Marine Science*, 4, 366.

de Mello, D. and de Oliveira, C., 2015. Biological matrices for sampling free-ranging cetaceans and the implications of their use for reproductive endocrine monitoring. *Mammal Review*, 46, 77–91. doi:10.1111/mam.12055

Douhard, M., Plard, F., Gaillard, J.M. et al., 2014. Fitness consequences of environmental conditions at different life stages in a long-lived vertebrate. *Proceedings of the Royal Society B: Biological Sciences*, 281, 20140276. doi:10.1098/rspb.2014.0276

Fazio, A., Argüelles, M. and Bertellotti, M., 2015. Change in southern right whale breathing behavior in response to gull attacks. *Marine Biology*, 162, 267–73.

Fernández Ajó, A., Hunt, K.E., Giese, A.C. et al., 2020. Retrospective analysis of the lifetime endocrine response of southern right whale calves to gull wounding and harassment: a baleen hormone approach. *General and Comparative Endocrinology*, 296, 113536.

Fernández Ajó, A.A., Hunt, K.E., Uhart, M. et al., 2018. Lifetime glucocorticoid profiles in baleen of right whale calves: potential relationships to chronic stress of repeated wounding by Kelp Gulls. *Conservation Physiology*, 6, coy045.

Fortune, S.M., Trites, A.W., Perryman, W.L. et al., 2012. Growth and rapid early development of North Atlantic right whales (*Eubalaena glacialis*). *Journal of Mammalogy*, 93, 1342–54.

Fowler, C.W., 1981. Density dependence as related to life history strategy. *Ecology*, 62, 602–10.

Frasier, T., Hamilton, P., Brown, M. et al., 2007. Patterns of male reproductive success in a highly promiscuous whale species: the endangered North Atlantic right whale. *Molecular Ecology*, 16, 5277–93.

Frasier, T. R., Hamilton, P. K., Brown, M. et al., 2009. Sources and rates of errors in methods of individual identification for North Atlantic right whales. *Journal of Mammalogy*, 90, 1246–55.

Gambell, R., 1993. International management of whales and whaling: a historical review of the regulation of commercial and aboriginal subsistence whaling. *Arctic*, 46, 97–107.

Goymann, W., 2012. On the use of non-invasive hormone research in uncontrolled, natural environments: the problem with sex, diet, metabolic rate and the individual. *Methods in Ecology and Evolution*, 3, 757–65.

Hamilton, P.K., Knowlton, A.R., and Marx, M.K., 2007. Right whales tell their own stories: the photo-identification catalog. In S.D. Kraus and R.M. Rolland, eds. *The Urban Whale: North Atlantic Right Whales at the Crossroads*, pp. 75–104. Harvard University Press, Cambridge, MA.

Hamilton, P.K., Knowlton, A.R., Marx, M.K., and Kraus, S.D., 1998. Age structure and longevity in North Atlantic right whales *Eubalaena glacialis* and their relation to reproduction. *Marine Ecology Progress Series*, 171, 285–92.

Harcourt, R., van der Hoop, J., Kraus, S., and Carroll, E.L., 2019. Future directions in *Eubalaena spp.*: comparative research to inform conservation. *Frontiers in Marine Science*, 5, 530. doi:10.3389/fmars.2018.00530.

Hayes, S., Josephson, E., Maze-Foley, K., and Rosel, P., 2017. US Atlantic and Gulf of Mexico Marine Mammal Stock Assessments—2016. NOAA Technical Memorandum NMFS-NE 241. NOAA, Silver Spring, MD.

Hogg, C., Rogers, T., Shorter, A., et al., 2009. Determination of steroid hormones in whale blow: It is possible. *Marine Mammal Science*, 25, 605–18.

Howle, L., Kraus, S., Werner, T., and Nowacek, D., 2019. Simulation of the entanglement of a North Atlantic right whale (*Eubalaena glacialis*) with fixed fishing gear. *Marine Mammal Science*, 35, 760–78.

Hunt, K., Lysiak, N., Matthews, C. et al., 2018. Multi-year patterns in testosterone, cortisol and corticosterone in baleen from adult males of three whale species. *Conservation Physiology*, 6, coy049. doi:10.1093/conphys/coy049

Hunt, K., Lysiak, N., Moore, M., and Rolland, R.M., 2016. Longitudinal progesterone profiles from baleen of female North Atlantic right whales (*Eubalaena glacialis*) match recent calving history. *Conservation Physiology*, 4, cow014. doi:10.1093/conphys/cow014

Hunt, K., Lysiak, N., Moore, M., and Rolland, R., 2017. Multi-year longitudinal profiles of cortisol and corticosterone recovered from baleen of North Atlantic right whales (*Eubalaena glacialis*). *General and Comparative Endocrinology*, 254, 50–9. doi:10.1016/j.ygcen.2017.09.009

Hunt, K., Moore, M., Rolland, R. et al., 2013. Overcoming the challenges of studying conservation physiology in large whales: a review of available methods. *Conservation Physiology*, 1, cot006.

Hunt, K., Robbins, J., Buck, C. et al., 2019. Evaluation of fecal hormones for noninvasive research on reproduction and stress in humpback whales (*Megaptera novaeangliae*). *General and Comparative Endocrinology*, 280, 24–34.

Hunt, K., Rolland, R., and Kraus, S., 2014a. Detection of steroid and thyroid hormones via immunoassay of North Atlantic right whale (*Eubalaena glacialis*) respiratory vapor. *Marine Mammal Science*, 30, 796–809. doi:10.111/mms.12073

Hunt, K., Rolland, R., Kraus, S., and Wasser, S., 2006. Analysis of fecal glucocorticoids in the North Atlantic right whale (*Eubalaena glacialis*). *General and Comparative Endocrinology*, 148, 260–72.

Hunt, K., Stimmelmayr, R., George, C. et al., 2014b. Baleen hormones: a novel tool for retrospective assessment of stress and reproduction in bowhead whales (*Balaena mysticetus*). *Conservation Physiology*, 2, cou030. doi:10.1093/conphys/cou030

International Whaling Commission, 2012. Report of the IWC Workshop on the Assessment of Southern Right Whales, 13–16 September 2011, Buenos Aires, Argentina. IWC Scientific Committee SC/64/rep5.

Katona, S. and Kraus, S., 1999. Efforts to conserve the North Atlantic right whale. In J.R. Twiss and R.R. Reeves, eds. *Conservation and Management of Marine Mammals*, pp. 311–31. Smithsonian Institution Press, Washington, DC.

Kellar, N., Keliher, J., Trego, M.L. et al., 2013. Variation of bowhead whale progesterone concentrations across demographic groups and sample matrices. *Endangered Species Research*, *22*, 61–72.

Kellar, N., Trego, M., Marks, C. et al, 2009. Blubber testosterone: a potential marker of male reproductive status in short-beaked common dolphins. *Marine Mammal Science*, *25*, 507–22.

Kenney, R., 2018. Right whales: *Eubalaena glacialis, E. japonica*, and *E. australis*. In B. Würsig, J. Thewissen, and K. Kovacs, eds. *Encyclopedia of Marine Mammals*, 3rd edition, pp. 962–72. Academic Press, London.

Knowlton, A.R., Hamilton, P.K., Marx, M.K. et al., 2012. Monitoring North Atlantic right whale (*Eubalaena glacialis*) entanglement rates: a 30 yr retrospective. *Marine Ecology Progress Series*, *466*, 293–302.

Knowlton, A.R., Kraus, S.D., and Kenney, R.D., 1994. Reproduction in North Atlantic right whales (*Eubalaena glacialis*). *Canadian Journal of Zoology*, *72*, 1297–305.

Knowlton, A.R., Marx, M.K., Hamilton, P.K. et al., 2017. Task 2: Final Report on 2017 Right Whale Entanglement Scar Coding Efforts. Report submitted to: National Marine Fisheries Service, 2017.

Knowlton, A.R., Robbins, J., Landry, S. et al., 2015. Effects of fishing rope strength on the severity of large whale entanglements. *Conservation Biology*, *30*, 318–28. doi:10.1111/cobi.12590

Knowlton, A.R., Sigukjósson, J., Ciano, J.N., and Kraus, S.D., 1992. Long-distance movements of North Atlantic right whales (*Eubalaena glacialis*). *Marine Mammal Science*, *8*, 397–405.

Kraus, S., Fasick, J., Werner, T., and McFarron, P., 2014. Enhancing the visibility of fishing ropes to reduce right whale entanglements. Report to the Bycatch Reduction Engineering Program (BREP), National Marine Fisheries Service, Office of Sustainable Fisheries, pp. 67–75.

Kraus, S. and Hagbloom, M., 2016. Project 4 Report: Assessments of vision to reduce right whale entanglements. Consortium for Wildlife Bycatch Reduction. New England Aquarium, Boston, MA. 15 pp. Available at: https://www.bycatch.org/ (accessed 6 July 2020).

Kraus, S. and Rolland, R., 2007. *The Urban Whale: North Atlantic Right Whales at the Crossroads*. Harvard University Press, Cambridge, MA.

Kraus, S.D., Hamilton, P.K., Kenney, R.D. et al., 2001. Reproductive parameters of the North Atlantic right whale. *Journal of Cetacean Research and Management (Special Issue)*, *2*, 231–6.

Kraus, S.D., Kenney, R., Mayo, C.A. et al., 2016. Recent scientific publications cast doubt on North Atlantic right whale future. *Frontiers in Marine Science*, *3*, 137. doi:10.3389/fmars.2016.00137

Kraus, S.D., Pace, R.I., and Frasier, T., 2007. High investment, low return: the strange case of reproduction in *Eubalaena glacialis*. In S.D. Kraus and R.M. Rolland, eds. *The Urban Whale: North Atlantic Right Whales at the Crossroads*, pp. 172–99. Harvard University Press, Cambridge, MA.

Leaper, R., Cooke, J., Trathan, P. et al., 2006. Global climate change drives southern right whale (*Eubalaena australis*) population dynamics. *Biology Letters*, *2*, 289–92.

Lindström, J., 1999. Early development and fitness in birds and mammals. *Trends in Ecology & Evolution*, *14*, 343–8.

Lysiak, N., 2008. Investigating the migration and foraging ecology of North Atlantic right whales with stable isotope geochemistry of baleen and zooplankton. PhD thesis, Boston University, Boston, MA.

Lysiak, N., Trumble, S., Knowlton, A., and Moore, M., 2018. Characterizing the duration and severity of fishing gear entanglement on a North Atlantic right whale (*Eubalaena glacialis*) using stable isotopes, steroid and thyroid hormones in baleen. *Frontiers in Marine Science*, *5*, 168. doi:10.3389/fmars.2018.00168

Madliger, C.L., Love, O.P., Hultine, K.R., and Cooke, S.J., 2018. The conservation physiology toolbox: status and opportunities. *Conservation Physiology*, *6*, coy029.

Malik, S., Brown, M., Kraus, S., and White, B., 2006. Analysis of mitochondrial DNA diversity within and between North and South Atlantic right whales. *Marine Mammal Science*, *16*, 545–58.

Marcil-Ferland, D., Festa-Bianchet, M., Martin, A., and Pelletier, F., 2013. Despite catch-up, prolonged growth has detrimental fitness consequences in a long-lived vertebrate. *American Naturalist*, *182*, 775–85. doi:10.1086/673534

Marón, C., Baltramino, L., Di Martino, M. et al., 2015a. Increased wounding of southern right whale (*Eubalaena australis*) calves by kelp gulls (*Larus dominicanus*) at Península Valdés, Argentina. *PLoS ONE*, *10*, e0139291. doi:10.1371/journal.pone.0139291

Marón, C., Di Martino, M., Chirife, A. et al., 2018. No evidence of malnutrition in dead southern right whale calves off Argentina as inferred from blubber thickness measurements and lipid content analysis. SC/67b/CMP/03 report presented to the International Whaling Commission Scientific Committee, Slovenia. Available at: https://iwc.int/home (accessed 6 July 2020).

Marón, C., Rowntree, V., Sironi, M. et al., 2015b. Estimating population consequences of increased calf mortality in the southern right whales of Argentina. SC/66a/BRG/1 report presented to the International Whaling Commission Scientific Committee, San Diego, CA. Available at: https://iwc.int/home (accessed 6 July 2020).

McCauley, D., Pinsky, M., Palumbi, S. et al. 2015. Marine defaunation: animal loss in the global ocean. *Science*, *347*, 1255641.

Metcalfe, N. and Monaghan, P., 2001. Compensation for a bad start: grow now, pay later? *Trends in Ecology & Evolution*, *16*, 254–60.

Meyer-Gutbrod, E. and Green, C., 2014. Climate-associated regime shifts drive decadal-scale variability in recovery of North Atlantic right whale population. *Oceanography*, *27*, 148–53.

Meyer-Gutbrod, E. and Greene, C., 2017. Uncertain recovery of the North Atlantic right whale in a changing ocean. *Global Change Biology*, *24*, 455–64. doi:10.1111/gcb.13929

Miller, C.A., Best, P.B., Perryman, W.L. et al., 2012. Body shape changes associated with reproductive status, nutritive condition and growth in right whales *Eubalaena glacialis* and *E. australis*. *Marine Ecology Progress Series*, *459*, 135–56.

Miller, C.A., Reeb, D., Best, P.B. et al., 2011. Blubber thickness in right whales *Eubalaena glacialis* and *Eubalaena australis* related with reproduction, life history status and prey abundance. *Marine Ecology Progress Series*, *438*, 267–83.

Monsarrat, S., Pennino, M., Smith, T. et al., 2015. A spatially explicit estimate of the prewhaling abundance of the endangered North Atlantic right whale. *Conservation Biology*, *30*, 783–91. doi:10.1111/cobi.12664

Moore, M., McLellan, W., Daoust, P.Y. et al., 2007. Right whale mortality: a message from the dead to the living. In S.D. Kraus and R.M. Rolland, eds. *The Urban Whale: North Atlantic Right Whales at the Crossroads*, pp. 358–79. Harvard University Press, Cambridge, MA.

Moore, M.J., Miller, C.A., Morss, M.S. et al., 2001. Ultrasonic measurement of blubber thickness in right whales. *Journal of Cetacean Research and Management*, *2*, 301–9.

Moore, M.J. and Van der Hoop, J.M., 2012. The painful side of trap and fixed net fisheries: chronic entanglement of large whales. *Journal of Marine Biology*, *2012*, 230653. doi:10.1155/2012/230653

Nijs, G. and Rowntree, V., 2016. Rare sightings of southern right whales (*Eubalaena australis*) on a feeding ground off the South Sandwich Islands, including a known individual from Península Valdés, Argentina. *Marine Mammal Science*, *33*, 342–9. doi:10.1111/mms.12354

North Atlantic Right Whale Consortium, 2019. North Atlantic Right Whale Consortium 2019 Database. Anderson Cabot Center for Ocean Life at the New England Aquarium, Boston, MA. Available at: https://www.narwc.org/ (accessed 6 July 2020).

Nowacek, D.P., Johnson, M.P., Tyack, P.L. et al., 2001. Buoyant balaenids: the ups and downs of buoyancy in right whales. *Proceedings of the Royal Society B*, *268*, 1811–16. doi:10.1098/rspb.2001.1730

Ohsumi, S. and Kasamatsu, F., 1986. Recent off-shore distribution of the southern right whale in summer. *Report of the International Whaling Commission (Special Issue)*, *10*, 177–85.

Pace, R.M., Corkeron, P.J., and Kraus, S.D., 2017. State–space mark–recapture estimates reveal a recent decline in abundance of North Atlantic right whales. *Ecology and Evolution*, *7*, 8730–41.

Palme, R., 2019. Non-invasive measurement of glucocorticoids: advances and problems. *Physiology & Behavior*, *199*, 229–43.

Payne, R., 1986. Long term behavioral studies of the southern right whale (*Eubalaena australis*). *Report of the International Whaling Commission (Special Issue)*, *10*, 161–7.

Payne, R., Brazier, O., Dorsey, E. et al., 1983. External features in southern right whales (*Eubalaena australis*) and their use in identifying individuals. In R. Payne, ed. *Communication and Behavior of Whales, AAAS Selected Symposia Series 76*. Westview Press, Boulder, CO.

Payne, R., Rowntree, V., Perkins, J. et al, 1990. Population size, trends and reproductive parameters of right whales, *Eubalaena australis*, off Península Valdés, Argentina. *Report of the International Whaling Commission (Special Issue)*, *12*, 271–8.

Pettis, H.M. and Hamilton, P.K., 2015. North Atlantic Right Whale Consortium 2015 Annual Report Card. Available at: https://www.narwc.org/report-cards.html (accessed 6 July 2020).

Pettis, H.M. and Hamilton, P.K. 2017. North Atlantic Right Whale Consortium 2017 Annual Report Card. Available at: https://www.narwc.org/report-cards.html (accessed 6 July 2020).

Pettis, H.M. and Hamilton, P.K. 2018. North Atlantic Right Whale Consortium 2018 Annual Report Card. Available at: https://www.narwc.org/report-cards.html (accessed 6 July 2020).

Pettis, H.M., Rolland, R.M., Hamilton, P.K. et al., 2004. Visual health assessment of North Atlantic right whales (*Eubalaena glacialis*) using photographs. *Canadian Journal of Zoology*, *82*, 8–19.

Pettis, H.M., Rolland, R.M., Hamilton, P.K. et al., 2017. Body condition changes arising from natural factors and fishing gear entanglements in North Atlantic right whales *Eubalaena glacialis*. *Endangered Species Research*, *32*, 237–49. doi:10.3354/esr00800

Reeves, R. and Smith, T., 2006. A taxonomy of world whaling. In D.P. DeMaster, D.F. Doak, T.M. Williams, and R.L. Brownell Jr, eds. *Whales, Whaling, and Ocean Ecosystems*, pp. 82–101. University of California Press, Berkeley, CA.

Reeves, R.R., Rolland, R.M., and Clapham, P.J., 2000. Causes of Reproductive Failure in North Atlantic Right Whales: New Avenues of Research. Workshop Report. 26–8 April 2000, Falmouth, MA.

Reilly, S., Bannister, J., Best, P., et al. 2008. *Eubalaena australis* (Chile-Peru subpopulation). IUCN Red List of Threatened Species. Version 2010.4. Available at: www.iucnredlist.org (accessed 6 July 2020).

Rolland, R., Hunt, K., Kraus, S. and Wasser, S., 2005. Assessing reproductive status of right whales (*Eubalaena glacialis*) using fecal hormone metabolites. *General and Comparative Endocrinology*, 142, 308–17.

Rolland, R., McLellan, W.A., Moore, M.J. et al., 2017. Fecal glucocorticoids and anthropogenic injury and mortality in North Atlantic right whales (*Eubalaena glacialis*). *Endangered Species Research*, 34, 417–29.

Rolland, R., Parks, S., Hunt, K. et al., 2012. Evidence that ship noise increases stress in right whales. *Proceedings of the Royal Society B: Biological Sciences*, 279, 2363–8. doi:10.1098/rspb.2011.2429

Rolland, R., Schick, R., Pettis, H. et al., 2016. Health of North Atlantic right whales *Eubalaena glacialis* over three decades: from individual health to demographic and population health trends. *Marine Ecology Progress Series*, 542, 265–82. doi:10.3354/meps11547

Rosenbaum, H.C., Brownell, R., Brown, M. et al., 2000. World-wide genetic differentiation of *Eubalaena*: questioning the number of right whale species. *Molecular Ecology*, 9, 1793–802.

Rowntree, V., McGuinness, P., Marshall, K. et al., 1998. Increased harassment of right whales (*Eubalaena australis*) by kelp gulls (*Larus dominicanus*) at Península Valdés, Argentina. *Marine Mammal Science*, 14, 99–115. doi:10.1111/j.1748-7692.1998.tb00693.x

Rowntree, V.J., Uhart, M.M., Sironi, M. et al., 2013. Unexplained recurring high mortality of southern right whale (*Eubalaena australis*) calves at Península Valdés, Argentina. *Marine Ecology Progress Series*, 493, 275–89. doi:10.3354/meps10506

Schipper, J., Chanson, J., Chiozza, F. et al., 2008. The status of the world's land and marine mammals: diversity, threat and knowledge. *Science*, 322, 225–30.

Sharp, S.M., McLellan, W.A., Rotstein, D.S. et al., 2019. Gross and histopathologic diagnoses from North Atlantic right whale (*Eubalaena glacialis*) mortalities between 2003 and 2018. *Diseases of Aquatic Organisms*, 135, 1–31. doi:10.3354/dao03376

Sironi, M., Alzugaray, L., Saez, A. et al., 2019. The use of underwater viewing deck of the semi-submersible whale-watching vessel Yellow Submarine at Península Valdés, Argentina, as a platform for opportunity for scientific research with the southern right whale, *Eubalaena australis*. *Latin American Journal of Aquatic Mammals*, 13, 2–8. doi:10.5597/lajam00242

Sironi, M., Rowntree, V., Di Martino, M.D. et al., 2014. Updated information for 2012–2013 on southern right whale mortalities at Península Valdés, Argentina.

SC/65b/BRG/06 report presented to the International Whaling Commission Scientific Committee, Portugal. Available at: https://iwc.int/home (accessed 6 July 2020).

Sironi, M., Rowntree, V., Di Martino, M. et al., 2018. Southern right whale mortalities at Península Valdés, Argentina: updated information for 2016–2017. SC/67B/CMP/06 report presented to the International Whaling Scientific Committee, Slovenia. Available at: https://iwc.int/home (accessed 6 July 2020).

Sironi, M., Rowntree, V., Snowdon, C. et al., 2009. Kelp gulls (*Larus dominicanus)* feeding on southern right whales (*Eubalaena australis*) at Península Valdés, Argentina: updated estimates and conservation implications. SC/61/BRG/19 report presented to the International Whaling Commission Scientific Committee, Portugal. Available at: https://iwc.int/home (accessed 6 July 2020).

Sorochan, K.A., Plourde, S.P., Morse, R. et al., 2019. North Atlantic right whale (*Eubalaena glacialis*) and its food: (II) interannual variations in biomass on western North Atlantic shelves. *Journal of Plankton Research*, 41, 687–708.

Szesciorka, A.R., Calambokidis, J., and Harvey, J.T., 2016. Testing tag attachments to increase the attachment duration of archival tags on baleen whales. *Animal Biotelemetry*, 4, 18. doi:10.1186/s40317-016-0110-y

Thomas, P., 1988. Kelp gulls, *Larus dominicanus*, are parasites on flesh of the right whale, *Eubalaena australis*. *Ethology*, 79, 89–103.

Thomas, P., Reeves, R., and Brownell, R., 2016. Status of the world's baleen whales. *Marine Mammal Science*, 32, 682–734.

Thomas, P., Uhart, M., McAloose, D. et al., 2013. Workshop on the southern right whale die-off at Península Valdés, Argentina. SC/61/BRG19, International Whaling Commission, Portugal. Available at: https://iwc.int/home (accessed 6 July 2020).

Tormosov, D., Mikhaliev, Y., Best, P. B. et al., 1998. Soviet catches of southern right whales *Eubalaena australis*, 1951–1971. Biological data and conservation implications. *Biological Conservation*, 86, 185–97. doi:10.1016/S0006-3207(98)00008-1

Trumble, S., Norman, S., Crain, D. et al., 2018. Baleen whale cortisol levels reveal a physiological response to 20th century whaling. *Nature Communications*, 9, 4587.

Tulloch, V., Plagányi, É., Brown, C. et al., 2019. Future recovery of baleen whales is imperiled by climate change. *Global Change Biology*, 25, 1263–81. doi:10.1111/gcb.14573

Van der Hoop, J., Corkeron, P., and Moore, M., 2017. Entanglement is a costly life-history stage in large whales. *Ecology and Evolution*, 7, 92–106. doi:10.1002/ece3.2615

Van der Hoop, J., Nousek-McGregor, A.E., Nowacek, D.P. et al., 2019. Foraging rates of ram-filtering North Atlantic right whales. *Functional Ecology*, 33, 1290–306.

Weisbrod, A., Shea, D., Moore, M., and Stegeman, J., 2009. Organochlorine exposure and bioaccumulation in the endangered North Atlantic right whale (*Eubalaena glacialis*) population. *Environmental Toxicology and Chemistry*, 19, 654–66. doi:10.1002/etc.5620190318

Whitehead, H. and Payne, R., 1981. New techniques for assessing populations of right whales without killing them. *Mammals in the Seas (Report of the Food and Agriculture Organization of the United Nations, Working Party on Marine Mammals)*, 3, 189–211.

Wikelski, M. and Cooke, S.J., 2006. Physiological assessment of environmental stressors. *Trends in Ecology & Evolution*, 1, 38–46.

Zerbini, A., Rosenbaum, H., Mendez, M. et al., 2016. Tracking southern right whales through the southwest Atlantic: an update on movements, migratory routes and feeding grounds. Paper SC/66b/BRG/26 presented to the IWC Scientific Committee, Slovenia.

Zerbini, A.N., Fernández Ajó, A., Andriolo, A. et al., 2018. Satellite tracking of Southern right whales (*Eubalaena australis*) from Golfo San Matías, Rio Negro Province, Argentina. Paper SC/67B/CMP/17 presented to the IWC Scientific Committee, Slovenia.

CHAPTER 13

Weathering the impacts of climate change: methods for measuring the environment at scales relevant to conservation physiology

Brian Helmuth

> ⮌ **Take-home message**
>
> Most plants and animals experience their environments very differently to humans. Ecomechanical approaches provide insights that can enhance the efficacy of conservation measures such as the identification of climate refugia.

13.1 Introduction

For many organisms, the world is an exceedingly small place. Plants, seaweeds, and sessile invertebrates spend the entirety of their existence affixed to one tiny bit of real estate, wholly dependent on the motion of the outside world to deliver food, nutrients, and oxygen, remove wastes, and to disperse their gametes, larvae, or seeds. Small leaf-dwelling arthropods have body temperatures that are strongly affected by the heterogeneity on and within leaves (Caillon et al. 2014). Some reptiles spend their lives in areas covering only tens to hundreds of square metres (Christie et al. 2012) and the home range size of some limpets is less than 1 m^2 (Fenberg 2013). To such creatures, the environmental conditions that ultimately drive their physiology (their microclimates) can depend as much or more on the physical characteristics of their immediate physical habitat (i.e. their microhabitat) as

they do on much larger-scale weather conditions (Kearney 2006; Potter et al. 2013). Even for larger and/or more mobile organisms such as insects, fish, and birds, the presence of habitat-forming species ('ecosystem engineers'; *sensu* Jones et al. 1994) such as trees, grasses, kelps, and bed-forming bivalves such as oysters and mussels can strongly dictate the environments that these organisms experience through effects on shading (Jurgens and Gaylord 2017) or modifications of wind (Jucker et al. 2018) and water flow (Hurd 2015). Yet such details are often forgotten or ignored by ecological forecasts of the impacts of global climate change, especially when focusing on processes such as species ranges, species extinctions, and biodiversity (Woods et al. 2015; Pincebourde et al. 2016). This serves as a major impediment in our understanding of how physiological performance under a changing climate ultimately may drive these larger-scale ecological responses.

Brian Helmuth, *Weathering the impacts of climate change: methods for measuring the environment at scales relevant to conservation physiology*
In: *Conservation Physiology: Applications for Wildlife Conservation and Management.* Edited by: Christine L. Madliger, Craig E. Franklin, Oliver P. Love, and Steven J. Cooke, Oxford University Press (2021). © Oxford University Press. DOI: 10.1093/oso/9780198843610.003.0013

This chapter explores the disconnect between the methods we typically use to measure and model environmental conditions and the mechanisms through which global environmental change ultimately impacts organisms through their physiology. Drawing on the use of short case studies, it examines why a consideration of what may appear superficially to be needlessly pedantic details may offer critical insights into the role of physiology in conservation and management. Specifically, this chapter examines the growing body of literature that suggests that small temporal- and spatial-scale processes may play a much larger role in ecosystem-level and biogeographic processes than previously recognized. If we are to capitalize on the promise of physiological research for informing conservation and management, then these precise studies of physiological mechanisms must be matched with equally careful measurements and models of the environmental conditions that drive them. While the primary focus of this chapter is on the body temperature of organisms—a driver of many if not most physiological processes (Somero 2012)—similar principles apply to a range of other biophysical processes such as oxygen and nutrient exchange (Patterson 1992).

13.2 What scales matter for conservation physiology?

13.2.1 Temporal variability

At a proximal level, individual organisms—including humans—care very little about global climate or mean annual conditions, except to the extent that these ultimately have cascading, teleconnected effects on local weather patterns (Forchhammer and Post 2004). Predictions of physiological performance based only on changes in climatic averages thus have little direct relevance in the real world (Helmuth et al. 2014). While increases in the order of 1.5–3°C (or more) in mean global temperature are forecast in coming decades, changes of this magnitude are already being observed over shorter time scales. Estimates of thermal performance based on weather, rather than on long-term changes in climate thus yield fundamentally different predictions of physiological performance (Helmuth et al. 2014).

The American lobster (*Homarus americanus*) fishery provides a pertinent case study. American lobster is the most economically valuable single-species fishery in the north-eastern United States and Canada (Caputi et al. 2013). The preferred water temperature of *H. americanus* is 5–20°C, with physiological performance declining rapidly at temperatures above 20°C (Aiken and Waddy 1986). Heat waves at the southern end of the species range (in New York, Rhode Island, and Massachusetts south of Cape Cod) have been implicated in mass mortality events (Pearce and Balcom 2005). For example, in 1999 bottom temperatures in Long Island Sound reached 21–23.5°C, corresponding to infection by a parasitic paramoeboid (Caputi et al. 2013). Subsequent incidents have resulted in the collapse of the fishery in much of this region. In contrast, in the Gulf of Maine (GOM), north of Cape Cod, where waters are cooler, catches have surged. In the spring of 2012 water temperatures in the GOM were higher than normal, and by the end of the summer reached levels that were ~3°C warmer than the climatology (Mills et al. 2017). This anomaly lasted for several months, and led to record catches by the commercial lobster fishery in Maine (Mills et al. 2017). Presumably a lot of these increases are the result of increased physiological performance, when the warmer waters were closer to the species' thermal optimum. Importantly, however, the GOM is warming faster than 99 per cent of the global ocean (Pershing et al. 2015). In the future, anomalies similar to those observed in 2012, when added on top of increases in the baseline climatology, will likely result in mortality similar to that observed in the south. Whether any particular year will be a 'boom' or 'bust' thus depends as much on interannual variation in temperatures as it does on climatology. Adaptive conservation measures, such as preserving the stock's reproductive potential by releasing large females can potentially serve as a partial buffer (Le Bris et al. 2018). However, management strategies such as reductions in catch limits during years when temperatures are suboptimal, and allowing fishers to shift to catching other species, are also needed (Caputi et al. 2016; Mills et al. 2017). Critically, these 'early warning systems' could be directly informed by an understanding of the target species' physiology (Cooke and O'Connor 2010; Caputi et al. 2016; Mills et al. 2017).

In some cases, short-term, extreme events such as heat waves (Pansch et al. 2018) and cold snaps (Winkler et al. 2013) can have consequences that last for decades. Wethey et al. (2011) showed, for example, that an extremely cold winter in Europe in 1962–1963 caused changes in the biogeographic distribution of species from which populations were still recovering. As a result, neither contemporary environmental conditions at range edges nor longer-term climatic trends were predictive of observed species range limits. In this case, distributions were attributable to a single extreme winter that had occurred almost 50 years prior.

Even higher frequency (sub-seasonal) temporal variability can have significant physiological and ecological effects. Baumann and Smith (2018) recorded daily variations in coastal pH of nearly one unit, and measurements in streams show similarly high temporal variability in pH, oxygen, and nitrate (Rode et al. 2016). Kearney et al. (2012) and Montalto et al. (2014) used energetics modelling (Dynamic Energy Budgets) to show that exposure to high frequency, fluctuating (hourly) temperature conditions yielded different predictions of reproductive output than did exposure to constant conditions with the same overall mean temperature. In the scenario modelled by Kearney et al. (2012), lizards in the fluctuating environment were exposed to shorter-term, highly stressful conditions punctuated by periods of recovery; under constant conditions animals experienced chronic, low-level stress with no intermittent recovery.

Today, an increasing number of environmental datasets provide high temporal frequency terrestrial data at hourly to daily time scales. Downscaled model forecasts can include stochastically generated 'weather' that statistically represents likely future patterns of daily variability (Dell'Aquila et al. 2012). Used in tandem, these datasets and models facilitate explorations of the importance of temporal data resolution (Mislan and Wethey 2011; Montalto et al. 2014). While fewer databases at these high frequencies exist for the marine environment (Bates et al. 2018), new methods are being adopted to bridge this gap (Canonico et al. 2019). The further development of high-resolution forecasts of

weather conditions under ongoing climate change thus offer considerable promise for the nascent field of conservation physiology (Wikelski and Cooke 2006; Mangano et al. 2020). For example, in the lobster case study above, anomalously warm conditions in 2012 were detected early in the year, and the subsequent impacts on the stock were potentially predictable given a basic understanding of lobster thermal physiology (Mills et al. 2017).

13.2.2 Spatial heterogeneity

Perhaps a larger challenge than the ability to obtain high temporal frequency data lies in recording and modelling environmental data at the high spatial resolutions relevant to an organism's physiology, that is, its microclimate. Remote sensing from satellites still remains the most reliable method of maintaining continuous, near-global coverage of surface conditions (Canonico et al. 2019), but variability in factors such as temperature, soil moisture, and rainfall within each pixel can be very high (Rahaghi et al. 2019). In alpine boulder fields, for example, differences in surface temperature of 18°C have been recorded over the scale of centimetres (Graham et al. 2012). Similarly, Denny et al. (2011) showed that temperature variation within a single intertidal mussel bed rivalled the differences in average temperatures recorded along 14° of latitude. In terrestrial environments, much of this heterogeneity is the result of shading, either by elements of the landscape such as complex surface topography (Zhang et al. 2017, Choi et al. 2019), or by canopy-forming organisms (Jurgens and Gaylord 2017).

When combined with variation in physiological sensitivity to environmental conditions, this spatial heterogeneity can lead to even higher levels of variability in physiological performance, especially when combined with inter-individual differences in thermal tolerance. Dong et al. (2017) found that intraspecific variation in physiological sensitivity among individual snails exceeded differences observed among species. Studies such as these point to the necessity of considering the combined influence of physiological response (sensitivity) and exposure to local environmental conditions when estimating overall vulnerability of populations or species (Maxwell et al. 2019), a concept drawn from the sustainability science literature (Turner et al. 2003).

An increasing number of studies have begun to explore the consequences of microhabitats and microclimates to much larger-scale ecological and biogeographic consequences of climate change (Potter et al. 2013; Pincebourde et al. 2016). In particular, several authors have pointed out how the cumulative effects of processes such as behavioural thermoregulation (Sunday et al. 2014), small-scale refugia (Dong et al. 2017), and genetic variance in tolerance to environmental change (Pearson et al. 2009) may ultimately drive the responses of populations, assemblages, and species distributions in nature. A core message of these studies is that the scale of environmental driver that we must consider cannot be assumed to be the same as that of the ecological response that we are attempting to explain.

A case in point with this issue lies in assumptions, often unsupported, of latitudinal 'gradients' in environmental factors such as temperature (Torossian et al. 2016). Gradients are a ubiquitous feature of biological systems. Gradients in oxygen drive rates of gas exchange over respiratory surfaces, and the larger the difference in concentration across a layer of epithelium, the higher the rate of gas transfer through Brownian motion. Similarly, gradients in temperature at the surface of an animal's skin determine rates of heat gain and loss through convection and conduction; when an animal's body temperature approximates that of its immediate environment, heat flux slows and the organism reaches steady state. Localized concentrations of pollen, seeds, gametes, and larvae can affect rates of spread of invasive species, and, ultimately, weather patterns are driven by gradients in temperature and pressure. True gradients can serve as useful natural laboratories for exploring the influence of environmental stressors.

Yet, increasingly assumptions are made about the presence of gradients that belie the true underlying spatial and/or temporal heterogeneity in the system (De Frenne et al. 2013). At best the term is used imprecisely, especially with studies purporting to study latitudinal gradients in temperature (Torossian et al. 2016). If measured at a sufficiently coarse resolution, geographic gradients indeed do appear to clearly occur. Certainly no one would argue against the observation that the poles are colder than the equator. But crucially, the observed *trend* in decreasing land or sea surface temperature with increasing latitude is not necessarily indicative of a gradient. By definition, a gradient implies a motive force—for example, heat moves along temperature gradients. But simply demonstrating a positive correlation between space or time and a response variable such as temperature using a regression model is not sufficient evidence of a gradient, especially when the number of data points is limited. When such assumptions of spatial patterns are tested using temporally coarse metrics, for example, annual average temperatures, the idea of a gradient tends to hold. But when one examines the actual metrics that are most likely responsible for physiological and ecological responses, such as daily maximum or minimum temperature, the idea of a gradient becomes much more difficult to justify (Helmuth et al. 2006; Seabra et al. 2015).

Temperatures in intertidal zones—the regions between the high and low tidelines along the world's coastlines—provide a case study. Intertidal invertebrates and algae (seaweeds) evolved from marine ancestors, yet daily they are exposed to terrestrial conditions during low tide. In many places, the tides controlling exposure are 'mixed semidiurnal,' meaning that there are effectively two low tides and two high tides that vary in amplitude. Each day, the timing of each tide moves later by ~50 min, and the 'lower low' tide gradually transitions into the 'higher low' tide and vice versa. The end result is that in any given season, the lowest tides, when maximum exposure to air occurs, can tend to occur at specific times of day (Mislan et al. 2009). On the Pacific coast of North America, summer-time low tides tend to occur primarily at night at many sites in southern California, so that even searing midday weather conditions have very little effect on mid- and low intertidal organisms. In contrast, low tides in summer occur midday at many northern sites in Oregon and Washington (Helmuth et al. 2006). Subsequently, mid- and low intertidal animals at these northern sites experience higher body temperatures and thus greater levels of physiological stress than do animals at more southern sites where, overall, weather conditions are hotter (Place et al. 2008). These 'geographic mosaics' have since been shown to occur in many locations (Dong et al. 2017) and habitats (Ackerly et al. 2015) and not only for temperature but also in factors such as ocean pH and phytoplankton availability (Krause-Jensen et al. 2015; Kroeker et al. 2016).

The take-home message here is that our assumptions of how physiological response changes over large geographic gradients cannot be blindly based on our perceptions of how factors such as average temperature change with latitude, but must instead be related to whatever environmental 'signal' is most likely to serve as the proximal driver of the response being measured. Similar arguments against gradients in time can be made when one considers that climate change (a trend in global temperature) often does not mean that, especially at any particular location, each year is warmer than the preceding year. How selection that occurs during an anomalously cold year may affect resilience of a population or community during a subsequent hot year and vice versa remains an active and critical area of investigation (Dillon and Lozier 2019).

A practical consequence for conservation is the ability to identify areas of low vulnerability, that is, 'rescue sites' or conversely, when prioritizing conservation efforts, sites of high vulnerability which may be past saving. Rescue sites (or climate refugia) are increasingly being considered in conservation management strategies (Maclean et al. 2015), e.g., in marine spatial planning (Rilov et al. 2019). For example, Woodson et al. (2019) evaluated microclimatic conditions at two sites with fishing cooperatives in Baja California, Mexico, and showed that they were robust (i.e. showed very little environmental change) to large-scale climate forcing. An important consideration in efforts such as these is a recognition that optimal locations need not occur in the centre of species distributions and that the most stressful conditions may not occur only at range edges (Bozinovic et al. 2011). Instead, they suggest that rescue sites and stepping stones (Hannah et al. 2014), which interact within a metapopulation or metacommunity context to drive overall resilience (Lawson et al. 2012), and rates of range expansion under climate change (Wang et al. 2020), are key to effective design and management (Woodson et al. 2019).

13.2.3 Niche (organism)-level environmental conditions

Ultimately, physiological performance (and hence survival, growth, and reproduction) is driven by cellular and subcellular processes, and so a consideration of how external environmental conditions drive the temperature and chemistry of tissues is critical (Patterson 1992). In other words, while at larger scales we often focus on aspects of the environment such as air, water, and surface temperature, it is an organism's body temperature that drives physiological performance—and these temperatures can be quite different from the temperature of the surrounding habitat. Kearney (2006) describes drivers such as organismal body temperature as niche-level processes, as distinguished from conditions of the habitat/microhabitat or broader environment. This section examines why these distinctions are critical, and why knowledge of broad environmental conditions may not be sufficient to predict physiological stress and ecological function.

As humans, we often fail to recognize the fundamental disconnect between the ways in which we experience our environment, and the ways that most other organisms on the planet experience those same conditions (Helmuth 2002; Pincebourde et al. 2016). For example, like other vertebrates, we are capable of maintaining gas exchange over the surface of our lungs through ventilation, a process that is largely unaffected by external air flow. In stark contrast, many invertebrates and all plants do not have this capability. Instead, their gas exchange surfaces rely on ambient fluid movement (convection) to drive gas exchange. Thus, when flow stops, aquatic animals can experience hypoxia even when there is plenty of oxygen in the surrounding water, due to limitations of mass flux through the diffusion boundary layer overlying the animal's surface (Patterson 1992). For example, during dark respiration many corals can experience anoxia under conditions of restricted water flow (Shashar et al. 1993). In contrast, during peak levels of photosynthesis the reverse can occur and levels of oxygen can become so high in coral tissues that it causes oxidative stress (Finelli et al. 2006). Because levels of turbulence and gas exchange are affected by morphology, two corals exposed to identical conditions of water flow and ambient oxygen can experience markedly different rates of gas exchange: a coral colony with a shape that enhances turbulence can thrive in low flows even while an adjacent, streamlined colony is suffocating (Finelli et al. 2006). Similar processes occur with the uptake of nutrients such as dissolved nitrogen and phosphorus (Thomas and Atkinson 1997). In such cases, simple measurements of ambient oxygen or nutrients in the overlying water column are not sufficient to understand or predict levels of physiological stress; instead, the interactions of the organism with its environment must be considered.

Similar processes drive heat exchange, but as endotherms with high levels of behavioural and physiological control of our body temperatures, the impacts of heat exchange on ectothermic organisms can be difficult for humans to conceptualize. Plants and animals (including homeotherms) exchange heat with their surroundings via the mechanisms of conduction, convection, evaporation, and radiation. However, while for a homeotherm these mechanisms of heat gain and loss are compensated for by decreases and increases in metabolic heat production (van Ooijen et al. 2004), for an ectotherm they drive body temperature in ways that can make assumptions about heat exchange, body temperature, and thermal physiology difficult to intuit (Lewis and Linn 1994).

An understanding of how heat exchange mechanisms affect organism temperature is important for recognizing why measuring environmental conditions can be so difficult, and why seemingly intuitive assumptions can lead us astray (Lewis and Linn 1994). An important first step is understanding that heat (energy) and temperature (average kinetic energy) are not synonymous, and two objects (or organisms) with identical heat content can have very different temperatures and vice versa. The total amount of heat energy required to raise a fixed mass by 1K (equivalent to an increase of 1°C) is termed specific heat capacity (c_p); denser (more mass per volume) materials with a high c_p (such as water) will require more heat energy to raise their temperature and thus have a higher 'thermal inertia'. Seastars will increase the percentage of water in their bodies preceding periods of extreme temperature (Pincebourde et al. 2009). Surface area to volume

ratios have an enormous impact, so that all other factors being equal, larger organisms will take much longer to heat or cool under the same environmental conditions compared with a smaller organism. For example, the large body size of pythons can help them to maintain high and stable temperatures (Shine and Madsen 1996).

By far the largest heat source to most ectothermic organisms is short-wave solar radiation (Kingsolver 1985; Scheffers et al. 2017). Shading by canopy species (Wood et al. 2014; Faye et al. 2017; Scheffers et al. 2017) and by topography (Zhang et al. 2017; Choi et al. 2019) can thus be a significant determinant of ectotherm temperature. But an equally important consideration is the shape, surface area, colour, and surface properties of the organism itself, which determine the heat gain through radiation. The colour of a butterfly's wings can have huge effects on its heat gain through an influence on surface absorptivity (the proportion of incident short-wave radiation absorbed as heat energy), as can its orientation to the sun (Kingsolver 1985), which alters the projected surface area subject to solar radiation. Ore dust pollution has thus been shown to significantly alter animal temperature in intertidal organisms (Erasmus and De Villiers 1982), although the role of colour polymorphism remains unclear (Seuront et al. 2018). Largely because of the influence of heat gain by solar radiation, the temperatures of sun-exposed organisms can be significantly higher—in some cases by 15°C or more—than the temperature of the surrounding air and can vary by 25°C over the course of a few hours (Helmuth 2002).

The rapid heat gain from solar radiation can, to a varying extent, be offset by other mechanisms of heat loss. An ectothermic organism tightly adhered to a rock will readily exchange heat with that surface as the result of direct contact, and its body temperature can rapidly equilibrate with rock temperature via conduction (Wethey 2002). Heat energy flows along a temperature gradient: if the organism is hotter than its underlying substrate it will lose heat and vice versa when it is cooler. Many organisms living in close contact with the ground thus have body temperatures that closely approximate that surface temperature; for example, limpets (Harley et al. 2009), barnacles (Wethey 2002), and small arthropods (Pincebourde et al. 2016). Notably, however, when those rock surfaces are also being heated by solar radiation they too will be considerably hotter than the overlying air, as will the organisms attached to that rock.

Wind and water movement also drive heat exchange with the fluid (air or water) surrounding an organism. Convective heat transfer describes the exchange of heat energy between an organism's outer surface and the fluid in which it is immersed, and is a special form of conduction. Here again heat exchange occurs along temperature gradients and under conditions of high fluid velocity and elevated turbulence, the temperature of an organism can quickly equilibrate with that of the air moving over its surface (Helmuth et al. 2011). Organisms that initially start out cooler than the nearby air can rapidly heat to air temperature through convective heat gain, and organisms hotter than air will cool to the same. In order to fully compensate for heat gain from solar radiation, however, wind velocities and turbulence levels need to be quite high in air (Helmuth et al. 2011). The rate of heat exchange can be strongly affected by organism morphology, and animals and plants that have shapes that enhance turbulence and/or have a large surface area relative to their volume will exchange heat more quickly with the air (Harley et al. 2009). Therefore, during the day air temperature typically sets the lower limits of body temperature for many plants and animals, and the amount of direct solar radiation impacting the surface then increases temperatures above that lower threshold. To some extent these mechanisms also apply under water. Fabricius (2006) and Jimenez et al. (2008) showed that in low flow environments, the tissues of darkly pigmented corals can heat to temperatures higher than the temperature of the surrounding water due to heat gain from solar radiation coupled with decreased heat efflux from convection.

For animals with wet surfaces, heat can also be rapidly dissipated through the phase change of water from liquid to vapour state (evaporation). When liquid undergoes a phase change—a process driven not only by temperature differentials but also via mass flux to the surrounding air—it releases a large amount of heat energy (the latent heat of vaporization), which is then removed from the organism's surface through convection. Depending on the relative humidity of the surrounding air (a measure of the air's water content), an animal or plant's body temperature can drop below that of air (Broitman et al. 2009); the lower the relative humidity the higher the differential.

Finally, heat is also exchanged with the organism's surroundings and with the sky via infrared heat exchange. All surfaces are constantly absorbing and emitting infrared (long-wave) heat energy at a rate proportional to the fourth power of their temperature. Especially at night in the absence of direct solar radiation, infrared heat loss to cloud-free skies (cloud cover increases the emission of infrared heat from the sky to the ground) can be substantial. Via this mechanism, frost can form on surfaces even when local air temperature is above freezing.

For mobile animals, the ability to behaviourally thermoregulate can have highly significant effects on their exposure to potentially damaging environmental conditions (Adolph 1990; Sunday et al. 2014). For example, 'mushrooming' behaviour by intertidal limpets can reduce their body temperatures relative to the underlying rock surface (Williams and Morritt 1995). More commonly, however, mobile organisms behaviourally thermoregulate by moving through heterogeneous environments in order to occupy refuges (Woods et al. 2015). The behavioural rules governing the movement of organisms in thermally heterogeneous environments are often complex (Scheffers et al. 2013; Sears et al. 2016) and for many organisms are not well understood (Monaco et al. 2017). For example, variation among sites in the temporal autocorrelation of environmental conditions—the predictability of future environments given current conditions—is likely to play an important but only minimally explored role (Dong et al. 2017).

The end result of all of these mechanisms of heat exchange is that air temperature can be a very dubious proxy for the temperature of ectothermic plants and animals, and at a minimum the relationship between air and organism temperature needs to be validated before air temperature can be used as a meaningful indicator of physiological stress. It also means that the instruments that we use to record temperature in the field may not reflect the organisms that we care about (Lima et al. 2011). Just as the colour, size, and shape of an organism affect its temperature, so do these same characteristics affect the temperature that an instrument records (Judge et al. 2018). In summary, these studies illustrate that the concept of a single 'site temperature' is flawed, and that sun-exposed environments with any level of substratum complexity can present organisms with extremely high heterogeneity in thermal microenvironments (Choi et al. 2019).

13.2.4 Implications for conservation

A practical consequence of understanding how the environment (and environmental change) drives niche-level conditions and ultimately physiological processes is in how we enact conservation strategies such as interventions, indicators, and early warning systems (Gsell et al. 2016; Mills et al. 2017). For example, while artificial shading can, under some circumstances, be viewed as an anthropogenic disturbance that can alter community composition (Pardal-Souza et al. 2017), it also can be used as part of interventions to reduce the degree of feminization through temperature sex determination (TSD) in turtle eggs (Hill et al. 2015; Esteban et al. 2018). Incubation temperature has long been known to influence sex ratio in sea turtles, and with warmer temperatures under climate change, very high levels of feminization are being observed (Fuentes et al. 2010). Esteban et al. (2018) measured reductions in nest temperatures under different materials, and then, combined with variation among beaches in nest temperature, designed a mitigation matrix to inform nest shading and relocation to help balance nestling sex ratios of three species of sea turtles. Comparably, Wood et al. (2014) investigated the viability of different methods of natural shading, such as tree planting, to reduce the negative effects of increased nest temperatures.

Other novel forms of intervention depend on understanding both physiology and environmental exposure. Farrell et al. (2008) used models of aerobic scope to forecast the success of spawning populations of salmon. This understanding provided a mechanistic approach to explain historical mortality observed during migration. Danner et al. (2012) and Caldwell et al. (2015) modelled the required water flow in rivers needed to maintain oxygen and temperature within acceptable physiological ranges for salmon. This model was then used to inform dam openings and closures during otherwise lethally extreme conditions, offering a solution that maximized water use by both humans and fish.

Habitat complexity can be used as a conservation mechanism, not only by reducing stressors such as temperature (Choi et al. 2019) but also through provision of refugia from predators (Magel et al. 2016). For example, the maintenance and restoration of structuring species such as oyster beds can have spill-over effects on biodiversity and community stability (McAfee et al. 2017) and can facilitate fisheries (Scyphers et al. 2011). At smaller scales, on artificial structures such as seawalls, the enhancement of surface complexity can affect patterns of biodiversity (Chapman and Underwood 2011; Lai et al. 2015; Loke and Todd 2016); however, the presence of such structures can also enhance the spread of invasive species (Prosser et al. 2018).

13.3 Conclusions

The application of a detailed, mechanistic understanding of how organisms physiologically respond to their environment is emerging as a vital approach in climate change adaptation strategies (Chown et al. 2004; Chown and Gaston 2008; Gunderson et al. 2016). Too often, however, the value of this detailed understanding of organismal sensitivity is significantly degraded by combining these measurements with spatially and/or temporally coarse environmental data that may have little direct relevance to the organism or assemblage in question. The use of data such as annual means or climatic averages must be avoided if we are to capitalize on the promise of conservation physiology (Wikelski and Cooke 2006). At a minimum, we need to understand the degree to which these commonly used metrics are correlated with actual drivers of physiological performance, and how those relationships may be changing (Di Cecco and Gouhier 2018). To this end, detailed measurements of environmental conditions at the level of organisms provide windows of insight that otherwise are unavailable. A mechanistic understanding of the combined influence of physiological sensitivity and environmental exposure will continue to play an increasingly important role as conservation biologists attempt to contend with ongoing environmental change.

References

Ackerly, D.D., Cornwell, W.K., Weiss, S.B. et al., 2015. A geographic mosaic of climate change impacts on terrestrial vegetation: which areas are most at risk? *PLoS ONE*, *10*(6), e0130629.

Adolph, S.C., 1990. Influence of behavioral thermoregulation on microhabitat use by 2 *Sceloporus* lizards. *Ecology*, *71*(1), 315–27.

Aiken, D.E. and Waddy, S.L., 1986. Environmental influence on recruitment of the American lobster, *Homarus americanus*: a perspective. *Canadian Journal of Fisheries and Aquatic Sciences*, *43*, 2258–70.

Bates, A.E., Helmuth, B., Burrows et al., 2018. Biologists ignore ocean weather at their peril. *Nature*, *560*, 299–301.

Baumann, H. and Smith, E.M., 2018. Quantifying metabolically driven pH and oxygen fluctuations in US nearshore habitats at diel to interannual time scales. *Estuaries and Coasts*, *41*(4), 1102–17.

Bozinovic, F., Calosi, P., and Spicer, J.I., 2011. Physiological correlates of geographic range in animals. *Annual Review of Ecology, Evolution and Systematics*, *42*, 155–79.

Broitman, B.R., Szathmary, P.L., Mislan, K.A.S. et al., 2009. Predator-prey interactions under climate change: the importance of habitat vs body temperature. *Oikos*, *118*, 219–24.

Caillon, R., Suppoa, C., Casas, J. et al., 2014. Warming decreases thermal heterogeneity of leaf surfaces: Implications for behavioural thermoregulation by arthropods. *Functional Ecology*, *28*, 1449–58.

Caldwell, J., Rajagopalan, B., and Danner, E., 2015. Statistical modeling of daily water temperature attributes on the Sacramento River. *Journal of Hydrological Engineering*, *20*(5), 04014065.

Canonico, G., Buttigieg, P.L., Montes, E. et al., 2019. Global observational needs and resources for marine biodiversity. *Frontiers in Marine Science*, *6*, 367).

Caputi, N., de Lestang, S., Frusher, S., and Wahle, R.A., 2013. The impact of climate change on exploited lobster stocks. In: Phillips, B.E., ed. Lobsters: Management, Aquaculture and Fisheries, second edition, pp. 84–112. Wiley-Blackwell, Chichester

Caputi, N., Kangas, M., Denham et al., 2016. Management adaptation of invertebrate fisheries to an extreme marine heat wave event at a global warming hot spot. *Ecology and Evolution*, *6*(11), 3583–93.

Chapman, M.G. and Underwood, A.J., 2011. Evaluation of ecological engineering of 'armoured' shorelines to improve their value as habitat. *Journal of Experimental Marine Biology and Ecology*, *400*(1–2), 302–13.

Choi, F., Gouhier, T.C., Lima, F. et al., 2019. Mapping physiology: biophysical mechanisms define scales of climate change impacts. *Conservation Physiology*, *7*(1), coz028. doi:10.1093/conphys/coz028.

Chown, S.L. and Gaston, K.J., 2008. Macrophysiology for a changing world. *Proceedings of the Royal Society B*, *275*, 1469–78.

Chown, S.L., Gaston, K.J., and Robinson, D., 2004. Macrophysiology: large-scale patterns in physiological traits and their ecological implications. *Functional Ecology*, *18*, 159–67.

Christie, K., Craig, M.D., Stokes, V.L., and Hobbs, R.J., 2012. Home range size and micro-habitat density requirements of *Egernia napoleonis*: implications for restored Jarrah forest of South Western Australia. *Restoration Ecology*, *20*(6), 740–6.

Cooke, S.J. and O'Connor, C.M., 2010. Making conservation physiology relevant to policy makers and conservation practitioners. *Conservation Letters*, *3*, 159–66.

Danner, E.M., Melton, F.S., Pike, A. et al., 2012. River temperature forecasting: a coupled-modeling framework for management of river habitat. *IEEE Journal of Selected Topics in Applied Earth Observations and Remote Sensing*, *5*(6), 1752–60.

De Frenne, P., Rodriguez-Sanchez, F., Coomes, D.A. et al., 2013. Microclimate moderates plant responses to macroclimate warming. *Proceedings of the National Academy of Sciences of the United States of America*, *110*(46), 18561–5.

Dell'Aquila, A., Calmanti, S., Ruti, P. et al., 2012. Impacts of seasonal cycle fluctuations in an A1B scenario over the Euro-Mediterranean. *Climate Research*, *52*, 135–57.

Denny, M.W., Dowd, W.W., Bilir, L., and Mach, K.J., 2011. Spreading the risk: small-scale body temperature variation among intertidal organisms and its implications for species persistence. *Journal of Experimental Marine Biology and Ecology*, *400*, 175–90.

Di Cecco, G.J. and Gouhier, T.C., 2018. Increased spatial and temporal autocorrelation of temperature under climate change. *Scientific Reports*, *8*, 14850.

Dillon, M.E. and Lozier, J.D., 2019. Adaptation to the abiotic environment in insects: the influence of variability on ecophysiology and evolutionary genomics. *Current Opinion in Insect Science*, *36*, 131–9.

Dong, Y.-W., Li, X.-X., Choi, F.M.P. et al., 2017. Untangling the roles of microclimate, behaviour and physiological polymorphism in governing vulnerability of intertidal snails to heat stress. *Proceedings of the Royal Society B*, *284*, 20162367.

Erasmus, T. and De Villiers, A.F., 1982. Ore dust pollution and body temperatures of intertidal animals. *Marine Pollution Bulletin*, *13*(1), 30–2.

Esteban, N., Laloe, J.O., Kiggen, F. et al., 2018. Optimism for mitigation of climate warming impacts for sea turtles through nest shading and relocation. *Scientific Reports, 8*, 17625.

Fabricius, K.E., 2006. Effects of irradiance, flow, and colony pigmentation on the temperature microenvironment around corals: Implications for coral bleaching? *Limnology and Oceanography, 51*(1), 30–7.

Farrell, A.P., Hinch, S.G., Cooke, S.J. et al., 2008. Pacific salmon in hot water: applying aerobic scope models and biotelemetry to predict the success of spawning migrations. *Physiological and Biochemical Zoology, 81*(6), 697–708.

Faye, E., Rebaudo, F., Carpio, C. et al., 2017. Does heterogeneity in crop canopy microclimates matter for pests? Evidence from aerial high-resolution thermography. *Agriculture Ecosystems & Environment, 246*, 124–33.

Fenberg, P.B., 2013. Intraspecific home range scaling: a case study from the owl limpet (*Lottia gigantea*). *Evolutionary Ecology Research, 15*, 103–10.

Finelli, C.M., Helmuth, B.S.T., Pentcheff, N.D., and Wethey, D.S., 2006. Water flow controls oxygen transport and photosynthesis in corals: potential links to coral bleaching. *Coral Reefs, 25*, 47–57.

Forchhammer, M.C. and Post, E., 2004. Using large-scale climate indices in climate change ecology studies. *Population Ecology, 46*, 1–12.

Fuentes, M.M.P.B., Hamann, M. and Limpus, C.J., 2010. Past, current and future thermal profiles of green turtle nesting grounds: implications from climate change. *Journal of Experimental Marine Biology and Ecology, 383*, 56–64.

Graham, E.A., Rundel, P.W., Kaiser, W. et al., 2012. Fine-scale patterns of soil and plant surface temperatures in an alpine fellfield habitat, White Mountains, California. *Arctic Antarctic and Alpine Research, 44*(3), 288–95.

Gsell, A.S., Scharfenberger, U., Özkundakci, D. et al., 2016. Evaluating early-warning indicators of critical transitions in natural aquatic ecosystems. *Proceedings of the National Academy of Sciences of the United States of America, 113*(50), E8089–95.

Gunderson, A.R., Armstrong, E.J., and Stillman, J.H., 2016. Multiple stressors in a changing world: The need for an improved perspective on physiological responses to the dynamic marine environment. *Annual Review of Marine Science, 8*, 12.1–22.

Hannah, L., Flint, L., Syphard, A.D. et al., 2014. Fine-grain modeling of species' response to climate change: holdouts, stepping-stones, and microrefugia. *Trends in Ecology & Evolution, 29*(7), 390–7.

Harley, C.D.G., Denny, M.W., Mach, K.J. and Miller, L.P., 2009. Thermal stress and morphological adaptations in limpets. *Functional Ecology, 23*, 292–301.

Helmuth, B., 2002. How do we measure the environment? Linking intertidal thermal physiology and ecology through biophysics. *Integrative and Comparative Biology, 42*(4), 837–45.

Helmuth, B., Broitman, B.R., Blanchette, C.A. et al., 2006. Mosaic patterns of thermal stress in the rocky intertidal zone: implications for climate change. *Ecological Monographs, 76*(4), 461–79.

Helmuth, B., Russell, B.D., Connell, S.D. et al., 2014. Beyond long-term averages: making biological sense of a rapidly changing world. *Climate Change Responses, 1*, 10–20.

Helmuth, B., Yamane, L., Lalwani, S. et al., 2011. Hidden signals of climate change in intertidal ecosystems: what (not) to expect when you are expecting. *Journal of Experimental Marine Biology and Ecology, 400*, 191–9.

Hill, J.E., Paladino, F.V., Spotila, J.R., and Tomillo, P.S., 2015. Shading and watering as a tool to mitigate the impacts of climate change in sea turtle nests. *PLoS ONE, 10*(6), e0129528.

Hurd, C., 2015. Slow-flow habitats as refugia for coastal calcifiers from ocean acidification. *Journal of Phycology, 51*, 599–605.

Jimenez, I.M., Kühl, M., Larkum, A.W.D. and Ralph, P.J., 2008. Heat budget and thermal microenvironment of shallow-water corals: do massive corals get warmer than branching corals? *Limnology and Oceanography, 53*, 1548–61.

Jones, C.G., Lawton, J.H., and Shachak, M., 1994. Organisms as ecosystem engineers. *Oikos, 69*, 373–86.

Jucker, T., Hardwick, S.R., Both, S. et al., 2018. Canopy structure and topography jointly constrain the microclimate of human-modified tropical landscapes. *Global Change Biology, 24*(11), 5243–58.

Judge, R., Choi, F., and Helmuth, B., 2018. Recent advances in data logging for intertidal ecology. *Frontiers in Ecology and Evolution, 6*, 213.

Jurgens, L.J. and Gaylord, B., 2017. Physical effects of habitat-forming species override latitudinal trends in temperature. *Ecology Letters, 21*(2), 190–6.

Kearney, M., 2006. Habitat, environment and niche: what are we modelling? *Oikos, 115*(1), 186–91.

Kearney, M.R., Matzelle, A., and Helmuth, B., 2012. Biomechanics meets the ecological niche: the importance of temporal data resolution. *Journal of Experimental Biology, 215*, 922–33.

Kingsolver, J.G., 1985. Thermal ecology of *Pieris* butterflies (Lepidoptera: Pieridae): a new mechanism of behavioral thermoregulation. *Oecologia, 66*, 540–5.

Krause-Jensen, D., Duarte, C.M., Hendriks, I.E., et al., 2015. Macroalgae contribute to nested mosaics of pH variability in a sub-Arctic fjord. *Biogeosciences Discussions, 12*, 4907–45.

Kroeker, K.J., Sanford, E., Rose, J.M. et al., 2016. Interacting environmental mosaics drive geographic variation in mussel performance and species interactions. *Ecology Letters*, 19, 771–9.

Lai, S., Loke, L.H.L., Hilton, M.J. et al., 2015. The effects of urbanisation on coastal habitats and the potential for ecological engineering: a Singapore case study. *Ocean & Coastal Management*, 103, 78–85.

Lawson, C.R., Bennie, J.J., Thomas, C.D. et al., 2012. Local and landscape management of an expanding range margin under climate change. *Journal of Applied Ecology*, 49(3), 552–61.

Le Bris, A., Mills, K.E., Wahle, R.A. et al., 2018. Climate vulnerability and resilience in the most valuable North American fishery. *Proceedings of the National Academy of Sciences of the United States of America*, 115(8), 1831–6.

Lewis, E.L. and Linn, M.C., 1994. Heat energy and temperature concepts of adolescents, adults and experts: implications for curricular improvements. *Journal of Research in Science Teaching*, 31(6), 657–78.

Lima, F.P., Burnett, N.P., Helmuth, B., Aveni-Deforge, K. et al., 2011. Monitoring the intertidal environment with bio-mimetic devices. In George, A., ed. *Advances in Biomimetics*. INTECH publishing, London.

Loke, L.H.L. and Todd, P.A., 2016. Structural complexity and component type increase intertidal biodiversity independently of area. *Ecology*, 97(2), 383–93.

Maclean, I.M.D., Hopkins, J.J., Bennie, J. et al., 2015. Microclimates buffer the responses of plant communities to climate change. *Global Ecology and Biogeography*, 24, 1340–50.

Magel, J.M.T., Pleizier, N., Wilson, A.D.M. et al., 2016. Do physical habitat complexity and predator cues influence the baseline and stress-induced glucocorticoid levels of a mangrove-associated fish? *Comparative Biochemistry and Physiology, Part A*, 203, 281–7.

Mangano, M.C., Mieszkowska, N., Helmuth, B., et al., 2020 Moving toward a strategy for addressing climate displacement of marine resources: a proof-of-concept. *Frontiers in Marine Science*, 7, 408.

Maxwell, S.L., Butt, N., Maron, M. et al., 2019. Conservation implications of ecological responses to extreme weather and climate events. *Diversity and Distributions*, 25(4), 613–25.

McAfee, D., O'Connor, W.A., and Bishop, M.J., 2017. Fast-growing oysters show reduced capacity to provide a thermal refuge to intertidal biodiversity at high temperatures. *Journal of Animal Ecology*, 86(6), 1352–62.

Mills, K.E., Pershing, A.J., and Hernandez, C.M., 2017. Forecasting the seasonal timing of Maine's lobster fishery. *Frontiers in Marine Science*, 4, 337.

Mislan, K.A.S. and Wethey, D.S., 2011. Gridded meteorological data as a resource for mechanistic macroecology

in coastal environments. *Ecological Applications*, 21(7), 2678–90.

Mislan, K.A.S., Wethey, D.S., and Helmuth, B., 2009. When to worry about the weather: role of tidal cycle in determining patterns of risk in intertidal ecosystems. *Global Change Biology*, 15(12), 3056–65.

Monaco, C.J., McQuaid, C.D., and Marshall, D.J., 2017. Decoupling of behavioural and physiological thermal performance curves in ectothermic animals: a critical adaptive trait. *Oecologia*, 185(4), 583–93.

Montalto, V., Sará, G., Ruti, P. et al., 2014. Testing the effects of temporal data resolution on predictions of bivalve growth and reproduction in the context of global warming. *Ecological Modelling*, 278, 1–8.

Pansch, C., Scotti, M., Barboza, F.R. et al., 2018. Heat waves and their significance for a temperate benthic community: a near-natural experimental approach. *Global Change Biology*, 24(9), 4357–67.

Pardal-Souza, A.L., Dias, G.M., Jenkins, S.R. et al., 2017. Shading impacts by coastal infrastructure on biological communities from subtropical rocky shores. *Journal of Applied Ecology*, 54(3), 826–35.

Patterson, M.R., 1992. A chemical engineering view of cnidarian symbioses. *American Zoologist*, 32(4), 566–82.

Pearce, J. and Balcom, N., 2005. The 1999 Long Island Sound Lobster mortality event: findings of the Comprehensive Research Initiative. *Journal of Shellfish Research*, 24(3), 691–7.

Pearson, G.A., Lago-Leston, A., and Mota, C., 2009. Frayed at the edges: selective pressure and adaptive response to abiotic stressors are mismatched in low diversity edge populations. *Journal of Ecology*, 97, 450–62.

Pershing, A.J., Alexander, M.A., Hernandez, C.M. et al., 2015. Slow adaptation in the face of rapid warming leads to collapse of the Gulf of Maine cod fishery. *Science*, 350(6262), 809–12.

Pincebourde, S., Murdock, C.C., Vickers, M., and Sears, M.W., 2016. Fine-scale microclimatic variation can shape the responses of organisms to global change in both natural and urban environments. *Integrative and Comparative Biology*, 56(1), 45–61.

Pincebourde, S., Sanford, E., and Helmuth, B., 2009. An intertidal sea star adjusts thermal inertia to avoid extreme body temperatures. *The American Naturalist*, 174(6), 890–7.

Place, S.P., O'Donnell, M.J., and Hofmann, G.E., 2008. Gene expression in the intertidal mussel *Mytilus californianus*: physiological response to environmental factors on a biogeographic scale. *Marine Ecology Progress Series*, 356, 1–14.

Potter, K.A., Woods, H.A., and Pincebourde, S., 2013. Microclimatic challenges in global change biology. *Global Change Biology*, 19, 2932–9.

Prosser, D.J., Jordan, T.E., Nagel, J. et al., 2018. Impacts of coastal land use and shoreline armoring on estuarine ecosystems: an introduction to a special issue. *Estuaries and Coasts*, 41(Suppl 1), S2–18.

Rahaghi, A.I., Lemmin, U., and Barry, D.A., 2019. Surface water temperature heterogeneity at subpixel satellite scales and its effect on the surface cooling estimates of a large lake: airborne remote sensing results from Lake Geneva. *Journal of Geophysical Research—Oceans*, 124(1), 635–51.

Rilov, G., Mazaris, A.D., Stelzenmüller, V. et al., 2019. Adaptive marine conservation planning in the face of climate change: what can we learn from physiological, genetic and ecological studies? *Global Environmental Change*, 17, e00566.

Rode, M., Wade, A.J., Cohen, M.J. et al., 2016. Sensors in the stream: the high-frequency wave of the present. *Environmental Science & Technology*, 50(19), 10297–307.

Scheffers, B.R., Brunner, R.M., Ramirez, S.D. et al., 2013. Thermal buffering of microhabitats is a critical factor mediating warming vulnerability of frogs in the Philippine biodiversity hotspot. *Biotropica*, 45(5), 628–35.

Scheffers, B.R., Edwards, D.P., Macdonald, S.L. et al., 2017. Extreme thermal heterogeneity in structurally complex tropical rain forests. *Biotropica*, 49(1), 35–44.

Scyphers, S.B., Powers, S.P., Heck, K.L., and Byron, D., 2011. Oyster reefs as natural breakwaters mitigate shoreline loss and facilitate fisheries. *PLoS ONE*, 6(8), e22396.

Seabra, R., Wethey, D.S., Santos, A.M., and Lima, F.P., 2015. Understanding complex biogeographic responses to climate change. *Scientific Reports*, 5, 12930.

Sears, M.W., Angilleta, M.J., Schuler S. et al., 2016. Configuration of the thermal landscape determines thermoregulatory performance of ectotherms. *Proceedings of the National Academy of Sciences of the United States of America*, 113(38), 10595–600.

Seuront, L., Ng, T.P.T., and Lathlean, J.A., 2018. A review of the thermal biology and ecology of molluscs, and of the use of infrared thermography in molluscan research. *Journal of Molluscan Studies*, 84, 203–32.

Shashar, N., Cohen, Y., and Loya, Y., 1993. Extreme diel fluctuations of oxygen in diffusive boundary layers surrounding stony corals. *Biological Bulletin*, 185, 455–61.

Shine, R. and Madsen, T., 1996. Is thermoregulation unimportant for most reptiles? An example using water pythons (*Liasis fuscus*) in tropical Australia. *Physiological Zoology*, 69(2), 252–69.

Somero, G.N., 2012. The physiology of global change: linking patterns to mechanisms. *Annual Review of Marine Science*, 4, 39–61.

Sunday, J.M., Bates, A.E., Kearney, M.R. et al., 2014. Thermal-safety margins and the necessity of thermoregulatory behavior across latitude and elevation. *Proceedings of the National Academy of Sciences of the United States of America*, 111(15), 5610–15.

Thomas, F.I.M. and Atkinson, M.J., 1997. Ammonium uptake by coral reefs: effects of water velocity and surface roughness on mass transfer. *Limnology and Oceanography*, 42(1), 81–8.

Torossian, J.L., Kordas, R.L., and Helmuth, B., 2016. Cross-scale approaches to forecasting biogeographic responses to climate change. *Advances in Ecological Research*, 55, 371–433.

Turner, B.L., Kasperson, R.E., Matson, P.A. et al., 2003. A framework for vulnerability analysis in sustainability science. *Proceedings of the National Academy of Sciences of the United States of America*, 100(14), 8074–9.

van Ooijen, A.M.J., van Marken Lichtenbelt, W.D., van Steenhoven, A.A., and Westerterp, K.R., 2004. Seasonal changes in metabolic and temperature responses to cold air in humans. *Physiology & Behavior*, 82, 545–53.

Wang, W., Wang, J., Choi, F.M.P. et al., 2020. Global warming and artificial shorelines reshape seashore biogeography. *Global Ecology and Biogeography*, 29, 220–31.

Wethey, D.S., 2002. Biogeography, competition, and microclimate: the barnacle *Chthamalus fragilis* in New England. *Integrative and Comparative Biology*, 42, 872–80.

Wethey, D.S., Woodin, S.A., Hilbish, T.J. et al., 2011. Response of intertidal populations to climate: Effects of extreme events versus long term change. *Journal of Experimental Marine Biology and Ecology*, 400(1–2), 132–44.

Wikelski, M. and Cooke, S.J., 2006. Conservation physiology. *Trends in Ecology & Evolution*, 21(2), 38–46.

Williams, G.A. and Morritt, D., 1995. Habitat partitioning and thermal tolerance in a tropical limpet, *Cellana grata*. *Marine Ecology Progress Series*, 124(1–3), 89–103.

Winkler, D.W., Luo, M.K., and Rakhimberdiev, E., 2013. Temperature effects on food supply and chick mortality in tree swallows (*Tachycineta bicolor*). *Oecologia*, 173(1), 129–38.

Wood, A., Booth, D.T., and Limpus, C.J., 2014. Sun exposure, nest temperature and loggerhead turtle hatchlings: Implications for beach shading management strategies at sea turtle rookeries. *Journal of Experimental Marine Biology and Ecology*, 451, 105–14.

Woods, H.A., Dillon, M.E., and Pincebourde, S., 2015. The roles of microclimatic diversity and of behavior in mediating the responses of ectotherms to climate change. *Journal of Thermal Biology*, 54, 86–97.

Woodson, C.B., Micheli, F., Boch, C. et al., 2019. Harnessing marine microclimates for climate change adaptation and marine conservation. *Conservation Letters*, 12(2), e12609.

Zhang, Y.L., Chang, X.L., and Liang, J., 2017. Comparison of different algorithms for calculating the shading effects of topography on solar irradiance in a mountainous area. *Environmental Earth Sciences*, 76(7), 295.

PART V

Improving Wildlife Rehabilitation and Captive Management Programmes

A veterinary perspective on the conservation physiology and rehabilitation of sea turtles

Charles Innis and Kara Dodge

> ⮑ **Take-home message**
>
> Understanding the physiological dysfunction caused by human-made and natural stressors improves the veterinary management and rehabilitation of imperilled sea turtle species and provides resource managers with a deeper understanding of risks to sea turtle populations.

14.1 Introduction: the conservation challenge

Turtles have existed for approximately 250 million years. The first known marine turtle, *Odontochelys*, appeared approximately 220 million years ago, and by 100 million years ago, four families of marine turtles existed. Today, only two of those families survive. Family *Dermochelidae* has only one remnant species, the leatherback turtle (*Dermochelys coriacea*). The leatherback evolved approximately 100 million years ago, and is the largest living turtle species, with exceptionally large adults approaching 900 kg and over 1 m in length. Family *Cheloniidae*, the 'hard-shelled' sea turtles, has six species: green turtle (*Chelonia mydas*), hawksbill turtle (*Eretmochelys imbricata*), loggerhead turtle (*Caretta caretta*), olive ridley turtle (*Lepidochelys olivacea*), Kemp's ridley turtle (*Lepidochelys kempii*), and flatback turtle (*Natator depressus*).

The life history of turtles is forgiving of the loss of eggs and juveniles to environmental variables and predation, but turtle populations are negatively impacted by premature adult mortality (Congdon et al. 1993). It may take decades for a turtle to reach adult size and sexual maturity. Historically, upon reaching adulthood, turtles had a very high annual survival rate, reproduced for decades, and had a high chance of self-replacement prior to death. But human population growth and activities have added many new threats for adults, and turtles in general have suffered from habitat loss, road mortality, and overcollection (in many cases for the illegal wildlife trade). Sea turtle populations, specifically, have declined globally due to fisheries interactions, egg harvest, coastal development, direct hunting, recreational and commercial boating, marine debris, and pollution. The consequences of climate change for sea turtle populations are not yet clear but are predicted to be adverse. Sea turtles are categorized using terms such as 'critically endangered', 'endangered', 'threatened', and 'vulnerable' by various countries and global conservation organizations.

Charles Innis and Kara Dodge, *A veterinary perspective on the conservation physiology and rehabilitation of sea turtles* In: *Conservation Physiology: Applications for Wildlife Conservation and Management*. Edited by: Christine L. Madliger, Craig E. Franklin, Oliver P. Love, and Steven J. Cooke, Oxford University Press (2021). © Oxford University Press. DOI: 10.1093/oso/9780198843610.003.0014

Sea turtles are often found ill or injured. In many countries, marine rehabilitation facilities provide veterinary care for marine mammals, sea birds, and sea turtles. With proper intervention, many injured or ill sea turtles can be returned to the wild. For example, in the United States, at least 11 000 sea turtles have been released to the wild after successful rehabilitation (Figure 14.1) (Innis et al. 2019). Despite these successes, the role of rehabilitation in conservation remains unclear and controversial, in general. Pundits question whether rehabilitated animals resume normal behaviour and contribute to the breeding population, whether these animals represent a disease risk, and whether resources used for rehabilitation are better used for other aspects of conservation (Caillouet et al. 2016). Proponents show evidence that at least some rehabilitated individuals have been later documented as alive and breeding, that the public outreach and education provided by rehabilitation

programmes is significant, and that our understanding of species biology is enhanced by studying turtles during rehabilitation (Innis et al. 2019).

In parallel (and often in collaboration) with sea turtle biologists, veterinarians and rehabilitation personnel are adding substantially to our understanding of sea turtle biology and conservation. In this chapter, we describe physiological insights that have been gained during veterinary assessment and rehabilitation of sea turtles, and the value of this work for the conservation of sea turtles. Specifically, physiological assessments of sea turtles involved in fisheries interactions, cold-stunning events, oil spills, and watercraft trauma have been important in identifying physiological derangements, improving triage and treatment of affected animals, and improving rehabilitation outcomes. Physiological assessment methods including point-of-care acid–base and blood gas analysis, cardiac monitoring (electrocardiogram [ECG], echocardi-

Figure 14.1 Kemp's ridley sea turtles are released to the ocean after being rehabilitated following a cold-stunning event. The physiological response of these turtles to long-distance transport was investigated (Hunt et al. 2019, 2020). Courtesy of New England Aquarium.

ography, Doppler), haematological and plasma bio-chemical analysis, endocrine assessment (e.g. adrenocortical and thyroid status), and diagnostic imaging modalities such as radiology, sonography, and computed tomography are often used. This chapter will provide evidence that documenting the physiological dysfunction caused by various anthropogenic and natural stressors can be useful for resource managers in developing management plans for endangered species. Sea turtles, in particular, have been cited as an example of specific and appropriate use of physiological data in establishing conservation recovery plans (Mahoney et al. 2018).

14.2 Physiological assessment of sea turtle threats

14.2.1 Fisheries interactions

Fisheries that result in the intentional capture and killing of sea turtles for consumption have been prohibited in many countries but are a persistent concern for some species and regions (Wallace et al. 2011). However, incidental capture (bycatch) of sea turtles in various fisheries remains one of the greatest threats to sea turtle populations (Wallace et al. 2013). Fisheries that involve trawls, gillnets, driftnets, long lines, purse seines, pot/trap buoy lines, weirs/pound nets, and dredges can cause prolonged forced submergence and physical trauma. Mitigation measures have been implemented in some fisheries (e.g. the use of 'turtle excluder devices' in trawl nets), yet many turtles continue to drown or suffer fatal trauma. Of particular interest are turtles that remain alive at the time of capture but are so physiologically deranged or physically traumatized that they will later die. Such 'post-release mortality' is the subject of robust debate among resource managers. In some jurisdictions, protocols for estimating post-release mortality have been developed, such that annual mortality can be estimated (Stacy et al. 2016). While laudable, such estimates often rely largely on expert opinion and limited empirical data. Due to the challenges of studying ocean-going animals and inferring outcomes from satellite telemetry (Swimmer and Gilman 2012), there is a lack of robust studies

involving physical and physiological assessment and post-release monitoring of variably affected individuals. Clearly, when a management agency does not 'count' interactions as fatal when they truly are fatal, the impact of those interactions is underestimated.

The effects of forced submergence in trawls and gillnets have been studied experimentally and under natural conditions for several sea turtle species, often involving point-of-care analysers to document the turtles' physiological status. Such analysers are generally small, portable, and somewhat rugged, and can be safely and effectively utilized in real time in the field. In general, and as expected for an air-breathing vertebrate, forced submersion can result in respiratory and metabolic acidosis, hypoxia, hyperkalaemia, hyperphospha-taemia, hyperglycaemia, high corticosterone, and evidence of muscle exertion (i.e. elevated plasma concentrations of creatine kinase and lactate; Stabenau et al. 1991; Harms et al. 2003; Stabenau and Vietti 2003; Snoddy et al. 2009; Phillips et al. 2015). Snoddy et al., (2009) found that derangements in blood biochemistry (increased blood lactate, lactate dehydrogenase, creatine kinase, phosphorus, and glucose) were associated with a decline in health status as indicated by physical examination. While duration appears to play an important role in determining the outcome of forced submergence (Sasso and Epperly 2006; Snoddy et al. 2009), factors such as the capture depth and severity of entanglement may also affect post-release survival. Such data have compelled resource managers, for example, to implement soak time limits and seasonal prohibition of unattended recreational gillnets (Snoddy et al. 2009).

Only recently, veterinarians in Europe documented decompression sickness (DCS) during clinical evaluation of loggerhead turtles involved in fisheries interactions (García-Párraga et al. 2014; Fahlman et al. 2017). While sea turtles move between great depths without incident during their normal behavioural repertoire, they appear to be prone to DCS during conditions of forced submergence at depth, exertion (i.e. escape efforts), and rapid ascent during net retrieval. Affected turtles develop severe gas embolism, with gas apparent in the heart, major blood vessels, brain, kidneys, spleen, and liver. Diagnosis of DCS can be made by several modalities including radiography, sonography, computed tomography, gas composition analysis, and positive response to hyperbaric oxygen therapy (HBOT). Spirometry data (i.e. measuring the speed and volume of respiratory air flow) indicate that expiratory and inspiratory flow, and tidal volume are reduced by DCS, but improve after HBOT (Portugues et al. 2018). This major discovery has clear implications for post-release mortality estimates. If DCS is occurring worldwide, many forcibly submerged turtles that are released alive due to externally 'good' appearance may later die due to the combined effects of DCS and physiological dysfunction. The physiological derangements specifically associated with DCS in sea turtles have not been documented to date, but investigation is clearly warranted.

In the western North Atlantic, leatherback turtles seasonally forage off south-eastern Canada and the north-eastern United States, where they commonly become entangled in vertical ropes that are used to connect pots (traps) to surface buoys (Innis et al. 2010; Hamelin et al. 2017). Entanglement in fixed pot-rope gear can result in severe physical trauma and physiological dysfunction for these turtles (Innis et al. 2010), and a high percentage of leatherbacks show evidence of scarring from past rope interactions (Archibald and James 2018). Affected turtles have a stress response (elevated corticosterone), metabolic changes that may eventually lead to negative energy balance (elevated thyroxine and beta-hydroxybutyrate [BHB], and depressed blood urea nitrogen [BUN]) (Figure 14.2) (Innis et al. 2010; Hunt et al. 2016b). While post-entanglement survival has been documented for some leatherback turtles (Innis et al. 2010), the outcome for more severely injured or physiologically deranged individuals warrants further study.

Overall, while obvious mortality of sea turtles in fisheries has been noted for decades, recognition of the significance of cryptic post-release mortality for sea turtles and other animals is more recent. Accurate estimation of losses resulting from fisheries bycatch requires an informed, comprehensive understanding of the physiological effects of these interactions beyond simply counting the observed live and dead.

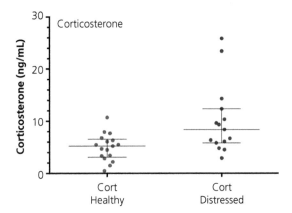

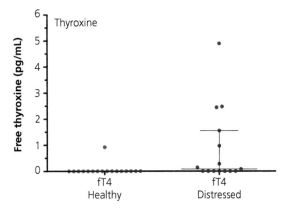

Figure 14.2 Corticosterone (Cort; top panel) and free thyroxine (fT4; bottom panel) in 17 healthy leatherback turtles and 15 distressed leatherback turtles (entangled or stranded). Scatterplots show all data points; horizontal bars are medians and whiskers are interquartiles. Corticosterone and fT4 concentrations were both significantly higher in distressed turtles than in healthy turtles. Courtesy of *Conservation Physiology*, Hunt et al. (2016b).

14.2.2 Cold-stunning

Natural 'cold-stunning' of free-ranging sea turtles in the United States and Europe has been reported for many years (e.g. Roberts et al. 2014). The origin of the term 'cold-stunning' is unclear, but it has come into general use to describe the lethargic to moribund condition of sea turtles during exposure to relatively cold environmental temperatures. Cold-stunned sea turtles that are found stranded on beaches are often transferred to rehabilitation facilities for treatment.

Being at the lower threshold of survivable body temperatures, cold-stunned turtles may be affected by fatal to near-fatal physiological disruption. Physiological assessment of affected turtles often indicates severe bradycardia, apnoea, respiratory and metabolic acidosis, hypoxia, hyperkalaemia, severe dehydration, and coagulopathy (Innis et al. 2007, 2009; Keller et al. 2012; Stacy et al. 2013; Barratclough et al. 2019). Assessment of renal function by traditional plasma biochemical analysis (e.g. uric acid), as well as novel evaluation by iohexol clearance testing (measurement of glomerular filtration rate, GFR) indicates reduced renal function, which generally improves during rehabilitation, but may evolve into renal failure in some cases (Innis et al. 2009, 2016; Kennedy et al. 2012).

Endocrine assessment has also been fruitful for assessing the status of cold-stunned turtles and influencing their clinical treatment. Cold-stunned Kemp's ridley turtles show very depressed thyroid function, while their plasma corticosterone values are among the highest documented for any sea turtle under any stressor (Hunt et al. 2012). These derangements lead to ongoing complications in rehabilitation such as persistent anorexia, and susceptibility to opportunistic infections such as pneumonia, sepsis, and osteomyelitis.

Retrospective analysis of physiological data derived from cold-stunned Kemp's ridley turtles led to the development of the first objective mortality prediction index for a turtle species (Figure 14.3) (Stacy et al. 2013). Using receiver-operating-characteristic analysis, a method that utilizes patient data to derive an optimal model for prognostic sensitivity and specificity, a highly specific and sensitive model for mortality prediction was derived, indicating that blood pH, pO_2, and potassium were highly relevant in predicting mortality. This model has subsequently been used at New England Aquarium to direct triage decisions when very large numbers of cold-stunned turtles are found stranded simultaneously (i.e. directing effort and resources towards turtles with better prognoses). The model has also been utilized for comparative analysis of turtles exposed to other stressors, such as turtles evaluated during the Deepwater Horizon (DWH) oil spill, as discussed later in this chapter (Stacy et al. 2013).

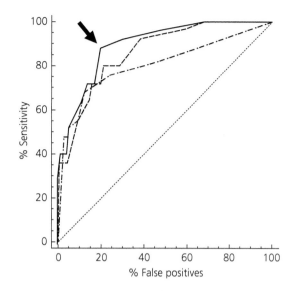

Figure 14.3 Diagnostic performance of three indices (dash-dotted, dashed, and continuous lines) for prediction of mortality in Kemp's ridley sea turtles based on blood analysis. The diagonal (dotted) line indicates the area under the curve = 0.50 (a non-informative test). Arrow shows the cut-off value for predicting mortality using the index with highest sensitivity = 88.0 and specificity = 80.5 per cent (solid line). This index relies on assessment of blood pH, pO_2, and potassium. Courtesy of *Conservation Physiology*, Stacy et al. (2013).

14.2.3 Oil spills

Sea turtles have been affected by a number of oil spills; however, physiological effects have only been intensively studied during the 2010 DWH oil spill in the Gulf of Mexico (Stacy et al. 2017). Previously, experimental exposure of loggerhead turtles to crude oil indicated possible haemolytic anaemia and salt gland dysfunction, but these conditions were not well characterized (Lutcavage et al. 1995). During the DWH spill, hundreds to thousands of turtles were likely oiled, many of which are believed to have died while mired in oil at sea (Wallace et al. 2017). Rescue efforts recovered several hundred oiled sea turtles at sea, many of which were juvenile Kemp's ridley and green turtles (Stacy et al. 2017). Upon transfer to a rehabilitation site, point-of-care analysers and haematologic and plasma biochemical analyses were used to serially assess their physiological status. Results were interpreted by experienced sea turtle veterinarians, and data were also evaluated using previously developed objective mortality prediction indices.

Many oiled turtles mounted a physiological stress response and were affected by acidosis and electrolyte abnormalities, likely caused by the combined effects of oiling, hyperthermia, and transport. Severe physiological disruption was diagnosed in 36 per cent of turtles at time of admission to rehabilitation, with 25 per cent exceeding the threshold for predicted mortality without clinical intervention (Stacy et al. 2017). These data were used to direct the clinical treatment of the turtles, and to follow their improvement during rehabilitation, resulting in >90 per cent survival rate (Stacy et al. 2017). These data were also used to estimate the magnitude of sea turtle losses resulting from the spill as part of the natural resource damage assessment (Wallace et al. 2017). Results indicated that large numbers of oiled turtles that remained at sea (i.e. those that were not hospitalized) likely died during the event (Wallace et al. 2017; NOAA 2019). This information was combined with data derived for other involved species (e.g. Venn-Watson et al. 2015) and contributed to a US$20 billion legal settlement, with portions of that settlement delegated to sea turtle conservation efforts (NOAA 2019).

14.2.4 Vessel strike

Sea turtles are often injured by collisions with commercial and recreational vessels (Hazel and Gyuris 2006; Denkinger et al. 2013; Mettee and Norton 2017; Dourdeville et al. 2018; Foley et al. 2019). Such collisions can result in various combinations of sharp and blunt trauma from impact with the hull and components of steering and propulsion systems. In many cases, the collision is quickly fatal, while other turtles sustain injuries that result in delayed mortality or morbidity (Foley et al. 2019). Carapace trauma may result in damage to the lungs and other coelomic viscera (e.g. intestinal transection). Traumatic brain injury, decapitation, enucleation (loss of one or both eyes), limb amputation, and even complete body transection may occur. As for other causes of premature adult mortality, there is potential for boat strike to remove large numbers of reproductively mature adults from the population when they aggregate in coastal breeding and feeding areas. In Massachusetts, between 2012 and 2017, at least 33 adult leatherback turtles were killed

by vessel strike (Massachusetts Audubon Wellfleet Bay Wildlife Sanctuary, unpublished data), including at least three females that had previously been documented nesting in the Caribbean (Dourdeville et al. 2018). A recent large-scale study of vessel strikes in Florida found evidence of injury from watercraft in as many as one third of stranded turtles of some species and reported the highest frequency of occurrence in reproductively active adults (Foley et al. 2019).

While such injuries are commonly managed by veterinarians, to date, there is no large, detailed case series that describes physiological and physical assessment of sea turtles after vessel strike. General discussions of their status indicate that inflammatory leukograms, anaemia, hypoproteinaemia, hypocalcaemia, and dehydration may be seen (Mettee and Norton, 2017). Leukograms may show heterophilia, lymphocytosis, monocytosis, left shift (immature heterophilia), and toxic cellular changes (Mettee and Norton 2017). It is expected that future studies will further define the extent of physiological derangement for such cases, and that terminal changes such as respiratory and metabolic acidosis and hyperkalaemia will be detected.

Based on the frequency and severity of boat strike, and its potential for severe adverse population effects, mitigation measures should be considered, including seasonal speed limits, seasonal closures of high use habitat, and enforcement of such regulations (Shimada et al. 2017; Foley et al. 2019).

14.3 Clinical studies of physiology enhance understanding of sea turtle general biology

As a result of these many threats, injured and ill sea turtles often receive veterinary care. Assessment of sea turtles during veterinary interventions has produced novel data regarding basic sea turtle biology. In some cases, aspects of sea turtle biology that were previously unstudied, or poorly studied, have been enhanced by clinical experiences. For example, clinical studies of renal function in convalescent cold-stunned Kemp's ridley turtles resulted in the largest dataset for glomerular filtration rates for any sea turtle species to date (Innis et al. 2016). Clinical evaluation of entangled leatherback turtles indicated that BHB may be an important metabolic indicator for sea turtles (Innis et al. 2010), leading to further investigation of BHB during fasting under controlled conditions (Price et al. 2013). The safety and success of various sea turtle ecology studies, laparoscopic validation of sex identification methods for determining sex ratios, and studies of sea turtle vision and hearing have been facilitated by collaboration of veterinarians and sea turtle biologists (e.g. Harms et al. 2007, 2014; Crognale et al. 2008; Dodge et al. 2014; Innis et al. 2014; Piniak et al. 2016; Shertzer et al. 2018). Such studies may aid in the development of acoustic and visual deterrents to fisheries interactions, and studies of sea turtle hearing may also be important in developing regulations and procedures regarding oceanic exploration and effects of seismic testing.

Physiological differences among various life stages, species, seasons, and populations of sea turtles have been documented by veterinary assessments and basic field research. Changes in haematologic and blood chemistry data have been recorded as sea turtles mature, including increasing packed cell volume, lymphocyte percentage, blood urea nitrogen, albumin, total protein, cholesterol, and decreasing glucose, aspartate aminotransferase, alkaline phosphatase, and heterophil percentage (Bradley et al. 1998; Hamann et al. 2006; Kakizoe et al. 2007; Fong et al. 2010; Labrada-Martagón et al. 2010; Delgado et al. 2011; Rousselet et al. 2013). These baseline data could be useful comparatively for future conservation studies that seek to assess sea turtle maturation and recruitment into the adult population. Failure of newly recruited turtles to show the expected physiological maturation could indicate resource limitation or other factors that are delaying maturation.

Sex-based and seasonal physiological differences may also be seen. While the mechanisms behind these variations are not entirely known, they are likely due to differing metabolic conditions related to migration, foraging, water temperature, muscle and fat mass, breeding condition, etc. Male green turtles in the Bahamas had significantly higher uric acid concentrations and lower cholesterol concentrations than females (Bolten and Bjorndal 1992).

Male green turtles in the Arabian Sea had higher concentrations of calcium, higher aspartate aminotranferase activity, lower concentrations of iron, and lower lactate dehydrogenase activity compared with females (Hasbún et al. 1998). For leatherback turtles, males show higher plasma alkaline phosphatase and sodium, and a lower monocyte percentage compared with females (Innis et al. 2010; Harris et al. 2011). Kelly et al. (2015) compared summer and autumn blood data for juvenile loggerhead turtles in North Carolina. In autumn, turtles had higher total leukocyte count, heterophils, monocytes, total protein, globulin, and blood urea nitrogen, with lower haematocrit and chloride values in comparison with summer. Inter-annual variation among blood biochemical data within the same location has been shown for green turtles (Labrada-Martagón et al. 2010).

Female sea turtles, in particular, appear to enter altered metabolic and physiological states during folliculogenesis, egg production, and nesting, including mobilization of lipid, protein, and calcium. Blood urea nitrogen may be an order of magnitude lower in nesting loggerhead and leatherback turtles in comparison with foraging females, but the reason and mechanism for this change are not known (Deem et al. 2009; Innis et al. 2010; Perrault et al. 2012). Casal et al. (2009) documented relatively high triglycerides, cholesterol, total protein, albumin, globulin, and calcium in nesting female loggerhead turtles. With the exception of increased calcium, Harris et al. (2011) noted decreased concentrations or activities of many biochemical parameters in nesting leatherbacks, including several tissue enzymes, total protein, globulin, cholesterol, potassium, and uric acid. Perrault et al. (2012) evaluated maternal health indices of nesting leatherback turtles and found correlations between egg hatching success, hatchling emergence, and plasma alkaline phosphatase activity, blood urea nitrogen, calcium, calcium:phosphorus ratio, carbon dioxide, cholesterol, creatinine, and phosphorus. Honarvar et al. (2011) found that haematocrit, calcium, potassium, sodium, phosphorus, total protein, albumin, and globulin concentrations decreased over the nesting season in female leatherbacks. Goldberg et al. (2013) found decreasing serum concentrations of leptin (an appetite-suppressing hormone) and increasing concentrations of ghrelin (a hunger-stimulating hormone) in female hawksbills through the nesting season, which may stimulate food intake towards the end of nesting or after the post-nesting migration. In addition to haematologic and bio-chemical evaluation, other medical tools have been used to document demographic variation in sea turtles. For example, ultrasonographic fat measure-ments have been defined in leatherback sea turtles, providing an objective assessment of body condi-tion and an additional tool for future comparative studies (Harris et al. 2016). Collectively, these stud-ies highlight the extensive physiological changes that may occur in sea turtles throughout their lives, and points to the importance of understanding such changes when interpreting and utilizing such data.

14.4 Improving capture, handling, and transportation of sea turtles

Sea turtles may be intentionally captured for eco-logical research, and sea turtles are often trans-ported long distances for treatment, rehabilitation, and subsequent release. Capture events may result in physical exertion and physiological changes, so monitoring is warranted to ensure the safety of the event. For example, assessment of healthy free-ranging leatherback turtles during capture events indicate a modest increase in plasma corticosterone, blood pH, and blood potassium concentrations (Figure 14.4) (Innis et al. 2014; Hunt et al. 2016b). While not documented in sea turtles to date, severe hyperkalaemia could lead to cardiac arrhythmias during prolonged or exertional capture events.

For clinics in northern latitudes, rehabilitated tur-tles may need to be transported hundreds of miles south for release at a warm-water beach. In the United States, sea turtles are commonly transported in individual boxes in trucks, often for longer than 24 h, while some turtles are transported by plane. Transportation exposes a sea turtle to numerous stimuli, including removal from water, handling, confinement in a small dry container, substantial increases in ambient noise and vibration, and pos-sibly temperature variation. Sea turtles, like other vertebrates, have a robust physiological stress response, exhibiting rises in corticosterone in response to stressors such as cold-stunning and

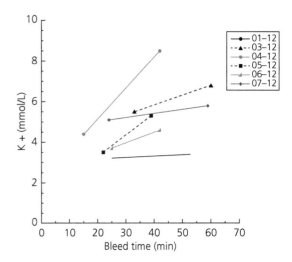

Figure 14.4 Post-capture and pre-release blood potassium concentrations vs. time for six leatherback turtles (*Dermochelys coriacea*) as determined during ecological field studies. The significant rise in potassium is likely secondary to exertion. Hyperkalaemia can cause life-threatening cardiac arrhythmias, justifying physiological monitoring during such studies. Courtesy of *Conservation Physiology*, Innis et al. (2014).

entanglement (Snoddy et al. 2009; Hunt et al. 2012). No standard criteria currently exist for best practices of transporting sea turtles of any species. Experience indicates that transport duration can be shortened appreciably by adjusting timing of release dates (e.g. waiting to release turtles until summer months when nearby beaches have warm water) or by adjusting logistics (e.g. adding staff/vehicles to allow rotating teams of drivers). During transport of Kemp's ridley and loggerhead turtles ranging up to 26 h in dur-ation, significant elevations in stress-related ana-lytes have been detected, including elevations in corticosterone, glucose, white blood cell count, and heterophil:lymphocyte ratio (Hunt et al. 2016a; Hunt et al. , 2020). However, the stress response was only modest in comparison with more severe stressors (e.g. cold-stunning, entanglement), and other physiological variables were considered normal (e.g. blood gas status, pH, electrolytes). As a result, transportation of sea turtles under well-planned, controlled conditions for up to 24 h can be considered safe. These results were used comparatively to dem-onstrate that the deranged physiological status of turtles affected by the DWH oil spill was not likely attributable to transport alone (Stacy et al. 2017).

Preliminary data suggest that it could be of benefit to briefly recover transported sea turtles in temporary saltwater tanks near the release site. Hunt et al. (2019) showed that the physiological stress response following transportation of Kemp's ridley turtles largely dissipates if they are allowed to recover in a holding pool for 6 h, or overnight. Whether this effort is truly of measurable benefit requires further study of post-release behaviour; however, it stands to reason that potentially adverse physiological changes should and likely can be minimized when possible.

Finally, research into the anaesthetic management of sea turtles has aided both clinical management and basic research (Moon and Stabenau 1996; Chittick et al. 2002; Harms et al. 2007; MacLean et al. 2008; Phillips et al. 2017). Turtles may need to be anaesthetized for some diagnostic procedures (e.g. bronchoscopy), and must be anaesthetized for surgical procedures (e.g. removal of fishing line from the intestine). Anaesthesia may also be needed for basic research involving turtle hearing and vision (Harms et al. 2007). Physiological assessment during anaesthetic events often involves determination of body temperature, heart rate, respiratory rate, venous blood gas analysis, and other blood biochemical data. Advances in anaesthesia protocols made in the course of managing sea turtle cases in rehabilitation (e.g. Chittick et al. 2002) increased the margin of safety and dramatically decreased anaesthesia recovery times, making it possible to use anaesthesia during field research, and facilitating permitting decisions for such studies in which turtles are released soon after full anaesthetic recovery (Harms et al. 2007; MacLean et al. 2008).

14.5 Other tools

Many other tools have been used to study sea turtle physiology, ecology, and behaviour. Complementary and collaborative studies between sea turtle biologists and veterinarians are seeking to understand post-release survival of medically managed turtles by combining data on physical and physiological conditions with subsequent telemetry data on turtle movements and survival. Satellite telemetry tags have been used extensively to study behaviours and habitat preferences of sea turtles, including turtles released from fishing gear and rehabilitation. Determination of the post-release fate of bycaught sea turtles is essential for evaluating the true impacts of different fisheries on populations, and different satellite tag models have been used to infer outcomes. Some studies have employed pop-up satellite archival tags (PSATs) (Swimmer et al. 2006, 2014; Sasso and Epperly 2007) while other studies have used time-depth-recorders (TDRs) to infer post-release outcomes of disentangled turtles (Snoddy and Williard 2010) and identify opportunistic fisheries-related mortality (Hays et al. 2003). However, there has been considerable debate as to the suitability of traditional satellite tags for inferring post-release mortality, since it is difficult to distinguish true mortality from tag failure (e.g. Chaloupka et al. 2004; Swimmer and Gilman 2012).

More recently, 'survivorship' pop-up archival satellite tags (sPATs) have been developed to overcome some of these limitations and improve confidence in post-release fate determinations. These tags have been used extensively in studies on sharks and fish and offer promise for future studies of post-release survival in sea turtles (Maxwell et al. 2018). Satellite tags have also been used to monitor the post-release fate of rehabilitated sea turtles. These tags can provide information on short-term survival, as well as movements, dive behaviour, and habitat preferences (Cardona et al. 2012; Mestre et al. 2014; Coleman et al. 2017). This information can be compared with data from wild conspecifics to assess reintegration of rehabilitated individuals and provides a basis on which to evaluate the conservation benefit of rehabilitation to sea turtle population recovery efforts. Post-release monitoring studies of sea turtle behaviour have shown that injured and rehabilitated turtles can be successfully reintroduced to the wild, and bycaught turtles can survive and resume normal behaviours after release from fishing gear. A major limitation of satellite telemetry in sea turtle ecology studies is their relatively short battery life relative to the lifespan of turtles. For example, a juvenile turtle released with a satellite tag may be effectively monitored for 1–2 years, whereas it may be decades before its contribution to the breeding population is realized.

Acoustic telemetry has also been used for post-release monitoring of sea turtles (Hart et al. 2012;

MacDonald et al. 2013). While this technology has been shown to be highly effective for monitoring post-release survival, movements, and site fidelity in fish, sharks, and pinnipeds, there have been comparatively few acoustic tracking studies for sea turtles. Acoustic tags have a longer battery life (up to 10 years) compared with satellite tags; however, externally attached tags are susceptible to premature shedding (Lamont et al. 2015). Surgically implanted acoustic tags are associated with longer tag detections and have been safely deployed in a range of non-turtle species (Horning et al. 2008; Kneebone et al. 2014; Bino et al. 2018). To improve our understanding of long-term post-release outcomes for bycaught and rehabilitated sea turtles, implanted acoustic tags could be a viable and cost-effective alternative to satellite tags.

While satellite and acoustic tags are useful for inferring outcomes, they offer no insight on individual condition prior to release. Integrating physical and physiological profiles of individual turtles with post-release telemetry data can improve future veterinary assessment and management of sea turtle entanglement cases, potentially enhancing post-entanglement survival. Integration of telemetry and physiology data can also improve our understanding of terminal injuries in sea turtles, which can be used to make humane euthanasia decisions.

Technological innovations and miniaturization of sensors over the past three decades have resulted in a suite of on-animal archival tags, or 'biologgers', that collect detailed behavioural, physiological, and environmental data (Rutz and Hays 2009). In sea turtles, physiological parameters such as body temperature (Southwood et al. 2005) and heart rate (Southwood et al. 1999) have been collected using biologgers. Accelerometers have been used to infer behaviours and energy expenditure (Wilson et al. 2008), and tags are being developed to understand sea turtle response to their acoustic environment (Tyson et al. 2017). Direct observations from animal-borne cameras (Reina et al. 2005), remotely operated vehicles (ROVs) (Patel et al. 2016), autonomous underwater vehicles (AUVs) (Dodge et al. 2018), and unmanned aerial vehicles (UAVs) (Rees et al. 2018) can give insight on physiological parameters (respiratory rate), as well as energy expenditure (inferred from stroke rate), behaviours (feeding,

social), and habitat. Since biologgers collect large quantities of high-resolution data, it is necessary to recover these devices. Future innovations in marine animal tagging should include remote data transmission capability for high-resolution archival tags, as well as continued miniaturization of tags for smaller taxa and rechargeable power sources for extending tag deployment durations. These features will revolutionize our understanding of sea turtle biology and biodiversity at a global scale.

14.6 Conclusions

Physiological assessment of sea turtles during their natural repertoire of behaviours and under the effect of various stressors has been productive. Data derived from such studies have already been used in informing management plans and damage assessment plans for these endangered species. Yet, much remains to be studied. It is important to better establish the long-term outcome for turtles that are adversely affected during various human interactions, in comparison with their physiological state at the time of those interactions. In some cases, physical injury and deranged physiological states will be fatal, while effects in other cases may be relatively minor. Circumstances that result in the death of large numbers of adult sea turtles will adversely affect the population, so such situations must be clearly recognized by resource managers. Future research by sea turtle biologists, veterinarians, and rehabilitation personnel should seek to understand cryptic post-release mortality, the contribution of rehabilitated turtles to the breeding population, and the effects of emerging stressors. Accomplishing such goals will require deployment of existing technologies, such as implanted telemetry systems, and the development of new technologies, such as miniaturized multi-faceted physiological sensors, and improved tag systems to detect delayed mortality. Resource managers and permitting agencies should encourage and facilitate such studies.

Acknowledgements

We thank Drs Brian Stacy, Nicole Stacy, and Craig Harms for critically reviewing and enhancing this chapter.

References

Archibald, D.W. and James, M.C., 2018. Prevalence of visible injuries to leatherback sea turtles *Dermochelys coriacea* in the Northwest Atlantic. *Endangered Species Research*, 37, 149–63.

Barratclough, A., Tuxbury, K., Hanel, R. et al., 2019. Baseline plasma thromboelastography in Kemp's ridley (*Lepidochelys kempii*), green (*Chelonia mydas*) and loggerhead (*Caretta caretta*) sea turtles and its use to diagnose coagulopathies in cold-stunned Kemp's ridley and green sea turtles. *Journal of Zoo and Wildlife Medicine*, 50(1), 62–68.

Bino, G., Kingsford, R.T., Grant, T., et al., 2018. Use of implanted acoustic tags to assess platypus movement behaviour across spatial and temporal scales. *Scientific Reports*, 8(1), 1–12.

Bradley, T.A., Norton, T.M., and Latimer, K.S., 1998. Hemogram values, morphologic characteristics of blood cells, and morphometric studies of loggerhead sea turtles, *Caretta caretta*, in the first year of life. *Bulletin of the Association of Reptilian and Amphibian Veterinarians*, 8(3), 8–16.

Bolten, A.B. and Bjorndal, K.A., 1992. Blood profiles for a wild population of green turtles (*Chelonia mydas*) in the southern Bahamas: size-specific and sex-specific relationships. *Journal of Wildlife Diseases*, 28, 407–13.

Caillouet, C.W., Putman, N.F., Shaver, D.J. et al., 2016. A Call for evaluation of the contribution made by rescue, resuscitation, rehabilitation, and release translocations to kemp's ridley sea turtle (*lepidochelys kempii*) population recovery. *Herpetological Conservation and Biology*, 11, 486–96.

Cardona, L., Fernández, G., Revelles, M., and Aguilar, A., 2012. Readaptation to the wild of rehabilitated loggerhead sea turtles (*Caretta caretta*) assessed by satellite telemetry. *Aquatic Conservation: Marine and Freshwater Ecosystems*, 22, 104–12.

Casal, A.B., Camacho, M., López-Juardo, L.F. et al., 2009. Comparative study of hematologic and plasma biochemical variables in Eastern Atlantic juvenile and adult nesting loggerhead sea turtles (*Caretta caretta*). *Veterinary Clinical Pathology*, 38, 213–21.

Chaloupka, M., Parker, D., and Balazs, G., 2004. Modelling post-release mortality of loggerhead sea turtles exposed to the Hawaii-based pelagic longline fishery. *Marine Ecology Progress Series*, 28, 285–93.

Chittick, E.J., Stamper, M.A., Beasley, J.F. et al., 2002. Medetomidine, ketamine, and sevoflurane for anesthesia of injured loggerhead sea turtles: 13 cases (1996–2000). *Journal of the American Veterinary Medical Association*, 221, 1019–25.

Coleman, A.T., Pitchford, J.L., Bailey, H., and Solangi, M., 2017. Seasonal movements of immature Kemp's ridley sea turtles (*Lepidochelys kempii*) in the northern gulf of Mexico. *Aquatic Conservation: Marine and Freshwater Ecosystems*, 27, 253–67.

Congdon, J.D., Dunham, A.E., and van Loben Sels, R.C., 1993. Delayed sexual maturity and demographics of Blanding's turtles (*Emydoidea blandingii*): implications for conservation and management of long-lived organisms. *Conservation Biology*, 7, 826–33.

Crognale, M.A., Eckert, S.A., Levenson, D.H., and Harms, C.A., 2008. Leatherback sea turtle *Dermochelys coriacea* visual capacities and potential reduction of bycatch by pelagic longline fisheries. *Endangered Species Research*, 5, 249–56.

Deem, S.L., Norton, T.M., Mitchell, M. et al., 2009. Comparison of blood values in foraging, nesting, and stranded loggerhead turtles (*Caretta caretta*) along the coast of Georgia, USA. *Journal of Wildlife Diseases*, 45, 41–56.

Delgado, C., Valente, A., Quaresma, I. et al., 2011. Blood biochemistry reference values for wild juvenile loggerhead sea turtles (*Caretta caretta*) from Madeira archipelago. *Journal of Wildlife Diseases*, 47, 523–9.

Denkinger, J., Parra, M., Muñoz, J.P. et al., 2013. Are boat strikes a threat to sea turtles in the Galapagos Marine Reserve? *Ocean and Coastal Management*, 80, 29–35.

Dodge, K.L., Galuardi, B., Miller, T.J., and Lutcavage, M.E., 2014. Leatherback turtle movements, dive behavior, and habitat characteristics in ecoregions of the Northwest Atlantic Ocean. *PLoS ONE*, 9, e91726.

Dodge, K.L., Kukulya, A.L., Burke, E., and Baumgartner, M.F., 2018. TurtleCam: a 'smart' autonomous underwater vehicle for investigating behaviors and habitats of sea turtles. *Frontiers in Marine Science*, 5, 90. doi:10.3389/fmars.2018.00090

Dourdeville, K.M., Wynne, S., Prescott, R. et al., 2018. Three-island, intra-season nesting leatherback (*Dermochelys coriacea*) killed by vessel strike off Massachusetts, USA. *Marine Turtle Newsletter*, 155, 8–11.

Fahlman, A., Crespo-Picazo, J.L., Sterba-Boatwright, B. et al., 2017. Defining risk variables causing gas embolism in loggerhead sea turtles (*Caretta caretta*) caught in trawls and gillnets. *Scientific Reports*, 7, 2739. doi:10.1038/s41598-017-02819-5

Foley A.M., Stacy, B.A., Hardy, R.F. et al., 2019. Characterizing watercraft-related mortality of sea turtles in Florida (USA). *Journal of Wildlife Management*, 83(5), 1057–72.

Fong, C.L., Chen, H.C., and Cheng, I.J., 2010. Blood profiles from wild populations of green sea turtles in Taiwan. *Journal of Veterinary Medicine and Animal Health*, 2(2), 8–10.

García-Párraga, D., Crespo-Picazo, J.L, de Quirós, Y.B. et al., 2014. Decompression sickness ('the bends') in sea turtles. *Diseases of Aquatic Organisms*, 111, 191–205.

Goldberg, D.W., Leitao, S.A.T., Godfrey, M.H. et al., 2013. Ghrelin and leptin modulate the feeding behaviour of the hawksbill turtle *Eretmochelys imbricata* during nesting season. *Conservation Physiology*, 1, cot016. doi:10.1093/conphys/cot016

Hamann, M., Schäuble, C.S., Simon, T., and Evans, S., 2006. Demographic and health parameters of green sea turtles *Chelonia mydas* foraging in the Gulf of Carpentaria, Australia. *Endangered Species Research*, 2, 81–8.

Hamelin, K.M., James, M.C., Ledwell, W. et al., 2017. Incidental capture of leatherback sea turtles in fixed fishing gear off Atlantic Canada. *Aquatic Conservation: Marine and Freshwater Ecosystems*, 27, 631–42.

Harms, C.A., Eckert, S.A., Kubis, S.A. et al., 2007. Field anaesthesia of leatherback sea turtles (*Dermochelys coriacea*). *Veterinary Record*, 161, 15–21.

Harms, C.A., Mallo, K.M., Ross, P.M., and Segars, A., 2003. Venous blood gases and lactates of wild loggerhead sea turtles (*Caretta caretta*) following two capture techniques. *Journal of Wildlife Diseases*, 39, 366–74.

Harms, C.A., Piniak, W.E.D., Eckert, S.A., and Stringer, E.M., 2014. Sedation and anesthesia of hatchling leatherback sea turtles (*Dermochelys coriacea*) for auditory evoked potential measurement in air and in water. *Journal of Zoo and Wildlife Medicine*, 45, 86–92.

Harris, H.S., Benson, S.R., Gilardi, K.V. et al., 2011. Comparative health assessment of Western Pacific leatherback turtles (*Dermochelys coriacea*) foraging off the coast of California, 2005–2007. *Journal of Wildlife Disease*, 47, 321–37.

Harris, H.S., Benson, S.R., James, M.C. et al., 2016. Validation of ultrasound as a noninvasive tool to measure subcutaneous fat depth in leatherback sea turtles (*Dermochelys coriacea*). *Journal of Zoo and Wildlife Medicine*, 47, 275–9.

Hart, K.M., Sartain, A.R., Fujisaki, I. et al., 2012. Home range, habitat use, and migrations of hawksbill turtles tracked from Dry Tortugas National Park, Florida, USA. *Marine Ecology Progress Series*, 457, 193–207.

Hasbún, C.R., Lawrence, A.J., Naldo, J. et al., 1998. Normal blood chemistry of free-living green sea turtles, *Chelonia mydas*, from the United Arab Emirates. *Comparative Haematology International*, 8, 174–7.

Hays, G.C., Broderick, A.C., Godley, B.J. et al., 2003. Satellite telemetry suggests high levels of fishing-induced mortality in marine turtles. *Marine Ecology Progress Series*, 262, 305–9.

Hazel, J. and Gyuris, E., 2006. Vessel-related mortality of sea turtles in Queensland, Australia. *Wildlife Research*, 33, 149–54. doi:10.1071/WR04097

Honarvar, S., Brodsky, M.C., Fitzgerald, D.B. et al., 2011. Changes in plasma chemistry and reproductive output of nesting leatherbacks. *Herpetologica*, 67, 222–35.

Horning, M., Haulena, M., Tuomi, P.A., and Mellish, J.A., 2008. Intraperitoneal implantation of life-long telemetry transmitters in otariids. *BMC Veterinary Research*, 4, 51. doi:10.1186/1746-6148-4-51

Hunt, K.E., Innis, C.J., Kennedy, A.E. et al., 2016a. Assessment of ground transportation stress in juvenile Kemp's ridley sea turtles (*Lepidochelys kempii*). *Conservation Physiology*, 4, cov071. doi:10.1093/conphys/cov071

Hunt, K.E., Innis, C., Merigo, C., and Rolland, R.M., 2016b. Endocrine responses to diverse stressors of capture, entanglement and stranding in leatherback turtles (*Dermochelys coriacea*). *Conservation Physiology*, 4, cow022. doi:10.1093/conphys/cow022

Hunt, K.E., Innis, C., Merigo, C. et al., 2019. Ameliorating transport-related stress in endangered Kemp's ridley sea turtles (*Lepidochelys kempii*) with a recovery period in saltwater pools. *Conservation Physiology*, 7(1), coy065.

Hunt, K.E., Innis, C. and Rolland, R.M., 2012. Corticosterone and thyroxine in cold-stunned Kemp's ridley sea turtles (*Lepidochelys kempii*). *Journal of Zoo and Wildlife Medicine*, 43, 479–93.

Hunt, K.E., Merigo, C., Burgess, E.A. et al., 2020. Effects of ground transport in Kemp's ridley (*Lepidochelys kempii*) and loggerhead (*Caretta caretta*) turtles. *Integrative Organismal Biology*, 2(1). doi.org/10.1093/iob/obaa012

Innis, C.J., Finn, S., Kennedy, A. et al., 2019. A summary of sea turtles released from rescue and rehabilitation programs in the United States, with observations on re-encounters. *Chelonian Conservation and Biology*, 18, 3–9.

Innis, C.J., Kennedy, A., McGowan, J.P. et al., 2016. Glomerular filtration rates of naturally cold-stunned Kemp's ridley turtles (*Lepidochelys kempii*): comparison of initial vs. convalescent values. *Journal of Herpetological Medicine and Surgery*, 26, 100–3.

Innis, C.J., Merigo, C., Cavin, J.M. et al., 2014. Serial assessment of the physiological status of leatherback turtles (*Dermochelys coriacea*) during direct capture events in the northwestern Atlantic Ocean: comparison of post-capture and pre-release data. *Conservation Physiology*, 2, cou048. doi:10.1093/conphys/cou048

Innis, C., Merigo, C., Dodge, K. et al., 2010. Health evaluation of leatherback turtles (*Dermochelys coriacea*) in the northwestern Atlantic during direct capture and fisheries gear disentanglement. *Chelonian Conservation and Biology*, 9, 205–22.

Innis, C.J., Ravich, J.B., Tlusty, M.F. et al., 2009. Hematologic and plasma biochemical findings in cold-stunned Kemp's ridley sea turtles (*Lepidochelys kempii*). *Journal of the American Veterinary Medical Association*, 235, 426–32.

Innis, C., Tlusty, M., Merigo, C., and Weber, E.S., 2007. Metabolic and respiratory status of cold-stunned

Kemp's ridley sea turtles (*Lepidochelys kempii*). *Journal of Comparative Physiology B, 177,* 623–30.

Kakizoe, Y., Sakaoka, K., Kakizoe, F. et al., 2007. Successive changes of hematologic characteristics and plasma chemistry values of juvenile loggerhead turtles (*Caretta caretta*). *Journal of Zoo Wildlife Medicine, 38,* 77–84.

Keller, K., Innis, C., Tlusty, M.F. et al., 2012. Metabolic and respiratory derangements associated with death in cold-stunned Kemp's ridley turtles (*Lepidochelys kempii*): 32 cases (2005–2009). *Journal of the American Veterinary Medical Association, 240,* 317–23.

Kelly, T.R., Braun McNeill, J., Avens, L. et al., 2015. Clinical pathology reference intervals for an in-water population of juvenile loggerhead sea turtles (*Caretta caretta*) in Core Sound, North Carolina, USA. *PLoS ONE, 10*(3), e0115739. doi:10.1371/journal.pone.0115739

Kennedy, A., Innis, C. and Rumbeiha, W., 2012. Determination of glomerular filtration rate in juvenile Kemp's ridley turtles (*Lepidochelys kempii*) using iohexol clearance, with preliminary comparison of clinically healthy turtles vs. those with renal disease. *Journal of Herpetological Medicine and Surgery, 22,* 25–9.

Kneebone, J., Chisholm, J. and Skomal, G., 2014. Movement patterns of juvenile sand tigers (*Carcharias taurus*) along the east coast of the USA. *Marine Biology, 161,* 1149–63.

Labrada-Martagón, V., Méndez-Rodríguez, L.C., Gardner, S.C. et al., 2010. Health indices of the green turtle (*Chelonia mydas*) along the Pacific coast of Baja California Sur, Mexico. I. Blood biochemistry values. *Chelonian Conservation and Biology, 9,* 162–72.

Lamont, M.M., Fujisaki, I., Stephens, B.S., and Hackett, C., 2015. Home range and habitat use of juvenile green turtles (*Chelonia mydas*) in the northern Gulf of Mexico. *Animal Biotelemetry, 3,* 53. doi:10.1186/s40317-015-0089-9

Lutcavage, M.E., Lutz, P.L., Bossart, G.D., and Hudson, D.M., 1995. Physiologic and clinicopathologic effects of crude oil on loggerhead sea turtles. *Archives of Environmental Contamination and Toxicology, 28,* 417–22.

MacDonald, B.D., Madrak, S.V., Lewison, R.L. et al., 2013. Fine scale diel movement of the east Pacific green turtle, *Chelonia mydas*, in a highly urbanized foraging environment. *Journal of Experimental Marine Biology and Ecology, 443,* 56–64.

MacLean, R.A., Harms, C.A., and Braun-McNeill, J., 2008. Propofol anesthesia in loggerhead (*Caretta caretta*) sea turtles. *Journal of Zoo and Wildlife Medicine, 44,* 143–50.

Mahoney, J.L., Klug, P.E., and Reed, W.L., 2018. An assessment of the US endangered species act recovery plans: using physiology to support conservation. *Conservation Physiology, 6*(1), coy036. doi:10.1093/conphys/coy036

Maxwell, S.M., Witt, M.J., Abitsi, G. et al., 2018. Sea turtles and survivability in demersal trawl fisheries: do coma-

tose olive ridley turtles survive post-release? *Animal Biotelemetry, 6,* 11. doi:10.1186/s40317-018-0155-1

Mettee, N.S. and Norton, T.M., 2017. Trauma and wound care. In C.A. Manire, T.M. Norton, B.A. Stacy et al., eds. Sea Turtle Health and Rehabilitation, pp. 657–74. J. Ross Publishing, Plantation, FL.

Mestre, F., Bragança, M.P., Nunes, A., and dos Santos, M.E., 2014. Satellite tracking of sea turtles released after prolonged captivity periods. *Marine Biology Research, 10,* 996–1006.

Moon, P.F. and Stabenau, E.K., 1996. Anesthetic management of sea turtles. *Journal of the American Veterinary Medical Association, 208,* 720–6.

NOAA, 2019. Three ways we are restoring sea turtles after Deepwater Horizon. Available at: https://www.fisheries.noaa.gov/feature-story/three-ways-we-are-restoring-sea-turtles-after-deepwater-horizon (accessed 15 July 2019).

Patel, S.H., Dodge, K.L., Haas, H.L., and Smolowitz, R.J., 2016. Videography reveals in-water behavior of loggerhead turtles (*Caretta caretta*) at a foraging ground. *Frontiers in Marine Science, 3,* 254. doi:10.3389/fmars.2016.00254

Perrault, J.R., Miller, D.L, Eads, E. et al., 2012. Maternal health status correlates with nest success of leatherback sea turtles (*Dermochelys coriacea*) from Florida. *PLoS ONE, 7*(2), e31841. doi:10.1371/journal.pone.0031841

Phillips, B.E., Cannizzo, S.A., Godfrey, M.H. et al., 2015. Exertional myopathy in a juvenile green sea turtle (*Chelonia mydas*) entangled in a large mesh gillnet. *Case Reports in Veterinary Medicine,* 604320. http://dx.doi.org/10.1155/2015/604320

Phillips, B.E., Posner, L.P., Lewbart, G.A. et al., 2017. Effects of alfaxalone administered intravenously to healthy yearling loggerhead sea turtles (*Caretta caretta*) at three different doses. *Journal of the American Veterinary Medical Association, 250,* 909–17.

Piniak, W.E.D., Mann, D.A., Harms, C.A. et al., 2016. Hearing in the juvenile green sea turtle (*Chelonia mydas*): a comparison of underwater and aerial hearing using auditory evoked potentials. *PLoS ONE, 11*(10), e0159711.

Portugues, C., Crespo-Picazo, J.L., García-Párraga, D. et al., 2018. Impact of gas emboli and hyperbaric treatment on respiratory function of loggerhead sea turtles (*Caretta caretta*). *Conservation Physiology, 6*(1), cox074. doi:10.1093/conphys/cox074

Price, E.R., Jones, T.T., Wallace, B.P., and Guglielmo, C.G., 2013. Serum triglycerides and ß-hydroxybutyrate predict feeding status in green turtles (*Chelonia mydas*): evaluating a single blood sample method for assessing feeding/fasting in reptiles. *Journal of Experimental Marine Biology and Ecology, 439,* 176–80.

Rees, A., Avens, L., Ballorain, K. et al., 2018. The potential of unmanned aerial systems for sea turtle research and conservation: a review and future directions. *Endangered Species Research*, 35, 81–100.

Reina, R.D., Abernathy, K.J., Marshall, G.J., and Spotila, J.R., 2005. Respiratory frequency, dive behaviour and social interactions of leatherback turtles, *Dermochelys coriacea* during the inter-nesting interval. *Journal of Experimental Marine Ecology and Biology*, 316, 1–16. doi:10.1016/j.jembe.2004.10.002

Roberts, K., Collins, J., Paxton, C.H. et al., 2014. Weather patterns associated with green turtle hypothermic stunning events in St. Joseph Bay and Mosquito Lagoon, Florida. *Physical Geography*, 35, 134–50.

Rousselet, E., Stacy, N.I., La Victoire, K. et al., 2013. Hematology and plasma biochemistry of immature, captive-reared loggerhead sea turtles (*Caretta caretta*). *Journal of Zoo and Wildlife Medicine*, 44, 859–74.

Rutz, C. and Hays, G.C., 2009. New frontiers in biologging science. *Biology Letters*, 5, 289–92.

Sasso, C.R. and Epperly, S.P., 2006. Seasonal sea turtle mortality risk from forced submergence in bottom trawls. *Fisheries Research*, 81(1), 86–8.

Sasso, C.R. and Epperly, S.P., 2007. Survival of pelagic juvenile loggerhead turtles in the open ocean. *Journal of Wildlife Management*, 71, 1830–5.

Shertzer, K.W., Avens, L., McNeill, J.B. et al., 2018. Characterizing sex ratios of sea turtle populations: a Bayesian mixture modeling approach applied to juvenile loggerheads (*Caretta caretta*). *Journal of Experimental Marine Biology and Ecology*, 504, 10–19.

Shimada, T., Limpus, C., Jones, R., and Hamann, M., 2017. Aligning habitat use with management zoning to reduce vessel strike of sea turtles. *Ocean and Coastal Management*, 15, 163–72.

Snoddy, J.E., Landon, M., Blanvillain, G., and Southwood, A., 2009. Blood biochemistry of sea turtles captured in gillnets in the lower Cape Fear River, North Carolina, USA. *Journal of Wildlife Management*, 73, 1394–401.

Snoddy, J.E. and Southwood Williard, A., 2010. Movements and post-release mortality of juvenile sea turtles released from gillnets in the lower Cape Fear River, North Carolina, USA. *Endangered Species Research*, 12, 235–47.

Southwood, A.L., Andrews, R.D., Lutcavage, M.E. et al., 1999. Heart rates and diving behavior of leatherback sea turtles in the eastern Pacific Ocean. *Journal of Experimental Biology*, 202, 1115–25.

Southwood, A.L., Andrews, R.D., Paladino, F.V., and Jones, D.R., 2005. Effects of diving and swimming behavior on body temperatures of Pacific leatherback turtles in tropical seas. *Physiological and Biochemical Zoology*, 78, 285–97.

Stabenau, E.K., Heming, T.A., and Mitchell, J.F., 1991. Respiratory, acid-base and ionic status of Kemp's ridley sea turtles (*Lepidochelys kempii*) subjected to trawling. *Comparative Biochemistry and Physiology*, 99A, 107–11.

Stabenau, E.K. and Vietti, K., 2003. The physiological effects of multiple forced submergences in loggerhead sea turtles (*Caretta caretta*). *Fishery Bulletin*, 101, 889–99.

Stacy, B.A., Keene, J.L., and Schroeder, B.A., 2016. Report of the Technical Expert Workshop: Developing National Criteria for Assessing Post-Interaction Mortality of Sea Turtles in Trawl, Net, and Pot/Trap Fisheries. US Department of Commerce, NOAA. NOAA Technical Memorandum NMFS-OPR-53, 110 pp.

Stacy, N.I., Field, C.L., Staggs, L. et al., 2017. Clinicopathological findings in sea turtles assessed during the Deepwater Horizon oil spill response. *Endangered Species Research*, 33, 25–37.

Stacy, N.I., Innis, C.J., and Hernandez, J.A., 2013. Development and evaluation of three mortality prediction indices for cold-stunned Kemp's ridley sea turtles (*Lepidochelys kempii*). *Conservation Physiology*, 1(1), cot003. doi: 10.1093/conphys/cot003

Swimmer, Y., Arauz, R., McCracken, M. et al., 2006. Diving behavior and delayed mortality of olive ridley sea turtles *Lepidochelys olivacea* after their release from longline fishing gear. *Marine Ecology Progress Series*, 323, 253–61.

Swimmer, Y., Empey Campora, C., McNaughton, L. et al., 2014. Post-release mortality estimates of loggerhead sea turtles (*Caretta caretta*) caught in pelagic longline fisheries based on satellite data and hooking location. *Aquatic Conservation Marine and Freshwater Ecosystems*, 24, 498–510.

Swimmer, Y. and Gilman, E., 2012. Report of the Sea Turtle Longline Fishery Post-release Mortality Workshop, 15–16 November 2011. US Department of Commerce. NOAA Technical Memorandum, NOAA-TM-NMFS-PIFSC-34, 31 pp.

Tyson, R.B., Piniak, W.E.D., Domit, C. et al., 2017. Novel bio-logging tool for studying fine-scale behaviors of marine turtles in response to sound. *Frontiers in Marine Science*, 4, 219. doi:10.3389/fmars.2017.00219

Venn-Watson, S., Colegrove, K.M., Litz, J. et al., 2015. Adrenal gland and lung lesions in Gulf of Mexico common bottlenose dolphins (*Tursiops truncatus*) found dead following the Deepwater Horizon oil spill. *PLoS ONE*, 10(5), e0126538. doi:10.1371/journal.pone.0126538

Wallace, B.P., Brosnan, T., McLamb, D. et al., 2017. Effects of the Deepwater Horizon oil spill on protected marine species. *Endangered Species Research*, 33, 1–7.

Wallace, B.P., DiMatteo, A.D., Bolten, A.B. et al., 2011. Global conservation priorities for marine turtles. *PLoS ONE*, 6, e24510. doi:10.1371/journal.pone.0024510

Wallace, B.P., Kot, C., DiMatteo, A.D. et al., 2013. Impacts of fisheries bycatch on marine turtle populations worldwide: toward conservation and research priorities. *Ecosphere*, 4, 40. doi:10.1890/ES12-00388.1

Wilson, R.P., Shepard, E.L.C., and Liebsch, N., 2008. Prying into the intimate details of animal lives: use of a daily diary on animals. *Endangered Species Research*, 4, 123–37. doi:10.3354/esr00064

Applications of minimally invasive immune response and glucocorticoid biomarkers of physiological stress responses in rescued wild koalas (*Phascolarctos cinereus*)

Edward J. Narayan and Renae Charalambous

> ⇨ **Take-home message**
>
> Stress evaluation through quantification of glucocorticoids and immune response biomarkers in minimally invasive biological samples can boost conservation management and rehabilitation programmes for Australasian wildlife species.

15.1 Introduction: physiological stress in Australasian wildlife under environmental change

Ultimate and proximate sources of stress are affecting native wildlife species in Australia predominately through anthropogenic factors such as land clearing and habitat fragmentation, and infectious diseases (Narayan and Williams 2016; McAlpine et al. 2017; Finn and Stephens 2017; Narayan 2019; Narayan and Vanderneut 2019). Environmental stressors are responsible for the current plight of iconic small mammal or marsupial species such as koalas (*Phascolarctos cinereus*), with their conservation status currently listed as vulnerable to extinction with an alarmingly decreasing population trend (Woinarski and Burbidge 2016).

The definition of stress is highly contentious; however, it can be broadly defined as a change in the psychological, physiological, and/or physical well-being of a living organism as a result of exposure to any biological and/or environmental factor that acts as a stressor (challenges regular physiological capacity) (Hing et al. 2016). Currently, haematology (blood cell profiling) and glucocorticoid (stress hormone) monitoring are increasingly applied in wildlife conservation programmes to assess stress levels of animals both in the wild and in captivity. This is made possible through access to minimally invasive blood and/or other biological samples such as faeces, urine, and hair (Palme 2019). These techniques have been used across various animal groups for clinical diagnostic studies; however, there seems to be limited published studies in Australasian small mammals associated with exposure to known environmental stressor(s) (for examples, see Table 15.1) (Rishniw et al. 2012; Brearley et al. 2013;

Edward J. Narayan and Renae Charalambous, *Applications of minimally invasive immune response and glucocorticoid biomarkers of physiological stress responses in rescued wild koalas* (Phascolarctos cinereus) In: *Conservation Physiology: Applications for Wildlife Conservation and Management.* Edited by: Christine L. Madliger, Craig E. Franklin, Oliver P. Love, and Steven J. Cooke, Oxford University Press (2021). © Oxford University Press. DOI: 10.1093/oso/9780198843610.003.0015

King and Bradshaw 2010; Phelps and Kingston 2018; McMichael et al. 2017). Australasian mammalian fauna include diverse species of marsupials or pouched small mammals that are also facing an eminent extinction crisis through climate change and other human-induced threats. There are various *in situ* and *ex situ* conservation programmes in Australasia that will benefit from conservation physiology tools.

15.2 Acute and chronic stress responses to environmental challenges

The physiological stress response is a highly coordinated neuro-endocrine reaction to an unpredictable, uncontrollable, and/or aversive stimulus (also known as a stressor) (Beehner and Bergman 2017). Wildlife stress responses can involve activation of the hypothalamic–pituitary–adrenal (HPA) axis, which is a complex and essential negative-feedback system involving glucocorticoids among other neuroendocrine mediators (Hing et al. 2016). When the stress response is activated, the HPA axis begins releasing a cascade of hormones tasked with managing the response to the stressor (Narayan and Williams 2016). This process begins in the higher brain centres including the limbic areas of the hippocampus followed by activation of the paraventricular nucleus (PVN) of a small neuroendocrine structure called the hypothalamus, which is situated above the brainstem, secreting corticotropin-releasing hormone (CRH) (Tasker and Herman 2011; Narayan and Williams 2016). The secretion of CRH stimulates production of a protein proopiomelanocortin (POMC). POMC serves as the basis for a number of stress-related hormones, including adrenocorticotropic hormone (ACTH), β-lipotropin (β-LPH), and β-endorphin (Tasker and Herman 2011; Narayan and Williams 2016). CRF ACTH acts on the adrenal cortices (also known as the adrenal glands) to result in an increased synthesis and output of glucocorticoids (Narayan and Williams 2016). Glucocorticoids then act to assist metabolically and behaviourally by diverting the storage of glucose to fat to instead partition energy by supplying it to all

parts of the body for the upcoming challenge (Narayan and Williams 2016). Additionally, the production of glucocorticoids will assist in balancing pH after the challenge and acts as a chemical blocker within the negative-feedback process to CRH secretion and HPA axis synergy (Narayan and Williams 2016). The function of the HPA axis in response to stress comes at the cost of diverting energy away from other bodily functions (Narayan and Williams 2016).

If the stressor experienced is acute (short-lived), then the body can return to homeostasis through a negative-feedback mechanism (Narayan and Williams 2016; Beehner and Bergman 2017). On the contrary, if the stressor experienced is chronic, then resources that are integral for survival are continually diverted in an attempt to keep the HPA axis activated (Narayan and Williams 2016; Beehner and Bergman 2017). The stress response operates like a classic life history trade-off, prioritizing present energy usage over future energy storage (Wingfield and Sapolsky 2003). For this reason, the stress response is adaptive over the short term but unsustainable over long-term activation (Beehner and Bergman 2017).

Activation of the HPA axis and exposure to chronic stress has been proven to subsequently decrease animal health and longevity (Juster et al. 2010). This is because chronic stress has been shown to be connected with a number of adverse health effects and disorders, most likely the result of the reallocation of resources during the stress response, as the body gives higher priority to survival over maintenance (Sapolsky 2004; Whirledge and Cidlowski 2010). Some of the major negative health implications associated with chronic stress include the suppression of growth, reproduction, and function of the immune system (Lattin and Romero 2014).

Chronic stress has been shown to reduce the growth of *in utero* koala offspring and their body weight after birth through the suppression of growth factors such as insulin-like growth factor (IGF-1) (Emack et al. 2008). Reduced growth is consistently obvious in *in utero* males compared with females with chronically stressed mothers (Emack et al. 2008). It is presumed that this is due to

Table 15.1 Examples of published studies applying minimally invasive biomarkers to assess physiological stress in Australian wildlife species.

Study species	Environmental trauma and/or disease	Physiological biomarker	Physiological response	Outcome	Conclusion	Reference
Squirrel glider: sp. *Petaurus norfolcensis*	Urbanization: land clearing and urban edges	Hair cortisol levels	Edge contrast had high cortisol levels Edge contrast where hollows or tree over storeys were present had a lesser effect on hair cortisol Edges caused a potential physiological stress to sugar gliders	Low-contrast minor roads (roads that can be safely crossed/residential areas) are the most suitable edge type in urban landscapes. High-contrast roads have the least conservation value for long-term conservation	Hair cortisol measurements were able to make inferences about physiological demands of living near urban landscapes; however, further investigation is required	Brearley et al. 2012
Barrow Island kangaroo. sp. *Macropus robustus isabellinus*	High environmental temperatures Low water availability (drought)	Hormone assays (blood serum) Osmolarity (blood serum) Haematology (blood serum)	Red blood cell to whole blood ratio was significantly lower in the island species in comparison with mainland species Total red blood cell counts were also much lower Significant negative correlation between total body water content and lysine vasopressin in Barrow Island kangaroos Significant negative correlation between total leukocytes and blood plasma	The Barrow Island euros had significantly elevated plasma osmolarity (285.9 ± 4.4 mOsm/kg), plasma lysine vasopressin (16.7 ± 4.6 pg/ml) and plasma cortisol (1.71 ± 0.22 lg/dl) Eosinophil counts were significantly depressed and negatively correlated with plasma cortisol levels, suggesting that the exposure was chronic and compromising the euros' immune capabilities	The extremely challenging conditions of prolonged drought reveal Barrow Island euro species to suffer from a non-regenerative normocytic hypochromic anaemia that can compromise respiratory capacities and activity limits. The low genetic diversity of the island and the anaemia will cause less resilience to climatic changes in the long term. Climate change scenarios are predicting major changes in weather and rainfall patters for biodiversity hotspots, making Barrow Island at a greater risk	King and Bradshaw 2010

Table 15.1 Continued.

Study species	Environmental trauma and/or disease	Physiological biomarker	Physiological response	Outcome	Conclusion	Reference
Cave-roosting bat. sp. *Hipposideros diadema*	Human disturbance in and around caves	Leukocyte profiles	Direct cave disturbance was identified as the primary driver of neutrophil-to-lymphocyte ratios Where direct cave disturbance is absent, bat abundance and cave complexity increase also caused stress responses The abundance of confamilial species increased total leukocytes	Environmental and biological contexts play a significant role in modulating physiological stress responses of individuals to multiple gradients of human disturbance	Both environmental and biological factors can influence an individual's ability to cope with a stress response to human disturbance. Mitigating human disturbance allows population persistence	Phelps and Kingston 2018
Flying foxes: *Pteropus alecto* (the black flying fox) *P. conspicillatus* (the spectacled flying fox)	Disease: Hendra virus	Urinary cortisol concentrations Hendra virus RNA-prevalence	Mean urinary cortisol concentration in Hendra virus RNA-positive pooled urine samples was significantly higher than in Hendra virus RNA-negative samples Females showed a rise in urinary cortisol concentrations during birthing season in comparison with males at the same time of year	Urinary cortisol excretion is modulated by both life cycle and ecological factors	Association between low winter temperatures and increased Hendra virus infection and excretion, putatively mediated by the physiological cost of thermoregulation	McMichael et al. 2017

males exhibiting an overall higher growth rate than females, making it possible that males are more vulnerable to insult during gestation (Emack et al. 2008). Furthermore, reduced body weight after gestation as a result of maternal stress can be critical to the survival of offspring through increased evolutionary disadvantages such as a decrease in species-specific biological fitness (Emack et al. 2008). The negative consequences of excessive glucocorticoid production on growth in koalas can cause stunted development through the production of IGF-1 (Emack et al. 2008).

Activation of the HPA axis in response to stress has also been shown to influence immune function, leading to increased morbidity and mortality (Brearley et al. 2013). This is concerning as stress exacerbates the impact of disease on animals who may already be struggling for survival (Hing et al. 2016). Higher prevalence of disease in response to stress can be explained through the production of glucocorticoids in response to activation of the HPA axis (Hing et al. 2016). Glucocorticoid production can have profound physiological effects on immunological processes via the receptors on immune cells and changes in immune gene expression in target tissues (Hing et al. 2016). Glucocorticoids influence the trafficking of leukocytes (white blood cells responsible for fighting infection) and suppress the secretion of proinflammatory cytokines (regulators of inflammation as a response to infection to heal and repair) (Chrousos 2009). The pathogenesis of chronic stress-related disease can be summarized through the sustained, excessive secretion of multiple homeostatic systems (Chrousos 2009). These diseases represent the effects of two physiological processes whose mediators are supposed to be secreted in a quantity-limited and time-limited fashion, but have instead gone awry (Chrousos 2009). The negative consequences of excessive glucocorticoid production on disease in koalas include inflamed tissues or systemic infection, anti-chlamydial antibodies as a sign of infection, organ dysfunction, and clinical signs of chlamydiosis (fatal if left untreated) (Grogan et al. 2018).

Additional to increased morbidity and mortality and decreased growth, long-term activation of the HPA axis in response to stress has been shown to inhibit reproduction (Chrousos 2009). Precise levels of glucocorticoids are required for proper reproductive function, and if this balance is disrupted, so is fertility (Wingfield and Sapolsky 2003; Whirledge and Cidlowski 2010). The impact of glucocorticoid production on reproduction results in an extended follicular stage, making the overall reproductive cycle length longer and/or more irregular (Wingfield and Sapolsky 2003). Furthermore, uterine maturation can become impaired during stress by at least two mechanisms (Wingfield and Sapolsky 2003). First, stress in some species can be associated with a decline in levels of progesterone (progesterone mediates preparation of the uterine wall for implantation during the luteal phase) (Wingfield and Sapolsky 2003). Second, stress in most species decreases proactive female behaviours designed to increase the likelihood of sex (i.e. proceptivity), as well as responsiveness to proceptive behaviours on the part of a male (i.e. receptivity) (Wingfield and Sapolsky 2003). When stress is encountered in males, the production of glucocorticoids precedes a decline in testosterone concentration (Whirledge and Cidlowski 2010). Leydig cells are responsible for the production of testosterone, the hormone required to regulate male fertility. Elevated glucocorticoids are known to decrease testosterone biosynthesis by Leydig cells, as well as induce Leydig cell apoptosis, and spermatogonia apoptosis within the seminiferous tubules (Whirledge and Cidlowski 2010). The negative consequences of excessive glucocorticoid production on reproduction in koalas can therefore cause infertility through the imbalance and inhibition of important hormones required for reproduction in both males and females (Chrousos 2009).

15.3 Biomarkers indicative of stress-related disease

Three common biomarkers of immune function include leukocytes, neutrophil-to-lymphocyte ratios (N/L), and urea. Urea is a colourless crystalline compound and the main nitrogenous breakdown product of protein metabolism in mammals, which is excreted in urine (Beier et al. 2011; Zhao et al. 2018). Leukocytes, also known as white blood cells, are the first component of the innate immune system, which mediate a first-line of defence against a microbial attack (Gordon-Smith 2013). The primary role of leukocytes is to recognize, ingest foreign or degraded cells or proteins, kill pathogens, and to present specific pathogen antigens to assist in the immune response (Gordon-Smith 2013). Additional to leukocytes, N/L ratios are components of white blood cells that assist in the immune response to stress (Gordon-Smith 2013). Neutrophils divide and enter the bloodstream to combat infection by eating foreign bacteria, whereas lymphocytes work to make two types of antibodies to fight foreign bacteria, T-type and B-type antibodies (Gordon-Smith 2013). High blood N/L ratios are indicative of a disease condition that is acting to compromise immunity (Gordon-Smith 2013).

Chlamydia is an infectious disease associated with compromised immunity in koalas, and measuring levels of leukocytes and N/L ratios in the blood of koalas admitted into veterinary clinics is useful for disease detection (Gordon-Smith 2013; Fabijan et al. 2019). Furthermore, increased levels of urea in blood are indicative of diseases that could be compromising kidney function (Bellomo et al. 2012; Speight et al. 2014). Oxalate nephrosis and renal failure are both diseases associated with kidney dysfunction in koalas. Thus, measuring levels of urea in the blood of koalas admitted into veterinary clinics is useful for disease detection (Speight et al. 2014). Further, measuring glucocorticoids is important for determining the lethal and sublethal effects of stress (Narayan et al. 2013). When measured, glucocorticoids can represent physiological stress, which can provide insight into wildlife well-being such as any permanent negative changes relating to inhibition of growth, reproduction, and immune function (Sheriff et al. 2011).

In the remaining portion of this chapter, we will highlight, using rescued koalas, the measurement of immune and minimally invasive hormonal stress biomarkers for assessing physiological stress in an Australasian mammal of high conservation importance. Within each case study, we provide

application(s) of specific biomarker(s) that can be useful indices of acute and/or chronic stress and therefore aid the clinical care and management of koalas. The end-goal of clinical care for wild koalas is treatment and release; however, the outcome is dependent on each admission and the nature/extent of trauma and disease.

15.4 Case study 1: haematological indicators of rescued koalas

Haematological biomarkers can be useful to track the physiological status of rescued koalas and other Australasian mammals while receiving clinical care, leading to improved prognosis of trauma and disease, more specialized treatment and care, and eventually increasing the chances of koala rehabilitation and release back to the wild. Koalas are rescued and admitted into clinical care on a regular basis in Australia due to the high incidences of environmental trauma such as bushfires, vehicle collision impacts, habitat disturbance and land clearance, and diseases such as chlamydia (Narayan 2019). The specific environmental trauma and disease-related conditions found in rescued koalas at one such clinic, Adelaide Koala and Wildlife Hospital, South Australia (SA), Australia are shown in Figure 15.1.

Neutrophils and lymphocytes are two components of white blood cells, which when analysed as a ratio, are a clear representation of physiological stress (Narayan and Hero 2011). The normal ratio, represented as a percentage of neutrophils to lymphocytes (N/L) in koalas is 40:55 (Dickens 1975). Any reading of neutrophils that is higher than 40 per cent, in addition to any reading of lymphocytes that is lower than 55 per cent, can be classed as abnormal (Dickens 1975). N/L ratios reflect physiological stress as they indicate a temporary redistribution of white blood cells to areas of the body where they are most needed during the stress response (Davis and Maerz 2011). Examples of locations where white blood cells redistribute include the epidermis to either fight an infection or close a wound (Davis and Maerz 2011). The result of this redistribution includes the increased production of glucocorticoids, which increases the proportion of

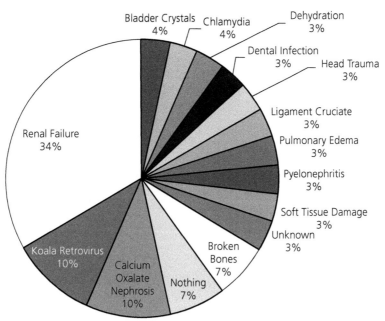

Figure 15.1 Examples of environmental trauma and diseases in rescued wild koalas. The most common diagnosis in adult wild koalas (*n* = 30) admitted to clinical care over an 11-month period (January–November 2017) was renal failure at 34 per cent. Following renal failure, the next most common diagnoses were retrovirus at 10 per cent and calcium oxalate nephrosis at 10 per cent.

neutrophils and decreases the proportion of lymphocytes in circulation, making the N/L ratio useful for assessing the degree of stress encountered by small mammals (Davis and Maerz 2011).

In particular, leukocytes are a white blood cell, which functions to protect the body against infectious diseases and foreign invaders (Davis et al. 2004, 2008). It is common to use leukocyte counts to assess immune function, as the irregular production of white blood cells is indicative of host defence mechanisms (Davis et al. 2004). Rescued wild koalas diagnosed with retrovirus-KoRV with more advanced stages leading to leukaemia, lymphoma, malignant tumours, and immune deficiency disorders can have either substantially low or high leukocyte count (Denner and Young 2013). In an earlier study, wild house finches (*Haemorhous mexicanus*) injected with a bacterial disease were found to have excessively high leukocyte counts compared with finches that had not been injected with the same bacterial disease (Davis et al. 2004). Furthermore, those finches that had been injected with the bacterial dis-

ease were susceptible to dramatic and irregular changes in their leukocyte readings once exposed to stress including capture and release (Davis et al. 2004).

Blood urea nitrogen is determined by the complex balance between urea production, urea metabolism, and urea excretion, and is moderated by a number of renal and non-renal dependent factors (Seleno 2011). Where blood urea nitrogen is not a direct factor in system dysfunction, it is associated with an increased severity of renal systematic illnesses (Seleno 2011). For example, elevated blood urea nitrogen is correlated with increased mortality in critically ill human patients already suffering with heart-related issues (Seleno 2011). Elevated levels of urea in blood profiles are an indicator of failing renal function (Lanyon et al. 2012), and koalas diagnosed with bladder crystals show high urea readings indicating renal dysfunction.

Overall, evaluating biomarkers of the immune system such as leukocytes, N/L ratios, and urea using blood sampling is an effective method of assessing physiological stress that can aid clinical care, rehabilitation, and conservation programmes for Australasian wildlife species. In the case of koalas, availability of minimally invasive haematological biomarkers allows for the rapid assessment of the physiological status of each koala patient, which assists in more specialized clinical prognosis and treatment to increase the chances of faster recovery and rehabilitation.

15.5 Case study 2: faecal glucocorticoid (FGM) monitoring in rescued koalas

In comparison with eutherian mammals, koalas have a low adrenal weight: body weight ratio (~45 mg/kg), and adrenocortical histological changes to disease has been similar to that of mammals (Booth et al. 1990). Cortisol is the major glucocorticoid present in the blood of koalas (Blanshard and Bodley 2009; Vogelnest and Woods 2009). It shows irregular pulsatile changes in blood making it difficult to use a single-point blood sample as a stress biomarker (McDonald et al. 1990). Koala FGM assay has been biologically validated and shown to be a reliable measure of biologically relevant changes in FGM concentrations in response to environmental stressors (Buchanan and

Goldsmith 2004; Narayan et al. 2013; Webster et al. 2017). Due to the fairly large and complex digestive tract of the koala, FGMs are excreted periodically. Generally, there are two peaks over an approximately 9-day period, the first at 24 h (female) and 48 h (male), and then another peak roughly 9–10 days after the first peak (Narayan et al. 2013). This later peak is highly dependent upon the individual's rate of metabolism. These two peaks could be explained by the theory stating that larger particles are excreted more rapidly by passing straight through the proximal colon. Smaller particles are retained in the caecum for approximately 9 days. Below, we discuss some cases looking at individual koalas in care for changes in FGM levels.

As the first example, an adult wild male koala with a body condition score of 2 (1 = emaciated, 2 = poor, 3 = fair, 4 = good, and 5 = excellent; after A. Reiss in Jackson et al. 2003) showed increased FGM levels due to surgery and levels remained high over 10 days. This may be due to infection placing continuous stress on the individual. These changes in FGM levels may also be indicative of a change in care or routine (Figure 15.2).

In a second example, we explored chloramphenicol intervention in clinical care, which is a strongacting antibiotic used to treat chlamydia bacterial infections in koalas. As can be seen in Figure 15.3,

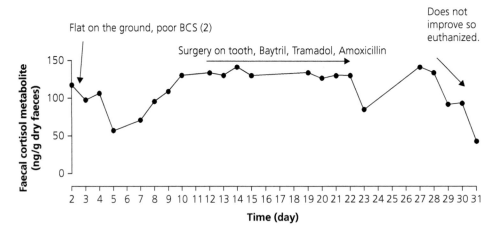

Figure 15.2 Faecal glucocorticoid metabolite (FGM) changes in response to clinical intervention in a koala patient. FGM levels were quite high (above maximum baseline levels for wild healthy males, which is 46.44 ng/g; Narayan, 2019) in clinic. Thus, despite the extensive clinical intervention, this koala patient had very high levels of physiological stress from the beginning due to high exposure to environmental trauma (potentially heat stress). White blood cell (leukocytes) counts were also incredibly high. Prognosis: conjunctivitis, mandibular infection, and dehydration. Outcome: euthanized.

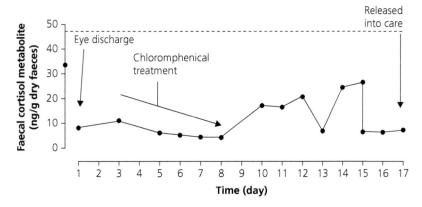

Figure 15.3 Successful treatment of infectious disease matches with reduction in faecal glucocorticoid metabolite (FGM) levels. A 2-year old male koala weighing 5.2 kg that was admitted in 2016 with a good body condition score (BCS) of 4. Presented with slightly red eyes and dried discharge. Due to the discharge he had a swab taken to be tested for chlamydia, which returned positive. Treated with chloramphenicol. Prognosis: chlamydia, successfully treated, released into care. As shown, FGM levels subsided during treatment indicating the koala responded well to the clinical care.

chloramphenicol intervention led to a gradual decrease in FGM levels. This potentially demonstrates positive response of the koalas to the clinical intervention, which could be indicative of successful treatment and stress management.

One of the major implications of stress evaluation in this way is that it allows the clinic staff to assess and identify a physiological stress response in the koalas. An application of these findings has been acknowledged in the latest version of the Koala Rehabilitation Manual that koalas are highly responsive to stress and excessive handling should be avoided, especially with young koalas (after E. Narayan in Flanagan 2019).

15.6 Case study 3: chronic stress in rescued koalas using fur cortisol analysis

Although cortisol can be measured in blood plasma, blood serum, saliva, urine, or faeces (Mastromonaco et al. 2014), evidence suggests that hair is the most effective sample medium for capturing chronic stress (Mastromonaco et al. 2014). This is because hair is thought to incorporate blood-borne hormones during its growth phase, it is relatively stable, and any cortisol detected in hair reflects physiological stress experienced over the period of hair growth, which can be weeks to months (Mastromonaco et al. 2014). Acquiring samples to

measure cortisol is not always a stress-free process due to the pressure associated with capture and handling of the animal (Mastromonaco et al. 2014). However, any stress experienced during capture and handling to acquire hair is not likely to impact glucocorticoid levels in hair samples so long as they are collected concurrently, thus preventing further growth and integration (Cattet et al. 2014; Mastromonaco et al. 2014).

There are multiple studies that use a range of techniques to detect cortisol in hair (Macbeth et al. 2010; Ashley et al. 2011; Dettmer et al. 2012; Cattet et al. 2014; Di Francesco et al. 2017). The production of excessive glucocorticoids such as cortisol can be indicative of some degree of chronic stress, as cortisol is produced when the body has reached overexertion and the duration of a particular stressor extends beyond what an organism can handle (Beehner and Bergman 2017). However, the body retains a baseline level of cortisol, which must first be considered when determining if levels of cortisol are excessive in response to particular stressors (Narayan et al. 2013). Since there have been no previous studies that analyse cortisol in hair in koalas, there are no current baseline levels to compare the results of this research with.

Koala fur cortisol can be readily measured using methods described recently by Charalambous and Narayan (2019). A total of 45 wild rescued koalas were admitted to clinical care between December 2017 and September 2018. Each koala was given a prognosis relating to why they were admitted into care. The varying prognoses for the koalas included suffering with a disease (i.e. chlamydia, retrovirus, etc.), being attacked by a dog, being caught in a bushfire, being hit by a car (HBC), and being in an unsuitable environment (i.e. being displaced by land clearing). A fur sample (0.2 g) taken from each koala was analysed for cortisol. The prognosis with the highest median cortisol reading was disease at 1370.99 ng/g (Figure 15.4). Following disease, HBC had the next highest median cortisol reading at 1285.48 ng/g, with unknown prognosis following at 1258.9 ng/g (Figure 15.4). There were no significant differences between each prognosis and the associated median cortisol reading, which could be explained by low sample size.

It is notable that disease was the prognosis with the highest median cortisol reading (Figure 15.4) as the presence of chronic stress does not allow the HPA axis to reach a recovery phase, resulting in dysfunction of the negative-feedback mechanism and the onset of subsequent health impairments (Bonier et al. 2009). In terms of disease, the cost of the HPA axis not reaching a recovery phase involves increased susceptibility to disease, shedding of infectious agents, and a shift in host–parasite equi-

librium (Martin 2009). Furthermore, the prolonged exposure of glucocorticoids in the body can affect immunological processes via receptors on immune cells and changes in immune gene expression in target tissues (Martin 2009). Therefore, it is not surprising that those wild rescued koala patients who were given a disease prognosis had the highest median cortisol reading.

Wild rescued koalas who were given the prognosis of HBC, being in an unsuitable environment, being attacked by a dog, and being caught in a bushfire all returned cortisol readings, although they were not as elevated as disease (Figure 15.4). This may indicate that these events and the consequences associated with them possibly did not allow the HPA axis to reach a recovery phase, resulting in dysfunction of the negative-feedback mechanism under chronic stress (Bonier et al. 2009). Rapid land development, urbanization, and human population growth are dramatically altering the habitats in which native wildlife dwell (Donnelly and Marzluff 2006). The change in suitable habitat such as the decline of forest cover dramatically expands the likelihood of mortality and morbidity in koalas (Narayan and Williams 2016). Koalas living in fragmented habitats are forced to increase their movement in search for food (Davies, Gramotvev, McAlpine et al. 2013; Davies, Gramotvev,

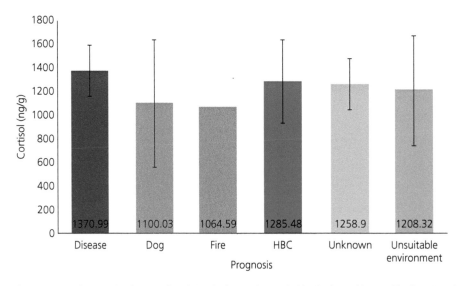

Figure 15.4 Chronic stress evaluation using fur cortisol analysis in koalas. Median cortisol levels of 45 wild rescued koala patients based on their prognosis when admitted to clinical care during December of 2017 through to September of 2018.

Seabrook et al. 2013). This increases their vulnerability to predation, in particular to dogs, who view koalas as prey and therefore koalas are at a high risk at being attacked (Davies, Gramotvev, McAlpine et al. 2013; Davies, Gramotvev, Seabrook et al. 2013). Furthermore, as koalas are forced to move further for food, they increase the probability of falling victim to vehicle trauma as roads are being placed between interconnecting koala habitats (Davies, Gramotvev, McAlpine et al. 2013; Davies, Gramotvev, Seabrook et al. 2013). As koalas are increasingly needing to search for suitable food sources in locations further away, this species is finding itself in areas outside of its niche (Donnelly and Marzluff 2006).

Whether koalas are sustaining injuries from disease, HBC, being attacked by dogs, or whether unsuitable habitat or bushfires are the source of morbidity or mortality, koalas are exposed to an alarming number of stressors. Thus, by measuring physiological stress levels of wild koalas we can tap into their physiological resilience and adaptive capacity towards environmental change. This physiological data can be highly powerful and useful for the on-ground monitoring and management of wild koalas such as during habitat and population regeneration programmes, translocation, and long-term eco-physiological monitoring of koala populations with on-going national conservation efforts.

15.7 Conclusions and future directions

We have used koalas as an example of an Australian native small mammal species to demonstrate the usefulness of physiological biomarkers of stress evaluation in rescued wildlife in clinical care. These studies have been made possible with the outstanding work of wildlife rescuers and veterinarian staff who provide daily care to recuperating koala patients in clinical care. Both traditional blood sampling and other minimally invasive biological sampling such as faeces and hair (fur) could provide valuable data related to physiological stress in rescued koala patients. It can be accepted that blood biochemistry profiles are effective in the assessment of physiological stress and can be used in disease diagnosis on an individual/case needs basis. In particular, leukocyte counts are effective in assessing

immune function, N/L ratios are a clear representation of physiological stress, and urea readings are associated with an increased severity of renal systematic illnesses. These biological samples and robust analytical hormone quantification methods also provide glucocorticoid data in real time that can be used to make objective assessments of physiological stress in koalas under clinical care.

We propose that haematological methods such as leukocyte counts, N/L ratios, and urea readings can provide rapid diagnosis of clinical issues related to animal health and well-being. Using faecal sampling, the acute stress responses of koalas can be obtained to gauge how they are faring in response to management intervention and clinical care. Furthermore, fur cortisol analysis provides a robust biological sampling technique to assess chronic stress. Together, these conservation physiology tools can be highly useful and boost on-the-ground management efforts and overarching conservation goals to benefit Australia's iconic small mammals, with techniques being transferred to study species other than koalas such as Tasmanian devils and wombats.

References

Ashley, N.T., Barboza, P.S., Macbeth, B.J. et al., 2011. Glucocorticosteroid concentrations in feces and hair of captive caribou and reindeer following adrenocorticotropic hormone challenge. *General and Comparative Endocrinology*, 172(3), 382–91.

Beehner, J.C. and Bergman, T.J., 2017. The next step for stress research in primates: to identify relationships between glucocorticoid secretion and fitness. *Hormones and Behavior*, 91, 68–83.

Beier, K., Eppanapally, S., Bazick, H.S. et al., 2011. Elevation of blood urea nitrogen is predictive of long-term mortality in critically ill patients independent of 'normal' creatinine. *Critical Care Medicine*, 39(2), 305–13. doi:10.1097/CCM.0b013e3181ffe22a

Bellomo, R., Kellum, J.A., and Ronco, C., 2012. Acute kidney injury. *The Lancet*, 380(9843), 756–66. doi:10.1016/s0140-6736(11)61454-2

Blanshard, W. and Bodley, K., 2009. Koalas. In L. Vogelnest and R. Woods, eds. *Medicine of Australian Mammals*, pp. 227–327. CSIRO Publishing, Melbourne.

Bonier, F., Martin, P.R., Moore, I.T. et al., 2009. Do baseline glucocorticoids predict fitness? *Trends in Ecology & Evolution*, 24(11), 634–42.

Booth, R.J., Carrick, F.N., and Addison, P.A., 1990. The structure of the koala adrenal gland and the morphological changes associated with the stress of disease. In A.K. Lee, K.A. Handasyde, and G.D. Sanson, eds. *Biology of the Koala*, pp. 281–8. Surrey Beatty & Sons, Sydney.

Brearley, G., McAlpine, C., Bell, S., and Bradley, A., 2012. Influence of urban edges on stress in an aboreal mammal: a case study of squirrel gliders in southeast Queensland Australia. *Landscape Ecology*, 21, 1407–19.

Brearley, G., Rhodes, J., Bradley, A. et al., 2013. Wildlife disease prevalence in human-modified landscapes. *Biological Reviews*, 88(2), 427–42.

Buchanan, K.L. and Goldsmith, A.R., 2004. Noninvasive endocrine data for behavioural studies: the importance of validation. *Animal Behaviour*, 67(1), 183–5.

Cattet, M., Macbeth, B.J., Janz, D.M. et al., 2014. Quantifying long-term stress in brown bears with the hair cortisol concentration: a biomarker that may be confounded by rapid changes in response to capture and handling. *Conservation Physiology*, 2(1).

Charalambous, R. and Narayan, E., 2019. Cortisol measurement in koala (*Phascolarctos cinereus*) fur. *JoVE (Journal of Visualized Experiments)*, 150, e59216.

Chrousos, G.P., 2009. Stress and disorders of the stress system. *Nature Reviews Endocrinology*, 5(1), 374–81.

Davies, N.A., Gramotvev, G., McAlpine, C. et al., 2013. Physiological stress in koala populations near the arid edge of their distribution. *Public Library of Science One, 8,* 1–12.

Davies, N.A., Gramotvev, G., Seabrook. L. et al., 2013. Movement patterns of an arboreal marsupial at the edge of its range: a case study of the koala. *Movement Ecology*, 1, 1–15.

Davis, A.K., Cook, K.C., and Altizer, S., 2004. Leukocyte profiles in wild house finches with and without mycoplasmal conjunctivitis, a recently emerged bacterial disease. *EcoHealth*, 1(4), 362–73. doi:10.1007/s10393-004-0134-2

Davis, A.K. and Maerz, J.C., 2011. Assessing stress levels of captive-reared amphibians with hematological data: implications for conservation initiatives. *Journal of Herpetology*, 45(1), 40–4.

Davis, A.K., Maney, D.L., and Maerz, J.C., 2008. The use of leukocyte profiles to measure stress in vertebrates: a review for ecologists. *Functional Ecology*, 22(5), 760–72. doi:10.1111/j.1365-2435.2008.01467.x

Denner, J. and Young, P.R., 2013. Koala retroviruses: characterization and impact on the life of koalas. *Retrovirology*, 10(1), 108.

Dettmer, A.M., Novak, M.A., Suomi, S.J. et al., 2012. Physiological and behavioral adaptation to relocation stress in differentially reared rhesus monkeys: hair cortisol as a biomarker for anxiety-related responses. *Psychoneuroendocrinology*, 37(2), 191–9.

Dickens, R.K., 1975. The koala (*Phascolarctos cinereus*) past, present and future. *Australian Veterinary Journal*, 51(10), 459–63.

Di Francesco, J., Navarro-Gonzalez, N., Wynne-Edwards, K. et al., 2017. Qiviut cortisol in muskoxen as a potential tool for informing conservation strategies. *Conservation Physiology*, 5(1), cox052.

Donnelly, R. and Marzluff, J.M., 2006. Relative importance of habitat quantity, structure, and spatial pattern to birds in urbanizing environments. *Urban Ecosystems*, 9(2), 99–117.

Emack, J., Kostaki, A., Walker, C.D. et al., 2008. Chronic maternal stress affects growth, behaviour and hypothalamo–pituitary–adrenal function in juvenile offspring. *Hormones and Behavior*, 54(4), 514–20.

Fabijan, J., Caraguel, C., Jelocnik, M. et al., 2019. Chlamydia pecorum prevalence in South Australian koala (*Phascolarctos cinereus*) populations: identification and modelling of a population free from infection. *Scientific Reports*, 9(1), 6261.

Finn, H.C. and Stephens, N.S., 2017. The invisible harm: land clearing is an issue of animal welfare. *Wildlife Research*, 44(5), 377–91. doi:10.1071/wr17018

Flanagan, C., 2019. Koala rehabilitation manual. Koala Hospital. doi:10.13140/RG.2.2.23429.29924

Gordon-Smith, T., 2013. Structure and function of red and white blood cells. *Medicine*, 41(4), 193–9. doi:10.1016/j.mpmed.2013.01.023

Grogan, L.F., Peel, A.J., Kerlin, D., et al., 2018. Is disease a major causal factor in declines? An evidence framework and case study on koala chlamydiosis. *Biological Conservation*, 221, 334–44.

Hing, S., Narayan, E.J., Thompson, R.C.A. et al., 2016. The relationship between physiological stress and wildlife disease: consequences for health and conservation. *Wildlife Research*, 43(1), 51–60. doi:10.1071/wr15183

Jackson, S., Reid, K., Spittal, [sic] D., and Romer, L., 2003. Koalas. In S. Jackson, ed. *Australian Mammals: Biology and Captive Management*, pp. 145–81. CSIRO Publishing, Melbourne.

Juster, R.P., McEwan, B.S., and Lupien, S.J., 2010. Allostatic load biomarkers of chronic stress and impact on health and cognition. *Neuroscience and Biobehavioral Reviews*, 35, 2–16.

King, J.M. and Bradshaw, S.D., 2010. Stress in an island kangaroo? The Barrow Island euro, *Macropus robustus isabellinus*. *General and Comparative Endocrinology*, 167(1), 60–7.

Lanyon, J. M., Sneath, H. L., and Long, T., 2012. Evaluation of exertion and capture stress in serum of wild dugongs

(*Dugong dugon*). *Journal of Zoo and Wildlife Medicine*, *43*(1), 20–32. doi:10.1638/2010-0178.1

Lattin, C.R. and Romero, L.M., 2014. Chronic stress alters concentrations of corticosterone receptors in a tissue-specific manner in wild house sparrows (*Passer domesticus*). *Journal of Experimental Biology*, *217*, 2601–8.

Macbeth, B.J., Cattet, M.R.L., Stenhouse, G.B. et al., 2010. Hair cortisol concentration as a noninvasive measure of long-term stress in free-ranging grizzly bears (*Ursus arctos*): considerations with implications for other wildlife. *Canadian Journal of Zoology*, *88*(10), 935–49.

Martin, L.B., 2009. Stress and immunity in wild vertebrates: timing is everything. *General and Comparative Endocrinology*, *163*, 70–6.

Mastromonaco, G.F., Gunn, K., McCurdy-Adams, H. et al., 2014. Validation and use of hair cortisol as a measure of chronic stress in eastern chipmunks (*Tamias striatus*). *Conservation Physiology*, *2*(1), cou055, https://doi.org/10.1093/conphys/cou055

McAlpine, C., Brearley, G., Rhodes, J. et al., 2017. Time-delayed influence of urban landscape change on the susceptibility of koalas to chlamydiosis. *Landscape Ecology*, *32*(3), 663–79. doi:10.1007/s10980-016-0479-2

McMichael, L., Edson, D., Smith, C. et al., 2017. Physiological stress and Hendra virus in flying-foxes (*Pteropus spp.*), Australia. *PLoS ONE*, *12*(8), e0182171.

Narayan, E., 2019. Physiological stress levels in wild koala sub-populations facing anthropogenic induced environmental trauma and disease. *Scientific Reports*, *9*(1), 6031.

Narayan, E.J., Webster, K., Nicolson, V. et al., 2013. Non-invasive evaluation of physiological stress in an iconic Australian marsupial: the koala (*Phascolarctos cinereus*). *General and Comparative Endocrinology*, *187*, 39–47.

Narayan, E. and Hero, J.M., 2011. Urinary corticosterone responses and haematological stress indicators in the endangered Fijian ground frog (*Platymantis vitiana*) during transportation and captivity. *Australian Journal of Zoology*, *59*(1), 79–85.

Narayan, E. and Vanderneut, T., 2019. Physiological stress in rescued wild koalas are influenced by habitat demographics, environmental stressors, and clinical intervention. *Frontiers in Endocrinology*, *10*.

Narayan, E.J. and Williams, M., 2016. Understanding the dynamics of physiological impacts of environmental stressors on Australian marsupials, focus on the koala (*Phascolarctos cinereus*). *BMC Zoology*, *1*(2), 1–13. doi:10.1186/s40850-016-0004-8

Palme, R., 2019. Non-invasive measurement of glucocorticoids: advances and problems. *Physiology & Behavior*, *199*, 229–43.

Phelps, K.L. and Kingston, T., 2018. Environmental and biological context modulates the physiological stress response of bats to human disturbance. *Oecologia*, *1–2*, 1–12.

Rishniw, M., Pion, P.D. and Maher, T., 2012. The quality of veterinary in-clinic and reference laboratory biochemical testing. *Veterinary Clinical Pathology*, *41*(1), 92–109. doi:10.1111/j.1939-165X.2011.00386.x

Sapolsky, R.M., 2004. Social status and health in humans and other animals. *Annual Review of Anthropology*, *33*, 393–418.

Seleno, N., 2011. Elevation of blood urea nitrogen is predictive of long-term mortality in critically ill patients independent of 'normal' creatinine: Beier K, Eppanapally S, Bazick H, et al. Crit Care Med 2011; 39: 305–13. *Journal of Emergency Medicine*, *40*(6), 724.

Sheriff, M.J., Dantzer, B., Delehanty, B. et al., 2011. Measuring stress in wildlife: techniques for quantifying glucocorticoids. *Oecologia*, *166*(4), 869–87.

Speight, K.N., Haynes, J.I., Boardman, W. et al., 2014. Plasma biochemistry and urinalysis variables of koalas (*Phascolarctos cinereus*) with and without oxalate nephrosis. *Veterinary Clinical Pathology*, *43*(2), 244–54. doi:10.1111/vcp.12145

Tasker, J.G. and Herman, J.P., 2011. Mechanisms of rapid glucocorticoid feedback inhibition of the hypothalamic-pituitary-adrenal axis. *Stress*, *14*, 398–406.

Vogelnest, L. and Woods, R., 2009. *Medicine of Australian Mammals: An Australian Perspective*. CSIRO Publishing, Melbourne.

Webster, K., Narayan, E., and De-Vos, N., 2017. Fecal glucocorticoid metabolite response of captive koalas (*Phascolarctos cinereus*) to visitor encounters. *General and Comparative Endocrinology*, *244*, 157–63.

Whirledge, S. and Cidlowski, J.A., 2010. Glucocorticoids, stress, and fertility. *Minerva Endocrinologica*, *35*, 109–25.

Wingfield, J.C. and Sapolsky, R.M., 2003. Reproduction and resistance to stress: when and how. *Journal of Neuroendocrinology*, *15*, 711–24.

Woinarski, J. and Burbidge, A.A., 2016. *Phascolarctos cinereus*. The IUCN Red List of Threatened Species. Available at: www.iucnredlist.org/ (accessed 26 June 2020).

Zhao, J., Zhu, L., Fan, C. et al., 2018. Structure and function of urea amidolyase. *Bioscience Reports*, *38*(1), 1–12. doi:10.1042/BSR20171617

How thermal ecophysiology assists the conservation of reptiles: case studies from New Zealand's endemic fauna

Alison Cree, Kelly M. Hare, Nicola J. Nelson, Christian O. Chukwuka, and Jo Virens

> ⊃ **Take-home message**
>
> Understanding the constraints and opportunities that environmental temperature places on different life-history stages of reptiles, including embryos in nests and those that develop within live-bearing females, can allow for better-planned translocations, improved captive management, and stronger predictions about risks from changes in climate.

16.1 Introduction

Conservation physiology has a short history—little more than a decade—as an explicitly recognized discipline (Cooke et al. 2013). However, herpetologists have long appreciated the relevance of thermal biology to the conservation of (the non-avian) reptiles (Spellerberg 1975), a major paraphyletic group of vertebrates accounting for over 10 000 species (Uetz et al. 2019). Like most of the world's animals, reptiles are ectotherms and thus profoundly influenced by the variations in body temperature that occur with environmental conditions. As an ectotherm's body temperature increases, rates of ecologically important functions such as locomotion, growth, and embryonic development increase in a curvilinear fashion to a maximum performance

level then decline rapidly towards the upper limits for survival (Paaijmans et al. 2013; Sinclair et al. 2016). This relationship between body temperature and performance in ectotherms is widely illustrated in the form of a thermal performance curve (also described as a thermal dependence curve or thermal reaction norm; Figure 16.1).

Parameters that define the thermal performance curves in reptiles and other ectotherms can differ in value among lineages as a result of evolutionary change (e.g. Araújo et al. 2013). However, such curves do not provide a complete indication of fitness consequences for multiple reasons (Sinclair et al. 2016). For one, key values do not always take into account time-dependent effects (e.g. from duration of exposure, or from prior acclimatization in the wild or acclimation in the laboratory;

Alison Cree, Kelly M. Hare, Nicola J. Nelson, Christian O. Chukwuka, and Jo Virens, *How thermal ecophysiology assists the conservation of reptiles: case studies from New Zealand's endemic fauna* In: *Conservation Physiology: Applications for Wildlife Conservation and Management*. Edited by: Christine L. Madliger, Craig E. Franklin, Oliver P. Love, and Steven J. Cooke, Oxford University Press (2021). © Oxford University Press. DOI: 10.1093/oso/9780198843610.003.0016

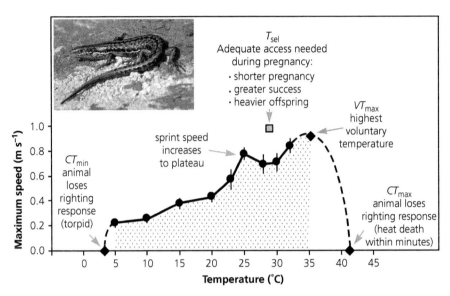

Figure 16.1 Thermal performance curve for the viviparous New Zealand skink *Oligosoma maccanni*. At the lower limit of thermal tolerance (critical thermal minimum, or CT_{min}), coordinated movement is not possible, although skinks are not at immediate risk of death. Voluntary activity is constrained to body temperatures between about 5°C and 35°C (stipple). Within this range, sprint speed increases with temperature, until the optimal range where speed is high and stable. Thermal preference (selected temperature or T_{sel}, shown here by day for females in early pregnancy, at an arbitrary sprint speed) also falls within this range, and pregnancy success is reduced if access is low. At the voluntary thermal maximum (VT_{max}) animals attempt movement to cooler conditions. If body temperature continues to rise, skinks will eventually reach their critical thermal maximum (CT_{max}), where coordinated movement is not possible and death is likely within minutes. Sprint speed values below 5°C and above 32°C are speculative, as indicated by dashed lines. Data sources: activity temperature range, Hare et al. (2009); Virens and Cree (unpubl. data); sprint speed between 5°C and 32°C for non-pregnant females, Gaby et al. (2011); daytime thermal preference for pregnant females, (J) Hare et al. (2009); effects of access to T_{sel} during pregnancy, (K) Hare and Cree (2010); VT_{max} and CT_{max} for males, Virens and Cree (2019). Values shown are mean ± SE; some error bars are too small to be visible. Inset photo: Alison Cree. Adapted by permission from Springer, Switzerland. *New Zealand Lizards*. D.G. Chapple, ed. (Thermal and metabolic physiology of New Zealand lizards, Hare, K.M. and Cree, A. 2016. pp. 239–247.) Copyright. Sprint speed data by permission from Springer Nature. *Journal of Comparative Physiology A* (Thermal dependence of locomotor performance in two cool-temperate lizards. Gaby et al. 2011). Copyright.

Kingsolver and Woods 2016). Of particular relevance to this chapter, key values can also vary among life-history stages; notably, thermal tolerances of embryos can be lower than those of adults and thus set limits on the distributions of species (Levy et al. 2015). Nonetheless, thermal performance curves remain a useful heuristic for exploring potential effects of variation in environmental temperatures on ectotherms (Sinclair et al. 2016).

Given that reptile species are exquisitely adapted to operate within thermal limits, an understanding of the constraints imposed by temperature is highly relevant to conservation. For example, we can improve conservation outcomes by recognizing the thermal properties of natural habitats at risk of potential degradation, by providing more suitable

conditions in captive management, and by estimating limits to species distribution (e.g. Pike et al. 2010; Kearney et al. 2018). These applications are of course becoming ever more relevant as global warming continues at an alarming rate (often in conjunction with changes in other environmental conditions that can affect thermal tolerance, including water availability). To date, much attention has focused on the vulnerability of tropical ectotherms, given how closely the mean habitat temperatures for these taxa already approach mean optimal temperatures for performance (i.e. tropical taxa have a low thermal safety margin; Kingsolver et al. 2013). Ectotherms from temperate regions, with generally higher thermal safety margins, may experience some initial benefit from warmer climates in certain aspects of performance (Kingsolver et al. 2013). Nonetheless, overall effects on mid-latitude species are difficult to predict; given the wide variation in environmental temperatures already experienced, some taxa are at risk of acute heat stress under the temperature extremes predicted for future decades (Kingsolver et al. 2013; Paaijmans et al. 2013).

The archipelago of New Zealand (also known as Aotearoa New Zealand, hereafter NZ), provides habitats for over 100 proposed species of endemic, terrestrial reptiles (Chapple 2016). The climates occupied by NZ reptiles range from subtropical in the north to cool-temperate in the south (latitudes of 34 to 47°S), and habitats range from coastal to alpine (sea level to >2000 m; Hare et al. 2016). In most areas the climate has a maritime influence with relatively cool summers and mild winters, but semi-arid parts of inland South Island or Te Waipounamu approach near-continental conditions (and heatwaves—marine and terrestrial—are beginning to be experienced; Salinger et al. 2019). The terrestrial endemic reptiles fall within the two sister-groups of lepidosaurs: the rhynchocephalians, represented by the world's only surviving species, the tuatara (*Sphenodon punctatus*), and endemic squamates, represented by members of two families of lizards, Scincidae and Diplodactylidae (there are no terrestrial snakes). There is much cryptic genetic diversity within the lizards; at least 61 species of skinks and 43 species of geckos have been proposed, although not all are formally described (Chapple 2016; Van Winkel et al. 2018). Within NZ, these undescribed taxa are often recognized with temporary 'tag names' that allow treatment as entities for conservation purposes (Hitchmough et al. 2016), and we follow this approach here.

Among NZ's endemic and terrestrial reptiles, two major ecological distinctions have the potential to influence thermal preferences and tolerances. First, many species are at least partly nocturnal (sometimes described as diurno-nocturnal) whereas others are strictly diurnal (Cree 2014; Hare et al. 2016). Second, and even more unusual globally, most species are viviparous (live-bearing), with only two (the tuatara and one skink) laying eggs (Cree and Hare 2016a). Slow life histories are commonplace, especially among the nocturnal species.

For example, life spans can reach about a century in tuatara and over 50 years in some geckos (values within the top 3 per cent for lepidosaurs), and reproduction is less than annual in several species (Cree and Hare 2016a).

Reptiles in NZ have been profoundly affected by human settlement, which began about 730 years ago. The resulting habitat alterations and introduction of numerous exotic predators (especially mammals) have contributed to about 85 per cent of terrestrial reptiles being rated, using the NZ Threat Classification System, as either already threatened with extinction or at risk of becoming so (Hitchmough et al. 2016). Although NZ lizards, especially, were frequently overlooked in the past, awareness is growing of their ecological, evolutionary, and cultural significance, including to Māori, the indigenous people of NZ (Chapple 2016; Towns et al. 2016). Reptiles have thus begun to receive wider attention in conservation management over recent decades, including in formal recovery plans (Nelson et al. 2015; Towns et al. 2016).

Here, we summarize the current understanding of the thermal ecophysiology, including effects of temperature on embryos, for the most studied endemic reptiles of New Zealand. In addition, we emphasize how this information is relevant to, or already applied in, conservation. Where information exists, we also consider the influence of water availability; in nature, temperature and moisture are likely to co-vary in ways that influence survival. We consider case studies in three areas: (1) thermal effects on embryos of two egg-laying species (the tuatara and a skink, both diurno-nocturnal); (2) thermal effects on embryos of two live-bearing species (a diurnal skink and a diurno-nocturnal gecko); and (3) upper thermal tolerances—voluntary and/or critical, including in pregnant females (for the same viviparous species as in 2). We then synthesize our findings, making recommendations for further research and for more effective integration of physiology into conservation. Although space limitations prevent us discussing invasive pest reptiles (one skink has established and a turtle has potential to establish; Nelson et al. 2015), we note that knowledge of thermal biology is relevant to predicting the potential distribution of exotic species as well (e.g. Chapple et al. 2016).

16.2 Case study 1: influences of temperature and water availability on embryonic development in egg-laying species—tuatara and the egg-laying skink

Globally and in NZ, we know more about the effects of temperature and water availability on embryonic development of oviparous reptiles than for live-bearing species. One reason is surely the sheer abundance of egg-laying species (about 80 per cent of reptiles globally), although this abundance is not true of NZ species. Other reasons, however, probably apply to NZ taxa. Eggs are discrete, immobile structures laid in the ground and are thus directly exposed to external factors. Their eggshells can provide visible evidence of failed development (Thompson et al. 1996), unlike the situation in viviparous species in which aborted embryos may be consumed by mothers or otherwise disappear without trace (Cree and Hare 2010; Caldwell et al. 2018). Eggs are also more easily collected and studied in the laboratory, with fewer logistical demands and welfare concerns. Finally, oviparous reptiles are more likely than viviparous species to exhibit temperature-dependent sex determination (TSD), a phenomenon that attracts considerable interest. Below, we discuss effects of temperature and water availability on embryonic development on two species individually, followed by a combined section on conservation implications and applications.

16.2.1 The tuatara (*Sphenodon punctatus*)

The evolutionary status of tuatara as the last living rhynchocephalian has contributed to this species receiving the most scientific attention among all reptiles within NZ. The tuatara is also of high significance as a taonga, or treasure, among Māori iwi (tribes), many of whom identify as kaitiaki (guardians) of local populations. Additional factors favouring attention include the species' relatively large body size (females reach up to ~500 g; eggs ~4 g at laying) and conspicuous nesting behaviour. The tuatara genus has also experienced a large reduction in distribution that is obvious from the fossil record, leading to a conservation status today

of 'at risk—relict' for the extant species (for an overview, see Cree 2014; Hitchmough et al. 2016).

Most information about environmental influences on embryonic development in tuatara comes from the abundant population on Stephens Island (Takapourewa) in Cook Strait (40°S). This population lies at the cool end of the surviving range, although tuatara survived to at least 46°S on South Island until human arrival (Cree 2014). On Stephens Island, about nine eggs (range 1–18) are laid in shallow nests (top-most eggs 10–230 mm below soil surface) in open ground during the austral spring, at an early (gastrula) stage of development (Thompson et al. 1996; Nelson et al. 2004d; Cree 2014). Nesting begins earlier in years with warmer springs (Nelson et al. 2018). Development proceeds over about 10–16 months, ceasing for several months over winter; eggs in warmer locations probably hatch earliest (Thompson et al. 1996). Temperatures in natural nests fluctuate daily (by up to 16°C; Thompson et al. 1996), and also seasonally between extremes of 1.6 and 38.4°C in successful nests (Nelson et al. 2004d). The parchment-shelled eggs swell as water is absorbed from the soil during development. Desiccation resulting from low soil water potential contributes to egg mortality, which can exceed 50 per cent in some study-years (Thompson et al. 1996).

The effects of several constant temperatures on embryonic development of tuatara from Stephens Island have been examined (Figure 16.2A). Although constant temperatures do not reflect natural conditions, they provide valuable evidence that thermal effects exist. Under moist conditions (high substrate water potential; see Figure 16.2A for details) the lowest constant temperature for successful hatching lies between 15 and 18°C; between 18 and 24°C hatching success is high (≥87.5 per cent), although in one study it was significantly lower at 18°C than at 21 or 22°C (Nelson et al. 2004b). The upper thermal limit is unclear; hatching after a constant 25°C is clearly possible (Thompson 1990; A. Cree unpubl obs.) and embryos can survive exposure to a similar constant-temperature equivalent in natural nests (Nelson et al. 2004b). Nonetheless, the temperatures yielding success in tuatara are low compared with those for many squamates (Andrews and Schwarzkopf 2012).

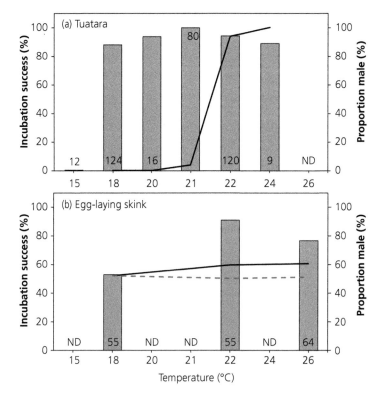

Figure 16.2 Effects of temperature on egg incubation success and offspring sex for two egg-laying species of New Zealand reptiles. (A) Tuatara, *Sphenodon punctatus* (at water potentials of −120 to −170 kPa). (B) The egg-laying skink, *Oligosoma suteri* (at −120 to −270 kPa). Vertical bars show incubation success for the sample sizes indicated within or above the bars. Solid or dashed lines indicate the percentage of hatchlings that were male from the same or related studies (note, sample sizes may differ). ND = no data. Note categorical (non-linear) *x*-axis. Over the temperature ranges tested, the tuatara exhibits an FM pattern of TSD, whereas the egg-laying skink shows no evidence for TSD. For the skink, sexes were assigned from cloacal morphology (including presence of inferred hemipenes) at hatching (dashed line), and reassessed for 22°C and 26°C based on evidence from recaptures of some individuals when adult (solid line). Although some males appear to have been incorrectly assigned as female at hatching, the inference that TSD does not exist seems unlikely to change. Data sources for tuatara: Thompson (1990); Nelson et al. (2004a, b); Mitchell et al. (2006); Besson et al. (2012); Alison Cree (unpubl. data). Data sources for *O. suteri*: Hare et al. (2002); Cree & Hare (2016a); Kelly Hare (unpubl. data).

Hatching success is not affected by relatively moist water potentials spanning −90 to −400 kPa (Thompson 1990).

Within the range of temperatures yielding high incubation success, temperature has large consequences for incubation time in tuatara (e.g. the duration of ~280 days at 18°C is halved at 24°C; Cree 2014). This variation could have important consequences under field conditions given the fitness effects from season of hatching observed in other reptiles (e.g. Warner and Shine 2007). Incubation temperature also has a dramatic effect on offspring sex. The TSD pattern under temperatures tested to date with eggs from Stephens Island

stock across a range of studies is FM; that is, only females (F) are produced from low incubation temperatures (18–20°C); only males (M) are produced from high temperatures (~22–24°C); and both sexes are produced over a relatively narrow range of intermediate temperatures (~21–22°C; see Figure 16.2 for references). Nests on Stephens Island also show sex ratios consistent with an FM pattern (Nelson et al. 2004a). Nonetheless, research with warmer temperatures is needed to confirm FM rather than FMF (i.e. a possible switch back to females at higher temperatures than tested thus far). To date, the FM pattern appears unique among reptiles to tuatara, and the possibility that it forms an 'eco-evolutionary trap' resulting in extreme male bias under global warming has been raised (Pezaro et al. 2017; see further below).

Effects of incubation conditions on aspects of phenotype other than sex seem modest in tuatara. In one study, temperatures between 18 and 22°C did not influence hatchling mass or snout-vent length, and the amount of water absorbed by the eggs had only a weak effect on hatchling size (Thompson 1990). However, another study suggested that offspring from 18°C are smaller than those from warmer incubation temperatures at 10 months, and that natural nests produce hatchings in which snout-vent length is maximized when constant-temperature equivalents during incubation are between 21 and 23°C (Nelson et al. 2004b). Selected temperature (the temperature preferred on a thermal gradient) in offspring at 3 months of age was not affected by incubation temperatures between 18 and 22°C (Nelson et al. 2017), and metabolic rate at 2–3 years of age was not affected by incubation at 20°C compared with 23 or 25°C (Jarvie et al. 2018; note that in some of these comparisons, sex and incubation temperature are unavoidably confounded). More dramatic differences in phenotype were seen when comparing offspring from constant incubation conditions in the laboratory with those from naturally incubated eggs; the latter experience wider variation in temperatures as well as in water potential. The naturally incubated hatchlings were more aggressive, with a greater incidence of tail loss and faster sprint speeds at 10 months of age under laboratory conditions (Nelson et al. 2004b, 2006). The production of tuatara offspring that have high survival and reproduction when released to the wild after artificial incubation may be improved by investigating variable incubation regimes in the future.

16.2.2 The egg-laying skink *(Oligosoma suteri)*

The egg-laying skink lives on rocky beaches in northern NZ (34–37°S) where it overlaps in distribution with the tuatara (van Winkel et al. 2018). As with the tuatara, it has the conservation status of 'at risk—relict' (Hitchmough et al. 2016). However, the egg-laying skink is smaller (~16 g) and has a more semi-marine lifestyle: it possesses nasal salt glands,

and dives in saltwater pools to forage and escape predators (Hare et al. 2016 and references within). Like most other squamates (and unlike tuatara), the egg-laying skink retains eggs *in utero* for a substantial period of embryonic development (up until stage 32 out of a 40-point staging scheme; Hare et al. 2002); nothing is known about environmental influences during this phase of development. At oviposition, eggs are laid in clutches of one to five (each ~0.9 g) in sand or soil beneath large stones in early summer; females nest communally and wild-laid nests take about 3 months to incubate (Cree and Hare 2016a and references therein). Substrate warmth influences nest-site selection (Chapple et al. 2017), and nests can experience a diel temperature range between 18°C and 32°C (V. Stenhouse pers. comm. in Chapple et al. 2017).

Laboratory incubation studies, similar to some previously performed for tuatara, used three constant incubation temperatures (18, 22, and 26°C) and two water potentials (−120 and −270 kPa). Water potential had no influence on incubation duration or sex (Hare et al. 2002), and although hatchlings were slightly larger for all size measures when from the wetter substrate, this effect diminished over time and water potential had no influence on hatchling speed or survival (Hare et al. 2004, 2008). Temperature, in contrast, has a major impact; in particular, duration of incubation increased as temperature decreased (Hare et al., 2002), with the lowest incubation temperature being near the lower lethal limit for incubation (Hare et al. 2002; Figure 16.2B). Hatchlings from 18°C had morphological abnormalities, atypical colouration, smaller size, reduced growth, slower sprint speeds, and lower survival in captivity (Hare et al. 2002, 2004, 2008). However, there was no evidence of TSD (Figure 16.2B). After release to the wild, skinks from only the two warmer incubation temperatures were recaptured 5–7 years later, exhibiting similar sizes and sprint speeds as wild-incubated-and-raised individuals (Hare et al. 2020, but much lower rates of diving (~27 per cent captive-incubated-and-raised chose to dive vs. 78 per cent of wild-incubated-and-raised skinks; Hare et al. 2020; Miller et al. 2010). Thus, a cool incubation temperature of 18°C seems more detrimental to the egg-laying skink than to tuatara.

16.2.3 Conservation implications and applications: embryonic development in egg-laying species

To date, information from egg incubation studies with tuatara from Stephens Island stock has been applied to conservation in several ways. These include: (1) guidance on management of captive populations to produce viable offspring of known sex ratios (Blanchard and the Tuatara Recovery Group 2002); (2) assessments of the thermal suitability of potential sites for reintroduction (e.g. to Orokonui Ecosanctuary in the southern South Island; Besson et al. 2012; Jarvie et al. 2014); and (3) production of viable offspring of known sex ratios for conservation translocations (e.g. Jarvie et al. 2015); such offspring have often been produced from research projects that have simultaneously helped to refine aspects of TSD (e.g. Nelson et al. 2004a, 2010; Keall et al. 2010). More generally, laboratory findings have helped explain current nesting patterns (e.g. why eggs are not laid in forested locations on Stephens Island) and with predicting potential effects of vegetation changes from human disturbance (Thompson et al. 1996; Cree 2014). Using data from Stephens Island, studies are underway to develop species-distribution models for tuatara to assess potential threats of extinctions, and opportunities for reintroductions, including under climate change (Carter 2015; Jarvie 2016).

Studies on tuatara have also been extended to the much smaller population on the rocky islet (4 ha) of North Brother in Cook Strait. Unlike the Stephens Island population, which has a wide range of nesting habitats and nest characteristics and is not thought to be at immediate risk from a strong male bias under global warming (Nelson et al. 2004d), the situation on North Brother seems much grimmer. Nest sites on North Brother are in shallow, north-facing soils covered in low vegetation only, and the adult population is already male-biased (Grayson et al. 2014). Modelling of soil temperatures at typical nest depths, combined with information on development rates and TSD pattern, suggest that under a maximum warming scenario hatchlings will be exclusively male by 2080 (Grayson et al. 2014). When these results are combined with low egg production rates and other demographic

patterns, population extinction becomes likely within a few centuries of warming (Grayson et al. 2014). Short-term management could provide cooler incubation conditions, but the value is complicated by the already-low genetic variation within this population (for discussion, see Cree 2014). Even without TSD, the vulnerability of tuatara eggs and hatchlings on small islands to higher temperatures and possible drought is of serious concern (especially in warmer, more northerly locations, and with rising sea levels potentially reducing available habitat; Cree 2014).

For the egg-laying skink, incubation data have provided the basis for species-distribution modelling. The results indicate that embryos could successfully develop further south than populations currently extend (an absence that could reflect previous extirpations), and that even greater extension is possible under predicted warming scenarios (Stenhouse et al. 2018). Such predictions are timely, given that the conservation status is unclear for mainland populations of this species and that dispersal is limited by the fragmented nature of suitable habitat. Translocations of egg-laying skinks may be needed to restore the species to safe havens on the mainland, and knowledge of where egg incubation should be successful will inform these decisions.

16.3 Case study 2: influence of temperature on embryonic development in live-bearing species—McCann's skink and the Otago/Southland gecko

Unlike the oviparous species above, live-bearing lizards must provide their embryos with suitable conditions *in utero* throughout the entire period of embryonic development. Most live-bearing lizards (including all NZ species examined to date) are lecithotrophic, meaning that most nutrients for the embryo have been committed, as yolk, at ovulation. Nonetheless, nutrients in small amounts may be supplied via placental tissues during pregnancy, and water uptake by the embryo, as well as gas exchange, are necessary for successful development (Stewart and Blackburn 2015). Female lizards also often adjust their thermoregulatory behaviour when pregnant in ways that may enhance survival and viability of offspring (Stewart and Blackburn 2015).

16.3.1 McCann's skink (*Oligosoma maccanni*) and the Otago/Southland gecko (*Woodworthia* 'Otago/Southland')

Among NZ's viviparous lizards, thermal effects during pregnancy have been examined in detail for only two species. The smaller-bodied species is McCann's skink (~4 g), a diurnal, sun-basking (heliothermic) lizard. The larger is the Otago/Southland gecko (~10 g), a nocturnal forager that has been revealed by trail cameras to also cryptically bask (Gibson et al. 2015). Populations of both species have been studied from subalpine tussock grassland near Macraes in the southern South Island (~45°S, ~500–710 masl) where the lizards occupy crevices in schist rock outcrops. McCann's skink, with a conservation status of 'not threatened' (Hitchmough et al. 2016), has been suggested as a surrogate or model species, with appropriate caution, for the sympatric skinks *O. grande* and *O. otagense*; the latter species are also diurnal and heliothermic, but larger-bodied and 'nationally endangered' (Hitchmough et al. 2016) and the subjects of captive management (e.g. Connolly and Cree 2008). The Otago/Southland gecko has the conservation status of 'at risk—declining', but is locally abundant at Macraes and may provide a model for gecko species at greater conservation risk. Given that these two species are sympatric and have often been studied in parallel, we consider them together in this section.

Laboratory studies of thermal effects during pregnancy have been grounded in knowledge of normal reproductive patterns and thermal conditions in the wild. For McCann's skink, pregnancies at Macraes last about 4–5 months from ovulation in spring until birth in mid–late summer (Cree and Hare 2016a and references therein). Birth dates are later (by days) for skinks from the highest elevations, where temperatures are probably coolest (Hare and Cree 2011). Females in early pregnancy select mean daytime body temperatures of 28.9°C (Hare et al. 2009). However, field body temperatures and microhabitat temperatures vary greatly in the changeable subalpine environment at Macraes, and thus access of pregnant females to their selected temperature is infrequent (Hare et al. 2009).

Laboratory studies with McCann's skink therefore proceeded by offering varied access to basking opportunity rather than constant temperatures (basking opportunity is avidly used by females when opportunity exists; Caldwell et al. 2018). Females with low access to basking opportunity (heat lamps available for a nominal 28 h/week) were less likely to have successful pregnancies than those with heat lamps for 40 h/week (which approximated conditions at Macraes at the time of the study) or 56 h/week (Figure 16.3B; Cree and Hare 2010). Gestation length was strongly affected by basking opportunity, averaging 51 days longer at the coolest regime than the warmest (Figure 16.3A; Cree and Hare 2010). Those pregnancies that were successful under the coolest regime produced offspring that were smaller and/or grew more slowly than those from other regimes, albeit some of these effects were sex-specific (Hare and Cree 2010). Although offspring sex was not influenced by thermal conditions during pregnancy, some behavioural and colouration differences were apparent at 3 months of age (Hare and Cree 2010). Overall, these observations for McCann's skink show that thermal conditions during pregnancy influence pregnancy success, offspring phenotypes, and gestation length (and hence date of birth, which in other viviparous species can influence survival in the wild; Wapstra et al. 2010).

In contrast with McCann's skink, the Otago/Southland gecko at Macraes has a remarkably long pregnancy of ~14 months (Cree and Hare 2016a). Elsewhere in the world gestation lengths of a year or more are seen in only a few unrelated lizards, often but not always from cool climates (Stewart and Blackburn 2015). Embryonic development in the Macraes population of Otago/Southland geckos occurs from ovulation in spring until autumn about 6–7 months later; females then retain fully developed offspring *in utero* over winter, a phenomenon inferred to benefit survival of offspring by delaying birth until warm, spring conditions (Cree and Hare 2016a and references therein). The geckos at

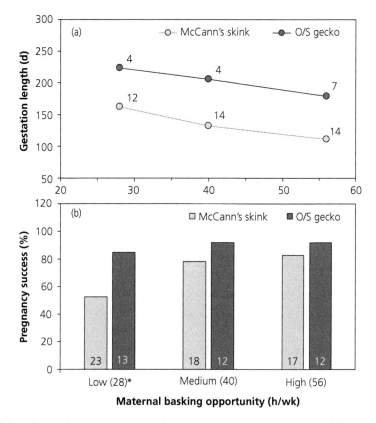

Figure 16.3 Effect of thermal regime during pregnancy on developmental outcomes for two viviparous lizards from Macraes, New Zealand: McCann's skink (*Oligosoma maccanni*) and the Otago/Southland gecko (*Woodworthia* 'Otago/Southland'; 'O/S gecko' in figure). Thermal regimes varied in terms of access to a basking lamp; thus, opportunity to reach maternal selected temperature was low (28 h/week*; nominal for skinks), intermediate (40 h/week, approximately natural conditions at Macraes), or high (56 h/week). During the remaining time, females were held at 15–18°C by day and 12°C overnight. (A) Gestation length for females that delivered spontaneously, assuming an ovulation date of 1 October. Sample sizes are shown beside symbols (standard errors are smaller than the size of the symbols). (B) Pregnancy success as the percentage of females producing at least one viable offspring. Note that for skinks only, the low regime was a nominal 28 h/week; after many failed pregnancies, basking opportunity was increased (after about 91 per cent of gestation) to 40 h/week, yielding some successful pregnancies from the remaining females. For geckos, pregnancy success is a measure of developmental success only, as some females were induced to deliver (thus, whether offspring would have been viable if females had been left to deliver naturally is unknown). Sample sizes are shown at base of bars. Sources: for skinks, data redrawn from Cree and Hare (2010). For geckos, data redrawn from Cree and Hare (2016b).

Macraes also show remarkably large fluctuations (for lizards) in thermal preference during pregnancy. Daytime preference is clearly elevated (reaching a mean of about 28°C) during at least early–mid-pregnancy (Rock et al. 2000; Cree and Hare 2016b); it can then fall to a mean of 21°C during the final few weeks of pregnancy, a response predicted to depress metabolism and thereby enhance survival of fully developed offspring *in utero* (Cree and Hare 2016b).

Initial work investigating thermal effects during pregnancy in Otago/Southland geckos confirmed that success under indoor conditions was feasible (Cree et al. 2003). However, it also identified a high risk of failure when females were unable to reach what was subsequently identified as selected temperature during early–mid-pregnancy (Rock and Cree 2003). More recent studies, performed in parallel with McCann's skinks, were better informed by knowledge of microhabitat and body temperatures in the wild and by the selected temperatures of pregnant females (Rock et al. 2000, 2002). When female geckos were offered basking that allowed selected body temperature to be achieved for 28, 40, or 56 h per week, viable embryos developed equally well under all regimes (Figure 16.3B; noting that, for unknown reasons, captive female geckos sometimes require hormonal induction to reduce the risk of 'over-gestation' and stillbirths; Cree and Hare 2010, 2016b). Gestation length in the geckos was substantially longer than in McCann's skinks and, as expected, was greatest when basking opportunity was lowest. Spontaneously delivered offspring from the coolest regime were smaller and grew more slowly in mass than those from warmer regimes, but they had similar high survival (to 2 months) and were equally fast at running (Cree and Hare 2016b). Effects on offspring sex of Otago/Southland geckos are under study.

16.3.2 Conservation implications and applications: embryonic development in live-bearing species

Overall, our research examining thermal effects on McCann's skink and the Otago/Southland gecko demonstrates several points relevant to conservation. Pregnancies can be successful indoors with appropriate vitamin supplementation, lighting, and other key aspects of husbandry (e.g. control of ectoparasites; Hare et al. 2010). However, the thermal regime (especially basking opportunity) affects pregnancy outcomes in multiple ways. In particular, females must have sufficient access to selected temperature for pregnancy to be successful, and the degree of access has a strong effect on date of birth. Although neither species is currently in need of captive management for conservation purposes, our results are relevant to future modelling of climate-change impacts (e.g. changes in cloud cover; Hare and Cree 2010), and were used by Department of Conservation staff when holding skinks of high conservation concern at Macraes prior to translocation and long-term captive holding (Lesley Judd, Department of Conservation, pers. comm.). They also have relevance to captive-management programmes underway for other lizards (see discussion below).

16.4 Case study 3: Effects of life-history stage on upper thermal tolerances of live-bearing lizards—McCann's skink and the Otago/Southland gecko

Biomarkers that provide information about an animal's physiological status are a key aspect of the conservation physiologist's toolbox (Madliger et al. 2018). As in many vertebrates, plasma concentrations of glucocorticosteroids have been explored as a potential biomarker of 'stress' in tuatara and a few NZ lizards large enough to be readily blood-sampled (e.g. Barry et al. 2010; Anderson et al. 2017). Although secretion of corticosterone clearly rises acutely with capture, baseline concentrations of this metabolic hormone can be challenging to interpret given the natural fluctuations that occur with factors such as season, female reproductive condition, and body temperature within the activity range (Cree et al. 2003; Anderson et al. 2017). Body temperature, on the other hand, has a profound and obvious influence on physiological processes, can be measured rapidly and with minimal contact in small animals (Figure 16.4), and can reach values that are unequivocally harmful in the short- or very short term (Figure 16.1).

Given the context of climate change, it is unsurprising that measures of upper thermal tolerance in lizards have seen a global resurgence of interest. One measure, critical thermal maximum (CT_{max}), has been reported for at least 350 species of lizards worldwide (Bennett et al. 2018; Diele-Viegas et al. 2018). The typical testing procedure involves heating an individual lizard at a constant rate until an animal loses the ability to right itself when turned on its back. At this point, fitness is effectively reduced to zero as the animal is unable to escape the warming environment. Provided that the animal is immediately returned to cooler temperatures (and before the possibility of involuntary muscle spasms; Huang et al. 2006; see also Camacho and Rusch 2017), survival without obvious ill-effects seems high, although more explicit reporting of this is desirable.

From ecological and ethical perspectives, one could nonetheless argue that an animal's well-being and activity have already been restricted by body temperatures several degrees below CT_{max}, that is,

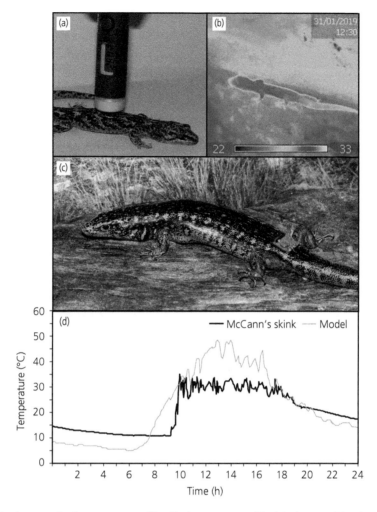

Figure 16.4 Recent developments allowing measurement of lizard body temperatures with minimal contact. (A) An inexpensive and portable infrared thermometer allows skin temperature to be measured from a short distance in this adult gecko (*Woodworthia* 'Otago/Southland', ~12 g). Skin temperatures measured with infrared thermometers can be comparable with values from a thermocouple inserted into the cloaca, provided that the size of the lizard, and distance and orientation of the device, eliminate any influence of background temperature (Hare et al. 2007; Chukwuka et al. 2019). Photo: Alison Cree. (B) Image from a thermal imaging camera of a juvenile Otago/Southland gecko (~2 g) at Macraes on a sunny day in summer. The gecko has been exposed within its retreat by tilting the rock above. The gecko's skin (29.6°C over dorsal pelvis) is close to the temperature of the warm rock above, and also close to the voluntary thermal maximum for juveniles of this species. Photo: Christian Chukwuka. (C) An Otago skink (*Oligosoma otagense*, ~40 g) with a biologger weighing ~0.34 g attached to its pelvis. The device, a miniaturized modification of a commercial temperature logger (DS1922L Thermocron iButton; Maxim Integrated, San Jose, CA, USA), enables near-continuous records of skin temperature to be recorded over several days (Virens and Cree 2018). Photo: Alison Cree. (D) Temperature trace obtained from a pregnant McCann's skink (~6 g) on a summer's day at Macraes. The skink was fitted with a temperature biologger similar to that in (C). Once the skink has emerged to bask, its body temperature rises rapidly to reach voluntary thermal maximum for this species; the skink then thermoregulates around a slightly lower set-point for the rest of the day. Also shown is the temperature inside a painted copper model, recorded using an unmodified iButton. The model, which was calibrated to have similar heating and cooling rates to a live skink, was placed on a nearby rock surface in an unshaded position. Clearly, the live skink would have exceeded the CT_{max} of this species if it had not retreated to shade (Virens and Cree, unpubl. data).

at and above the voluntary thermal maximum (VT_{max}). Although a variety of ways of assessing VT_{max} in lizards have been proposed, a recent recommended definition is 'a well-defined movement to avoid a warming...environment' (Camacho and Rusch 2017). As with CT_{max}, values can vary somewhat with testing procedure (e.g. heating rate); biological variables probably also influence VT_{max}, although current information is limited (for a review, see Camacho and Rusch 2017).

16.4.1 McCann's skink (*Oligosoma maccanni*) and the Otago/Southland gecko (*Woodworthia* 'Otago/Southland')

Current studies examining CT_{max} and/or VT_{max} provide the first explicit values for two of NZ's viviparous lizards (extending earlier information available on lethal temperatures; see Hare and Cree 2016). We focus on sex and life-history stage as aspects of within-species variation. Of particular interest, we included two stages that have been largely or completely overlooked for lizards: pregnancy and the neonatal period. For McCann's skinks, an important finding is that CT_{max} was lower during late pregnancy than in postpartum females or males, but not neonates. Reassuringly, all tested stages (adults, neonates, and embryos *in utero*) survived with no deleterious effects detected (Virens and Cree 2019). Effects on embryos during early pregnancy, when developmental responses might be detrimental, have yet to be tested. Mean values overall were similar to mean values for reptile species worldwide, consistent with an analysis suggesting relatively little evolution in CT_{max} among reptiles compared with that for critical thermal minimum (Araújo et al. 2013). The value for McCann's skink of maintaining a high CT_{max} is apparent when we consider that, even in the cool climate of Macraes, lizard models on rock surfaces reach well above CT_{max} on cloudless days (Figure 16.4).

We also examined values for VT_{max} in both McCann's skinks and Otago/Southland geckos. In McCann's skinks, VT_{max} was, as expected, about 5–6°C lower than for CT_{max}. Once again, significant variation among sex and life-history groups was detected, but with some differences in the pattern compared with CT_{max}. Pregnant female and male skinks had relatively low values for VT_{max}, whereas neonates had a relatively high value (Virens and Cree 2019). For the Otago/Southland gecko, preliminary indications are that VT_{max} is about 4–5°C lower than in McCann's skinks, with some influence from prior acclimatization. Although little if any variation in VT_{max} temperature exists among the sex and life-history groups tested to date, the time that geckos tolerate exposure to heating before an escape response is initiated can vary among groups (Chukwuka 2020). Thus, for both species, VT_{max} is somewhat plastic, and within-species variation in voluntary thermal tolerance (both the temperature and the duration of tolerance) should be considered.

16.4.2 Conservation implications and applications: upper thermal tolerances in live-bearing species

For both McCann's skinks and Otago/Southland geckos, field body temperatures measured in summer at Macraes already sometimes reach the VT_{max} measured in the laboratory (e.g. Figure 16.4). Thus, at the very least, further climate warming has the potential to alter species-specific patterns of activity and microhabitat use. We envisage values for upper thermal tolerance, especially VT_{max}, being used to create biophysical models, with different life-history stages (e.g. pregnant females and neonates) included. Elsewhere, the value of mechanistic niche approaches for predicting (and ideally, ameliorating) changes in distribution of lizard species with changing climates is apparent (e.g. Kearney et al. 2018).

Simultaneously, our studies with McCann's skinks and Otago/Southland geckos have demonstrated approaches for assessing thermal parameters that could readily be used with species of higher conservation concern. These include procedures for monitoring thermal microhabitats using data loggers (Hare et al. 2009) and for recording thermoregulatory behaviour in relation to weather conditions using trail cameras (Gibson et al. 2015). They also include methods for measuring field body temperatures with minimal disturbance, such as spot temperatures using infrared cameras or highly portable and inexpensive infrared thermometers (Hare et al. 2007; Chukwuka et al. 2019), or as continuous records of skin temperature over days using miniaturized biologgers weighing only ~0.3 g (Figure 16.4; Virens and Cree 2018). Such approaches may contribute to greater recognition of the value of natural vegetation and microhabitat shelters for shade, and to improved captive management for lizards generally (see discussion below).

16.5 Synthesis, future directions, and recommendations

16.5.1 Synthesis

The case studies described above illustrate several features central to the mission of conservation physiology. These include the use of an experimental approach to establish cause and effect (Seebacher and Franklin 2012; Cooke et al. 2017), the setting of measurable thresholds that can be detected before a population is destabilized (Madliger et al. 2017), and the development and/or application of novel methods and repurposed technologies as part of the discipline's toolbox (Madliger et al. 2018).

However, conservation physiology ideally moves beyond the (relatively simple) identification of potential benefits and is applied in ways that yield successful conservation outcomes (Madliger 2017). The research described above on captive egg incubation for tuatara comes closest to this goal, yielding offspring that have contributed to the reintroduction or reinforcement of tuatara populations on offshore islands, and to reintroductions to parts of the former range on the two main islands of NZ (Nelson et al. 2019). It is appropriate to acknowledge that many additional elements contributed to these successes, including the development of techniques for eradication and/or fencing out of introduced mammals, collegial support from numerous sectors of the conservation community including Māori kaitiaki, zoos, and community ecosanctuaries, and the existence of a formal recovery planning process that set a public agenda for coordinated action (Cree 2014; Nelson et al. 2019). A legacy of more than a century's prior biological study on tuatara undoubtedly helped, as did awareness that the eyes of the international community were on efforts to assist the world's only surviving rhynchocephalian reptile.

16.5.2 Future directions and recommendations

The immediate challenges for conserving NZ's 100+ species of lizards, almost all viviparous, are much greater than for tuatara. Many species exist only on the mainland where predatory mammals and habitat modification (including urban spread) pose substantial challenges; already, 16 taxa are rated as 'nationally endangered' or 'nationally critical' (Hitchmough et al. 2016). The public profile for lizards is lower than for tuatara and, historically, funding has been limited (Towns et al. 2016).

Conservation physiology could make an immediate contribution towards NZ's lizards by assisting the development and monitoring of captive-management programmes. Several captive-management programmes combining salvage from the wild (for sites where threats are considered unmanageable) with an out-of-range 'breed-for-release' component have already been initiated by conservation authorities. One, involving western populations of the grand and Otago skinks (both 'nationally endangered'), did not achieve the hoped-for outcomes. In 2014, 85 skinks were transferred in an emergency salvage to outdoor enclosures at several locations (some distant from the species' current or historical range). The programme ended earlier than planned in 2018 when 67 individuals (including surviving wild-caught and captive-born individuals) were released into a predator-resistant ecosanctuary closer to the collection site (Hare et al. 2019). A dozen or so younger animals have been held back for later release (R. Gibson, Auckland Zoo, pers. comm.). To date, no formal evaluation of factors contributing to poor outcomes in captivity is available; however, unsuitable climates at some locations are thought to have contributed, as well as higher-than-expected levels of intraspecific aggression within some adult pairs (R. Gibson, pers. comm.). Salvage and breed-for-release programmes are now underway for several other poorly known taxa of *Oligosoma* (R. Gibson, pers. comm.; Towns et al. 2016). We recommend comparisons of thermal microhabitats and body temperatures between wild habitats and captive populations, as well as regular assessments of pregnancy status in captivity, to help increase adult survival and the production of healthy offspring.

Captive management is often considered an option of 'last resort'. However, it is worth emphasizing that, as with the aggression noted above for western populations of grand and Otago skinks, unexpected challenges often appear. Even when conducting the research described in this chapter, involving healthy, non-senescent animals from genetically diverse populations, difficulties arose. For tuatara, the challenges included compromised egg quality or egg size resulting from premature induction of oviposition (Thompson 1990; Nelson et al. 2004c). In the egg-laying skink, desiccation resulted when small individual containers were initially used for egg incubation, and difficulties with accurate sex assignment at hatching have since been recognized (Hare et al. 2002; Cree and Hare 2016a; Hare et al. unpublished data). In the viviparous McCann's skink, the challenges included an association between uncontrolled ectoparasitic mites and pregnancy failure in captivity (Hare et al. 2010), and the potential for maternal ingestion of aborted embryos to conceal evidence of pregnancy failure (Caldwell et al. 2018). In Otago/Southland geckos, pregnancy success was initially low when thermal regimes were inadvertently too cold for females to reach selected temperature (Rock and Cree 2003), and there remains potential, for unknown reasons, for some pregnant geckos in captivity to over-gestate, or delay delivery beyond the point of embryo survival (Cree and Hare 2010, 2016a). Furthermore, for most NZ reptiles, techniques are lacking for accurate sex assignment at hatching or birth. All of these experiences point to the value of starting research into captive management early, while healthy eggs or pregnant females are available in sufficient numbers to study (see also Connolly and Cree 2008).

At the same time as the needs are growing for NZ's lizards, and as institutional processes that authorize live-animal work and fieldwork become increasingly complex, the pressures of environmental change threaten to become overwhelming (Hare et al. 2019). Although largely buffered to date by maritime conditions, terrestrial NZ is beginning to experience heatwaves (Salinger et al. 2019), and climatic conditions (not just in temperature) will continue to change over the coming decades and beyond (Ministry for the Environment 2016). It is common practice to suggest areas for future research, and indeed these exist: topics where conservation physiologists can help include assessing the simultaneous effects on reptiles (including embryos) of combined factors (e.g. temperature, water availability, and wind), especially in areas where summer temperatures are already hot and predicted to rise rapidly, such as in central Otago (Ministry for the Environment 2016; see also Hare and Cree 2016). We also emphasize the value of developing mechanistic niche models to help predict areas of suitable habitat for reptiles under different climate-change scenarios, and the importance of working more closely and effectively with conservation managers and practitioners, including Māori kaitiaki and other human communities, to help set research priorities and identify where 'big wins' might still be possible. Crucially, however, conservation physiologists need to engage with an overarching imperative: the urgency for society globally to end an addiction to unsustainable practices and transform towards lifestyles that ensure a liveable planet (Steffen et al. 2018). Globally, ample evidence exists from physiologists of the high-risk stakes for biodiversity from uncontrolled climate change (Stillman 2019). With humility, therefore, we must all endeavour, as humans and not just scientists, to develop and demonstrate the changed behaviours necessary to sustain a liveable planet.

Acknowledgements

Our thanks to landowners for access to field sites, to Richard Gibson of Auckland Zoo, and Lesley Judd of the Department of Conservation for discussion and information, to Nancy Longnecker for discussion, and to the book's editors for insights. Recent studies described here for Otago lizards were assisted by the Miss E. L. Hellaby Indigenous Grasslands Research Trust and by a J.S. Watson Grant from the Royal Forest and Bird Protection Society of New Zealand. We warmly acknowledge the New Zealand conservation community, including the Department of Conservation, community ecosanctuary groups and iwi including Kāti Huirapa Rūnaka ki Puketeraki, Ngāti Koata, Ngāti Maru, Ngāti Whanaunga and Te Ātiawa o Te Waka-a-Māui; we appreciate their ongoing contributions to the conservation of New Zealand reptiles and the opportunities that we have had to be involved.

References

Anderson, L., Nelson, N., and Cree, A., 2017. Glucocorticoids in tuatara (*Sphenodon punctatus*): some influential factors, and applications in conservation management. *General and Comparative Endocrinology*, 244, 54–9.

Andrews, R.M. and Schwarzkopf, L., 2012. Thermal performance of squamate embryos with respect to climate, adult life history, and phylogeny. *Biological Journal of the Linnean Society*, 106, 851–64.

Araújo, M.B., Ferri-Yanez, F., Bozinovic, F. et al., 2013. Heat freezes niche evolution. *Ecology Letters*, 16, 1206–19.

Barry, M., Cockrem, J.F., and Brunton, D.H., 2010. Seasonal variation in plasma corticosterone concentrations in wild and captive adult Duvaucel's geckos (*Hoplodactylus duvaucelii*) in New Zealand. *Australian Journal of Zoology*, 58, 234–42.

Bennett, J.M., Calosi, P., Clusella-Trullas, S. et al., 2018. GlobTherm, a global database on thermal tolerances for aquatic and terrestrial organisms. *Scientific Data*, 5, 180022.

Besson, A.A., Nelson, N.J., Nottingham, C.M., and Cree, A., 2012. Is cool egg incubation temperature a limiting factor for the translocation of tuatara to southern New Zealand? *New Zealand Journal of Ecology*, 36, 90–9.

Blanchard, B. and the Tuatara Recovery Group, 2002. Tuatara captive management plan and husbandry manual. In Threatened Species Occasional Publication 21, p. 75. Department of Conservation, Wellington, New Zealand.

Caldwell, A.J., Cree, A., and Hare, K.M., 2018. Parturient behaviour of a viviparous skink: evidence for maternal cannibalism when basking opportunity is low. *New Zealand Journal of Zoology*, 45, 359–70.

Camacho, A. and Rusch, T.W., 2017. Methods and pitfalls of measuring thermal preference and tolerance in lizards. *Journal of Thermal Biology*, 68, 63–72.

Carter, A., 2015. Modelling the incubation microclimate to predict offspring sex ratios and hatching phenology in tuatara (*Sphenodon punctatus*). Unpublished PhD thesis. Victoria University of Wellington, Wellington, New Zealand.

Chapple, D.G., ed., 2016. *New Zealand Lizards*. Springer, Cham, Switzerland.

Chapple, D.G., Keall, S.N., Daugherty, C.H., and Hare, K.M., 2017. Nest-site selection and the factors influencing hatching success and offspring phenotype in a nocturnal skink. *Amphibia-Reptilia*, 38, 363–9.

Chapple, D.G., Reardon, J.T., and Peace, J.E., 2016. Origin, spread and biology of the invasive plague skink (*Lampropholis delicata*) in New Zealand. In D.G. Chapple, ed. *New Zealand Lizards*, pp. 341–59. Springer, Cham, Switzerland.

Chukwuka, C.O. 2020. Microhabitat use by the nocturnal, cool-climate gecko *Woodworthia* "Otago/Southland" in the context of global climate change. Unpublished PhD thesis, University of Otago, Dunedin, New Zealand.

Chukwuka, C.O., Virens, J., and Cree, A., 2019. Accuracy of an inexpensive, compact infrared thermometer for measuring skin surface temperature of small lizards. *Journal of Thermal Biology*, 84, 285–91.

Connolly, J.D. and Cree, A., 2008. Risks of a late start to captive management for conservation: phenotypic differences between wild and captive individuals of a viviparous endangered skink (*Oligosoma otagense*). *Biological Conservation*, 141, 1283–92.

Cooke, S.J., Birnie-Gauvin, K., Lennox, R.J. et al., 2017. How experimental biology and ecology can support evidence-based decision-making in conservation: avoiding pitfalls and enabling application. *Conservation Physiology*, 5, cox043. doi:10.1093/conphys/cox043

Cooke, S.J., Sack, L., Franklin, C.E. et al., 2013. What is conservation physiology? Perspectives on an increasingly integrated and essential science. *Conservation Physiology*, 1, cot001. doi:10.1093/conphys/cot001.

Cree, A., 2014. *Tuatara: Biology and Conservation of a Venerable Survivor*. Canterbury University Press, Christchurch, New Zealand.

Cree, A. and Hare, K.M., 2010. Equal thermal opportunity does not result in equal gestation length in a cool-climate skink and gecko. *Herpetological Conservation and Biology*, 5, 271–82.

Cree, A. and Hare, K.M., 2016a. Reproduction and life history of New Zealand lizards. In D.G. Chapple, ed. *New Zealand Lizards*, pp. 169–206. Springer, Cham, Switzerland.

Cree, A. and Hare, K.M., 2016b. Maternal basking regime has complex implications for birthdate and offspring phenotype in a nocturnally foraging, viviparous gecko. *Journal of Experimental Biology*, 219, 2934–43.

Cree, A., Tyrrell, C.L., Preest, M.R. et al., 2003. Protecting embryos from stress: corticosterone effects and the corticosterone response to capture and confinement during pregnancy in a live-bearing lizard (*Hoplodactylus maculatus*). *General and Comparative Endocrinology*, 134, 316–29.

Diele-Viegas, L.M., Vitt, L.J., Simervo, B. et al., 2018. Thermal physiology of Amazonian lizards (Reptilia: Squamata). *PLoS ONE*, 13(3), e0192834.

Gaby, M.J., Besson, A.A., Bezzina, C.N. et al., 2011. Thermal dependence of locomotor performance in two cool-temperate lizards. *Journal of Comparative Physiology A*, 197, 869–75.

Gibson, S., Penniket, S., and Cree, A., 2015. Are viviparous lizards from cool climates ever exclusively nocturnal? Evidence for extensive basking in a New Zealand gecko. *Biological Journal of the Linnean Society, 115,* 882–95.

Grayson, K.L., Mitchell, N.J., Monks, J.M. et al., 2014. Sex ratio bias and extinction risk in an isolated population of tuatara (*Sphenodon punctatus*). *PLoS ONE, 9*(4), e94214. doi:10.1371/journal.pone.0094214

Hare, J.R., Holmes, K.M., Wilson, J.L., and Cree, A., 2009. Modelling exposure to selected temperature during pregnancy: the limitations of squamate viviparity in a cool-climate environment. *Biological Journal of the Linnean Society, 96,* 541–52.

Hare, J.R., Whitworth, E., and Cree, A., 2007. Correct orientation of a hand-held infrared thermometer is important for accurate measurement of body temperatures in small lizards and tuatara. *Herpetological Review, 38,* 311–15.

Hare, K.M., Borrelle, S.B., Buckley, H.L. et al., 2019. Intractable: species in New Zealand that continue to decline despite conservation efforts. *Journal of the Royal Society of New Zealand, 49,* 301–319.

Hare, K.M., Chapple, D.G., Towns, D.R., and van Winkel, D., 2016. The ecology of New Zealand's lizards. In D.G. Chapple, ed. *New Zealand Lizards*, pp. 133–68. Springer, Cham, Switzerland.

Hare, K.M. and Cree, A., 2010. Exploring the consequences of climate-induced changes in cloud cover on offspring of a cool-temperate viviparous lizard. *Biological Journal of the Linnean Society, 101,* 844–51.

Hare, K.M. and Cree, A., 2011. Maternal and environmental influences on reproductive success of a viviparous grassland lizard. *New Zealand Journal of Ecology, 35,* 254–60.

Hare, K.M. and Cree, A., 2016. Thermal and metabolic physiology of New Zealand lizards. In D.G. Chapple, ed. *New Zealand Lizards*, pp. 239–67. Springer, Cham, Switzerland.

Hare, K.M., Daugherty, C.H., and Cree, A., 2002. Incubation regime affects juvenile morphology and hatching success, but not sex, of the oviparous lizard *Oligosoma suteri* (Lacertilia: Scincidae). *New Zealand Journal of Zoology, 29,* 221–9.

Hare, K.M., Hare, J.R., and Cree, A., 2010. Parasites, but not palpation, are associated with pregnancy failure in a captive viviparous lizard. *Herpetological Conservation and Biology, 5,* 563–70.

Hare, K.M., Longson, C.G., Pledger, S., and Daugherty, C.H., 2004. Size, growth, and survival are reduced at cool incubation temperatures in the temperate lizard *Oligosoma suteri* (Lacertilia: Scincidae). *Copeia, 2004,* 383–90.

Hare, K.M., Pledger, S., and Daugherty, C.H., 2008. Low incubation temperatures negatively influence locomotor performance and behaviour of the nocturnal lizard *Oligosoma suteri* (Lacertidae: Scincidae). *Copeia, 2008,* 16–22.

Hare, K.M., Schumann, N., Hoskins, A.J. et al., 2020. Predictors of translocation success in captive-reared lizards: implications for their captive management. *Animal Conservation, 23,* 320–9.

Hitchmough, R., Barr, B., Lettink, M. et al., 2016. Conservation status of New Zealand reptiles, 2015. New Zealand Threat Classification Series 17, p. 14. Department of Conservation, Wellington, New Zealand.

Huang, S.P., Hsu, Y., and Tu, M.C., 2006. Thermal tolerance and altitudinal distribution of two *Sphenomorphus* lizards in Taiwan. *Journal of Thermal Biology, 31,* 378–85.

Jarvie, S., 2016. Reintroduction biology of tuatara (*Sphenodon punctatus*): identifying suitable founder animals and conservation translocation sites. Unpublished PhD Thesis, p. 294. University of Otago, Dunedin, New Zealand.

Jarvie, S., Besson, A.A., Seddon, P.J., and Cree, A., 2014. Assessing the thermal suitability of translocation release sites for egg-laying reptiles with temperature dependent sex-determination: a case study with tuatara (*Sphenodon punctatus*). *Animal Conservation, 17*(Suppl. 1), 48–55.

Jarvie, S., Jowett, T., Thompson, M.B. et al., 2018. Effects of warm temperatures on metabolic rate and evaporative water loss in tuatara, a cool-climate rhynchocephalian survivor. *Physiological and Biochemical Zoology, 91,* 950–66.

Jarvie, S., Senior, A.M., Adolph, S.C. et al., 2015. Captive rearing affects growth but not survival in translocated juvenile tuatara. *Journal of Zoology, 297,* 184–93.

Keall, S.N., Nelson, N.J., and Daugherty, C.H., 2010. Securing the future of threatened tuatara populations with artificial incubation. *Herpetological Conservation and Biology, 5,* 555–62.

Kearney, M.R., Munns, S.L., Moore, D., et al., 2018. Field tests of a general ectotherm niche model show how water can limit lizard activity and distribution. *Ecological Monographs, 88,* 672–93.

Kingsolver, J.G., Diamond, S.E., and Buckley, L.B., 2013. Heat stress and the fitness consequences of climate change for terrestrial ectotherms. *Functional Ecology, 27,* 1415–23.

Kingsolver, J.G. and Woods, H.A., 2016. Beyond thermal performance curves: modeling time-dependent effects of thermal stress on ectotherm growth rates. *The American Naturalist, 187,* 283–94.

Levy, O., Buckley, L.B., Keitt, T.H. et al., 2015. Resolving the life cycle alters expected impacts of climate change. *Proceedings of the Royal Society B, 282,* 20150837.

Madliger, C.L., 2017. Validation, values and vision: ways early-career researchers can help propel the field of conservation physiology. *Conservation Physiology, 5,* cox045.

Madliger, C.L., Franklin, C.E., Hultine, K.R. et al., 2017. Conservation physiology and the quest for a 'good' Anthropocene. *Conservation Physiology, 5,* cox003. doi:101093/conphys/cox003

Madliger, C.L., Love, O.P., Hultine, K.R., and Cooke, S.J., 2018. The conservation physiology toolbox: status and opportunities. *Conservation Physiology, 6*(1), coy029.

Miller, K.A., Hare, K.M., and Nelson, N.J., 2010. Do alternate escape tactics provide a means of compensation for impaired performance ability? *Biological Journal of the Linnean Society, 99,* 241–9.

Ministry for the Environment, 2016. Climate change projections for New Zealand: atmospheric projections based on simulations undertaken for the IPCC 5th Assessment, p. 127. Ministry for the Environment, Wellington, New Zealand.

Mitchell, N.J., Nelson, N.J., Cree, A. et al., 2006. Support for a rare pattern of temperature-dependent sex determination in archaic reptiles: evidence from two species of tuatara (*Sphenodon*). *Frontiers in Zoology, 3,* 9.

Nelson, N.J., Briskie, J.V., Constantine, R. et al., 2019. The winners: species that have benefited from 30 years of conservation action. *Journal of the Royal Society of New Zealand. 49:* 281–300.

Nelson, N.J., Cree, A., Thompson, M.B. et al., 2004a. Temperature-dependent sex determination in tuatara. In N. Valenzuela and V. Lance, eds. *Temperature-dependent Sex Determination in Vertebrates,* pp. 53–8. Smithsonian, Washington, DC.

Nelson, N.J., Hitchmough, R., and Monks, J.M., 2015. New Zealand reptiles and their conservation. In A. Stow, N. Maclean, and G.I. Holwell GI, eds. *Austral Ark: the State of Wildlife in Australia and New Zealand,* pp. 382–404. Cambridge University Press, Cambridge.

Nelson, N.J., Keall, S.N., and Hare, K.M., 2017. Temperature selection by juvenile tuatara (*Sphenodon punctatus*) is not influenced by temperatures experienced as embryos. *Journal of Thermal Biology, 69,* 261–6.

Nelson, N.J., Keall, S.N., Refsnider, J.M., and Carter, A.L., 2018. Behavioral variation in nesting phenology may offset sex-ratio bias in tuatara. *Journal of Experimental Zoology A, 2018,* 1–9.

Nelson, N.J., Moore, J.A., Pillai, S., and Keall, S.N., 2010. Thermosensitive period for sex determination in the tuatara. *Herpetological Conservation and Biology, 5,* 324–9.

Nelson, N.J., Thompson, M.B., Pledger, S. et al., 2004b. Egg mass determines hatchling size, and incubation temperature influences post-hatching growth, of tuatara *Sphenodon punctatus. Journal of Zoology, London, 263,* 77–87.

Nelson, N.J., Thompson, M.B., Pledger, S. et al., 2004c. Induction of oviposition produces smaller eggs in tuatara (*Sphenodon punctatus*). *New Zealand Journal of Zoology, 31,* 283–9.

Nelson, N.J., Thompson, M.B., Pledger, S. et al., 2004d. Do TSD, sex ratios, and nest characteristics influence the vulnerability of tuatara to global warming? *International Congress Series, 1275,* 250–7.

Nelson, N.J., Thompson, M.B., Pledger, S. et al., 2006. Performance of juvenile tuatara depends on age, clutch, and incubation regime. *Journal of Herpetology, 40,* 399–403.

Paaijmans, K.P., Heinig, R.L., Seliga, R.A. et al., 2013. Temperature variation makes ectotherms more sensitive to climate change. *Global Change Biology, 19,* 2373–80.

Pezaro, N., Doody, S.J., and Thompson, M.B., 2017. The ecology and evolution of temperature-dependent reaction norms for sex determination in reptiles: a mechanistic conceptual model. *Biological Reviews, 92,* 1348–64.

Pike, D.A., Croak, B.M., Webb, J.K., and Shine, R., 2010. Subtle—but easily reversible—anthropogenic disturbance seriously degrades habitat quality for rock-dwelling reptiles. *Animal Conservation, 13,* 411–18.

Rock, J., Andrews, R.M., and Cree, A., 2000. Effects of reproductive condition, season and site on selected temperatures of a viviparous gecko. *Physiological and Biochemical Zoology, 73,* 344–55.

Rock, J. and Cree, A., 2003. Intraspecific variation in the effect of temperature on pregnancy in the viviparous gecko *Hoplodactylus maculatus. Herpetologica, 59,* 8–22.

Rock, J., Cree, A., and Andrews, R.M., 2002. The effect of reproductive condition on thermoregulation in a viviparous gecko from a cool climate. *Journal of Thermal Biology, 27,* 17–27.

Salinger, M.J., Renwick, J., Behrens, E. et al., 2019. The unprecedented coupled ocean-atmosphere summer heatwave in the New Zealand region 2017/18: drivers, mechanisms and impacts. *Environmental Research Letters, 14,* 044023.

Seebacher, F. and Franklin, C.E., 2012. Determining environmental causes of biological effects: the need for a mechanistic physiological dimension in conservation biology. *Philosophical Transactions of the Royal Society B—Biological Sciences, 367,* 1607–14.

Sinclair, B.J., Marshall, K.E., Sewell, M.A. et al., 2016. Can we predict ectotherm responses to climate change using thermal performance curves and body temperatures? *Ecology Letters, 19,* 1372–1385. doi:10.1111/ele.12686

Spellerberg, I., 1975. Conservation and management of Britain's reptiles based on their ecological and behavioural requirements: a progress report. *Biological Conservation, 7,* 289–300.

Steffen, W., Rockstrom, J., Richardson, K. et al., 2018. Trajectories of the earth system in the Anthropocene. *Proceedings of the National Academy of Sciences,115,* 8252–9.

Stenhouse, V., Carter, A.L., Chapple, D.G. et al., 2018. Modelled incubation conditions indicate wider potential distributions based on thermal requirements for an oviparous lizard. *Journal of Biogeography, 2018*, 1–12.

Stewart, J.R. and Blackburn, D.G., 2015. Viviparity and placentation in lizards. In J. Rheubert, D. Siegel, and S. Trauth, eds. *Reproductive Biology and Phylogeny of Lizards and Tuatara*, pp. 448–563. CRC Press, Boca Raton, FL.

Stillman, J.H., 2019. Heat waves, the new normal: summertime temperature extremes will impact animals, ecosystems, and human communities. *Physiology, 34*, 86–100.

Thompson, M.B., 1990. Incubation of eggs of tuatara, *Sphenodon punctatus*. *Journal of Zoology, London, 222*, 303–18.

Thompson, M.B., Packard, G.C., Packard, M.J., and Rose, B., 1996. Analysis of the nest environment of the tuatara *Sphenodon punctatus*. *Journal of Zoology, London, 238*, 239–51.

Towns, D.R., Hitchmough, R.A., and Perrott, J., 2016. Conservation of New Zealand lizards: a fauna not forgotten but undervalued? In D.G. Chapple, ed. *New Zealand Lizards*, pp. 293–320. Springer, Cham, Switzerland.

Uetz, P., Freed, P., and Hošek, J., eds, 2019. The Reptile Database. Available at: www.reptile-database.org (accessed 10 October 2019).

van Winkel, D., Baling, M., and Hitchmough, R., 2018. *Reptiles and Amphibians of New Zealand: a Field Guide*. Auckland University Press, Auckland, New Zealand.

Virens, J. and Cree, A., 2018. Further miniaturisation of the Thermochron iButton to create a thermal bio-logger weighing 0.3 g. *Journal of Experimental Biology, 221*, jeb176354. doi:10.1242/jeb.176354

Virens, J. and Cree, A., 2019. Pregnancy reduces critical thermal maximum, but not voluntary thermal maximum, in a viviparous skink. *Journal of Comparative Physiology B, 189*, 611–21.

Wapstra, E., Uller, T., While, G.M. et al., 2010. Giving offspring a head start in life: field and experimental evidence for selection on maternal basking behaviour in lizards. *Journal of Evolutionary Biology, 23*, 651–7.

Warner, D.A. and Shine, R., 2007. Fitness of juvenile lizards depends on seasonal timing of hatching, not offspring body size. *Oecologia, 154*, 65–73.

CHAPTER 17

Using applied physiology to better manage and conserve the white rhinoceros (*Ceratotherium simum*)

Anna J. Haw, Andrea Fuller, and Leith C.R. Meyer

> ➲ **Take-home message**
>
> Safe chemical capture and transport of wild white rhinoceros are essential for the conservation of this threatened species. The white rhinoceros is particularly sensitive to the drugs used for chemical capture, resulting in high morbidity or mortality during rhino management procedures. Understanding the causes and consequences of capture-related pathophysiology, through the measurement of physiological variables in experimental trials, can facilitate safer methods of capture and transport and, by extension, facilitate the conservation of this critically endangered species.

17.1 Introduction

In protecting the rhino, we will be protecting their habitat, and in protecting their habitat we preserve the wild nature on which our humanity depends.

Ian Player

The white rhinoceros (*Ceratotherium simum*) is one of five remaining rhinoceros (or rhino) species that once roamed throughout Europe, Asia, and Africa. The other living rhinoceros species are the greater one-horned (*Rhinoceros unicornis*), Javan (*Rhinoceros sondaicus*), and Sumatran (*Dicerorhinus sumatrensis*) rhinoceros in southern Asia, and the black rhinoceros (*Diceros bicornis*), which like the white rhinoceros, occurs in sub-Saharan Africa. All of these species are threatened with extinction (International Rhino Foundation (IRF) 2020). At the start of the 20th century, there were an estimated 500 000 wild rhinoceros. By 1970, this worldwide population had fallen to 70 000, and today only about 29 000 remain in the wild (International Rhino Foundation (IRF) 2020). Recently, one of the two subspecies of white rhinoceros, the northern white rhinoceros (*Ceratotherium simum cottoni*), was declared extinct in the wild (World Wildlife Fund 2020). For simplicity, throughout this chapter we will refer to the other white rhinoceros subspecies, the southern white rhinoceros (*Ceratotherium simum simum*), as 'white rhinoceros'.

As early as the end of the 19th century, the white rhinoceros, which once roamed much of southern Africa, had been reduced to a single population of about 200 animals, largely because of hunting and habitat destruction (Rookmaaker 2000). The Hluhluwe-Imfolozi Park in South Africa, proclaimed in 1897, became the sanctuary for this last

Anna J. Haw, Andrea Fuller, and Leith C.R. Meyer, *Using applied physiology to better manage and conserve the white rhinoceros (*Ceratotherium simum*)*
In: *Conservation Physiology: Applications for Wildlife Conservation and Management.* Edited by: Christine L. Madliger, Craig E. Franklin, Oliver P. Love, and Steven J. Cooke, Oxford University Press (2021). © Oxford University Press. DOI: 10.1093/oso/9780198843610.003.0017

remaining population, and later became world renowned for its success in rhino conservation by breeding the species back from the brink of extinction. By the early 1960s, there were approximately 650 rhinoceros in, or adjacent to, this reserve (Hübschle 2015; Buss 2017). At this time, the growing rhino population in Hluhluwe-Imfolozi was facing increasing pressure from human settlers pushing into rhino habitat, and dwindling food supplies within the reserve (Player 1967). Merely ensuring successful reproduction was not sufficient for the long-term survival of the species. To safeguard this growing population, hundreds of rhinoceros needed to be captured and translocated to other reserves. Thus was born 'Operation Rhino' (Hluhluwe Game Reserve 2020), the first major wild mammal translocation effort, and the start of a journey towards better understanding of wildlife pharmacology, and the physiological effects of chemical capture (immobilization) and transport in the white rhinoceros.

At the inception of Operation Rhino, chemical immobilization of large herbivores was largely a difficult and risky procedure. The first drugs used for immobilization often had narrow safety margins between doses that produced immobilization and those that produced death (Player 1967). Also, the large volumes of drug needed and prolonged induction times (time from darting to immobilization) caused further complications. It was not until the development of a semi-synthetic opioid called etorphine, in 1963, that wildlife immobilization became a viable tool for conservation (Harthoorn and Bligh 1965). Indeed, it was primarily through the introduction of etorphine that hundreds of white rhinoceros were translocated to parks throughout South Africa during Operation Rhino, thereby saving the species from extinction. Today, etorphine remains the most commonly used opioid in the chemical immobilization of rhinoceros; although, it is often used in combination with a tranquilliser or sedative (Portas 2004; Burroughs et al. 2012a).

Etorphine is a highly potent full-agonist opioid, which is capable of inducing rapid immobilization, thus minimizing the risk of injury and pathologies related to prolonged physical exertion (Harthoorn and Bligh 1965; Blane et al. 1967; Burroughs et al.

2012b). Moreover, the effects of etorphine can be reversed with the administration of an opioid antagonist, making it an ideal agent for wildlife immobilization. However, together with etorphine's desirable immobilizing effects, this opioid drug also causes significant physiological dysfunction. Compared with other large herbivores, the white rhinoceros, for reasons not fully understood, is particularly sensitive to etorphine's negative effects (Fahlman 2008; Burroughs et al. 2012a). In white rhinoceros, these effects can include profound respiratory impairment, which leads to hypoxaemia, hypercapnia, and acidosis. Indeed, morbidity and mortality directly related to the capture event occur in etorphine-immobilized white rhinoceros, with worsening hypoxaemia noted as a risk factor for mortality (Kock et al. 1995). Apart from the respiratory effects, etorphine-induced immobilization also results in deleterious cardiovascular effects, namely hypertension and tachycardia, as well as significant muscle rigidity and tremors (Buss et al. 2016; De Lange et al. 2017).

Despite the risk of morbidity and mortality directly related to the capture event, Operation Rhino was successful in repopulating former white rhinoceros ranges, and today, the population stands at around 18 000 (International Rhino Foundation (IRF) 2020). Sadly, however, the success of Operation Rhino is now being jeopardized by the unprecedented rise in poaching, fuelled by a growing demand for rhino horn (Hübschle 2015; International Rhino Foundation (IRF) 2020). This demand stems largely from the growing middle classes of China and Vietnam, which are increasingly able to afford rhino horn (International Rhino Foundation (IRF) 2020). In these regions, rhino horn is illegally traded and used in traditional Chinese medicine, or kept as a status symbol. Although rhino horn, composed of keratin, has no known pharmacological effect, its current value per unit weight on the black market is believed to be greater than that of gold, diamonds, or cocaine (Biggs et al. 2013). The high value of rhino horn has attracted international criminal syndicates that link rhino horn source countries, like South Africa, to the demand in Asia, via a series of transit points and smuggling channels (Hübschle 2015; Christy 2016). These syndicates are well resourced and powerful, employing highly sophisticated techniques to outcompete anti-poaching efforts. In 2007, South Africa reported losing just 13 rhinoceros to poaching. By 2015, this number had risen to 1175 (Christy 2016). Between 2012 and 2017 poaching led to a 15 per cent decrease in white rhinoceros numbers (International Rhino Foundation (IRF) 2020). Although current anti-poaching and anti-horn-use efforts appear to be reducing the number of animals poached, additional rhinoceros conservation interventions, on a scale and intensity greater than that of Operation Rhino, are needed.

Many rhinoceros conservation efforts require hands-on interventions usually involving veterinary procedures. Success of these interventions requires a scientific understanding of the effects of the procedures and novel research developments in order to reduce their associated risks. Immobilization of white rhinoceros with etorphine remains central to the implementation of more advanced management and conservation strategies, such as animal identification, placing tracking or identification devices, dehorning, translocations, and procedures for the treatment of wounded animals. The goal now is to minimize, if not prevent, the negative effects of immobilization and transport such that any conservation-related intervention does not place additional risk of morbidity and mortality on individual animals.

By conducting controlled experimental trials, and using gold-standard physiological monitoring techniques, we are able to gain a greater understanding of the physiological consequences of capture and translocation. With this improved understanding, we can work towards keeping immobilized animals as physiologically stable as possible. It is no longer adequate to gauge the success of immobilization and transport purely on survival. Physiological dysfunction that occurs during these stressful interventions may place an animal at increased risk of death or disease in the future. Moreover, the need to immobilize physiologically compromised animals because of severe wounds resulting from poaching is increasing. Minimizing the deleterious effects of immobilization and transport is therefore of critical importance.

17.2 Physiological measurements in immobilized white rhinoceros

Implementing controlled experimental trials in an endangered population of wild animals can be extremely difficult, if not impossible. Therefore, many of the early treatments advocated to improve the safety of immobilized white rhinoceros were based on trial and error, and anecdotal evidence from opportunistic studies or clinical work. Field veterinarians have mostly relied on easy to measure and non-specific physiological parameters, such as respiratory rate, to assess the physiological status of immobilized white rhinoceros. While pulse oximetry (a non-invasive method of measuring arterial haemoglobin oxygen saturation via spectrophotometry) is now more widely used to help gauge the level of oxygenation in immobilized animals, most pulse oximeters are not calibrated for use in rhino and often perform poorly due to animal movement, poor peripheral perfusion, or bright ambient light (Fahlman 2008).

The gold standard for the assessment of blood oxygenation, ventilation, and acid–base balance is arterial blood gas analysis (Proulx 1999). Monitoring the partial pressure of arterial oxygen and carbon dioxide, as well as the acid–base status of the animal throughout immobilization, is a valuable means to evaluate physiological effects that different chemical immobilization protocols have on wild animals (Fahlman 2008). Indeed, arterial blood gas analysis is strongly advocated as an investigative approach to reliably examine the respiratory-depressant effects of opioids (Whiteside et al. 2016). In the field environment, blood gases are rarely measured to assess the physiological status of immobilized animals because of the additional equipment, expense, and time needed to perform these analyses. However, portable blood gas analysers can be quick and easy to use, and can give wildlife veterinarians highly valuable insight into the physiological status of the immobilized animal. Research studies on the responses of rhinoceros now routinely include analyses of arterial blood, drawn from the animal's auricular artery (e.g. Haw et al. 2014; Buss et al. 2018).

While arterial blood gases offer a valuable window into the respiratory status of immobilized animals, the mechanisms behind certain changes in blood gases are difficult to elucidate without further monitoring capabilities. Rhinoceros held in bomas (i.e. small enclosures to temporarily house wild animals) provide an ideal opportunity to employ more sophisticated monitoring techniques, such as the standard equipment used to assess human physiology during exercise (Figure 17.1). Such a system, adapted for use in immobilized white rhinoceros, can be used to measure a multitude of variables, including expired minute ventilation (VE_{BTPS}) using spirometry. The apparatus also allows for capturing expired gas, which can then be analysed for carbon dioxide and oxygen concentrations. With these additional variables, physiological equations can be used to provide further information such as oxygen consumption and carbon dioxide production, to improve our overall understanding of the physiological effects of immobilization.

Physiological stress can also be better assessed by analysing certain biochemical markers, which are especially useful if serial blood samples are taken, for example during transport. Recent studies have used the concentrations of adrenaline, noradrenaline, glucocorticoids, and glucose in the blood as a means to indicate the degree of the stress response during capture and transport (De Lange et al. 2017; Pohlin et al. 2020). Acute phase reactants (APRs) in blood serum can also be measured as indicators of acute tissue injury or inflammation. In white rhinoceroses, haptoglobin and serum amyloid A are positive APRs, which increase, while albumin and iron are negative APRs, which decrease, during an acute phase response (Hooijberg et al. 2020; Pohlin et al. 2020). By measuring these APRs we can begin to work out how stress affects the immune response in white rhinoceros. An acute phase response is also often accompanied by alterations in plasma oxidants and antioxidants, leading to an excessive production of free radicals with resultant oxidative stress (Cray et al. 2009; Pohlin et al. 2020). This oxidative stress may be implicated in disease processes and, thus, the measurement of oxidative parameters has been proposed as a means to identify animals at risk of developing disease (Lykkesfeldt and Svendsen 2007).

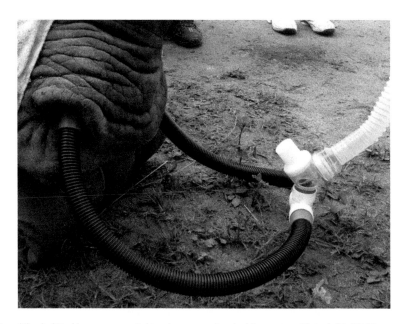

Figure 17.1 An immobilized white rhinoceros connected to a human exercise physiology system. Photo: Leith C.R. Meyer.

Below, we describe four case studies that illustrate how these tools can be used to better understand the physiological responses of white rhinoceros to management interventions, and how that knowledge can be applied to improve the conservation of the species.

17.3 Case study 1: improving oxygenation during chemical capture of white rhinoceros

Today, white rhinoceros are routinely immobilized with the potent opioid, etorphine, combined with the tranquillizer, azaperone (Burroughs et al. 2012a). Despite the recent progress in white rhinoceros capture techniques, this drug combination still leads to significant cardiorespiratory dysfunction in immobilized animals, and may be associated with mortalities (Kock et al. 1995; van Zijll Langhout et al. 2016). In an attempt to offset some of the negative effects, wildlife veterinarians have supplied various supportive treatments to immobilized white rhinoceros. Mixed opioid agonist-antagonists, such as butorphanol, as well as oxygen insufflation, are two such supportive treatments that were initially used largely based on anecdotal findings. Butorphanol is a synthetic mixed opioid agonist-antagonist, producing complex effects as a result of its varying affinity and efficacy at different opioid receptors. Veterinarians administer butorphanol to immobilized white rhinoceros in an attempt to reverse the respiratory-depressant effects of etorphine, while maintaining adequate immobilization. Oxygen administered intranasally or via nasotracheal intubation has been used to try increase blood oxygen levels in etorphine-immobilized white rhinoceros (Bush et al. 2004; Fahlman 2008).

In a large collaborative study in South Africa's Kruger National Park, researchers critically evaluated butorphanol and oxygen as supportive treatments in a controlled boma setting. Although the majority of managed white rhinoceros are darted from a helicopter in large reserves, this study was conducted in a boma setting to reduce confounding variables normally induced by stressors associated with a helicopter capture. In bomas, calm rhinoceros can be darted and weighed at each immobilization.

Food intake, defaecation, and demeanour can also be monitored throughout the study period. In this trial, the same eight white rhinoceros were immobilized on four occasions, each time receiving a different treatment post-immobilization. The treatments were butorphanol alone, butorphanol combined with oxygen insufflation, oxygen alone, and a control (Haw et al. 2014).

By measuring arterial blood gases, the researchers revealed that the respiratory-depressant effects of etorphine can be severe and life-threatening, with rhinoceros sometimes having arterial partial pressure of oxygen (P_aO_2) values below 25 mmHg, about a quarter of the normal value (98.2 ± 1.4 mmHg) (Citino and Bush 2007) in an awake animal (Figure 17.2). Although butorphanol administered post-induction attenuated this hypoxaemia, immobilized white rhinoceros remained critically hypoxaemic with P_aO_2 values still less than 55 mmHg. This finding from a controlled experiment is vitally important for field veterinarians. Butorphanol administration may give a veterinarian a false sense of security as it appears to increase the respiratory rate of immobilized rhinoceros. However, this improvement in respiratory rate is transient (<3 min) and does not translate into a profound improvement in oxygenation, such that immobilized rhinoceros may remain significantly compromised.

Another surprising finding from this study was that oxygen insufflation on its own did not improve blood oxygenation in boma-immobilized rhinoceros. The likely explanation is that these rhinoceros experienced a large degree of intrapulmonary shunting because of oxygen-induced absorption atelectasis (partial or complete collapse of the lung), as has been described in horses under general anaesthesia receiving a high fraction of inspired oxygen (Marntell et al. 2005; Staffieri et al. 2009). Oxygen in the alveoli is absorbed faster than nitrogen (Magnusson and Spahn 2003). With a high concentration of oxygen in the alveoli, the nitrogen, which normally acts as a 'stent' in the alveoli, is diluted, and the quick absorption of oxygen will result in the collapse of the alveoli. Furthermore, narrowed or compressed airways, because of pressure from abdominal organs and poor ventilation, will limit the ability for absorbed oxygen to be adequately

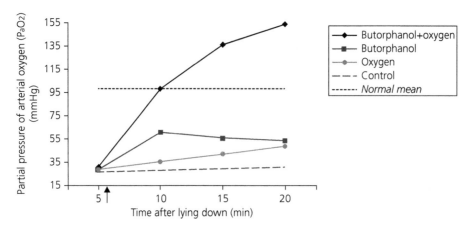

Figure 17.2 Arterial oxygenation in immobilized white rhinoceros. Changes in the partial pressure of arterial oxygen (P_aO_2) in immobilized white rhinoceros ($n = 8$) after receiving butorphanol, oxygen insufflation, butorphanol + oxygen insufflation, or sterile water (control). The black arrow at 6 min indicates the time at which the intervention was given. The dashed horizontal line indicates the average normal value in standing, unsedated white rhinoceros (Citino and Bush 2007). Figure redrawn with permission (Creative Commons Attribution 4.0 International License) from Haw et al. *BMC Veterinary Research* (2014), *10*, 253.

replaced. If atelectasis is severe enough to cause shunt fractions in excess of 50 per cent, oxygen supplementation will not be effective (Fahlman 2008).

However, when oxygen insufflation was combined with an intravenous injection of butorphanol, P_aO_2 values in etorphine-immobilized rhinoceros were entirely corrected, with P_aO_2 values reaching 154 ± 53 mmHg by the end of the recumbent period (Figure 17.2). This marked effect is difficult to explain, given the effects of oxygen and butorphanol alone, but is most likely a result of a transient improvement in tidal volume and chest expansion, and decreased metabolism induced by butorphanol. With a reduction in metabolism and subsequent higher oxygenation of blood entering the lungs, less oxygen will be absorbed from the alveoli, thus leaving more gas in the alveoli, and less risk of atelectasis. This study therefore led to the important finding that butorphanol combined with oxygen insufflation can correct immobilization-induced hypoxaemia in boma-immobilized white rhinoceros, thus limiting the risk of hypoxic damage to organs and muscles (Haw et al. 2014).

As most managed white rhinoceros are captured in the field (darted from a helicopter) it was important to test this supportive treatment of butorphanol + oxygen in field-captured rhinoceros. Field capture invariably leads to a number of physiological dysfunctions, even before capture drugs are administered. When the same protocol of butorphanol and oxygen was administered to field-captured white rhinoceros, oxygenation, although significantly improved to near-normal levels (81.2 mmol/l ± 23.7 mmHg), did not reach the same high values as in boma-immobilized white rhinoceros (Haw et al. 2015). In addition, these field-immobilized rhinoceros had a severe acidaemia of metabolic, rather than respiratory origin. The physical exertion during a helicopter chase likely resulted in anaerobic metabolism and the build-up of lactic acid, causing a metabolic acidosis.

This physical exertion, together with a fear response from fleeing the helicopter, may account for the lower oxygenation levels observed in field- compared with boma-immobilized rhino. Field-immobilized rhino likely had a higher sympathetic response compared with their habituated boma counterparts. Although not measured in that study, it is likely that an increased cardiac output resulted in increased blood transit time in pulmonary capillaries and pulmonary hypertension-induced oedema, reducing the ability of oxygen to diffuse from the alveoli into the arterial blood in field-immobilized rhino. Understanding the severe negative consequences of physical exertion during field immobilizations can help veterinarians modify capture techniques to reduce stress as far as possible. One method employed to reduce the severity of the metabolic derangements during induction is to administer butorphanol in the dart mixture, rather than after immobilization has occurred (Miller et al. 2013). However, this method may not be suitable for all field settings, as the rhinoceros usually remain standing, requiring a skilled ground team for safe handling of the immobilized rhinoceros (Miller et al. 2013).

17.4 Case study 2: tremors in immobilized white rhinoceros

Another, less studied side-effect of opioid-induced immobilization in white rhinoceros is muscle tremor. These tremors usually present as a conspicuous shaking and trembling, sometimes making it difficult to work with an immobilized animal and potentially exacerbating the pathophysiological effects of capture and immobilization. Tremors are difficult to objectively measure, hence the paucity of research on tremors in immobilized animals. However, with the use of accelerometer (activity) loggers secured to the legs of immobilized rhinoceros, it was possible to correlate objective tremor measurements with an observational scoring system (De Lange et al. 2017), opening the door to better investigate the causes and consequences of tremors.

As described above, butorphanol attenuates the severe etorphine-induced hypoxaemia in white rhinoceros. In addition, butorphanol administered with etorphine in the dart, or administered after immobilization, reduces the severity of tremors (Wenger et al. 2007; Burroughs et al. 2012b; Miller et al. 2013; De Lange et al. 2017). The exact mechanism underlying butorphanol's effect on tremors is unclear, but tremor intensity was found to be negatively correlated with arterial oxygenation and pH, and positively correlated with plasma catecholamine concentrations (De Lange et al. 2017).

A later study, using a human exercise physiology system, together with additional gas measuring devices, was conducted to better understand how butorphanol elicits its positive effects in immobilized white rhinoceros (Buss et al. 2018). This study was the first trial in rhinoceros in which accurate measurements of expired minute ventilation and expired oxygen fractions were obtained, enabling the researchers to calculate oxygen consumption. Oxygen consumption (VO_2) is a measure of metabolic activity and maintenance of homeostatic processes in mammals. The VO_2 in resting white rhinoceros is unknown, but metabolic rate per unit body mass tends to decrease with increasing mass (Porter and Brand 1995). Thus, the VO_2 of white rhinoceros should be less than that of horses. In this study, conducted in boma-housed animals, immobilized white rhinoceros were found to have elevated (8.2 ml/kg.min) oxygen consumption compared with that of horses at rest (3 ml/kg.min). This elevated oxygen consumption likely contributed significantly to hypoxaemia (Buss et al. 2018). In these immobilized white rhinoceros, muscle tremor scores and oxygen consumption were strongly correlated, suggesting that activity from skeletal muscle tremors is a significant contributing factor to the high oxygen consumption in immobilized rhinoceros (Buss et al. 2018). Surprisingly, this study demonstrated that butorphanol administered to immobilized white rhinoceros only briefly increased ventilation (<3 min), but the improved arterial oxygenation was sustained for a longer time. Thus, the improvement in arterial oxygenation following butorphanol administration appeared to be mainly attributed to a decrease in oxygen consumption, rather than an improvement in ventilation. This new insight reveals the severe repercussions of etorphine-induced tremors in rhinoceros, and the importance of taking action to minimize these tremors.

17.5 Case study 3: cardiovascular effects of immobilization in white rhinoceros

Arterial blood gas measurements and exercise physiology instruments have helped improve our understanding of respiratory physiology in immobilized white rhinoceros. In addition to the respiratory dysfunctions, however, etorphine also causes significant cardiovascular effects, namely increased arterial blood pressure (hypertension) and elevated heart rate (tachycardia), which together may compound the negative respiratory effects (Buss et al. 2016). The underlying mechanisms for increased arterial hypertension and tachycardia in etorphine-immobilized rhinoceros are not fully understood, but the etorphine-induced sympathetic response and hypoxia likely play a role (Buss et al. 2016).

To reduce the etorphine-induced hypertension, the tranquillizer azaperone is routinely added to the dart mixture. Azaperone counteracts etorphine-induced hypertension by antagonizing α_1-receptors in peripheral arterioles thus limiting vasoconstriction (Raath 1999; Boardman et al. 2014). To better understand the effects of azaperone when added to the dart mixture, researchers working in the Kruger National Park compared the effects of etorphine alone and etorphine combined with azaperone for the immobilization of white rhinoceros. This study was conducted in boma-housed rhinoceros to limit the confounding effects of stress and physical exertion on the cardiovascular system, which is likely pronounced in field captures. Arterial blood

pressures were recorded using a transducer secured at the level of the heart and connected to a physiological monitor (Buss et al. 2016).

This controlled study demonstrated that the inclusion of azaperone with etorphine in the immobilizing drug combination indeed reduced hypertensive blood pressures to values lower than those reported for unrestrained zoo animals. Heart rates were also lower with this combination of opioid and tranquillizer compared with the opioid alone; however, the animals remained tachycardic with heart rates more than double those of normal values in unrestrained awake animals (Buss et al. 2016). Although the rhinoceros immobilized with etorphine only were hypertensive at the beginning of the immobilization period, this hypertension resolved after 25 min of immobilization, but the tachycardia persisted (Buss et al. 2016) (Figure 17.3). Thus, whether etorphine is used alone, or in combination with azaperone, rhinoceros remain tachycardic throughout the immobilization period.

A possible explanation for the persistent tachycardia may be a reduction in the stroke volume of

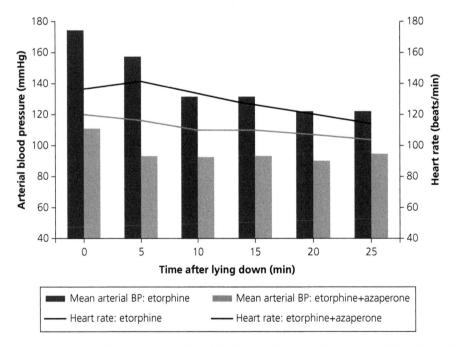

Figure 17.3 Mean blood pressure and heart rate in immobilized white rhinoceros. Mean arterial blood pressures (BP) and heart rate in white rhinoceros ($n = 6$) immobilized with etorphine, or etorphine + azaperone. Figure drawn from data published in Buss et al. *Journal of Zoo and Wildlife Medicine* (2016), *47*, 834–43.

the heart, or a reduction in total peripheral resistance (i.e. per heartbeat, less blood is pumped, or the blood flows through dilated vessels at a slower rate) (Buss et al. 2016). It is plausible that stroke volume is reduced in immobilized rhinoceros as a consequence of limited muscle activity in the limbs with subsequent blood pooling and decreased venous return to the heart. Also, etorphine-induced respiratory depression and chest wall rigidity could potentially decrease cardiac venous return by limiting the negative intra-thoracic pressure associated with inspiration (Buss et al. 2016). In addition, etorphine-induced hypoxaemia may lead to local vasodilation in response to tissue hypoxia. A persistent tachycardia with potential increased myocardial oxygen consumption in an immobilized rhinoceros is concerning due to the limited anaerobic capacity of the myocardium and the pronounced hypoxia commonly associated with etorphine-induced immobilization (Buss et al. 2016). Although no obvious adverse effects associated with the tachycardia were observed in the study rhinoceros, negative consequences could occur in animals compromised by age, blood loss from wounds, disease, or poor nutrition (Buss et al. 2016). With an increased understanding of the origins and consequences of

this tachycardia, we can begin to safeguard against potential negative effects in compromised animals. For example, efforts can be made to limit the time the animal is recumbent, and oxygenation can be improved by administering butorphanol with oxygen insufflation. Indeed, this same study demonstrated that butorphanol led to a rapid reduction in heart rate after it was administered to white rhinoceros immobilized with both etorphine, and etorphine + azaperone (Buss et al. 2016).

Apart from generalized hypertension, etorphine also has been implicated in causing pulmonary hypertension, as demonstrated in goats (Meyer et al. 2015) (Figure 17.4). In immobilized white rhinoceros, the alveolar–arterial oxygen gradient (A–a gradient) was found to be elevated, suggesting the presence of pulmonary hypertension (Buss et al. 2018). Pulmonary hypertension may reduce gas exchange across alveolar–capillary membranes because of pulmonary congestion and oedema, or a decrease in blood flow passage time through pulmonary vasculature (Meyer et al. 2015), thus leading to elevated A–a gradients. Indeed, in a preliminary study, where pulmonary pressures were measured in immobilized rhino for the first time, the authors found that supplementation of small doses of

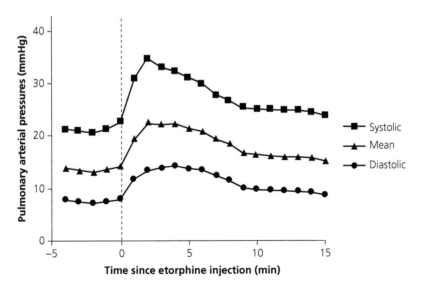

Figure 17.4 Effects of etorphine on pulmonary arterial pressures. Effects of etorphine injected intramuscularly (time = 0) on pulmonary arterial pressures in goats (*n* = 12). Figure redrawn with permission from Meyer et al. *BMC Veterinary Research* (2015) *11*, 18 under Creative Commons Attribution 4.0 International License.

etorphine resulted in increased pulmonary pressures with a mirrored decrease in blood oxygenation (Boesch et al. 2018).

17.6 Case study 4: transport of rhinoceros

As first demonstrated during Operation Rhino in the 1960s, safe translocation of large mammals is a fundamental component of species conservation. Although rhinoceros transport has improved over the years, morbidity and mortality rates directly related to transport are still unacceptably high. In South Africa and Namibia, the mortality rate for rhinoceros translocations is estimated to be 5 per cent (Miller et al. 2016), while the morbidity rate is likely much higher. Understandably, transport of a large wild mammal places significant stress on that animal, often leading to self-trauma, disease, and even death. Black rhinoceros commonly break their horns, fracture nasal bones, and experience muscle damage and heat stress (Morkel and Kennedy-Benson 2007), while white rhinoceros tend to develop post-capture anorexia, diarrhoea, and enterocolitis (Miller et al. 2016). Occasionally,

transport disasters can occur, as happened during 2018 in Kenya where all 11 translocated black rhinoceros died following the event (Balala 2018). According to the enquiry team, the cause of all the deaths was 'multiple stress syndrome intensified by salt poisoning and complicated by the following conditions: dehydration, starvation, proliferation of opportunistic bacteria in upper respiratory tract (*Pasteurella* species), gastric ulcers and gastritis' (Balala 2018). An improved understanding of the causes and consequences of transport-related stress can help us mitigate against these disasters in future.

During a recent translocation effort in southern Africa, researchers measured serum electrolyte, enzyme, and metabolite concentrations, as well as APRs and plasma oxidants in the rhinoceros. This study demonstrated that black and white rhinoceros transported hundreds of kilometres by vehicle experienced total body water loss, mobilization of energy reserves, muscular damage, and stress-induced immunomodulation (Pohlin et al. 2020) (Figure 17.5). With these new insights on the physiological changes of rhinoceros during transport, we can begin to make recommendations to improve the welfare of rhinoceros, for example through

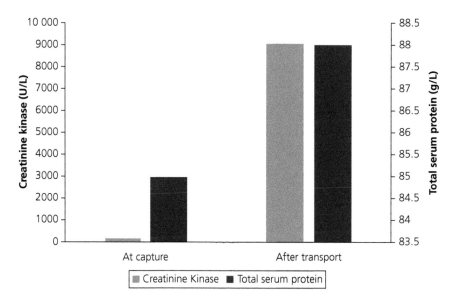

Figure 17.5 Changes in creatinine kinase and total serum protein in transported white rhinoceros. Mean concentrations of creatinine kinase and total serum protein in 24 white rhinoceros. The graph illustrates the changes in these variables after 1300 km road transportation. The increases in creatinine kinase and total serum protein suggest the presence of muscular damage and dehydration, respectively. Figure redrawn with permission from data published in Pohlin et al. *Journal of Wildlife Diseases* (2020), *52*, 2.

recommending limits for transport duration, or water and food deprivation duration (Pohlin et al. 2020). As another example, active measures can be implemented to prevent dehydration. Although rhinoceros are usually reluctant to drink during transport, we can reduce the risk of dehydration by avoiding transport during the hotter times of the day and year, and ensuring well-ventilated, cool crates. During transport of long duration, or over periods where environmental heat cannot be avoided, alternative methods, such as fluid administration, should be considered (Pohlin et al. 2020).

The physical stress of transport is accompanied by a fear response, as indicated by increases in plasma adrenaline and serum cortisol levels in transported white rhinoceros (Pohlin 2019). To reduce psychological stress during transport, most wild animals are given a tranquillizer. Azaperone is commonly used as a short-acting tranquillizer for rhinoceros transport (Morkel and Kennedy-Benson 2007). Midazolam, a benzodiazepine sedative, which is a relatively new drug used in wildlife capture and transport (Burroughs et al. 2012b), was considered as a possible useful anxiolytic to administer to transported rhinoceros. However, by comparing the effects of azaperone and midazolam in transported white rhinoceros, it was shown that midazolam had a greater negative influence on white blood cell responses to capture and transport, and could therefore represent a potential risk factor in the development of disease after transport (Pohlin et al. 2020). As wildlife translocation has already been linked with chronic stress and morbidity, extreme care should be taken with repeated use of midazolam, during capture and transport in white rhinoceros. In addition, as far as possible, rhinoceros should be closely monitored following translocation, so interventions to address physiological compromise can be implemented as early as possible if required.

17.7 Conclusions and future directions

Rhinoceros species are at the forefront of a brutal war against human greed and ignorance. As the demand for rhino horn continues to soar, poaching syndicates are using ever more powerful tools and weapons to obtain this sought-after commodity for the black market. Despite conservation efforts, rhino species are disappearing. In 2018, the northern white rhinoceros was declared extinct in the wild. There are only two known individuals of this subspecies in captivity, both of which are female. The Sumatran rhinoceros in Asia, which has an estimated population of fewer than 80 individuals, is likely the most endangered large mammal on earth, and could be the next to go extinct (International Rhino Foundation (IRF) 2020).

As conservationists attempt to stay ahead of the poaching syndicates, every effort should be made to limit the risk of morbidity and mortality directly related to rhino management interventions. Indeed, safe immobilization and transport of rhinoceros species will become more important as the fight against poaching intensifies further. Largely thanks to the actions of Operation Rhino in the 1960s, the white rhinoceros has become the most abundant and most extensively studied rhinoceros species, especially regarding its physiological responses to capture and transport. Early observations revealed this species' extreme sensitivity to capture drugs, making it even more important to implement research trials to help veterinarians manage the negative consequences of immobilization. Controlled trials and the use of more sophisticated monitoring devices have helped us to elucidate the physiological consequences of capture drugs and transport. Armed with this improved knowledge, we can confidently use the best available treatments to support white rhinoceros during immobilization and can begin to investigate novel approaches to improve capture and transport even further.

We now know that butorphanol administered to immobilized white rhinoceros leads to a reduction in oxygen consumption and heart rate, and an improvement in arterial oxygenation. Further improvements in oxygenation can be achieved by supplying oxygen via a nasotracheal tube. However, despite these supportive treatments, immobilized white rhinoceros remain physiologically compromised. While healthy animals may tolerate these physiological imbalances, diseased or injured animals may not. As rhino poaching escalates, with a significant impact on the white rhinoceros population, more critically injured animals will require opioid-induced immobilization for treatment and translocation.

Apart from poaching, changing climatic conditions, including increasing temperatures and irregular rainfall, occurring in much of Africa (Moise and Hudson 2008) also may result in more diseased and nutritionally compromised wild animals, including rhinoceros species. In an effort to prevent species from succumbing to the ills of rapidly changing environments, these physiologically stressed animals may require immobilization and transport to new environments. The compounding effects of opioid-induced negative effects on physiologically compromised animals will undoubtedly lead to an increased risk of capture-induced morbidity and mortality. Thus, it is imperative that we continue the quest to reduce the negative physiological effects of capture and transport.

Although we are now better equipped to treat opioid-induced pathophysiological events, the ultimate aim should be prevention, rather than treatment. Currently, azaperone is routinely co-administered with opioids during chemical immobilization of rhinoceros and other wildlife species. Azaperone enhances the efficacy of opioids, and reduces hypertension through its α_1-antagonistic effect, which blocks opioid-induced vasoconstriction (Bowdle 1998; Burroughs, et al. 2012b). Unfortunately, however, azaperone does not positively influence oxygenation or pH balance (Portas 2004). More recently, midazolam, a benzodiazepine, has been advocated as an alternative to azaperone for use in rhinoceros immobilization (van Zijll Langhout et al. 2016; Pohlin et al. 2020). Midazolam may reduce muscle tremors and therefore oxygen consumption, anaerobic metabolism, and lactic acidosis (van Zijll Langhout et al. 2016; Pohlin et al. 2020). Thus, midazolam may be a safer alternative to azaperone when combined with etorphine for the capture of wild white rhinoceros. However, further research is needed to definitively understand the beneficial, or possibly even negative, effects of these and other capture drugs.

Pulmonary hypertension is thought to be a significant contributing factor to the hypoxaemia and hypercapnia recorded in immobilized white rhinoceros (Buss et al. 2018). Understanding why this pulmonary hypertension and other cardiopulmonary side-effects occur during etorphine immobilization requires more in-depth physiological studies to better elucidate the mechanisms underlying the pathophysiology. Unravelling these mechanisms may shed greater insights into novel treatments that could be easier and safer to implement in the field.

Although attempts have been made to improve the welfare of rhinoceros during transport, it is difficult to advocate for specific treatments at this stage, given the lack of controlled trials. For logistical reasons, it is difficult to reduce confounding variables in transport trials. With more rigorous assessments of the short-term and long-term effects of transport on rhinoceros, we can better understand how best to limit deleterious physiological consequences, thereby reducing the risk of transport-related morbidity and mortality. Improved, long-term monitoring of rhinoceros post-transport, as well as after immobilization events, will help us gauge the more insidious (i.e. carryover) effects of capture and transport on eventual breeding success and vital rates. For example, do the physiological changes that occur during capture and transport affect fertility and, if so, how can we ensure optimal breeding rates following capture and transport events? As rhino poaching places ever more pressure on rhinoceros populations, ensuring normal fertility and conception after management interventions will be imperative.

Lastly, the dissemination of knowledge is a key factor in ensuring the survival of rhinoceros species. Research findings need to be converted into practical steps that practitioners can implement to improve the welfare of rhinoceros during immobilization and transport. Rhinoceros species likely differ in their physiological responses to capture and transport. However, the long history of white rhinoceros physiology research has set the stage for research into other rhinoceros species. Sharing methodologies and key insights can help rhinoceros researchers and managers across the world determine how best to support all rhinoceros species during capture and transport.

There is no one path for ensuring the survival of rhino species. However, by working across disciplines and implementing science-backed protocols, we can hope to give rhino a fighting chance. Multifaceted approaches are needed to tackle the challenges facing rhinoceros species. Conservation physiology will not save these iconic animals on its own, but it is certainly an important weapon in the overall fight against their extinction.

References

Balala, N., 2018. During the release of results of the independent inquiry into the deaths of rhinos translocated from Nairobi and Lake Nakuru National Parks to Tsavo East Rhino Sanctuary. Available at: https://www.facebook.com/MinistryOfTourismAndWildlifeKE/posts/press-statement-by-hon-najib-balala-egh-cabinet-secretary-ministry-of-tourism-wi/1774612095967991/ (accessed 8 July 2020).

Biggs, D., Courchamp, F., Martin, R., and Possingham, H. P., 2013. Legal trade of Africa's rhino horns. *Science*, 339(6123), 1038–9.

Blane, G.F., Boura, A.L.A., Fitzgerald, A.E., and Lister, R.E., 1967. Actions of etorphine hydrochloride, (M99): a potent morphine-like agent. *British Journal of Pharmacology and Chemotherapy*, 30(1), 11–22.

Boardman, W.S.J., Caraguel, C.G.B., Raath, J.P., and Langhout, M.V., 2014. Intravenous butorphanol improves cardiopulmonary parameters in game-ranched white rhinoceroses (*Ceratotherium simum*) immobilized with etorphine and azaperone. *Journal of Wildlife Diseases*, 50(4), 849–57.

Boesch, J.M., Gleed, R.D., Buss, P. et al., 2018. Effects of a supplemental etorphine dose on pulmonary artery pressure and cardiac output in immobilized, boma-habituated white rhinoceros (*Ceratotherium simum*): a preliminary study. *Journal of Zoo and Wildlife Medicine*, 49(4), 849–55.

Bowdle, T.A., 1998. Adverse effects of opioid agonists and agonist-antagonists in anaesthesia. *Drug Safety*, 19(3), 173–89.

Burroughs, R., Hofmeyr, M., Morkel, P. et al., 2012a. Chemical immobilization—individual species requirements. In: M. Kock and R. Burroughs, eds. *Chemical and Physical Restraint of Wild Animals. A Training and Field Manual for African Species*, 2nd edition, pp. 143–264. IWVS, Greyton.

Burroughs, R., Meltzer, D., and Morkel, P., 2012b. Applied pharmacology in chemical and physical restraint of wild animals. In: M. Kock and R. Burroughs, eds. *Chemical and Physical Restraint of Wild Animals. A Training and Field Manual for African Species*, 2nd edition, pp. 53–80. IWVS, Greyton.

Bush, M., Raath, J.P., Grobler, D., and Klein, L., 2004. Severe hypoxaemia in field-anaesthetised white rhinoceros (*Ceratotherium simum*) and effects of using tracheal insufflation of oxygen. *Journal of the South African Veterinary Association—Tydskrif Van Die Suid-Afrikaanse Veterinere Vereniging*, 75(2), 79–84.

Buss, P., 2017. *Understanding the Physiological Basis for Managing Anaesthetic Related Cardiopulmonary Side-effects in Wildlife.* University of the Witwatersrand, Johannesburg, South Africa.

Buss, P., Miller, M., Fuller, A. et al., 2016. Cardiovascular effects of etorphine, azaperone, and butorphanol combinations in chemically immobilized captive white rhinoceros (*Ceratotherium simum*). *Journal of Zoo and Wildlife Medicine*, 47(3), 834–43.

Buss, P., Miller, M., Fuller, A. et al., 2018. Postinduction butorphanol administration alters oxygen consumption to improve blood gases in etorphine-immobilized white rhinoceros. *Veterinary Anaesthesia and Analgesia*, 45(1), 57–67.

Christy, B., 2016. Special investigation: inside the deadly rhino horn trade. Here's how a pair of South Africans could undermine the international efforts to protect the vulnerable animals. *National Geographic*, 10. Available at: https://www.nationalgeographic.com/magazine/2016/10/dark-world-of-the-rhino-horn-trade/ (accessed 8 July 2020).

Citino, S.B. and Bush, M., 2007. Reference cardiopulmonary physiologic parameters for standing, unrestrained white rhinoceroses (*Ceratotherium simum*). *Journal of Zoo and Wildlife Medicine*, 38(3), 375–9.

Cray, C., Zaias, J., and Altman, N.H., 2009. Acute phase response in animals: a review. *Comparative Medicine*, 59(6), 517–26.

De Lange, S.S., Fuller, A., Haw, A. et al., 2017. Tremors in white rhinoceroses (*Ceratotherium simum*) during etorphine–azaperone immobilisation. *Journal of the South African Veterinary Association*, 88, a1466.

Fahlman, Å., 2008. *Advances in Wildlife Immobilisation and Anaesthesia. Clinical and Physiological Evaluation in Selected Species.* Swedish University of Agricultural Sciences, Uppsala, Sweden.

Harthoorn, A.M. and Bligh, J., 1965. The use of a new oripavine derivative with potent morphinelike activity for the restraint of hoofed wild animals. *Research in Veterinary Science*, 6(3), 290–300.

Haw, A., Hofmeyr, M., Fuller, A. et al., 2014. Butorphanol with oxygen insufflation corrects etorphine-induced hypoxaemia in chemically immobilized white rhinoceros (*Ceratotherium simum*). *BMC Veterinary Research*, 10(1), 253.

Haw, A., Hofmeyr, M., Fuller, A. et al., 2015. Butorphanol with oxygen insufflation improves cardiorespiratory function in field-immobilised white rhinoceros (*Ceratotherium simum*). *Journal of the South African Veterinary Association*, 86(1), a1276.

Hluhluwe Game Reserve, 2020. Saving the white rhino. Hluhluwe Game Reserve. Available at: https://www.hluhluwegamereserve.co.za/save-the-white-rhino/ (accessed 8 July 2020).

Hooijberg, E.H., Cray, C., Steenkamp, G. et al., 2020. Assessment of the acute phase response in healthy and injured southern white rhinoceros (*Ceratotherium simum simum*). *Frontiers in Veterinary Science*, 6, 475.

Hübschle, A.M., 2015. A game of horns. Transnational flows of rhino horn. Universität zu Köln. Available at: https://kups.ub.uni-koeln.de/6685/ (accessed 8 July 2020).

International Rhino Foundation (IRF), 2020. Rhinos are in crisis. International Rhino Foundation. Available at: https://rhinos.org/the-crisis/ (accessed 8 July 2020).

Kock, M.D., Morkel, P., Atkinson, M., and Foggin, C., 1995. Chemical immobilization of free-ranging white rhinoceros (*Ceratotherium simum simum*) in Hwange and Matobo National Parks, Zimbabwe, using combinations of etorphine (M99), fentanyl, xylazine, and detomidine. *Journal of Zoo and Wildlife Medicine*, 26(2), 207–19.

Lykkesfeldt, J. and Svendsen, O., 2007. Oxidants and antioxidants in disease: oxidative stress in farm animals. *The Veterinary Journal*, 173(3), 502–11.

Magnusson, L. and Spahn, D.R., 2003. New concepts of atelectasis during general anaesthesia. *British Journal of Anaesthesia*, 91(1), 61–72.

Marntell, S., Nyman, G., and Hedenstierna, G., 2005. High inspired oxygen concentrations increase intrapulmonary shunt in anaesthetized horses. *Veterinary Anaesthesia and Analgesia*, 32(6), 338–47.

Meyer, L.C.R., Hetem, R.S., Mitchell, D., and Fuller, A., 2015. Hypoxia following etorphine administration in goats (*Capra hircus*) results more from pulmonary hypertension than from hypoventilation. *BMC Veterinary Research*, 11(1), 18.

Miller, M., Buss, P., Joubert, J. et al., 2013. Use of butorphanol during immobilization of free-ranging white Rhinoceros (*Ceratotherium simum*). *Journal of Zoo and Wildlife Medicine*, 44(1), 55–61.

Miller, M., Kruger, M., Kruger, M. et al., 2016. A scoring system to improve decision making and outcomes in the adaptation of recently captured white rhinoceroses (*Ceratotherium simum*) to captivity. *Journal of Wildlife Diseases*, 52(2s), S78–85.

Moise, A.F. and Hudson, D.A., 2008. Probabilistic predictions of climate change for Australia and southern Africa using the reliability ensemble average of IPCC CMIP3 model simulations. *Journal of Geophysical Research: Atmospheres*, 113(D15).

Morkel, P. and Kennedy-Benson, A., 2007. Translocating black rhino. Current techniques for capture, transport, boma care, release and post-release monitoring. IUCN SCC African Rhino Specialist Group, 85 pp.

Player, I., 1967. Translocation of white rhinoceros in South Africa. *Oryx*, 9(2), 137–50.

Pohlin, F., 2019. Physiological responses to capture and transport in southern white rhinoceros (*Ceratotherium simum simum*) and southern-central black rhinoceros (*Diceros bicornis minor*). PhD Thesis. University of Pretoria, Pretoria, South Africa.

Pohlin, F., Hofmeyr, M., Hooijberg, E.H. et al., 2020. Challenges to animal welfare associated with capture and long road transport in boma-adapted black (*Diceros bicornis*) and semi-captive white (*Ceratotherium simum simum*) rhinoceroses. *Journal of Wildlife Diseases*, 56(2), 1–6.

Portas, T.J., 2004. A review of drugs and techniques used for sedation and anaesthesia in captive rhinoceros species. *Australian Veterinary Journal*, 82(8), 542–9.

Porter, R.K. and Brand, M.D., 1995. Cellular oxygen consumption depends on body mass. *American Journal of Physiology—Regulatory, Integrative and Comparative Physiology*, 269(1), R226–8.

Proulx, J., 1999. Respiratory monitoring: arterial blood gas analysis, pulse oximetry, and end-tidal carbon dioxide analysis. *Clinical Techniques in Small Animal Practice*, 14(4), 227–30.

Raath, J.P., 1999. Anesthesia in white rhinoceros. In: M.E. Fowler and R.E. Miller, eds. *Zoo and Wild Animal Medicine: Current Therapy*, 4th edition, pp. 556–61. W.B. Saunders, Philadelphia, PA.

Rookmaaker, K., 2000. The alleged population reduction of the southern white rhinoceros (*Ceratotherium simum simum*) and the successful recovery. *Säugetierkundliche Mitteilungen*, 45(2), 55–70.

Staffieri, F., Bauquier, S.H., Moate, P.J., and Driessen, B., 2009. Pulmonary gas exchange in anaesthetised horses mechanically ventilated with oxygen or a helium/oxygen mixture. *Equine Veterinary Journal*, 41(8), 747–52.

van Zijll Langhout, M., Caraguel, C.G.B., Raath, J.P., and Boardman, W.S.J., 2016. Evaluation of etorphine and midazolam anesthesia, and the effect of intravenous butorphanol on cardiopulmonary parameters in game-ranched white rhinoceroses (*Ceratotherium simum*). *Journal of Zoo and Wildlife Medicine*, 47(3), 827–33.

Wenger, S., Boardman, W., Buss, P. et al., 2007. The cardiopulmonary effects of etorphine, azaperone, detomidine, and butorphanol in field-anesthetized white rhinoceroses (*Ceratotherium simum*). *Journal of Zoo and Wildlife Medicine*, 38(3), 380–7.

Whiteside, G.T., Hummel, M., Boulet, J. et al., 2016. Robustness of arterial blood gas analysis for assessment of respiratory safety pharmacology in rats. *Journal of Pharmacological and Toxicological Methods*, 78, 32–41.

World Wildlife Fund, 2020. Species: rhino—white rhino. World Wildlife Fund. Available at: https://www.worldwildlife.org/species/white-rhino (accessed 8 July 2020).

PART VI

Increasing the Reach of Conservation Physiology

CHAPTER 18

Communication in conservation physiology: linking diverse stakeholders, promoting public engagement, and encouraging application

Taryn D. Laubenstein and Jodie L. Rummer

> ⮩ **Take-home message**
>
> Planning how research findings will be communicated with policy makers, stakeholders, and/or the general public, engaging stakeholders at various stages of the research process, and strategically choosing communication platforms are key elements that are critical to effective conservation outcomes.

18.1 Introduction

The preceding chapters have demonstrated how physiological concepts, tools, and knowledge can be applied to improving ecological conservation and management. Yet linking physiological data with real conservation action or changes in human behaviour can be difficult. Without a solid plan for communicating and engaging with policy makers, stakeholders, or the general public, even the most rigorous research findings can be overlooked or ignored.

In this chapter, we outline the benefits of communicating science beyond the 'ivory tower', provide guidance in navigating partnerships between researchers and practitioners, and outline the different modes of communication and stakeholder engagement that can suit a variety of conservation end-goals. In particular, we highlight knowledge

co-production, collaboration with social scientists, citizen science, and social media as four complementary ways of engaging with stakeholders. They are presented in order of most to least time- and resource-intensive, so that readers can work on incorporating effective communication and engagement into their work, regardless of career stage. We discuss the benefits and disadvantages of each method and give advice on how to successfully integrate them into a research programme. Finally, we look towards the future of communication and collaboration to see how the skills discussed here can be spread to the broader scientific community.

18.2 Why communicate?

Conservation physiologists are typically interested in achieving conservation action through their

---

Taryn D. Laubenstein and Jodie L. Rummer, *Communication in conservation physiology: linking diverse stakeholders, promoting public engagement, and encouraging application* In: *Conservation Physiology: Applications for Wildlife Conservation and Management.* Edited by: Christine L. Madliger, Craig E. Franklin, Oliver P. Love, and Steven J. Cooke, Oxford University Press (2021). © Oxford University Press. DOI: 10.1093/oso/9780198843610.003.0018 research. Communication is the bridge that can connect research with the people who can affect change, including decision makers, stakeholders, and the public at large. What we hope to emphasize in this chapter is that success in changing human behaviour goes beyond the quality of research or where it is published; rather, success will depend on research findings in combination with skills in collaborating and communicating with others.

When done properly and throughout the research process, communication and engagement can engender cooperation and support of stakeholders and promote meaningful stewardship of natural resources. There are also academic benefits that can be gained from communication and engagement, such as increased visibility and impact of research and the possibility to generate funding for future research. Indeed, funding bodies are increasingly recognizing the importance of collaboration and communication; some are even adding a communications section to grant applications and allotting for such expenditures in the budget. Finally, engaging in participatory research can ensure that research is relevant and useful to stakeholders and perhaps could even turn small projects into larger, more impactful collaborations.

18.3 Knowledge co-production

For many physiologists hoping to achieve conservation action, research is thought to be the first step towards reaching that goal. When a project has been designed, the data collected and analysed, and a paper written, then the findings can be disseminated. If findings are disseminated to policy makers, managers, and decision makers to inform legislation, change can happen when and where it is needed. While this is the traditional way of approaching conservation action, this one-way transfer of knowledge can be challenging and, at times, ineffective (Sturgis and Allum 2004). However, another method for achieving conservation success—knowledge co-production—is growing in popularity. This strategy engages researchers and multiple stakeholders spanning the science–policy–society

interface to contribute towards co-creating knowledge that will inform decision making (Lemos and Morehouse 2005).

With knowledge co-production, stakeholders are involved in the research process from the outset, often even initiating research projects. This early involvement means that research outputs from co-produced studies are often more relevant and useful for stakeholders (Meadow et al. 2015). Furthermore, stakeholders are more likely to perceive the results of co-produced studies as salient, credible, and legitimate, which in turn makes them more likely to incorporate results into the decision-making process (Cash et al. 2003). Not only is co-produced knowledge relevant and useful, it is also strengthened by incorporating multiple viewpoints. Local and cultural knowledge can provide examples of previous successes and failures (Fazey et al. 2006) and outline the most culturally appropriate ways to integrate research findings into conservation action (Naess 2013). Perhaps most importantly, co-produced research is based on the principles of democracy and social and environmental justice, meaning that researchers and stakeholders are placed on equal footing to achieve a mutually beneficial outcome (Cvitanovic et al. 2019).

To initiate a co-produced research project, the first step is to contact relevant stakeholders, unless they have already reached out to the research team. To ensure equity between all project members and improve uptake of project results, it is crucial that this step happens as early as possible. Determining the full range of relevant stakeholders can be tricky, but a starting place is to consider the primary users of the system or species of interest. Are they industry members, indigenous groups, managers, the general public, or some combination of these? Once a preliminary list has been collated, the next step is to determine the best ways to get in touch. Do the stakeholders frequently use and maintain a presence on social media (see Section 18.6)? One way to reach out to stakeholders is to work within social structures that already exist within the community (Djenontin and Meadow 2018). For instance, reaching out to well-connected leaders in a community can provide access to a wide sphere of stakeholders in a relatively short timeframe (Kirono et al. 2014).

However, this method can risk feeding into existing power imbalances (Djenontin and Meadow 2018). Therefore, seeking input from a diverse range of stakeholders is advised to democratize the process. It may also be possible to benefit from intermediates like knowledge brokers and boundary organizations to help establish a working relationship with key stakeholders (Reinecke 2015). Knowledge brokers are often embedded within research institutions, while boundary organizations are separate entities that can facilitate interactions between groups that may initially have trouble finding a common ground (Cvitanovic et al. 2015).

Once a team of researchers and stakeholders has been assembled, the project can be designed. This is when stakeholders can lay out their priorities, goals, and values to make sure they are incorporated into the study, and the team can then ensure that outcomes will be relevant. At this stage, collaborators can decide not only on the research questions to be answered, but also on the methods for answering those questions. Where will the study take place? What metrics will be used to gauge success? Local knowledge is crucial at this stage. In some cases, local experts can provide a detailed understanding of the ecosystem under investigation (see Section 18.3.1) or clarify end-goals that might differ from traditional scientific metrics of success. The most successful co-produced studies have started with all parties entering into the design discussions with open minds and a focus on listening (Armitage et al. 2011). Failing this, some stakeholders may disengage during early conversations if they perceive themselves to lack certain expert knowledge (Djenontin and Meadow 2018). The study design stage is also a key time to consider financial contributions of different stakeholders, as an equitable design process may also promote an equitable sharing of budget and resources (Podestá et al. 2013).

When a solid design is in place, the project can be implemented. However, just because the research process has commenced, this does not mean that communication with stakeholders should cease. Rather, continued engagement and clear communication with stakeholders at this time is critical. Barriers to communication can include language differences and jargon, all of which can be overcome using interpreters, communications specialists, or drawings and visual representations (Djenontin and Meadow 2018). Similarly, research outputs should be tailored to reach all stakeholders. For instance, instead of technical graphs and jargon-laden texts, elements of storytelling can be used to communicate results. Stories can use narrative devices like plot, characters, and descriptions to connect research findings with stakeholder values and interests (Young et al. 2016). Additionally, a formal dissemination plan can ensure that stakeholders are informed at regular intervals via appropriate channels (Castellanos et al. 2013). By following this overall format, the continued participation and satisfaction of all stakeholders is more certain.

In addition to the aforementioned steps, there are some intangible factors that can improve a co-produced research project, such as social capital and trust. Social capital is a term used to describe the networks and norms, like trust, that facilitate social engagement (Putnam 1995). Trust can be built through visibility in the field and in the community, for example, by hosting workshops, attending community meetings, and informally engaging with users in their element. These seemingly simple activities can build new social capital or even help to overcome a history of mistrust between stakeholders and outside researchers (Djenontin and Meadow 2018).

Though the above framework represents the current best practices for co-producing research, a variety of institutional factors can impede progress and need to be changed to promote further research of this nature. For instance, many institutions have inflexible structures, such as policies that limit data sharing, which can slow progress. Financial flexibility is also crucial, as it allows for improvements as the project proceeds, such as bringing on new hires to bolster the team's skill set (Djenontin and Meadow 2018). Over the long term, research institutions should provide training support in key skills that are needed for knowledge co-production, such as mediation, brokering, facilitation, and translation (Cvitanovic et al. 2019). Furthermore, given the growing role and importance of co-produced studies, institutions should recognize and reward researchers who take part in this type of research, as the diverse benefits and outputs are often not formally recognized through traditional pathways (Cvitanovic et al. 2019).

18.3.1 Case study: management of Greenland halibut

Biotelemetry—remote tracking of animal movements—has changed the way that scientists collect data about fish populations. With more accurate data that connect biological, environmental, and geographical factors to fish movements, managers can make informed decisions about fisheries stocks or marine protected areas (Crossin et al. 2017). Biotelemetry was used in a co-produced study on Greenland halibut in Cumberland Sound, Nunavut, Canada (Brooks et al. 2019).

Greenland halibut is a deep-water, circumpolar species that was primarily fished in Canadian waters by foreign fishing vessels until the 1980s (DFO 2006). Many of these quotas were then reallocated to coastal, indigenous communities to benefit local community economies (Brooks et al. 2019). For example, in 1994 the Pangnirtung community in Nunavut was allocated a 500-tonne quota. Initially, mark–recapture studies were undertaken to determine the geographical distribution of halibut across the region, but low tag returns resulted in suboptimal datasets (Treble 2003). Still, the tags that were returned suggested that there were two independent stocks of halibut: one offshore stock and one inshore stock. This prompted the Cumberland Sound Turbot Management Area (CSTMA) to be established such that the inshore stock could be specifically allocated to the Pangnirtung fishery (Figure 18.1).

Green halibut catches in the CSTMA declined throughout the 1990s and 2000s (Dennard et al. 2010), although catches were high in the offshore area just south of the CSTMA. These data supported Inuit Qaujimajatuqangit, or traditional knowledge, of fish movements, which suggested that the inshore halibut stock targeted by the Pangnirtung fishery was moving outside the CSTMA during the open season where they were fished by offshore vessels, thereby affecting quotas within the CSTMA (Brooks et al. 2019). Based on concerns voiced by Pangnirtung residents, the Pangnirtung Hunters

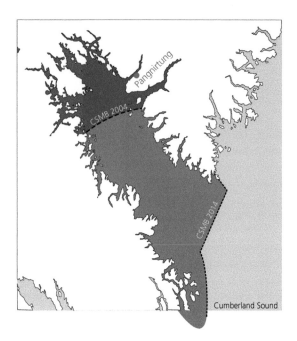

Figure 18.1 The Cumberland Sound Management Boundary (CSMB) shifted as a result of a co-produced study between Pangnirtung fishers, the Ocean Tracking Network, and the University of Windsor that demonstrated that Greenland halibut were moving out of the original management boundaries during the open season. Upon recommendation to Fisheries and Oceans Canada, the original boundary (CSMB 2004) was moved to 12 nautical miles offshore (CSMB 2014), ensuring that Pangnirtung fishers had access to the halibut stock. CSMB lines recreated from Brooks et al. (2019).

and Trappers Association (HTA), and other fishers, a research project was developed to determine whether Greenland halibut were migrating between the CSTMA and the offshore regions during the open season. To do this, a collaboration was established between the Ocean Tracking Network and the University of Windsor, Canada, organizations that had previously worked together, and it was determined that acoustic biotelemetry would be the best method to track the halibut, given the previous difficulties with mark–recapture studies (Brooks et al. 2019). Researchers presented their draft plans to the Pangnirtung HTA for feedback and used both Inuit traditional knowledge and fishery data to design the placement of biotelemetry receivers in the Sound.

The study results showed that Greenland halibut were, indeed, moving out of the CSTMA during the open season. This confirmed the suspicions of

Pangnirtung residents that their stock was vulnerable to commercial, offshore fishers. To remedy this, the study results and Inuit traditional knowledge from Pangnirtung fishers were presented to the Nunavut Wildlife Management Board in July 2013. A recommendation was made to Fisheries and Oceans Canada to move the CSTMA boundary. Consequently, the boundary was moved to 12 nautical miles offshore, ensuring that the Pangnirtung fishers had access to the inshore stock during the open season, and offshore fishers would not exploit the inshore stock while targeting the offshore stock.

The case of Greenland halibut in Cumberland Sound is a prime example of knowledge co-production in conservation physiology because local fishers and members of the community were involved from the onset. In fact, it was the concerns raised by Pangnirtung residents, the Pangnirtung HTA, and other fishers that prompted the research being funded. Similarly, Inuit traditional knowledge was valued equally alongside physiological data and used to design the placement of biotelemetry. Still, the study did encounter roadblocks. In 2011, after only one season of data collection, community elders raised concerns that the receivers were frightening ringed seals, a culturally and nutritionally valuable species for Pangnirtung residents. The research team tried to explain that the equipment would likely not affect the seals, but this did not convince the community, and the research was halted. This example highlights the importance of ongoing two-way communication to develop and maintain community trust and buy-in, as well as the challenges that can present in co-produced research projects. Ultimately, though, the project was successful in its aim to connect the Pangnirtung community with researchers to determine the movement patterns of halibut and thereby inform fisheries management to protect the livelihood of Pangnirtung fishers and their community.

18.4 Collaborating: social science

Findings from conservation physiology studies can provide critical information to decision makers so that conservation actions can be achieved. Yet the path from evidence-based recommendations to actions can be fraught with competing political,

social, and economic interests, meaning that even the most robust science may not be incorporated at the decision-making stage. Given the complexity of achieving conservation outcomes through human behavioural changes, it can be useful to collaborate with social scientists, who are experts in navigating this field.

Social science is a broad field, encompassing disciplines like sociology, economics, political science, and geography, to name a few. Together, these disciplines seek to understand social phenomena, such as culture and governance; social processes, such as decision making and social organization; and individual attributes, such as values and beliefs (Bennett et al. 2017). All of these factors contribute to conservation action and can be studied through the conservation social sciences. For instance, environmental psychologists can study how individual attitudes, beliefs, and norms shape people's responses to conservation actions, while environmental sociologists can reveal the patterns of influence among stakeholders and describe the relationships between stakeholders and their environment (for a complete guide to conservation social sciences, see Bennett et al. 2017).

Academics and practitioners alike have recognized the importance of incorporating the social sciences into the traditionally natural science-dominated field of conservation, highlighting the fundamental truth that conservation action cannot take place without human behavioural changes (Mascia et al. 2003; Schultz 2011; Hicks et al. 2016). Yet despite calls for social science to be mainstreamed into the conservation sciences (Bennett et al. 2016), collaborations between natural and social scientists are still not the norm. A survey of conservation experts across academia, government, and NGOs indicated that a host of barriers have prevented this type of collaboration from flourishing (Fox et al. 2006). Included among these barriers are insufficient funding for collaborative work, limited opportunities for interdisciplinary collaborations, a lack of support from the traditional academic rewards system for interdisciplinary work, and a mismatch in vocabulary between natural and social scientists. Indeed, beyond a difference in vocabulary, natural and social scientists can approach the same research problem with different ideologies and epistemologies about the natural world, resulting in difficulties blending their expertise into a coherent project (Bennett et al. 2016).

However, enthusiastic natural scientists should not let these barriers dissuade them from embarking on collaborations with social scientists. Interdisciplinary research is growing in popularity (Van Noorden 2015), and integration with the social sciences is becoming a greater priority in the conservation sciences. This is further evidenced by the formation of the Social Science Working Group within the Society for Conservation Biology, which has grown to over 700 members since its inception in 2003 (Mascia et al. 2003). Collaborations with social scientists can be accomplished in much the same manner as co-produced research, as described in the above section. In the same way, it is crucial to collaborate with social scientists from the inception of a research project so that their contributions can shape the methodology and design of the project (Viseu 2015). From there, an openness to different philosophies and modes of conducting research will be critical, but the rewards of collaboration will be great. Social scientists can ensure greater application of research findings through understanding the ways different social and cultural groups perceive the environment, improving management practices, facilitating higher social equity in conservation outcomes, and innovating new ways of thinking about conservation (Bennett et al. 2017).

18.4.1 Case study: stress in human–gorilla interactions

Wildlife tourism is a field that is often praised for promoting public awareness of conservation issues and funding conservation-focused research projects (Macfie and Williamson 2010). However, close human contact with wildlife has the potential to negatively affect the animals involved (Higginbottom et al. 2003). To investigate the factors that influence human–animal interactions in wildlife tourism, Dr Kathryn Phillips (née Shutt) utilized both physiological and social science methods.

Phillips travelled to the Dzanga–Sangha Gorilla Habituation and Ecotourism Project in the Central African Republic and chose western lowland gorillas as her study species. At this site, wild gorillas were being habituated to humans via daily exposure until they eventually demonstrated low levels of attention and aggression towards humans (MGVP and WCS 2009). Yet despite this outward appearance of habituation, Phillips wanted to measure the gorillas' physiological stress levels using faecal glucocorticoid metabolites (FGCMs). Glucocorticoids are hormones that are released from the adrenal cortex in response to stress in vertebrates (Selye 1955) and can be maladaptive when elevated over the long term (Cyr and Romero 2008). She also measured parasite infections in the same gorillas. Phillips found that the process of habituation was stressful for the gorillas, as evidenced by their FGCM levels being significantly higher than FGCM levels in unhabituated gorillas (Shutt et al. 2014). Moreover, she found that even habituated gorillas had elevated FGCMs after close encounters with humans, suggesting that the habituation process did not completely eliminate human-related stress responses. She also demonstrated a positive correlation between FGCMs and parasite infection, which could indicate that the immune system was being suppressed when FGCMs were high. This is particularly problematic in a wildlife tourism setting, as the gorillas experience close contact with researchers, tourists, and guides, and are susceptible to contracting human diseases, given their phylogenetic proximity to humans (Köndgen et al. 2008).

After Phillips determined that human contact could risk infecting the gorillas, she needed to identify the factors that increased the risk of infection to ensure they were mitigated. By employing social science methods including semi-structured interviews, questionnaires, and behavioural observations of tourists and staff, she learned that tourists had a modest, at best, understanding of the risks of disease transmission, and that this ignorance decreased tourists' motivation to follow regulations (Setchell et al. 2017). Encouragingly, tourists demonstrated a high willingness to follow regulations if they were informed as to the reasons why the rules were necessary. For instance, tourists said they would wear facemasks, declare illnesses, and provide evidence of vaccinations to decrease the risk of infecting the gorillas (Shutt 2014). As a result of Phillips' research, changes were made to a number of health procedures at the site, including a requirement for tourists and researchers to wear facemasks at all times and to disinfect their hands and boots before visiting the gorillas (K. Phillips, pers. comm.).

Phillips also noted that some tourists expressed a sense of unfairness pertaining to their vaccination requirements, given that local staff on-site had low access to healthcare and were predominantly unvaccinated. Phillips learned from senior management that the health of the staff was considered a low priority, as vaccinations were expensive and difficult to arrange logistically (Shutt 2014). Furthermore, senior management believed that staff and gorillas would have some immunity to local illnesses, and therefore would not need vaccinations. However, through the work of Phillips and another collaborator, the senior management grew to understand the risks of disease transmission by staff, consequently provided vaccinations for staff, and increased access to general healthcare (K. Phillips, pers. comm.).

Phillips' research demonstrates the value of collecting social science data to promote conservation action. While her physiological research revealed that human contact increased stress and possibly parasite infection in the habituated gorillas, it was her social science research that pinpointed the highest risks of infection and their causes. By identifying that tourists were uninformed about the risk of disease transmission, but willing to adhere to rules once informed, Phillips was able to suggest management actions that were successfully integrated into the programme. Similarly, her interviews with senior management revealed the misconceptions that led to the neglect of staff vaccinations and prompted management to prioritize staff health. Together, the physiological and social science data informed robust conservation decisions to protect the gorillas at this site.

18.5 Citizen science

For researchers keen to engage with the public, citizen science can be a useful tool. Citizen science, also known as participatory science, has a long history of bringing together members of the public to further scientific research. Though definitions for the term vary, here we define citizen science as research that involves non-professional scientists (i.e. members of the public) who take part in data collection and/or analysis. This differs from co-produced studies in that they tend to have deeper engagement with a targeted group of stakeholders who are involved from the outset in designing the research question, methods, and disseminating the results. Citizen science, on the other hand, can harness the power of numbers, drawing on the general public's enthusiasm to tackle huge datasets.

When designed correctly, citizen science projects can have major benefits for both research outputs and stakeholder engagement. By opening their research to public participation, researchers can save time and money while generating datasets at scales far greater than they could ever create on their own (Miller-Rushing et al. 2012). Additionally, citizen science gives researchers access to local knowledge that could be invaluable to a project's success (Kobori et al. 2016). For the general public, participating in research projects can increase scientific literacy (Cronje et al. 2011) as well as long-term environmental, civic, and research interests (Dickinson et al. 2012). Citizen science projects can also build social licence with local stakeholders to increase conservation action (Kelly et al. 2019).

The success of a citizen science project depends on careful planning. The first step is to decide how volunteers will be involved in data collection and/or analyses. Recent technological innovations have spurred an increase in citizen science projects, allowing researchers to easily disseminate information about their research while also broadening their pool of potential citizen scientists. Many popular projects have citizens collect environmental or wildlife data using emerging technology, such as smartphone apps, GPS, and photos (see Section 18.5.1). Other projects rely on citizens for analysis, asking them to classify photographs, videos, and sound recordings of plants or animals (Wiggins et al. 2014), which can then be used to create training sets for machine learning to classify the remaining data (Trouille et al. 2019). In either case, the project can be hosted on a pre-existing platform, such as Zooniverse, iNaturalist, or CitiSci.org, or, a new program can be created. Pre-existing platforms offer ease of use and affordability; whereas, new programs or interfaces are costly but can be tailored to suit unique projects.

Next, volunteer recruitment, engagement, and retention are crucial for successful citizen science projects (Locke et al. 2019). Recruiting through pre-existing platforms is fairly straightforward, as engaged citizens are already connected to the programme, but stand-alone projects can attract volunteers as well. It can be effective to reach out to local stakeholders through social media (see Section 18.6) or traditional media outlets like newspapers, TV, and radio. Magazine ads or flyers can also be created to post at conspicuous places used by potential stakeholders, such as community noticeboards. Once volunteers are recruited, they must be trained in proper methods for data collection and/or analysis. In the past, this step has made some researchers wary of using citizen science data, as researchers perceived the data to be less reliable than data produced by trained researchers. However, with proper training and oversight, volunteers can collect data of equal quality to data collected by professionals (Kosmala et al. 2016) The accuracy of citizen-collected data can be tested through expert validation and replication, while bias can be managed with high-performance computing and statistical programs (Bird et al. 2014). The training process should ideally be iterative, such that volunteers can give feedback to project staff about their experiences to improve protocols (Locke et al. 2019). Indeed, volunteer satisfaction is critical to retention and project completion. To retain volunteers, it can be useful to understand their motivations for participating in citizen science (Phillips et al. 2019), as volunteers whose citizen science experiences align with their motivations are more likely to continue participating (Clary et al. 1998) For instance, if volunteers are motivated by the prospect of contributing to scientific research, a series of regular communications about study outcomes can provide the spark to keep them engaged in the project (Locke et al. 2019). By ensuring volunteers remain motivated, citizen science projects can have long-term success in research outputs and stakeholder engagement.

18.5.1 Case study: Redmap (range extension database and mapping project)

Climate change is altering environmental conditions on a global scale, and many species have responded to these changes by shifting their geographical distributions to stay within their preferred environmental conditions (Chen et al. 2011). As species redistribute across the globe, this can impact biodiversity, ecosystem functioning, and human well-being (Pecl et al. 2017). Long-term monitoring programmes that are designed to document range shifts can be costly, particularly in the marine environment. Yet range shifts are occurring in marine ecosystems at nearly an order of magnitude faster than in terrestrial ecosystems (Sorte et al. 2010; Poloczanska et al. 2013), making monitoring programmes in marine ecosystems all the more urgent.

Professor Greta Pecl sought to address this knowledge gap when she founded Redmap in 2009. This citizen science project aimed to provide an early indication of range shifts in marine species by drawing from the knowledge of local fishers, divers, boaters, and other members of the public. To participate, citizens are encouraged to photograph marine species that they find living outside their normal range and submit those photographs to the Redmap website or upload them via the smartphone app. Species identifications are confirmed by one of more than 80 expert Australian scientists, and then the sighting is added to the dataset. Initially, the project was piloted in Tasmania, an area considered to be a 'hotspot' for ocean warming, as waters off the east coast are warming at nearly four times the global average (Johnson et al. 2011; Hobday and Pecl 2014). Based on the Redmap project's success in Tasmania, Redmap was expanded to encompass all Australian waters after 3 years.

Since the project was conceived, data generated by citizen scientists have already been incorporated into more than 20 scientific publications. The data have been used to parameterize habitat models to quantify shifts in habitat suitability (Champion et al. 2018), assess the likelihood of species undergoing range shifts (Robinson et al. 2015), and prompt scientific studies on data-poor species that may be undergoing range shifts (Ramos et al. 2015). The data may also be used in the future to reference historical distribution patterns and habitat ranges as they continue to shift with changing conditions.

Over the first decade of its existence, the Redmap programme logged more than 1900 unusual species sightings, but this does not necessarily mean that the public has learned about climate change in the process. Therefore, another goal of the Redmap programme has been to engage the public about the effects of climate change on marine ecosystems. To do this, Pecl collaborated with Melissa Nursey-Bray and Robert Palmer to assess the efficacy of Redmap in engaging with citizen scientists. Surveys revealed that Redmap users were learning about new range extension sightings, fish species, and what was happening in other parts of Australia (Nursey-Bray et al. 2018). However, surveys were unable to determine whether users connected the range extension sightings explicitly with the effects of climate change, indicating that a deeper enquiry into user knowledge of climate change will be necessary to evaluate this goal. Still, the survey did reveal that Redmap aligns well with many best practices of stakeholder participation in environmental management, such as early involvement of the public, integration of local and scientific knowledge, and a philosophy of equity, trust, and learning. Thus, Redmap can serve as a model for marine citizen science projects that contribute to science and improve community engagement with environmental issues.

18.6 Social media

Thanks to social media, it has never been easier to communicate than it is today. There were 2.62 billion social media users in the year 2018, and projections indicate that there could be over 3 billion users by 2021 (Clement 2018). The high prevalence of social media use can make it easier for researchers to reach out to the public, but these numbers can also seem daunting. How can one account stand out in the sea of content? As this section will reveal, deliberate, targeted use of social media can get information to the right people, create collaborations, and even launch grassroots campaigns.

Social media can help researchers to reach a broad audience by leveraging the power of networks and a special kind of relationship known as weak ties. Weak ties are low-investment relationships that are not based on personal relationships. Despite their casual nature, weak ties have been shown to be more useful than strong ties for reaching a broad network of people, as they foster the transfer of information across cultural and geographic boundaries (Granovetter 1973). This is particularly useful in the realm of social media, where most users are weakly connected, allowing for rapid dispersal of information to a wide audience (Zhao et al. 2010). This theory of weak ties can help researchers boost their media presence and build networks with journalists and decision makers (Evans and Cvitanovic 2018) while also using targeted messaging or groups to reach more specialized audiences (Shiffman 2018).

However, as with other forms of communication, social media has some limitations. Not everyone uses or has access to this technology; therefore, broad communication campaigns should incorporate components of both social and traditional media to ensure everyone gets the message. It is also wise to save sensitive topics for in-person meetings, as written communications strip away social cues such as body language and tone of voice, potentially leading to miscommunication. Keeping these limitations in mind, social media can be a valuable tool in a researcher's communications toolbox.

Twitter, Instagram, Facebook, LinkedIn, Reddit, YouTube, and Pinterest: the list of social media platforms can be dizzying, and each platform has its own nuances, benefits, and drawbacks. Here, we will focus on the 'Big Three' of social media: Twitter, Facebook, and Instagram. We outline the basics of each platform, their benefits and disadvantages, and the audiences that tend to congregate on each (Figure 18.2). For a more detailed explanation of the technical side of setting up each type of account, we recommend a number of excellent guides on social media for scientists (Bik and Goldstein 2013; Shiffman 2018).

18.6.1 Twitter

Twitter is a micro-blogging site that allows users to post messages of 280 characters or less, as well as photos, videos, and links to external websites. Users can search for topics or promote their work using hashtags (#). Twitter has emerged as one of the most-used social media platforms for scientists (Collins et al. 2016), serving as an online global

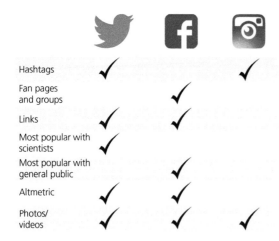

Figure 18.2 The social media focus includes the 'Big Three': Twitter, Facebook, and Instagram, respectively, across the top of the figure with checkmarks (✓) indicating relevance to various functions (e.g. hashtags, fan pages and groups, links, photos/videos), audience (e.g. scientists, general public), and benefits to altmetric scores along the left side of the figure.

faculty lounge that can connect far-flung researchers (Darling et al. 2013) and thereby facilitating collaborations and interdisciplinary research (Bik and Goldstein 2013).

One of the most-used features of Twitter among scientists is sharing and reading about the latest research (Collins et al. 2016). This makes sense, given that most scientists on Twitter follow and are followed by other scientists (Côté and Darling 2018). However, as scientists amass more followers, they can reach wider audiences; one study indicated that beyond a threshold of 1000 followers, scientists were able to reach a more diverse audience including journalists, policy makers, and the general public (Côté and Darling 2018).

Tweeting about a paper can increase its reach online and in academia. The alternative metric or 'altmetric' score of a paper quantifies its reach beyond traditional means (i.e., journal citations) through social and traditional media; an altmetric score can be increased by tweeting a link to the paper, so long as the associated website bears the digital object identifier (doi) of the paper. While a high altmetric score has inherent value, it can also affect academic impact. In some fields, highly tweeted papers are 11 times more likely to be highly cited (Eysenbach 2011). It is likely this combination

of broad and narrow outcomes—reaching wide audiences while also improving traditional academic metrics—that has led to the rise of Twitter within the scientific community.

18.6.2 Facebook

Facebook is the ubiquitous social media site, boasting a base of over 2.32 billion monthly users in December 2018 (Clement 2019). As such, many stakeholders will already have Facebook accounts set up, making this a convenient way to connect. Generally, Facebook interactions are more restricted than Twitter interactions because users must have mutually agreed to be 'friends' in order to communicate. However, a popular way for scientists to connect with stakeholders on Facebook without this step is through groups or fan pages. These are specialized features that allow people with similar interests to congregate and share ideas.

Groups or fan pages can be established for lab groups or for individual projects. These pages can be a useful jumping-off point for meeting conservation and science enthusiasts. For instance, researchers could establish a page for their lab group, then join related Facebook groups with similar topics to advertise the new page to quickly grow a following. Once a page has been established, it can be used not only to communicate directly with stakeholders, but also to promote other endeavours such as citizen science projects or crowd-funding opportunities, whereby researchers can source research funding directly from interested citizens (Hui and Gerber 2015). Finally, as with Twitter, links to papers shared on Facebook pages can increase altmetric scores.

18.6.3 Instagram

Instagram is a social media platform for sharing photos and videos. It may not be perceived as a space for academics; yet, research has shown that viewers pay more attention to pictures than text (Fahmy et al. 2014). This suggests that platforms like Instagram may have been overlooked for their potential to communicate science (Russmann and Svensson 2016).

Instagram is organized similarly to Twitter, where researchers can follow specific users or search for specific topics using hashtags (#). While accounts with professional photographs tend to accrue the most followers, niche accounts, such as science pages, can have success without the help of a professional photographer, so long as the images are particularly captivating. In the field of conservation physiology, photos or videos of experimental set-ups, charismatic study species, or research outcomes would transfer well to a platform like Instagram. Instagram can also be used to reach out to stakeholders who are frequent users of the platform and might not be reachable on Twitter or Facebook. The #keepemwet campaign is an excellent example of reaching out to a community on Instagram to promote conservation actions (see Section 18.6.4).

18.6.4 Case study: Keep 'em Wet (#keepemwet) fishing

Recreational fishing is a popular pastime, with estimates indicating that recreational fishers land approximately 47 billion fish annually (Cooke and Cowx 2006). Over 60 per cent of those fish are later returned to the wild in what is known as catch-and-release fishing (Cooke and Cowx 2006). While the goal of this method is to return fish back to the environment unharmed, scientific studies have found a range of negative effects on physiological performance that are associated with commonly used catch-and-release methods (Arlinghaus et al. 2007). Based on these findings, best-practice guidelines for catch-and-release fishing have been developed (Brownscombe et al. 2016). Yet conveying this information to the broad community of recreational fishers has proven difficult, as even state/provincial agencies sometimes provide inaccurate information about catch-and-release best practices (Pelletier et al. 2007).

In contrast to this top-down approach to communicating catch-and-release best practices, nature photographer Bryan Huskey decided to approach the issue from the bottom up. His grassroots conservation movement began in May 2013 when he coined the Instagram hashtag #keepemwet. This movement was born from his frustration with the online recreational fishing community, who frequently tagged images with the hashtag

#catchandrelease that depicted fish dry on banks or clearly being held out of the water for long periods of time. He was concerned that fishers were amassing followers with these photos that promoted harmful catch-and-release practices, and that such messages could potentially encourage others to take similar photos at the expense of the fish. Huskey started tagging his photos with #keepemwet, a phrase that was both catchy and directly related to conservation actions that fishers could take (Danylchuk et al. 2018). The tag caught on quickly and spread throughout the fly-fishing community. As a result, Huskey created official Instagram and Facebook pages, as well as the keepemwet.org website in 2015. The website was created to serve as a resource for learning science-backed methods for catch-and-release fishing. Huskey was able to drive traffic to the website by partnering with recreational fishing industry members, such as travel companies and gear manufacturers. Later, in 2016, Keepemwet Fishing launched an ambassador programme, which is a common practice in social media to use high-profile members of a community to promote a product or service. While ambassadors are commonly paid for their promotion of a product, the Keepemwet Fishing campaign only asked the angling professionals to promote science-backed catch-and-release guidelines.

The Keepemwet Fishing campaign has shown consistent growth in its reach across social media platforms, but measuring its impact is more difficult. Oftentimes, popular hashtags can take on a life of their own, and thus #keepemwet may be used on photos that do not promote catch-and-release best practices. Still, the organic origin and growth of this movement can serve as a useful case study for those looking to use social media for conservation physiology. Because Huskey was involved in the recreational fishing community, he knew that Instagram was a popular platform for other fishers, making it easy to reach out to them. As the Keepemwet Fishing campaign grew, it remained unaffiliated with government or corporate organizations, which may have contributed to its acceptance by the recreational fishing community (Hollenbeck and Zinkhan 2006). Then, by establishing an ambassador programme with high-profile anglers, the campaign was able to increase its reach without reducing its authenticity. Finally, the campaign may have succeeded because of its association with social pressure. Research from the social sciences has shown that anglers are willing to sanction other anglers' inappropriate catch-and-release practices, and this could increase as fishers learn more about science-based best practices for catch-and-release fishing (Guckian et al. 2018). Thus, the information provided by the Keepemwet Fishing campaign could increase social pressure and sanctioning within the recreational fishing community. Together, these strategies can be employed by other conservation physiologists looking to promote change in their stakeholder communities.

18.7 Conclusions and future directions

Effective communication and engagement are vital in translating research outputs into conservation action. However, there is no one-size-fits-all approach; the goals of the study will determine the audience, and the audience will dictate the best methods for engaging stakeholders and communicating the findings. Here, we have differentiated between a number of engagement and communication methods, but in reality, these methods can be mixed and matched in different combinations to best fit the needs of the researchers, the project, and the communities involved. For example, citizen science projects can be advertised on social media, and social scientists can join in on knowledge co-production teams. It is important to think outside the box in order to create a research programme with meaningful and effective impact.

Once these modes of communication and engagement have been mastered, it is key that researchers consider ways to enable other researchers to develop these skills. Avid social media users can help their colleagues set up Twitter accounts. Chief investigators on citizen science projects can share tips on volunteer recruitment and retention strategies. Researchers in co-produced studies can bring interested colleagues along to stakeholder meetings. Importantly, though, it is no one researcher's task to single-handedly up-skill their colleagues. Research institutions must facilitate improved engagement and communication skills in their researchers. This support can come in the form of

funding for workshops, external speakers, or short courses. It can also take the form of incentives for researchers to co-produce research and engage with stakeholders and the broader public. Institutions should recognize the key roles of communication and engagement in solving conservation problems and reward researchers that undertake non-traditional research projects. These skills are essential for researchers to produce high-quality, applicable outputs that are useable and useful to stakeholders, decision makers, and the broader community.

Acknowledgements

The authors thank Erin Walsh for figure illustrations. J.L.R. is funded by the Australian Research Council Centre of Excellence for Coral Reef Studies.

References

Arlinghaus, R., Cooke, S.J., Lyman, J. et al., 2007. Understanding the complexity of catch-and-release in recreational fishing: an integrative synthesis of global knowledge from historical, ethical, social, and biological perspectives. *Reviews in Fisheries Science*, 15, 75–167.

Armitage, D., Berkes, F., Dale, A. et al., 2011. Co-management and the co-production of knowledge: learning to adapt in Canada's Arctic. *Global Environmental Change*, 21, 995–1004.

Bennett, N.J., Roth, R., Klain, S.C. et al., 2016. Mainstreaming the social sciences in conservation. *Conservation Biology*, 31, 56–66.

Bennett, N.J., Roth, R., Klain, S.C. et al., 2017. Conservation social science: Understanding and integrating human dimensions to improve conservation. *Biological Conservation*, 205, 93–108.

Bik, H.M. and Goldstein, M.C., 2013. An introduction to social media for scientists. *PLoS Biology*, 11, e1001535.

Bird, T.J., Bates, A.E., Lefcheck, J.S. et al., 2014. Statistical solutions for error and bias in global citizen science datasets. *Biological Conservation*, 173, 144–54.

Brooks, J.L., Chapman, J.M., Barkley, A. et al., 2019. Biotelemetry informing management: case studies exploring successful integration of biotelemetry data into fisheries and habitat management. *Canadian Journal of Fisheries and Aquatic Sciences*, 76, 1238–1252.

Brownscombe, J.W., Danylchuk, A.J., Chapman, J.M. et al., 2016. Best practices for catch-and-release recreational fisheries—angling tools and tactics. *Fisheries Research*, 186, 693–705.

Cash, D.W., Clark, W.C., Alcock, F. et al., 2003. Knowledge systems for sustainable development. *Proceedings of the*

National Academy of Sciences of the United States of America, 100, 8086–91.

Castellanos, E.J., Tucker, C., Eakin, H. et al., 2013. Assessing the adaptation strategies of farmers facing multiple stressors: lessons from the Coffee and Global Changes project in Mesoamerica. *Environmental Science and Policy*, 26, 19–28.

Champion, C., Hobday, A.J., Tracey, S.R. et al., 2018. Rapid shifts in distribution and high-latitude persistence of oceanographic habitat revealed using citizen science data from a climate change hotspot. *Global Change Biology*, 24, 5440–53.

Chen, I.-C., Hill, J.K., Ohlemuller, R. et al., 2011. Rapid range shifts of species associated with high levels of climate warming. *Science*, 333, 1024–6.

Clary, E.G., Ridge, R.D., Stukas, A.A. et al., 1998. Understanding and assessing the motivations of volunteers: a functional approach. *Journal of Personality and Social Psychology*, 74, 1516–30.

Clement, J., 2018. Number of social network users worldwide from 2010 to 2021 (in billions. Available at: https://www.statista.com/statistics/278414/number-of-worldwide-social-network-users/ (accessed 8 July 2020).

Clement, J., 2019. Number of monthly active Facebook users worldwide as of 3rd quarter 2019 (in millions). Available at: https://www.statista.com/statistics/264810/number-of-monthly-active-facebook-users-worldwide/ (accessed 8 July 2020).

Collins, K., Shiffman, D., and Rock, J., 2016. How are scientists using social media in the workplace? *PLoS ONE*, 11, e0162680.

Cooke, S.J. and Cowx, I.G., 2006. The role of recreational fishing in global fish crises. *BioScience*, 54, 857–9.

Côté, I.M. and Darling, E.S., 2018. Scientists on Twitter: preaching to the choir or singing from the rooftops? *Facets*, 3, 682–94.

Cronje, R., Rohlinger, S., Crall, A. et al., 2011. Does participation in citizen science improve scientific literacy? A study to compare assessment methods. *Applied Environmental Education and Communication*, 10, 135–45.

Crossin, G.T., Heupel, M.R., Holbrook, C.M. et al., 2017. Acoustic telemetry and fisheries management. *Ecological Applications*, 27, 1031–49.

Cvitanovic, C., Hobday, A.J., Van Kerkhoff, L. et al., 2015. Improving knowledge exchange among scientists and decision-makers to facilitate the adaptive governance of marine resources: a review of knowledge and research needs. *Ocean and Coastal Management*, 112, 25–35.

Cvitanovic, C., Howden, M., Colvin, R.M. et al., 2019. Maximising the benefits of participatory climate adaptation research by understanding and managing the associated challenges and risks. *Environmental Science and Policy*, 94, 20–31.

Cyr, N.E. and Romero, L.M., 2008. Fecal glucocorticoid metabolites of experimentally stressed captive and free-living starlings: implications for conservation research. *General and Comparative Endocrinology*, 158, 20–8.

Danylchuk, A.J., Danylchuk, S.C., Kosiarski, A. et al., 2018. Keepemwet Fishing—an emerging social brand for disseminating best practices for catch-and-release in recreational fisheries. *Fisheries Research*, 205, 52–6.

Darling, E.S., Shiffman, D., Côté, I.M. et al., 2013. The role of Twitter in the life cycle of a scientific publication. *Ideas in Ecology and Evolution*, 6, 32–43.

Dennard, S.T., Macneil, M.A., Treble, M.A. et al., 2010. Hierarchical analysis of a remote, Arctic, artisanal longline fishery. *ICES Journal of Marine Science*, 67, 41–51.

DFO (Department of Fisheries and Oceans Canada), 2006. Fishery management plan Greenland halibut NAFO subarea O, 2006–2008. Fisheries and Oceans Canada, Central and Arctic Region, Winnipeg, Manitoba, Canada.

Dickinson, J.L., Shirk, J., Bonter, D. et al., 2012. The current state of citizen science as a tool for ecological research and public engagement. *Frontiers in Ecology and the Environment*, 10, 291–7.

Djenontin, I.N.S. and Meadow, A.M., 2018. The art of co-production of knowledge in environmental sciences and management: lessons from international practice. *Environmental Management*, 61, 885–903.

Evans, M.C. and Cvitanovic, C., 2018. An introduction to achieving policy impact for early career researchers. *Palgrave Communications*, 4, 88.

Eysenbach, G., 2011. Can tweets predict citations? Metrics of social impact based on Twitter and correlation with traditional metrics of scientific impact. *Journal of Medical Internet Research*, 13, e123.

Fahmy, S., Bock, M.A., and Wanta, W., 2014. *Visual Communication Theory and Research: a Mass Communication Perspective*. Palgrave Macmillan, New York.

Fazey, I., Fazey, J.A., Salisbury, J.G. et al., 2006. The nature and role of experiential knowledge for environmental conservation. *Environmental Conservation*, 33, 1–10.

Fox, H.E., Christian, C., Nordby, J.C. et al., 2006. Perceived barriers to integrating social science and conservation. *Conservation Biology*, 20, 1817–20.

Granovetter, M.S., 1973. The strength of weak ties. *American Journal of Sociology*, 78, 1360–80.

Guckian, M.L., Danylchuk, A.J., Cooke, S.J. et al., 2018. Peer pressure on the riverbank: assessing catch-and-release anglers' willingness to sanction others' (bad) behavior. *Journal of Environmental Management*, 219, 252–9.

Hicks, C.C., Levine, A., Agrawal, A. et al., 2016. Engage key social concepts for sustainability. *Science*, 352, 38–40.

Higginbottom, K., Green, R., and Northrope, C., 2003. A framework for managing the negative impacts of wildlife tourism on wildlife. *Human Dimensions of Wildlife*, 8, 1–24.

Hobday, A.J. and Pecl, G.T., 2014. Identification of global marine hotspots: sentinels for change and vanguards for adaptation action. *Reviews in Fish Biology and Fisheries*, 24, 415–25.

Hollenbeck, C.R. and Zinkhan, G.M., 2006. Consumer activism on the internet: the role of anti-brand communities. *Advances in Consumer Research*, 33, 479–85.

Hui, J. and Gerber, E., 2015. Crowdfunding science: sharing research with an extended audience. *Proceedings of the 2015 ACM International Conference on Computer-Supported Cooperative Work and Social Computing*, 31–43.

Johnson, C.R., Banks, S.C., Barrett, N.S. et al., 2011. Climate change cascades: shifts in oceanography, species' ranges and subtidal marine community dynamics in eastern Tasmania. *Journal of Experimental Marine Biology and Ecology*, 400, 17–32.

Kelly, R., Fleming, A., Pecl, G.T. et al., 2019. Social license through citizen science: a tool for marine conservation. *Ecology and Society*, 24, 16.

Kirono, D.G.C., Larson, S., Tjandraatmadja, G. et al., 2014. Adapting to climate change through urban water management: a participatory case study in Indonesia. *Regional Environmental Change*, 14, 355–67.

Kobori, H., Dickinson, J.L., Washitani, I. et al., 2016. Citizen science: a new approach to advance ecology, education, and conservation. *Ecological Research*, 31, 1–19.

Köndgen, S., Kühl, H., N'Goran, P.K. et al., 2008. Pandemic human viruses cause decline of endangered great apes. *Current Biology*, 18, 260–4.

Kosmala, M., Wiggins, A., Swanson, A. et al., 2016. Assessing data quality in citizen science. *Frontiers in Ecology and the Environment*, 14, 551–60.

Lemos, M.C. and Morehouse, B.J., 2005. The co-production of science and policy in integrated climate assessments. *Global Environmental Change*, 15, 57–68.

Locke, C.M., Anhalt-Depies, C.M., Frett, S. et al., 2019. Managing a large citizen science project to monitor wildlife. *Wildlife Society Bulletin*, 43, 4–10.

Macfie, E.J. and Williamson, E.A., 2010. *Best Practice Guidelines for Great Ape Tourism*. IUCN, Gland, Switzerland.

Mascia, M.B., Brosius, J.P., Dobson, T.A. et al., 2003. Conservation and the social sciences. *Conservation Biology*, 17, 649–50.

Meadow, A.M., Ferguson, D.B., Guido, Z. et al., 2015. Moving toward the deliberate coproduction of climate science knowledge. *Weather, Climate, and Society*, 7, 179–91.

MGVP (Mountain Gorilla Veterinary Project) and WCS (Wildlife Conservation Society), 2009. Conservation medicine for gorilla conservation. In T.S. Stoinski,

H.D. Steklis, and P.T. Mehlman, eds. *Conservation in the 21st Century: Gorillas as a Case Study*, pp. 57–78. Springer, Boston, MA.

Miller-Rushing, A., Primack, R., and Bonney, R., 2012. The history of public participation in ecological research. *Frontiers in Ecology and the Environment*, 10, 285–90.

Naess, L.O., 2013. The role of local knowledge in adaptation to climate change. *Wiley Interdisciplinary Reviews: Climate Change*, 4, 99–106.

Nursey-Bray, M., Palmer, R., and Pecl, G., 2018. Spot, log, map: assessing a marine virtual citizen science program against Reed's best practice for stakeholder participation in environmental management. *Ocean and Coastal Management*, 151, 1–9.

Pecl, G.T., Araújo, M.B., Bell, J.D. et al., 2017. Biodiversity redistribution under climate change: impacts on ecosystems and human well-being. *Science*, 355, eaai9214.

Pelletier, C., Hanson, K.C., and Cooke, S.J., 2007. Do catch-and-release guidelines from state and provincial fisheries agencies in North America conform to scientifically based best practices? *Environmental Management*, 39, 760–73.

Phillips, T.B., Ballard, H.L., and Lewenstein, B.V. et al., 2019. Engagement in science through citizen science: moving beyond data collection. *Science Education*, 103, 665–90.

Podestá, G.P., Natenzon, C.E., Hidalgo, C. et al., 2013. Interdisciplinary production of knowledge with participation of stakeholders: a case study of a collaborative project on climate variability, human decisions and agricultural ecosystems in the Argentine Pampas. *Environmental Science and Policy*, 26, 40–8.

Poloczanska, E.S., Brown, C.J., Sydeman, W.J. et al., 2013. Global imprint of climate change on marine life. *Nature Climate Change*, 3, 919–25.

Putnam, R.D., 1995. Bowling alone: America's declining social capital. *Journal of Democracy*, 6, 65–78.

Ramos, J.E., Pecl, G.T., Semmens, J.M. et al., 2015. Reproductive capacity of a marine species (*Octopus tetricus*) within a recent range extension area. *Marine and Freshwater Research*, 66, 999–1008.

Reinecke, S., 2015. Environmental science & policy knowledge brokerage designs and practices in four European climate services: A role model for biodiversity policies? *Environmental Science and Policy*, 54, 513–21.

Robinson, L.M., Gledhill, D.C., Moltschaniwskyj, N.A. et al., 2015. Rapid assessment of an ocean warming hotspot reveals 'high' confidence in potential species' range extensions. *Global Environmental Change*, 31, 28–37.

Russmann, U. and Svensson, J., 2016. Studying organizations on Instagram. *Information*, 7, 1–12.

Schultz, P.W., 2011. Conservation means behavior. *Conservation Biology*, 25, 1080–3.

Selye, H., 1955. Stress and disease. *Science, 122,* 625–31.

Setchell, J.M., Fairet, E., Shutt, K. et al., 2017. Biosocial conservation: integrating biological and ethnographic methods to study human–primate interactions. *International Journal of Primatology, 38,* 401–26.

Shiffman, D.S., 2018. Social media for fisheries science and management professionals: how to use it and why you should. *Fisheries, 43,* 123–9.

Shutt, K.A., 2014. Wildlife tourism and conservation: an interdisciplinary evaluation of gorilla ecotourism in Dzanga-Sangha, Central African Republic. PhD Thesis, Durham University, Durham.

Shutt, K., Heistermann, M., Kasim, A. et al., 2014. Effects of habituation, research and ecotourism on faecal glucocorticoid metabolites in wild western lowland gorillas: implications for conservation management. *Biological Conservation, 172,* 72–9.

Sorte, C.J.B., Williams, S.L., and Carlton, J.T., 2010. Marine range shifts and species introductions: comparative spread rates and community impacts. *Global Ecology and Biogeography, 19,* 303–16.

Sturgis, P. and Allum, N., 2004. Science in society: re-evaluating the deficit model of public attitudes. *Public Understanding of Science, 13,* 55–74.

Treble, M.A., 2003. Results of a Greenland halibut (*Reinhardtius hippoglossoides*) tagging project in Cumberland Sound, NAFO Division 0B, 1997–2000. Northwest Atlantic Fisheries Organization 345 Scientific Council meeting, SCR Doc. 03/41.

Trouille, L., Lintott, C.J., and Fortson, L.F., 2019. Citizen science frontiers: efficiency, engagement, and serendipitous discovery with human–machine systems. *Proceedings of the National Academy of Sciences of the United States of America, 116,* 1902–909.

Van Noorden, R., 2015. Interdisciplinary research by the numbers. *Nature, 525,* 306–7.

Viseu, A., 2015. Integration of social science into research is crucial. *Nature, 525,* 291.

Wiggins, A., Miller-Rushing, A.J., Parrish, J.K. et al., 2014. Next steps for citizen science. *Science, 343,* 1436–7.

Young, N., Nguyen, V.M., Corriveau, M. et al., 2016. Knowledge users' perspectives and advice on how to improve knowledge exchange and mobilization in the case of a co-managed fishery. *Environmental Science and Policy, 66,* 170–8.

Zhao, J., Wu, J., and Xu, K., 2010. Weak ties: subtle role of information diffusion in online social networks. *Physical Review E, 82,* 1–8.

CHAPTER 19

Optimism and opportunities for conservation physiology in the Anthropocene: a synthesis and conclusions

Steven J. Cooke, Christine L. Madliger, Jordanna N. Bergman, Vivian M. Nguyen, Sean J. Landsman, Oliver P. Love, Jodie L. Rummer, and Craig E. Franklin

19.1 Introduction

Conservation physiology arose as a 'discipline' based on the promise of using physiological knowledge, concepts, and tools to understand and solve conservation problems (Wikelski and Cooke 2006; Cooke et al. 2013). As such, the discipline is inherently mission-oriented. The success of conservation physiology should thus be assessed not just by the number of citations or other traditional measures of 'academic impact' but rather by the extent that conservation physiology delivers on its promise. Successes in conservation physiology are already being recognized (see Madliger et al. 2016); yet, there remain challenges in recognizing the success stories. Rather than waiting for the discipline to mature on its own, efforts have been taken to create a conceptual framework (Coristine et al. 2014) and to help build capacity within the conservation physiology community to ensure that research has impact (Cooke and O'Connor 2010; Madliger et al. 2017b).

Today, there is an urgency associated with conservation that likely extends beyond what Michael Soulé could have envisioned when he first described conservation science as a crisis discipline (Soulé 1985). We are in a biodiversity crisis unlike anything ever witnessed before in human history and with direct consequences on ecosystems, their functions, as well as the ecosystem services upon which humans depend (Cardinale et al. 2012). Amphibians (Beebee and Griffiths 2005) and other freshwater life are facing declines that have exceeded 80 per cent since 1970 (Reid et al. 2019). Novel stressors and threats continue to emerge and combine with existing ones to make life even more challenging for wildlife (Folt et al. 1999). Indeed, climate change is regarded as one of the major threats facing biodiversity and humanity today and for the coming decades (Bellard et al. 2012). Perhaps now, more than ever, there is urgent need for robust science to address these and other issues facing life on Earth.

Although it is easy to become despondent and frustrated about the threats to the natural world, it is also a time for optimism, given collective interest in rejecting a dystopian future and that changing attitudes and human behaviour is possible. For example, despite the fact that it is now accepted that we have entered the Anthropocene epoch (Lewis and Maslin 2015), there are efforts to identify what

Steven J. Cooke, Christine L. Madliger, Jordanna N. Bergman, Vivian M. Nguyen, Sean J. Landsman, Oliver P. Love, Jodie L. Rummer, and Craig E. Franklin,
Optimism and opportunities for conservation physiology in the Anthropocene: a synthesis and conclusions In: *Conservation Physiology: Applications for Wildlife Conservation and Management*. Edited by: Christine L. Madliger, Craig E. Franklin, Oliver P. Love, and Steven J. Cooke, Oxford University Press (2021).

is needed to achieve a 'good' Anthropocene and how to do so (Bennett et al. 2016; Dalby 2016; Madliger et al. 2017b). Similarly, rather than accepting the fact that biodiversity declines continue, some are advocating for strategies to 'bend the curve' and reverse this trend (Mace et al. 2018). To this end, in March of 2019, the United Nations Environment Programme (UNEP) announced the start of the 'Decade of Ecosystem Restoration'. There is also evidence of public support (e.g. climate change rallies and marches for extinction), which gives hope and suggests that the masses are ready for meaningful action. In that sense, the conservation science community needs to be poised to support and inform efforts to tackle these problems with the best available evidence (Ripple et al. 2017). We submit that conservation physiology has much to offer (as outlined in our introductory chapter) in this realm.

The chapters presented in this book span taxa, continents, tools, and issues that collectively provide a rich tapestry to explore and identify emergent themes. Here, we synthesize key themes that emerge from the case studies, providing an optimistic overview of future opportunities for conservation physiology. For each theme, we provide referenced commentary with the hope of providing today's conservation physiologists and those of the future with strategies and perspectives to help them deliver on the promise of conservation physiology. Finally, we consider what type of institutional and training changes are needed to build capacity for conservation physiology.

19.2 Emergent themes

19.2.1 Mechanisms matter in conservation

Simply documenting declines in wild populations through demographic studies often fails to identify the mechanistic basis for decline. An important aspect of conservation science is therefore to identify the threats that are negatively affecting the health, fitness, or survival of wild organisms. Only when threats are clearly identified and—ideally—mitigated, is it possible to expect populations to recover. Physiology can reveal the mechanisms underpinning population declines, changes in dis-

tribution patterns, alterations in health and fitness, and even drivers of mortality (Seebacher and Franklin 2012). When investigated within an experimental context, the mechanisms that are revealed are particularly powerful in that they contribute to understanding cause-and-effect relationships that are relevant in a regulatory context (Cooke and O'Connor 2010). Attempting to 'recover' a population without knowledge of the underlying mechanisms that are causing the declines can lead to wasted resources, as conservation efforts can be misdirected. As conservation physiology has matured, the field has become a trusted source of knowledge in the context of evidence-based conservation. These strengths have been highlighted repeatedly in the case studies presented in the preceding chapters. For example, the case study on Pacific salmon (Chapter 3) revealed the link between water temperature and disease and thus the interactive mechanisms driving migration failure during spawning migrations. It is clear that conservation science and practice have become far more mechanistic in the past decade or so, and conservation physiology has been a major driver of that trend.

19.2.2 Physiology is but one source of knowledge

When one settles down to read a book on conservation physiology, it might be assumed that the collective work will focus solely on conservation. That is not the case here, nor does that notion recognize the fact that conservation is best delivered from a holistic and integrated perspective. Although conservation science tends to be somewhat reductionist (e.g. consider subdisciplines such as conservation genetics, conservation medicine, and conservation physiology), at the end of the day, conservation is complex. So, applying diverse tools to identify solutions is essential. Consider a problem related to reproductive failure in a species. One approach might be to invest in genetic studies to determine if there is evidence of inbreeding. If that study takes 2 years, and it turns out that there is no evidence of inbreeding, then the community is no closer to being able to address the problem. However, if a problem is tackled from multiple dimensions using a diverse toolbox, it is possible to rapidly and accurately identify problems and therefore solutions. In this book, that concept was particularly relevant in the case study on Arctic fishes. Madigan et al. (Chapter 5) used both stable isotope analysis and biotelemetry to identify critical habitats and migration routes and predict population distribution change. Similarly, the research by Dzal and Willis (Chapter 9) involved applying tools from epidemiology and physiology to better understand how to respond to white-nose syndrome in bats. Finally, Ohmer et al. (Chapter 10) discussed how multiple metrics of immune function and stress physiology can be combined to understand disease susceptibility and improve management practices aimed at reversing population declines in some of the world's most imperilled amphibian species. Although it is quite common to take a reductionist approach in conservation, there is ample evidence that highlights the effectiveness and efficiency of bringing together multiple perspectives and approaches (i.e. interdisciplinarity) to problem solving when dealing with a crisis discipline (Balmford and Cowling 2006).

19.2.3 Physiology and behaviour are intertwined

When studying animals, it is impossible to consider behavioural or physiological aspects in isolation. Indeed, physiology and behaviour are inherently and intimately connected. Behaviour is underpinned by physiological mechanisms, processes, and systems. Consider locomotion. Moving from one site to another to avoid a disturbance represents a behavioural choice. Yet, the behaviour was prefaced by the sensory physiology apparatus identifying a relevant threat. Similarly, once the organism decided to move, say at a high speed, it was the physiological capacity of the organism that both enabled locomotion but also constrained it. And, after a high-speed retreat, there would have been a physiological recovery period during which behaviour would have been impaired. The same scenario can even be understood for sessile organisms, given that many organ systems, such as those related

to feeding/digestion and reproduction, involve aspects of physiology and behaviour. For those reasons, it is common for conservation studies on animals to include both physiological and behavioural components (Cooke et al. 2014). Cooke et al. (2014) advocated for better recognition of the intersection of behaviour, physiology, and conservation, which rang true in the case studies covered in this book. For example, Cree et al. (Chapter 16) explored the thermal biology of imperilled endemic reptiles in New Zealand, thinking about aspects of thermal stress as well as behavioural thermoregulation. In combination with many other case studies here and throughout the literature, it becomes clear that it is wise to couple behaviour and physiology when trying to solve conservation problems.

19.2.4 Embrace emerging tools and technologies

Conservation physiology is continually benefiting from novel developments in tools and technology. Some of these tools and technologies enable us to, for example, do more with less tissue, thus negating the need for lethal sampling. The work presented here by Hunt et al. (Chapter 12) was made possible by the rapid expansion of hormone measurement in non-traditional sample media such as whale blow. Other tools and technologies (e.g. point-of-care devices; Stoot et al. 2014; Harter et al. 2015; Talwar et al. 2017; Schwieterman et al. 2019) allow research to occur in remote locations, far from laboratory infrastructure. Some technologies, such as biotelemetry and biologging, allow us to study the behaviour and physiology of wild animals in their natural environment (Cooke et al. 2004; Wilson et al. 2015) to understand how animals respond to different stressors. For example, Tyson et al. (2017) used such technologies to understand how noise pollution affects sea turtles. In the laboratory, 'omics' technologies (e.g. genomics, proteomics, metabolomics, transcriptomics; see McMahon et al. 2014) are revolutionizing what we can do with small amounts of tissue. For instance, He et al. (2016) describe how transcriptomics can be used to inform how source populations are selected for species reintroduction programmes, and the case studies presented by Whitehead et al. (Chapter 7) here indicate that these novel techniques also contribute to pinpointing cause–effect relationships in species facing anthropogenic change, such as pollution. Even work on stable isotopes has evolved such that it is possible to understand not only what animals have been eating but also the environments that they encounter (Meier et al. 2017; Chapter 5).

Of course, technology is constantly evolving and improving. There have been new developments in nanosensors that could potentially be implanted into organisms to assess physiological state (e.g. blood biochemistry) in real time (Lee et al. 2018). A key message is that those working in the realm of conservation physiology are often at the frontier of biology, working to develop, refine, and apply new tools and approaches. Similarly, there are many efforts by conservation physiologists to refine their methods of interacting with animals to minimize welfare impacts and ensure that research does not impede conservation goals (Swaisgood 2007). As demonstrated by the case studies presented here, the conservation physiology toolbox is expanding rapidly (Madliger et al. 2018), but it is important to ensure that new tools and techniques are validated and ground-truthed along the way.

19.2.5 Physiology is relevant to conservation programmes in zoos and aquaria

Although a core aspect of conservation physiology emphasises 'field' research (i.e. field physiology; Costa and Sinervo 2004), that certainly does not preclude research on captive organisms, especially in zoos and aquaria. The concept of *ex situ* conservation (for a discussion, see Pritchard et al. 2012) embraces the idea that *in situ* conservation has failed or is otherwise insufficient. Most would agree that *ex situ* conservation means that a species is in an 'emergency state'; yet, the reality is such that *ex situ* opportunities are becoming more common, and we therefore need to embrace them and make them more efficient (Conde et al. 2013). Some have argued that zoos and aquaria have yet to fully recognize their potential for research and practice (Andrews and Kaufman 1994; Fa et al. 2014), and so the field

of conservation physiology has much scope to contribute to concepts such as 'rewilding' (Lorimer et al. 2015) and captive breeding programmes. In this book we included an entire subsection that focused on aspects of *ex situ* conservation and wildlife rehabilitation in captivity, including sea turtles (Chapter 14), koalas (Chapter 15), various New Zealand reptiles (Chapter 16), and rhinos (Chapter 17). Physiological approaches are particularly effective in identifying what organisms need to succeed (i.e. basic environmental and nutritional needs) while also providing objective tools for tracking the success of such activities. Increasingly, zoos and aquaria are employing experts with a physiological foundation (e.g. reproductive physiology, stress physiology), which is promising.

19.2.6 Conservation physiology extends across scales

The concept of 'scale' is intrinsically relevant in conservation physiology and its applications in management (Noss 1992). Various aspects of scale exist, with biological, spatial, temporal, allometric, and phylogenetic scales being the five most relevant to conservation physiology research (Cooke et al. 2014). Scale is critical to consider, as the scale at which we measure a biomarker may not be the same scale at which we are interested in its consequences, as Helmuth (Chapter 13) discussed here. With respect to biological scale, which refers to the hierarchy of biological organization (e.g. spanning genes, individuals, populations, and ecosystems), to understand causal mechanisms underlying demographic-level declines, physiological responses must first be assessed on an individual level. Although essential in designing effective conservation strategies, scaling physiology along the biological hierarchy from an individual- to a population level as a result of a specific environmental stressor is difficult to accomplish, as it requires multi-disciplinary expertise, longitudinal monitoring, and uninterrupted funding (Lindenmayer and Likens, 2018; Bergman et al. 2019). Furthermore, addressing temporal scale is important in interpreting acute versus chronic physiological responses. Understanding spatial scale is key in determining species distributions and physiological capacities with changing environmental conditions. Allometric scale (White et al. 2019) provides information on how traits scale with conservation implications (e.g. if certain size classes are more reproductively valuable or less vulnerable to exploitation). Finally, phylogenetic scale refers to genetic relationships between species shaped by evolutionary processes, offering information on adaptive physiological divergences between congeners. It is especially important to consider the various scales in policy application, including both upscaling and downscaling, to explore and determine best conservation practices and management strategies (Cooke et al. 2014). Synchronously investigating multiple physiological subdisciplines (e.g. reproductive physiology, stress physiology, genetics) may help reveal the mechanisms that are driving declines or changes in wild populations. Additionally, long-term datasets are needed, as they may provide a more comprehensive understanding of the physiological changes across scales, as different biomarkers vary in their response time to environmental perturbation (e.g. ranging from days to weeks), and could reflect seasonal variations or, for example, warming regimes.

19.2.7 Physiology can be incorporated into long-term monitoring programmes

Proactive conservation and management strategies, which rely on early identification and monitoring of potential threats, focus on ensuring demographic stability and are generally more cost- and time-effective for managing risks than reactive strategies (Drechsler et al. 2011). Recently, biomarkers (e.g. glucocorticoids, reproductive hormones, telomeres) have gained recognition as tools to measure organismal responses to environmental change with the potential to inform conservation policy. To use physiological indices for management strategies, it is essential to validate that they are both reflective of changing environmental conditions and predictive of population changes (Madliger and Love, 2014). Once the link between individual physiological responses and demographic changes as a result of an environmental perturbation is established, that biomarker can be incorporated into long-term monitoring programmes and used to

proactively develop and enforce recovery strategies prior to demographic collapse or extinction (Bergman et al. 2019). For example, Dupoué et al. (2017) identified a genetic biomarker, the telomere (i.e. specifically telomere length), as a reliable physiological parameter in predicting extinction risk in the common lizard (*Zootoca vivipara*). Telomere attrition (i.e. telomere shortening) is linked to repeated exposure to chronic life stress (Breuner et al. 2013) and can reflect biological age (i.e. in contrast to chronological age) and thus tell us a lot about reproductive status and capability (Monaghan and Haussmann 2006). The authors found that common lizard populations undergoing intense warming regimes due to climate change showed significantly shorter telomeres and higher risks of extinction when compared with their cooler habitat counterparts. By including this biomarker into long-term population monitoring, managers can determine when populations may be experiencing demographic-level declines so that they can proactively work to prevent extinction. Further, here, Crossin and Williams (Chapter 2) highlighted how longitudinal monitoring of energetic and stress physiology has assisted in determining predictors of breeding status and reproductive success in seabirds. As global biodiversity continues to decline, it is vital to develop strategies that prevent populations from reaching demographic instability or collapse. When monitored, biomarkers can indicate when populations are experiencing stress and undergoing declines, offering wildlife and resource managers the opportunity to implement recovery strategies before extinctions occur.

19.2.8 Conservation physiology is not just about vertebrates

A strong bias in conservation science exists, unfortunately, as high-level taxa, including charismatic mammals and other vertebrates, are disproportionately studied in comparison with invertebrate species and plants (Donaldson et al. 2016). The Red List of Threatened Species of the International Union for Conservation of Nature (IUCN) is a highly referenced, leading organization that monitors the status of species worldwide. Yet, even this international agency is still heavily biased towards vertebrates

324 CONSERVATION PHYSIOLOGY: APPLICATIONS FOR WILDLIFE CONSERVATION AND MANAGEMENT

(Eisenhauer et al. 2019). Although limited, conservation physiology research focused on invertebrate species has produced meaningful information. For example, physiological investigations in various invertebrate species have been relevant to a multitude of conservation-related questions including: How do reef-building scleractinian corals (i.e. *Agaricia agaricites* and *Agaricia tenuifolia*) respond (i.e. at the level of heat shock proteins) to high sea surface temperatures (Robbart et al. 2004)? How does the Baltic clam (*Macoma balthica*) respond (i.e. at the level of enzyme activities) to hypoxic conditions (Villnäs et al. 2019)? And, how do grasshoppers (*Chorthippus albonemus*) change gene regulation patterns in response to herbivore grazing intensity (Qin et al. 2017)? The case study by Alaux et al. (Chapter 4) illustrates how physiological information can allow for better management of bee populations, with great potential for expansion in this context. With invertebrates representing the most speciose and diverse group of animals globally, it is critical that physiological studies extend more often to underrepresented invertebrate species.

19.2.9 Conservation physiology informs sustainable resource management of non-imperilled species

The word 'conservation' inherently evokes connotations of science and practice that deals with imperilled species. Unfortunately, it is all too common to wait until organisms are imperilled before devoting resources or intellect. Yet, if management is successful, populations are sustained and ecosystems are left intact, such that no species or habitats become or remain at risk. In that context, a well-managed population or ecosystem can be a perfect example of conservation physiology in practice. That notion was represented throughout this book; indeed, many of the case studies did not focus on an imperilled organism. For example, Bouyoucos and Rummer (Chapter 11) herein discuss how combining ecophysiology techniques with community outreach and education are valuable steps towards conservation of shark populations predicted to be vulnerable to climate change in the future. Conservation 'wins' are best characterized by organisms and ecosystems that are not degraded to

the point of requiring emergency recovery plans. Overall, we all win if sustainable management leads to populations and ecosystems that are resilient to anthropogenic change.

19.2.10 Co-production increases likelihood of success

Co-production means working hand-in-hand with partners (i.e. stakeholders) from the idea-generation phase (i.e. before pen is put to paper, so to speak) right through to the project wrap-up (Chapter 18). Doing so ensures that the project has relevance, credibility, and legitimacy, while increasing the likelihood that the co-produced science results in responsible engagement, balanced, respectful knowledge exchange, and greater impact within the scientific community and the community at large (Nel et al. 2016). Co-production is simply the only way to ensure that the findings generated through this research will be embraced by stakeholders and other knowledge users. Co-production and effective knowledge mobilization hinge on sustained and iterative bidirectional communication (Young et al. 2016). In the context of conservation physiology, this means interacting continually with conservation practitioners and policy makers. Undertaking physiological research and then trying to 'feed it' to conservation practitioners is a recipe for failure but remains far too common. In this book, we highlighted numerous examples where co-production was clearly in practice (e.g. Chapter 3, Chapter 6, Chapter 8, Chapter 18). The concept of co-production is particularly important for conservation physiology given the common disconnect between knowledge generators and knowledge users, and should be of paramount importance heading into the future.

19.2.11 Shout from the rooftops—share our successes

Sharing successes contributes to solution-oriented narratives and offers positive outlooks to often complex and dreary conservation challenges. Focusing attention on success stories or 'brightspots' builds conservation optimism, which has been shown to underpin effective collaboration, drive creativity and innovation, and promote positive public perceptions—all of which are critical in mobilizing conservation research, education, and actions (Beever 2000; Bennett et al. 2016; Cvitanovic and Hobday 2018). Examples of successes in conservation physiology include identifying impacts of disturbance or environmental change, implementing disease control, and allowing managers to delineate and prioritize mitigation strategies because physiology offers mechanistic insights into the causes of change (Madliger et al. 2016). In a time of despair, where we as scientists and global citizens are seemingly constantly bombarded with dramatic and negative messages such as how the world is warming, how we are heading to the sixth mass extinction event, and that the biodiversity crisis is worse than climate change, we need optimism (Swaisgood and Sheppard 2010). We need to share success stories, not only to offer hope, but to communicate and share best practices so that these successes can rapidly spread across the globe in a time of urgency.

19.2.12 Don't be 'old school' when communicating your findings

Today's communication landscape is diverse in form (e.g. print media, online news sources, social media), highly fragmented (Bubela et al. 2009; Nisbet and Scheufele 2009), and requires the modern-day scientist to be flexible and creative if they are to be a successful communicator (Chapter 18). Furthermore, communication—like knowledge co-production—requires two-way dialogue (i.e. the 'dialogue model') as opposed to one-way information transfer from experts to non-experts (i.e. the 'deficit model'). It is also important to stay abreast of new developments in communication tools, many of which tap into more informal learning styles (National Research Council 2009). For example, social media is increasingly favoured by scientists as a method to disseminate information (e.g. Côté and Darling 2018), though successful implementation requires interaction among users (e.g. Bortree and Seltzer 2009; McClain 2017). Storytelling—and indeed conservation storytelling—is also being recognized for its ability to efficiently transfer information and its ease of implementation

(Leslie et al. 2013; Dahlstrom 2014; Veríssimo and Pais 2014; Green et al. 2018). Additionally, visual communication tools, such as graphic design (e.g. Rodriguez Estrada and Davis 2015) and videography/photography (e.g. Monroe et al. 2009), play critical roles in the modern-day science communicator's toolbox. Ultimately, effective communication strategies will require conservationists to use a blend of approaches, think outside the box, embrace dialogue, and be willing to 'adapt' to changing technologies.

19.3 Overcoming challenges that limit capacity for conservation physiology

Conservation physiology, although not new in terms of application (e.g. see discussion of *Silent Spring* by Rachel Carson; Wikelski and Cooke 2006), is still a relatively new discipline (i.e. first defined in detail by Wikelski and Cooke 2006; redefined by Cooke et al. 2013). As with all new and emerging disciplines, there are inherent challenges, especially when the goal is to deliver applied science to solve conservation problems. Doing the science alone is not enough for conservation physiology to succeed and evolve—if the science is ignored by practitioners and policy makers, conservation physiology will fail (Cooke and O'Connor 2010). Here, we discuss challenges that impede the development of the conservation physiologist as a valued team member, the training of the next generation of conservation physiologists, and the application of physiological knowledge to conservation problems by practitioners and policy makers.

If a scientist is so bold as to self-identify as a 'conservation physiologist', that scientist may face challenges. For example, institutions (i.e. especially universities) may fail to recognize conservation physiology as a valid research domain, which could impede the ability to secure tenure or promotion. Relatedly, there may be challenges with obtaining funding, if funding bodies are focused on resourcing more established disciplines. Fortunately, there are now a number of scholars around the globe who proudly identify as being conservation physiologists and an increasing number of success stories where entire research programmes, including long-term ones, have focused on conservation physiology.

Moreover, with an established journal (i.e. *Conservation Physiology*) there is further legitimacy to the field. Clearly there are links between conservation physiology being relevant to practitioners and policy makers (i.e. helping to solve problems) and the growing recognition of the value that conservation physiologists bring to a team. Nonetheless, as described above, it is important to continue to share success stories.

Another key challenge impeding conservation physiology is the development of effective training programmes. All too often, conservation scientists are not trained in physiology, and physiologists are not trained in conservation. Fortunately, there are a growing number of examples focusing on how this barrier is being surmounted (e.g. entire courses on conservation physiology; lectures on conservation physiology within conservation science, ecology, and physiology courses; development of texts such as this one). Other subdisciplines, such as conservation genetics and conservation behaviour, have benefited from the development of training frameworks that incorporate those subdisciplines into their core (e.g. Jacobson 1990). To date, we are unaware of any training frameworks that explicitly incorporate conservation physiology. We do not anticipate a time where there would be entire university programmes in the realm of conservation physiology, but rather, we hope that conservation physiology will be recognized as a valid and important aspect of conservation science and incorporated into broader training programmes. Beyond training the next generation, there are also opportunities to train and retrain conservation practitioners (e.g. through professional development courses at conferences) to understand what conservation physiology has to offer.

The final, and perhaps biggest, challenge facing conservation physiology is to have it embraced by practitioners and policy makers. This is not trivial (Cooke and O'Connor 2010). There are many complex reasons why practitioners may ignore science and perhaps especially novel information (Young et al. 2016). For example, it is well known that new knowledge is judged based on its legitimacy and relevance. Conservation physiology has struggled to demonstrate both. One of the biggest issues is that conservation physiology tends to focus on

molecules, cells, organs, and individuals, while conservation practitioners tend to care about populations, species, and ecosystems. This 'scalar' disconnect has been central to conservation physiology, with our findings therefore being regarded as 'interesting but not essential'. Conservation biology textbooks rarely cover and detail any physiology. Another key issue is that conservation practitioners and conservation physiologists rarely connect in formal settings such as conferences (Madliger et al. 2017a). Knowledge users and knowledge generators are rarely in the same space. This can be, of course, overcome with a co-production model, but that still requires knowledge generators and knowledge users to connect in some way. Fortunately, there is a growing number of examples where successes in conservation physiology have arisen because of meaningful partnerships with stakeholders. And, we anticipate this to become the norm over the next decade.

19.4 Conclusions

Conservation physiology is about generating an evidence base so that decisions can have meaningful impacts that benefit conservation. Doing so is an admirable task and one that is urgent, given the biodiversity crisis that exists today. Conservation physiology is increasingly being recognized for its ability to generate cause-and-effect relationships and understand mechanisms, which are essential for informing evidence-based conservation actions. The chapters in this book exemplify the many ways in which conservation physiology is relevant to stakeholders. We identified a number of themes that highlight both the challenges and opportunities in conservation physiology. For conservation physiology to continue to evolve and deliver on its promise requires concerted efforts from conservation physiologists, trainees, practitioners, policy makers, and other allies. Considering that conservation physiology is still a nascent discipline (Cooke et al. 2013), all of those working in this realm should be very proud of what they have collectively accomplished and be optimistic for the future (Cooke et al. 2020). Nonetheless, there is more work to do, and we hope that this chapter and this book in general will inspire others to rise to the challenge. There is no shortage of conservation problems that require the skills (see Cooke and O'Connor 2010) that well-trained conservation physiologists can bring to the table when partnered with those that will ensure that their research activities are relevant. Then, positive change can happen.

References

Andrews, C. and Kaufman, L., 1994. Captive breeding programmes and their role in fish conservation. In G. Mace, A. Feistner, and P. Olney, eds. *Creative Conservation*, pp. 338–51. Springer, Dordrecht.

Balmford, A. and Cowling, R.M., 2006. Fusion or failure? The future of conservation biology. *Conservation Biology*, 20, 692–5.

Beebee, T.J. and Griffiths, R.A., 2005. The amphibian decline crisis: a watershed for conservation biology? *Biological Conservation*, 125, 271–85.

Beever, E., 2000. Diversity: the roles of optimism in conservation biology. *Conservation Biology*, 14, 907–9.

Bellard, C., Bertelsmeier, C., Leadley, P. et al., 2012. Impacts of climate change on the future of biodiversity. *Ecology Letters*, 15, 365–77.

Bennett, E.M., Solan, M., Biggs, R. et al., 2016. Bright spots: seeds of a good Anthropocene. *Frontiers in Ecology and the Environment*, 14, 441–8.

Bergman, J.N., Bennett, J.R., Binley, A.D. et al. 2019. Scaling from individual physiological measures to population-level demographic change: case studies and future directions for conservation management. *Biological Conservation*, 238, 108242.

Bortree, D.S. and Seltzer, T., 2009. Dialogic strategies and outcomes: an analysis of environmental advocacy groups' Facebook profiles. *Public Relations Review*, 35(3), 317–19.

Breuner, C.W., Delehanty, B. and Boonstra, R., 2013. Evaluating stress in natural populations of vertebrates: total CORT is not good enough. *Functional Ecology*, 27, 24–36.

Bubela, T., Nisbet, M.C., Borchelt, R. et al., 2009. Science communication reconsidered. *National Biotechnology*, 27, 514–18.

Cardinale, B.J., Duffy, J.E., Gonzalez, A. et al., 2012. Biodiversity loss and its impact on humanity. *Nature*, 486, 59.

Conde, D.A., Colchero, F., Gusset, M. et al., 2013. Zoos through the lens of the IUCN Red List: a global metapopulation approach to support conservation breeding programs. *PLoS ONE*, 8, e80311.

Cooke, S.J. and O'Connor, C.M. 2010. Making conservation physiology relevant to policy makers and conservation practitioners. *Conservation Letters*, 3(3), 159–66.

Cooke, S.J., Hinch, S.G., Wikelski, M. et al., 2004. Biotelemetry: a mechanistic approach to ecology. *Trends in Ecology & Evolution*, 19(6), 334–43.

Cooke, S.J., Killen, S.S., Metcalfe, J.D. et al., 2014. Conservation physiology across scales: insights from the marine realm. *Conservation Physiology*, 2, cou024.

Cooke, S.J., Madliger, C.L. Cramp, R.L. et al., 2020. Reframing conservation physiology to be more inclusive, integrative, relevant and forward-looking: reflections and a horizon scan. *Conservation Physiology*, 8(1), coaa016.

Cooke, S.J., Sack, L., Franklin, C.E. et al., 2013. What is conservation physiology? Perspectives on an increasingly integrated and essential science. *Conservation Physiology*, 1, cot001. doi:10.1093/conphys/cot001

Coristine, L.E., Robillard, C.M., Kerr, J.T. et al., 2014. A conceptual framework for the emerging discipline of conservation physiology. *Conservation Physiology*, 2(1), cou033.

Costa, D.P. and Sinervo, B. 2004. Field physiology: physiological insights from animals in nature. Annual Reviews in Physiology, 66, 209–38.

Côté, I.M. and Darling, E.S., 2018. Scientists on Twitter: preaching to the choir or singing from the rooftops? *FACETS*, 3, 682–94.

Cvitanovic, C. and Hobday, A.J., 2018. Building optimism at the environmental science-policy-practice interface through the study of bright spots. *Nature Communications*, 9, 3466.

Dahlstrom, M.F., 2014. Using narratives and storytelling to communicate science with nonexpert audiences. *Proceedings of the National Academy of Sciences*, 111, 13614–20.

Dalby, S., 2016. Framing the Anthropocene: the good, the bad and the ugly. *The Anthropocene Review*, 3, 33–51.

Donaldson, M.R., Burnett, N.J., Braun, D.C. et al., 2016. Taxonomic bias and international biodiversity conservation research. *FACETS*, 1,105–13.

Drechsler, M., Eppink, F.V., and Wätzold, F., 2011. Does proactive biodiversity conservation save costs? *Biodiversity Conservation*, 20, 1045–55.

Dupoué, A., Rutschmann, A., Le Galliard, J.F. et al., 2017. Shorter telomeres precede population extinction in wild lizards. *Scientific Reports*, 7, 16976.

Duquesne, S. and Küster, E., 2010. Biochemical, metabolic, and behavioural responses and recovery of *Daphnia magna* after exposure to an organophosphate. *Ecotoxicology and Environmental Safety*, 73, 353–9.

Eisenhauer, N., Bonn, A., and Guerra, C.A., 2019. Recognizing the quiet extinction of invertebrates. *Nature Communications*, 10, 50.

Fa, J.E., Gusset, M., Flesness, N., and Conde, D.A., 2014. Zoos have yet to unveil their full conservation potential. *Animal Conservation*, 17, 97–100.

Farré, M. and Barceló, D., 2003. Toxicity testing of wastewater and sewage sludge by biosensors, bioassays and chemical analysis. *Trends in Analytical Chemistry*, 22, 299–310.

Folt, C.L., Chen, C.Y., Moore, M.V., and Burnaford, J., 1999. Synergism and antagonism among multiple stressors. *Limnology and Oceanography*, 44, 864–77.

Green, S.J., Grorud-Colvert, K., and Mannix, H., 2018. Uniting science and stories: perspectives on the value of storytelling for communicating science. *FACETS*, 3, 164–73.

Harter, T.S., Morrison, P.R., Mandelman, J.W. et al., 2015. Validation of the i-STAT system for the analysis of blood gases and acid-base status in juvenile sandbar shark. *Conservation Physiology*, 3, cov002.

He, X., Johansson, M.L., and Heath, D.D., 2016. Role of genomics and transcriptomics in selection of reintroduction source populations. *Conservation Biology*, 30, 1010–18.

Jacobson, S.K., 1990. Graduate education in conservation biology. *Conservation Biology*, 4, 431–40.

Lee, M.A., Nguyen, F.T., Scott, K. et al., 2018. Implanted nanosensors in marine organisms for physiological biologging: design, feasibility, and species variability. *ACS Sensors*, 4, 32–43.

Leslie, H.M., Goldman, E., Mcleod, K.L. et al., 2013. How good science and stories can go hand-in-hand: Science and stories. *Conservation Biology*, 27, 1126–9.

Lewis, S.L. and Maslin, M.A., 2015. Defining the anthropocene. *Nature*, 519, 171.

Lindenmayer, D.B. and Likens, G.E., 2018. Why monitoring fails. In *The Science and Application of Ecological Monitoring*, second edition, pp. 27–50. CSIRO Publishing, Clayton, Australia.

Lorimer, J., Sandom, C., Jepson, P. et al., 2015. Rewilding: science, practice, and politics. *Annual Review of Environment and Resources*, 40, 39–62.

Mace, G.M., Barrett, M., Burgess, N.D. et al., 2018. Aiming higher to bend the curve of biodiversity loss. *Nature Sustainability*, 1, 448.

Madliger, C.L. and Love, O.P. 2014. The need for a predictive, context-dependent approach to the application of stress hormones in conservation. *Conservation Biology*, 28(1), 283–7.

Madliger, C.L., Cooke, S.J., Crespi, E.J. et al., 2016. Success stories and emerging themes in conservation physiology. *Conservation Physiology*, 4(1), cov057 doi:10.1093/conphys/cov057

Madliger, C.L., Cooke, S.K., and Love, O.P., 2017a. A call for more physiology at conservation conferences. *Biodiversity and Conservation*, 26, 2507–15.

Madliger, C.L., Franklin, C.E., Hultine, K.R. et al., 2017b. Conservation physiology and the quest for a "good" Anthropocene. *Conservation Physiology*, 7, 1–10.

Madliger, C.L., Love, O.P., Hultine, K.R., and Cooke, S.J., 2018. The conservation physiology toolbox: status and opportunities. *Conservation Physiology*, 6, coy029.

McClain, C.R., 2017. Practices and promises of Facebook for science outreach: becoming a "Nerd of Trust." *PLoS Biology*, 15, e2002020.

McMahon, B.J., Teeling, E.C., and Höglund, J., 2014. How and why should we implement genomics into conservation? *Evolutionary Applications*, 7, 999–1007.

Meier, R.E., Votier, S.C., Wynn, R.B. et al., 2017. Tracking, feather moult and stable isotopes reveal foraging behaviour of a critically endangered seabird during the non-breeding season. *Diversity and Distributions*, 23, 130–45.

Monaghan, P. and Haussmann, M.F., 2006. Do telomere dynamics link lifestyle and lifespan? *Trends in Ecology & Evolution*, 21, 47–53.

Monroe, J.B., Baxter, C.V., Olden, J.D., and Angermeier, P.L., 2009. Freshwaters in the public eye: understanding the role of images and media in aquatic conservation. *Fisheries*, 34, 581–5.

National Research Council, 2009. *Learning Science in Informal Environments: People, Places, and Pursuits*. National Academies Press, Washington, DC.

Nel, J.L., Roux, D.J., Driver, A. et al., 2016. Knowledge co-production and boundary work to promote implementation of conservation plans. *Conservation Biology*, 30, 176–88.

Nisbet, M.C. and Scheufele, D.A., 2009. What's next for science communication? Promising directions and lingering distractions. *American Journal of Botany*, 96, 1767–78.

Noss R.F., 1992. Issues of scale in conservation biology. In P.L. Fieldler and S.K. Jain, eds. *Conservation Biology: The Theory and Practice of Nature Conservation, Preservation and Management*, pp. 240–1. Chapman and Hall, New York.

Pritchard, D.J., Fa, J.E., Oldfield, S., and Harrop, S.R., 2012. Bring the captive closer to the wild: redefining the role of ex situ conservation. *Oryx*, 46, 18–23.

Qin, X., Ma, J., Huang, X. et al., 2017. Population dynamics and transcriptomic responses of *Chorthippus albonemus* (Orthoptera: Acrididae) to herbivore grazing intensity. *Frontiers in Ecology and Evolution*, 5, 136.

Reid, A.J., Carlson, A.K., Creed, I.F. et al., 2019. Emerging threats and persistent conservation challenges for freshwater biodiversity. *Biological Reviews*, 94, 849–73.

Ripple, W.J., Wolf, C., Newsome, T.M. et al., 2017. 15,364 scientist signatories from 184 countries. World scientists' warning to humanity: a second notice. *BioScience*, 67, 1026–8.

Robbart, M.L., Peckol, P., Scordilis, S.P. et al., 2004. Population recovery and differential heat shock protein expression for the corals *Agaricia agaricites* and *A. tenuifolia* in Belize. *Marine Ecology Progress Series*, 283, 151–60.

Rodríguez Estrada, F.C. and Davis, L.S., 2015. Improving visual communication of science through the incorporation of graphic design theories and practices into science communication. *Science Communication*, 37, 140–8.

Schwieterman, G.D., Bouyoucos, I.A., Potgieter, K. et al., 2019. Analyzing tropical elasmobranch blood samples in the field: Blood stability during storage and validation of the HemoCue® haemoglobin analyser. *Conservation Physiology*, 7, coz081.

Seebacher, F. and Franklin, C.E., 2012. Determining environmental causes of biological effects: the need for a mechanistic physiological dimension in conservation biology. *Philosophical Transactions of the Royal Society B*, 367, 1607–14.

Soulé, M.E., 1985. What is conservation biology? *BioScience*, 35(11), 727–34.

Stoot, L.J., Cairns, N.A., Cull, F. et al., 2014. Use of portable blood physiology point-of-care devices for basic and applied research on vertebrates: a review. *Conservation Physiology*, 2(1), cou011.

Swaisgood, R.R., 2007. Current status and future directions of applied behavioral research for animal welfare and conservation. *Applied Animal Behaviour Science*, 102, 139–62.

Swaisgood, R.R. and, Sheppard, J.K., 2010. The culture of conservation biologists: show me the hope! *BioScience*, 60, 626–30.

Talwar, B., Bouyoucos, I.A., Shipley, O. et al., 2017. Validation of a portable, waterproof blood pH analyzer for elasmobranchs. *Conservation Physiology*, 5, cox012.

Tyson, R.B., Piniak, W.E., Domit, C. et al., 2017. Novel biologging tool for studying fine-scale behaviors of marine turtles in response to sound. *Frontiers in Marine Science*, 4, 219.

Veríssimo, D. and Pais, M.P., 2014. Conservation beyond science: scientists as storytellers. *Journal of Threatened Taxa*, 6, 6529–33.

Villnäs, A., Norkko, A., and Lehtonen, K.K., 2019. Multi-level responses of *Macoma balthica* to recurring hypoxic disturbance. *Journal of Experimental Marine Biology and Ecology*, 510, 64–72.

White C.R., Marshall D.J., Alton L.A. et al., 2019. Selection drives metabolic allometry. Nature *Ecology and Evolution*, 3, 598–603.

Wikelski, M. and Cooke, S.J., 2006. Conservation physiology. *Trends in Ecology & Evolution*, 21(1), 38–46.

Wilson, A.D., Wikelski, M., Wilson, R.P., and Cooke, S.J., 2015. Utility of biological sensor tags in animal conservation. *Conservation Biology*, 29, 1065–75.

Young, N., Nguyen, V.M., Corriveau, M. et al., 2016. Knowledge users' perspectives and advice on how to improve knowledge exchange and mobilization in the case of a co-managed fishery. *Environmental Science & Policy*, 66, 170–8.

Index

In all headings, 'conservation physiology' is abbreviated as 'CP'. Page numbers in *italics* refer to figures and tables.

locomotor traits 7–8
lymphocytes *132*, 167, 169, 171,
    260–1, 265

M

macrophysiology 58, 64
marine mammals:
    POP levels 111–12, 115
    sampling methods 115
    transcriptome profiling 115–16,
        117, 118
    see also whales (Cetacea); whales
        and non-traditional
        physiological tools study
marine predators, migratory:
    biotelemetry 69
    chemical tracers 70
    isoclock approach 71–3, *72*
        case studies:
            Pacific bluefin tuna 73–6, *75*
            Pacific salmon 76–8, *77*
        future applications 78–80, *79, 80*
    isotope turnover rates *79*
    management and conservation
        challenges 69–71
    stable isotope analysis (SIA) 70–1
    tag–recapture studies 69
    take-home message 69
    tissue sampling 79–80, *80*
marine protected areas (MPAs) 126,
    186, 187, 190, 306
metabolic acidosis 243, 245, 246
metabolic physiology:
    amphibians 170
    bats 151, 152, 153, 157
    birds, *see* yolk precursors studies
    contributions and tools 5, 6
    environmental impacts on immune
        function 170–1
    fish 42, 44, 46–7, 92–4, 100
    isoclock technique, *see* isoclock
        technique
    sharks 188, 192, *193*, 194
    whales 212, *213*
    see also bioenergetics; nutritional
        physiology
Mexico 73, 74, 231
microbiomes 155, 212–13
microclimates 157, 227, 229–31
microhabitats 227, 229–31, 277–8, 280–1
midazolam 297, 298
migration 29, 69, 206, 215, 306–7, *306*;
    see also barriers to fish
        movement, artificial; marine
        predators, migratory;
        salmon, Pacific
        (*Oncorhynchus* genus)
Millspaugh, J.J. 1–2

minnow, brassy (*Hybognathus
    hankinsoni*) 98
monitoring programmes, long-
    term 2, 126, 208, 310, 323
moose (*Alces alces*) 1
multiple competing hypothesis
    approach 23–4
murrelet, marbled (*Brachyramphus
    marmoratus*) 19, 21–4, *23, 24*

N

Natural Resources Damage
    Assessments (NRDAs) 115
Netherlands 62, 206
neurophysiology *5*, 198; *see also*
    sensory physiology
neutrophil-to-lymphocyte ratios
    (N:L) 169, *258*, 260–1, 265
newt, Eastern (*Notophthalmus
    viridescens*) 171
New Zealand 271; *see also* thermal
    ecophysiology reptile
    studies, New Zealand
niche (organism)-level environmental
    conditions 231–4
non-esterified fatty acid profiles 8, 131
noradrenaline 290
North America:
    American lobster 228–9
    Arctic-nesting waterfowl 24–5, *26*
    bats, *see* white-nose syndrome
        (WNS)
    bumblebees 62
    common eider 26–7
    fungal pathogens 143–4
    greater scaup 24–5
    harlequin duck 26, *27*
    marbled murrelet 21–4, *23, 24*
    see also Atlantic Ocean; Canada;
        Pacific Ocean; United States
Norway 206
nutritional physiology:
    Adverse Outcome Pathways
        (AOPs) 119, 121
    bats 157–8
    bees 61, *61*, 63, 64, 65
    birds, *see* yolk precursors studies
    kakapo 3
    killer whales 110
    overview 5
    Physioshark project 198
    stress physiology links 212
    whale reproductive health 211,
        211, 212, 214–15
    wildlife-provisioning tourism 126,
        129, 131, *132*
    see also bioenergetics; metabolic
        physiology

O

ocean acidification 7, 186, 188–9,
    194–5, 196, 198
ocean deoxygenation 188–9, 196
ocean warming 7, 187–90, 196, 198,
    228–9, 310
oestrogen 20, 114, 116
oil spills 110, 112–15, *113*, 245–6
OMICS technologies 7, 110, 321–2;
    see also genomic/
    transcriptomic biomarkers;
    transcriptome profiling
Operation Rhino, South Africa 288,
    297
ophidiomycosis 143
orca (*Orcinus orca*) 6, 8, 110
organochlorine compounds 29
owl, spotted (*Strix occidentalis*) 22
Oxleyan pygmy perch (OPP)
    (*Nannoperca oxleyana*) 95–6
oxygen consumption 46–7, 92–3, *93*,
    94, 170, 290, 293
oxygen insufflation 291–3, *292*
oyster, pearl (*Pinctada
    margaritifera*) 190

P

Pacific Ocean:
    North Pacific right whale
        (NPRW) 206, *207*
    Pacific bluefin tuna 73–6, *75*
    Pacific salmon, *see* salmon, Pacific
        (*Oncorhynchus* spp.)
Parvicapsula minibicornis 42
passive integrated transponder (PIT)
    tags 47, 98, 147, *149*, 176
pathogens 39, 40, 41–2, *42*, 45, 143–4;
    see also chytridiomycosis and
    amphibian declines;
    white-nose syndrome (WNS)
Pecl, Greta 310–11
penguins (*Spheniscidae*) 27–9
perch:
    golden (*Macquaria ambigua*) 100, *101*
    silver (*Bidyanus bidyanus*) 99, 100,
        101
    southern pygmy (*Nannoperca
        australis*) 95
    spangled (*Leiopotherapon
        unicolor*) 7, 100, *101*
performance traits 7–8
persistent organic pollutants
    (POPs) 111–12, 115–16,
    117, 118, 120
pesticides 1, *3*, 46, 63, 174; *see also*
    dichlorodiphenyltri-
    chloroethane (DDT)
pheromones 46